URBAN TRANSPORTATION PLANNING

McGraw-Hill Series in Transportation

URBAN TRANSPORTATION PLANNING

A Decision-Oriented Approach

Michael D. Meyer
Professor, School of Civil and Environmental Engineering
Georgia Institute of Technology

Eric J. Miller
Professor, Department of Civil Engineering
Director, Joint Program in Transportation
University of Toronto

Boston   Burr Ridge, IL   Dubuque, IA   Madison, WI   New York   San Francisco   St. Louis
Bangkok   Bogotá   Caracas   Kuala Lumpur   Lisbon   London   Madrid   Mexico City
Milan   Montreal   New Delhi   Santiago   Seoul   Singapore   Sydney   Taipei   Toronto

McGraw-Hill Higher Education ⚹

A Division of The **McGraw-Hill** Companies

URBAN TRANSPORTATION PLANNING: A DECISION-ORIENTED APPROACH,
SECOND EDITION

Published by McGraw-Hill, a business unit of The McGraw-Hill Companies, Inc. 1221 Avenue of the Americas, New York, NY 10020. Copyright © 2001, 1984, by The McGraw-Hill Companies, Inc. All right reserved. No part of this publication may be reproduced or distributed in any form or by any means, or stored in a database or retrieval system, without the prior written consent of The McGraw-Hill Companies, Inc., including, but not limited to, in any network or other electronic storage or transmission, or broadcast for distance learning.

Some ancillaries, including electronic and print components, may not be available to customers outside the United States.

This book is printed on acid-free paper.

2 3 4 5 6 7 8 9 0 DOC/DOC 0 9 8 7 6 5 4 3 2 1

ISBN 0–07–242332–3

Publisher: *Thomas E. Casson*
Executive editor: *Eric M. Munson*
Editorial coordinator: *Zuzanna Borciuch*
Senior marketing manager: *John Wannemacher*
Senior project manager: *Peggy J. Selle*
Media technology senior producer: *Phillip Meek*
Production supervisor: *Kara Kudronowicz*
Designer: *K. Wayne Harms*
Cover design and illustration by: *Rokusek Design*
Senior supplement producer: *David A. Welsh*
Compositor: *Lachina Publishing Services*
Typeface: *10/12 Times Roman*
Printer: *R. R. Donnelley & Sons Company/Crawfordsville, IN*

Library of Congress Cataloging-in-Publication Data

Meyer, Michael D.
    Urban transportation planning / Michael D. Meyer, Eric J. Miller.—2nd ed.
      p.   cm. — (McGraw-Hill series in transportation)
    Includes index.
    ISBN 0–07–242332–3
    1. Urban transportation–Planning. I. Miller, Eric J. II. Title. III. Series.

HE305 .M489   2001
388.4—dc21   00–048036
                                                       CIP

www.mhhe.com

CONTENTS

PREFACE

When we wrote the first edition of *Urban Transportation Planning: A Decision-Oriented Approach* in the early 1980s, the world was a very different place. Oil embargoes during the previous years had created heightened awareness of the precarious dependence of western economies on foreign sources of energy. Not surprisingly, energy contingency planning and the fuel-saving benefits of proposed transportation actions were prominently discussed in the book. The technology of transportation was viewed primarily as a given, and in many ways did not factor into the strategies for dealing with transportation problems. Land use was a critical input into travel demand modeling, but very few examples were available of how land-use policies could be used to influence transportation system performance. Formal environmental impact analysis had been born a mere 10 years prior to the book's publication, and thus the transportation profession was still defining/creating/ inventing ways for this to occur in a meaningful way.

It is symptomatic of the study of transportation and of the excitement that it holds for countless transportation professionals that much has changed since the book's first publication. In fact, our intent for this second edition was to provide a simple update of the material, while leaving the basic structure the same. As we began, we quickly learned that this approach could not be followed. Too much had changed. The incredible revolution in computing (e.g., the first edition's draft manuscript had been produced with an electric typewriter) and its ever increasing sophistication and availability have changed forever how transportation planning will be conducted. Terms such as GIS, GPS, IT, Internet, virtual reality, and e-commerce were unknown, at least outside of the science fiction literature. And today's students must know as much about the technology of transportation system operation as they do about the physical characteristics of transportation facilities. This new edition has incorporated this new context into an understanding of how transportation systems serve society.

The basic theme of the book, that the major purpose of transportation planning is to inform decision making, remains as the defining theme of this edition. This concept has, if anything, been reinforced over the past 15 years. The institutional framework for decision making has repeatedly been pointed to as one of the key characteristics influencing the effectiveness of planning. It has also been described as one of the important constraints limiting innovation and change. Given its central role in our description of transportation planning, decision making and its linkage to planning remains as one of the most important sections (Chapter 2) in this edition. The material has been updated to include legislation, regulations, and other contextual factors that influence the substance and form of transportation planning.

Several concepts not found in the first edition have been incorporated into this version. In most cases, they reflect the changes that have occurred over the past 15 years. However, in other cases, they represent ideas or principles that we strongly believe will characterize transportation planning over the next two decades, even though they are not widely recognized as such by most transportation professionals. These concepts include,

Metropolitan focus—The first edition discussed transportation planning in the context of *urban* areas. In this edition, we refer to transportation planning as it should occur in *metropolitan* areas. Most major cities of the world, and certainly those of North America, have evolved into regional concentrations of population and economic activity that, without a map showing jurisdictional boundaries, would be indistinguishable from one part of the region to another. Central cities, although still important as centers of business, government, and culture, have become just one part of a much larger metropolitan region. Many of the challenges of transportation planning now relate to travel and connections from one part of the region to another, bypassing the central city completely. Institutional structures have been established at the metropolitan level to guide transportation planning; modeling approaches have been expanded and enhanced to reflect this much broader context for planning; and the role of a metropolitan economy in the context of a global market has become an important focus of planning activities. Planning at the metropolitan level will be the focus of much of transportation planning in the future.

Systems perspective—Transportation has been thought of as a system for many years. However, this edition adds a new dimension to this "systems" perspective. The transportation system itself exists within and interacts with other much larger systems. This has important implications for problem definition, scale of analysis, methods and tools that can deal with these broader issues, and implementation strategies that transcend traditional institutional boundaries. Chapter 3 presents new material on the characteristics of a system and how these characteristics relate to transportation systems planning.

Linkage to development and land use—The first edition treated land use primarily as an important input to the technical planning process. Much has happened over the past 15 years. Many public officials have aggressively pursued public policies that have linked growth and transportation investment. Growth management, "smart" growth, and sustainable development have become important policy contexts in many jurisdictions for transportation investment decisions. This edition provides additional material on the important policy relationship between transportation systems and metropolitan development.

Linkage to environmental quality and ecosystem health—One of the trends identified in the first edition has continued to become more important over the past two decades and will likely be even more important in the future. A systems perspective leads to the observation that human activity will most likely have negative impacts on the natural environment. These impacts have been considered at higher scales of analysis, for example, regional air quality; watersheds; and, ultimately, global climate change. Incorporating environmental considerations at this level into transportation systems planning will be increasingly common in future years. We have laid the groundwork for this concept in this edition.

Linkage to community quality of life—The concept of a "community" and the "quality of life" experienced by its citizens has become more popular in recent years. This concern has led to new urban design principles (which in some cases are nothing more than urban design as it occurred 100 years ago), more attention given to the incidence of benefits and costs on different groups in society, and much stronger consideration to community impact analysis. This edition discusses the tools and techniques that can be used to establish a strong linkage between community quality of life and transportation system performance.

System performance orientation—Society has become more sensitive to the quality of service provided by schools, enforcement agencies, health organizations, and businesses. Citizens are asking for more accountability of public investment and continual monitoring of performance to identify opportunities for improvement. Transportation has not been immune from this trend. Many transportation agencies have developed measures of system performance that can be used to monitor the condition of the system and the quality of service being provided to the public. This edition adopts a performance-based planning approach to transportation planning. Chapter 2 describes how performance measures can be incorporated into the planning review.

Evolution in transportation system technology—The past decade has seen a significant addition to the toolbox of strategies that can be considered by transportation planners and decision makers. The application of advanced electronics, sensors, and network feedback technologies has made transportation system operation much more sophisticated. Intelligent transportation systems (ITS) technologies are now being considered, and many have been implemented, in most major metropolitan areas in North America and around the world. The vehicle itself will likely become much "smarter" in coming years with automated guidance and navigation capabilities, collision warning systems, self-monitoring diagnostics, and network connections to a central information management system becoming commonplace. The transportation planning process must consider such technologies, not only from the perspective of their impact on travel behavior, but also from the perspective of their relative costs and benefits as compared to traditional strategies.

Revolution in computer technology—Perhaps the most significant change that has occurred since the first edition relates to the rapid advances in computing power and the pervasiveness of computer use. Transportation planning very much relies on computer-based analysis, database management, and electronic presentation of results. The power of the computer now permits the simulation of individual travel behavior in the transportation system, changes in metropolitan development patterns, and the optimization of network operations. It seems likely that further advances in computing will add even greater capability to tomorrow's planners.

Adopting an implementation perspective—The first edition portrayed the planning process as including the implementation phase of plans and programs. Planning needed to incorporate in its approach concern for the feasibility of implementing strategies and actions. This edition continues this emphasis but adds a stronger focus on financial feasibility. Almost every conceivable action that results from a transportation planning process will require resources. For many years, transportation plans did not reflect closely the availability of these resources. In the United States, this came to an end in 1991 when the U.S. Congress required transportation plans and programs to be financially constrained; that is, they should reflect the level of resources that can reasonably be expected to be available over the plan or program time frame. This has led to increased emphasis on prioritization and on nontraditional sources of funding, both of which are covered in Chapter 9.

The first edition has been used in college courses around the world. As far as we can determine, many of this generation's transportation professionals have been exposed to the concepts in the first edition . . . and survived! We hesitated, therefore, in changing the structure of the book. However, we felt that the logic of information flow from one chapter to the next could be improved, and so we have made some changes from the first edition.

Chapter 1 introduces transportation to the reader and describes the significant changes that have occurred over the past 15 years. The relationship between transportation system performance and other systems, such as the economy or the environment, are highlighted. The evolution of the planning process toward greater environmental and community sensitivity is a basic point of departure for subsequent coverage of transportation planning.

Chapter 2, which was formerly Chapter 3, establishes the linkage between the transportation planning process and decision making. Different models of decision making are presented and their respective implications on planning discussed. The chapter ends by presenting a framework for transportation systems planning that incorporates many new concepts such as performance measures.

Chapter 3 is a new chapter that introduces the concept of systems as it pertains to transportation. Eight characteristics of systems are presented and their application in transportation planning highlighted. The chapter also describes the characteristics of urban travel that influence the substance and form of transportation planning.

Chapter 4 is the same chapter as in the first edition, only updated to reflect modern approaches to data collection and database management. The important influence of technology in these activities is noted, with special attention given to geographic information systems (GIS) and computer-based survey methodology. The role of performance measures and the concomitant need for data is a new addition to this chapter.

Chapter 5 introduces demand modeling. The reader should note that, unlike the first edition and other transportation texts, demand modeling is presented before the introduction of land-use modeling. Because land-use modeling is usually shown on process charts as occurring prior to demand modeling, textbooks have covered this material first. However, we have found in teaching our courses that if land use is presented first, we spend a great deal of time discussing its use in demand modeling and are thus required to present material on demand modeling. We have therefore reversed the order of presentation.

Chapter 6 begins by discussing the theory underlying urban development and how land-use models represent the evolution of changing metropolitan form over time. The chapter reviews several land-use models that are used throughout the world and critiques their use.

Chapter 7 provides an overview of transportation system performance and how this performance can be modeled. Known as transportation supply, system performance can be analyzed from the perspective of individual vehicle/person movement to network flows. The analysis of highway, transit, pedestrian, and bicycle facilities and services is described. A range of analysis tools, from simple heuristics to network simulation models, provides the reader with a sense of the many different types of tools that can be used to examine the performance of transportation systems.

Chapter 8 shows how the results of analysis can be synthesized into an evaluation framework for presentation to decision makers. In addition, evaluation includes the comparative evaluation of different alternatives, using such techniques as benefit–cost or cost-effectiveness methods. The assessment of impacts after project implementation, so-called ex post evaluation, is also discussed.

Chapter 9 describes different methods for establishing priorities among different projects, for developing a transportation investment program, and of identifying alternative financing strategies. The chapter also covers financial analysis and the relationship between system plans and programs. Characteristics of successful project implementation and of new institutional arrangements are presented.

The flow of information reflected in this structure of the book represents our latest thinking on the important issues associated with urban transportation planning and how they should be presented. This thinking has been greatly influenced by many individuals whom we have worked with and learned from. We are particularly grateful to the following individuals who took the time to provide feedback on earlier drafts of this book: Professors Adjo Amekudzi, Ralph Gakenheimer, Randy Guensler, Les Hoel, Tom Horan, John Leonard, Buzz Paaswell, Pete Parsonson, John Pucher, Ron Rice, Craig Roberts, Mike Rodgers, Ted Russell, Scott Rutherford, Richard Soberman, Gerry Steuart, Simon Washington, and Billy Williams; the following transportation professionals: Wayne Berman, Tom Brahms, Dan Brand, Sarah Campbell, Ann Canby, Janet D'Ignazio, Tom Deen, Don Emerson, Frank Francois, Randy Halvorsen, Kevin Heanue, Wayne Kober, Ken Kruckemeyer, Keith Lawton, Liz Levin, Ysella Llort, Steve Lockwood, Lance Neumann, Neil Pedersen, Alan Pisarski, John Poorman, Chuck Purvis, George Schoener, Jim Scott, Jim Shrouds, Robert Skinner, Jim Smedley, and Rich Steinmann; and the following students, Murtaza Haider, Dan Melcher, Matt Thornton, Elias Veith and Jean Wolf. Special thanks go to Dr. Lisa Rosenstein, who worked long hours making this manuscript readable, and to Lee Wilder and Lillie Brantley for making sure the work was done on time.

Finally, we would not be where we are today without the loving support of our family. To our wives, Heidi and Nancy, we can only express our gratitude and thanks for your endless support. One of the significant changes that did occur since the first edition was the addition of new family members. Given that this book is about the future, we dedicate this work to our next generation—Katie, Eric, Nathan, and Scott.

Michael D. Meyer, Ph.D., Atlanta

Eric J. Miller, Ph.D., Toronto

ABOUT THE AUTHORS

MICHAEL D. MEYER (Ph.D., Massachusetts Institute of Technology) is a Professor of Civil and Environmental Engineering and former Chair of the School of Civil and Environmental Engineering at the Georgia Institute of Technology. From 1983 to 1988, Dr. Meyer was Director of Transportation Planning and Development for Massachusetts where he was responsible for statewide planning, project development, traffic engineering, and transportation research. Dr. Meyer has written over 120 technical articles and has authored or coauthored numerous texts on transportation planning and policy. He is an active member of numerous professional organizations and has chaired committees relating to public transportation, transportation planning, environmental impact analysis, transportation policy, transportation education, and intermodal transportation. Dr. Meyer has consulted with many transportation organizations and has been involved with numerous expert review panels that have advised state and local officials on the most cost-effective investment in transportation. He is currently Senior Program Advisor for Multimodal Planning for Rizzo Associates, a Tetra Tech Company, a leading transportation consulting firm. He is the recipient of numerous awards, including the ASCE Harland Bartholomew Award, the TRB Pyke Johnson Award, and most recently, the Theodore M. Matson Award from six national transportation agencies recognizing his outstanding contributions in traffic engineering.

ERIC J. MILLER (Ph.D., Massachusetts Institute of Technology) is a Professor in the Department of Civil Engineering, University of Toronto, where he teaches courses in urban transportation planning, transportation demand analysis, and transportation and land use. He is also Director of the Joint Program in Transportation, the University of Toronto transportation research center. He is a member of the Transportation Research Board Passenger Travel Demand Forecasting Committee and past Associate Editor (Transportation) for the Canadian Journal of Civil Engineering. He has served on several travel demand modeling Peer Review Panels and consults widely throughout North America. Professor Miller's research interests include microsimulation modeling, integrated transportation–land-use modeling, modeling travel behavior, analyzing transportation–urban form interactions, and sustainable transportation systems.

1

Urban Transportation Planning: Definition and Context

1.0 INTRODUCTION

A metropolitan area's economic and social health depends to a large extent on the performance of its transportation system. Not only does the transportation system provide opportunities for the mobility of people and goods, but over the long term it influences patterns of growth and the level of economic activity through the accessibility it provides to land. In addition, it provides connections to other metropolitan areas, to the nation, and to the world. In recent years, changes to the urban transportation system have also been treated by many public officials as a means of meeting an assortment of national and community objectives. Such changes have been motivated in some cases by the desire to improve air quality, to enhance the viability of economic activity centers, to provide services to those needing mobility (e.g., low-income households, persons with disabilities, and the elderly), and to promote more sustainable community development. Planning for the development or maintenance of the urban transportation system is thus an important activity, both for promoting the efficient movement of people and goods in a metropolitan area and for providing a strong supportive role in attaining other community objectives.

The approach toward urban transportation planning presented in this book is very different from a transportation planning process that envisions a comprehensive and complete "plan" as the final product of the process. Rather, this approach recognizes that, to be effective, planning must be an integral and ongoing part of the *decision-making* process. The *product* of planning can be any form of communication with decision makers that provides useful information for understanding problems facing a metropolitan area, identifying alternative actions, selecting the best alternative, and developing successful implementation strategies.

This is not to say that plans are unimportant. Plans tell the public what kind of a system they can look forward to in the future and provide some sense of satisfaction that today's problems are being addressed. Plans establish the context within which subarea and project studies can be performed. Plans can anticipate where future development will occur (or through strategic investment, influence where this development should occur) and thus protect right-of-way for future transportation infrastructure. With funding constraints, plans tell a region what it can afford, and what it cannot, so that additional resources can be pursued where necessary. So, although the transportation plan is just one way of informing the decision-making process, it is an important part of defining a vision for the future and of establishing strategic transportation investment and system operations directions for the metropolitan area.

The first section of this chapter presents a definition of transportation planning that reflects the decision orientation of the process. The next section discusses a multimodal perspective on transportation planning that has responded to the changing context of planning, which is discussed in the final section.

1.1 A DEFINITION OF URBAN TRANSPORTATION PLANNING

There are many transportation planning processes underway in a metropolitan area at any given time, each defined at a different level of complexity and purpose. For example, while transit planners examine alternative service configurations, traffic engineers identify congestion-reducing alternatives for the highway network; regional planners look at urban development patterns and the provision of public services; individual employers consider alternative employee transportation programs; and social service agencies examine transportation options to improve delivery of their services to targeted population groups such as the elderly and disabled. With different groups and organizations (and thus different types of decisions) concurrently involved in planning activities, the requirements of these planning efforts will vary in important ways. However, the primary purpose of the planning effort is the same in each case—to generate information useful to decision makers for the specific types of decisions they are facing. Given that so many agencies and groups are involved with metropolitan-level transportation decision making, a regional perspective is needed on how these activities fit together. This is a major purpose of transportation planning.

The definition of urban transportation planning used in this book will focus on this basic purpose and on the following propositions suggested by Boulding [1974].

1. *The world moves into the future as a result of decisions (or the lack of decisions), not as a result of plans.* Planning can be effective only if it provides useful information to those who must make decisions. It must not only provide information that is desired by decision makers, but also provide information that is needed to understand fully the short- and long-term consequences of alternative choices.

2. *All decisions involve the evaluation of alternative images of the future and the selection of the most highly valued of feasible alternatives.* Decision making involves two major elements: an agenda consisting of alternative images of the future with some conception of the relationship between present action and future societal directions, and a valuation scheme that outlines preferences for the characteristics of likely decision outcomes. In the case of urban transportation, this valuation scheme is often intricately tied to societal values and goals as expressed in the political decision-making process.

3. *Evaluations and decisions are influenced by the degree of uncertainty associated with expected consequences.* Decisions regarding future actions are based on

implicit and explicit assumptions about the likely consequences of alternative decisions and the future state of the urban area in which the decisions will be implemented. Thus, the greater degree of uncertainty associated with these assumptions, the higher the value that should be placed on decisions that leave future options open.

4. *The products of planning should be designed to increase the chance of making better decisions.* Planning should examine a wide range of agendas, the values and objectives underlying the decision, past decisions that were not considered to be effective, failures of past predictions, and early warnings that the assumed state of the future is changing.

5. *The result of planning is some form of communication with decision makers.* The products of planning are only a small part of the information input to decision makers. To increase the usefulness of this planning information, planning products and processes should reflect the substantive and information-understanding requirements of those individuals who will use these products.

A decision-oriented approach to urban transportation planning should focus on the information needs of decision makers and should recognize the often limited capability of individuals unfamiliar with technical analysis to interpret the information produced. Importantly, planning should provide not only the information *desired* by decision makers but also information *needed* to provide a more complete understanding of the problem and of the implications of different solutions.

Since this textbook was first published, a few authors have adopted a similar perspective. Wachs, for example, stated in his book on ethics and planning:

> The purpose of planning tools is to provide systematic and neutral information to support decision-making, while the ethical content of planning is assumed to be in the definition of the problem and the weighing of information by decision makers [Wachs, 1985].

More recently, Banister argued that

> Transport planning must be seen as an integral part of a much wider process of decision making. Too often in the past transport solutions have been seen as the only way to resolve transport problems . . . transport planning must be seen as part of the land-use planning and development process, which requires an integrated approach to analysis and a clear vision of the type of city and society in which we wish to live [Banister, 1994].

The underlying assumption of a decision-oriented approach to planning is that the relevant decision makers can indeed be identified. In the context of urban transportation planning, decision makers are those individuals faced with the problem of allocating resources among competing needs to achieve certain ends. Decision makers can thus include elected officials who set general policies for resource allocation and who appropriate funds for the implementation of specific actions; transportation agency managers responsible for operating and maintaining components of the transportation system; private sector managers who must determine the most efficient routing of urban commodity shipments; and corporate officials concerned with employee transportation. It is also important to note that the decision-making structure in most metropolitan areas, consisting primarily of elected decision makers, is constantly changing. The transportation planning process needs to be flexible in responding to changing concerns and agendas, yet continue with a long-range (that is, beyond the next election cycle) perspective.

As input into the decision-making process, transportation planning outlines the strategic investments in facilities and services that are necessary to meet future system deficiencies. It also identifies the operational and technological changes to the existing network that will improve transportation service [Meyer, 1999]. The planning process is an opportunity to participate in and influence the decision-making process that allocates resources to achieve a desired future. It provides a sense of where society is heading and how transportation fits into this future. Planning can link the many individual decisions made by groups and organizations into a common vision of how each will help achieve a desired set of goals. Indeed, planning should help clarify and prioritize these goals.

At its simplest level, transportation planning is the process of answering four basic questions:

- *Where are we now* (such as trends and conditions relating to population, the transportation system, and the general state of the urban area)?

- *Where do we want to go* (major issues, public outreach results, obstacles, and opportunities)?

- *What will guide us* (mission statement, goals, objectives, public input, and performance measures)?

- *How will we get there* (revenue estimation, project and program implementation, public/private partnerships, and policy changes)? [Florida DOT, undated]

At its most challenging level, transportation planning focuses on balancing the many competing visions of what the future should look like and on developing an informed program of action among competing interests that will improve a community's quality of life and enhance transportation system performance. This latter perspective is particularly important in that the changing context of urban transportation planning over the past 40 years has introduced new issues into these visions and has opened up the process to numerous competing interest groups that never before participated in a meaningful way.

Given these considerations, urban transportation planning can be defined as follows:

Urban transportation planning is the process of

1. *Establishing* a vision of what a community wants to be and how the transportation system fits into this vision.

2. *Understanding* the types of decisions that need to be made to achieve this vision.

3. *Assessing* opportunities and limitations of the future in relationship to goals and desired system performance measures.

4. *Identifying* the near- and long-term consequences to the community and to transportation system users of alternative choices designed to take advantage of these opportunities or respond to these limitations.

5. *Relating* alternative decisions to the goals, objectives, or system performance measures established for an urban area, agency, or firm.

6. *Presenting* this information to decision makers in an understandable and useful form.

7. *Helping* decision makers establish priorities and develop an investment program.

Several concepts in this definition merit special attention. First, transportation planning is defined as a *process*. Such a process includes careful consideration of problem definition; incorporation of alternative viewpoints of analysis and evaluation; development of goals, objectives, and a statement of desirable transportation system performance; and completion of the technical analysis needed to determine impacts of alternative decisions. Technical analysis, considered by many to be synonymous with planning, is just one component of the planning process. Just as transportation planning occurs at the metropolitan level, so too such planning is being undertaken at higher levels of government. Thus, for example, the relationship between metropolitan and statewide transportation plans needs to be considered by both.

Second, transportation planning should assess opportunities as well as limitations of the future. The traditional approach to planning focuses most attention on identifying where problems or deficiencies will occur in the transportation system. Clearly, such a focus is an important element of planning. Often, however, opportunities to improve the existing operation of the transportation system exist even if no one perceives a problem. For example, even though bus routes in an urban transit network might be operating at acceptable performance levels, reorganizing the structure of the service could result in more efficient operations while maintaining, or even improving, service. Planning should thus include a *proactive* approach to transportation systems analysis.

Third, transportation planning should include a short-range and long-range perspective. The long-range planning component is a continuing activity that represents a statement of need and policy direction, thereby providing a context for periodic transportation decisions made in the near term. The long-range component of a transportation plan, to be relevant to decision makers, must be both flexible and responsive in scale (level of detail) and scope (alternatives and impacts considered) with respect to the kinds of decisions likely to be made. The short-range component takes into account the more immediate needs of transportation system performance and can focus on such things as operational changes or demand management (e.g., transportation system operations centers, ride sharing and variable work hours programs, and telecommunication substitutes for travel). Decisions in the short term, however, might change the context of future system design options and performance. Anticipated changes in the long term might strongly guide the desirability of actions to be implemented in the next 5 years. This temporal interrelationship is thus a critical consideration in transportation planning.

Fourth, as in Boulding's valuation scheme, goals, objectives, and performance measures form the basis of the measures of effectiveness used in evaluation. Goals and objectives have been part of transportation planning for many years.

Performance measures, however, have only recently been introduced into transportation planning. Performance measures are important because they can, in measurable ways, provide indications of the effectiveness of transportation planning and subsequent program implementation [Meyer, 1995; U.S. DOT, 1995; NCHRP, 1998]. In addition, they provide important benchmarks to the public and to political decision makers who are increasingly asking for greater accountability of public expenditures. Because goals and objectives are intricately tied to values, careful consideration should be given to whose values are represented in a statement of goals and objectives. In order to develop a representative goals set, opportunities for community input into the development of such a statement should be provided.

Finally, the most important decision makers for urban transportation planning are the elected and appointed government officials who provide and maintain a transportation system that meets the mobility and accessibility needs of their constituents. The types of transportation decisions facing these officials include expanding or modifying the existing transportation infrastructure (and thus raising the revenues to accomplish this); making changes to system operations that will enhance system performance; implementing policies that more closely link land use and transportation or that make the "price" of transportation more closely reflect the true "costs;" developing innovative approaches to transportation system management either through the application of advanced technologies or by influencing travel demand; and changing the institutional structure for transportation planning and program management. At the same time, officials must consider the equity implications of pursuing one program over another.

In addition to elected and appointed officials, numerous stakeholders, interest groups, community associations, business organizations, system users, and concerned citizens can also have a significant influence on the outcome. The planning process must provide opportunities for participation to all interested groups.

In summary, a decision-oriented approach to urban transportation planning focuses as much attention on the process of planning as it does on the techniques used. In some ways, this approach requires the planner to reverse the traditional sequence of planning (i.e., proceeding from problem definition to a final decision) by first understanding the requirements of the final decision and then by identifying the information and analyses needed to produce them. As a result, the information produced by this planning process is directly related to the needs of the decision-making process.

1.2   A MULTIMODAL PERSPECTIVE ON TRANSPORTATION PLANNING

Transportation planning has traditionally reflected the policy concerns and issues of the times in which it was occurring. Thus, the large-scale, regional transportation studies of the 1950s, which in many ways represent the early beginnings of today's metropolitan transportation planning process, focused almost exclusively on highway network expansion. Elected officials of the time were anxious to accommodate the tremendous increase in automobile use and to take advantage of federal aid that was available for highway construction. The motor vehicle industry lobbied successfully for dedicated gas taxes that could be used only for improvements to the road system, a legacy that still exists today. The transportation profession responded quickly in developing the theories, concepts, and tools that supported this focus. In the United States, the Bureau of Public Roads, the predecessor to the Federal Highway Administration (FHWA), published the first highway capacity manual in 1950. In 1965, the second edition was published by the Highway Research Board (now the Transportation Research Board) and included many new advances in analyzing the operational performance of the road system [Cambridge Systematics et al., 1998]. Subsequent versions were published in 1985, 1994, and 2000. In contrast, the first transit capacity manual in the United States was published in 1999.

As the policy environment for transportation planning began to change in the late 1960s and 1970s toward a more modally balanced metropolitan transportation system, the transportation planning profession lacked the technical tools, data, and process to adequately implement a multimodal transportation planning process. Plus, the structure for funding transportation projects still heavily favored highway investment. Recent transportation laws, regulations, and policies, however, have encouraged the development of a multimodal transportation planning process.

Multimodal transportation planning is defined as

The process of defining problems, identifying alternatives, evaluating potential solutions and selecting preferred actions that meet community goals in a manner that includes all feasible transportation modes [Meyer, 1993].

There are many transportation facilities and services available in a metropolitan area that provide opportunities for mobility and accessibility. Figure 1.1 shows one view of a multimodal transportation system. The system consists of different modal transportation networks (including the information/communications network) that by themselves allow a traveler to move from one location to another. Intermodal connections provide the ability to transfer from one modal network to another. In addition, metropolitan land-use patterns and the institutional structure for providing transportation service affect the overall performance of the system.

At its most fundamental level, multimodal transportation planning recognizes the fact that there is no single solution to the transportation problems facing a metropolitan area. A coordinated program of action is necessary to deal with the complex nature and interactions of the transportation phenomenon. Figure 1.2 illustrates the concept of a coordinated strategy that includes three components—supply management, demand management, and land-use management. Note that a key word used to describe each component is *management*. Because much of the U.S. transportation system is in place, the decisions of what additional capacity to provide, what types of operational improvements to make, how to influence demand for the purpose of reducing the impact of traffic, how to develop compatible land use, and how to provide the institutional and funding structure that supports the program are all in essence system management decisions.[1]

| [1]   Much of the following discussion in this section is based on [Meyer, 1998].

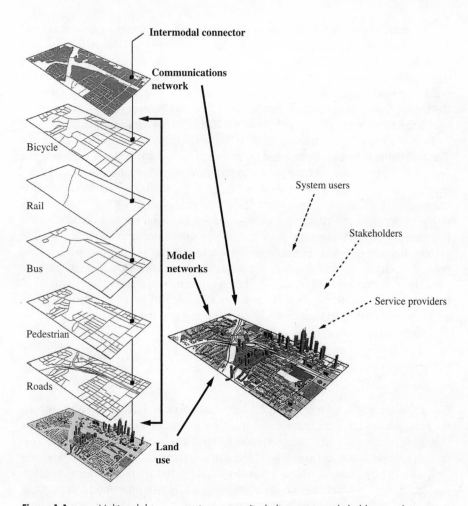

Figure 1.1   Multimodal transportation system (including users, stakeholders, and service providers)

SOURCE: Courtesy of Dr. William Bachman, Center for Geographic Information Systems, Georgia Institute of Technology.

Managing Transportation System Supply   Managing the transportation system by adding new facilities or by making operational changes to improve system performance has been the most common response to transportation problems for many years. Typical actions include new highways and transit facilities; improved traffic signalization schemes; traffic engineering improvements such as turn lanes, one-way streets, reversible lanes, and turn prohibitions; new or improved transit services; preferential treatment for those who use multioccupant vehicles; and ramp metering. Increasingly, transportation professionals have become interested in strategies that minimize the effects of accidents and other nonrecurring incidents on traffic flow, including incident detection programs, motorist information systems, and towing/enforcement efforts. The application of information processing, com-

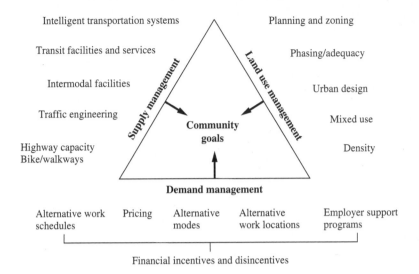

Figure 1.2 Components of a multimodal transportation program
| SOURCE: Meyer, 1998

munications technologies, advanced control strategies, and electronics, known as intelligent transportation systems (ITS), to improve the safety and efficiency of the transportation system has also become a part of many metropolitan transportation strategies [Smith, 1999; FHWA, 2000]. Table 1.1 shows the types of transportation strategies that would have been considered historically for different types of problems, as compared to an operations-oriented approach with ITS. The operations approach suggests a very different implementation environment than that existing for highway or transit facility expansion. Indeed, two of the major difficulties in the early development and deployment of ITS technologies have been a misunderstanding of what these types of strategies can accomplish and the mismatch between organizations geared toward building infrastructure and the ITS need for innovative service provision.

Because planners and engineers have had many years of experience with the supply side of the transportation system, there is more evidence in the literature on the resulting impact of these actions. In the unusual case where new capacity can be continually added to accommodate demand, these actions can significantly reduce congestion levels. In the long term, however, this additional capacity, if assigned to highway improvements only (e.g., additional lanes), would continue a heavy reliance on the automobile that could have serious implications for the provision of metropolitan mobility. In heavily urbanized areas, the construction of these actions (especially major highway improvements) can be costly, their implementation met with strong opposition, and even if feasible, they might take a long time to complete. It is for these reasons that other actions need to be considered.

Managing Transportation Demand In its broadest sense, demand management is any action or set of actions intended to influence the intensity, timing, and spatial

Table 1.1 Transportation problems, conventional approaches, and operational approaches with ITS

Problem	Solutions	Conventional Approach	Operational Approach with ITS
Traffic congestion	• Increase roadway throughput	• New roads • New lanes	• Advanced traffic control • Incident management • Corridor management • Advanced vehicle systems
	• Increase passenger throughput	• HOV lanes • Carpooling • Fixed route transit	• Real-time ride matching • Integrate transit and feeder services • Flexible route transit • New personalized public transit
	• Reduce demand	• Flex time programs	• Telecommuting • Transportation pricing
Lack of mobility and accessibility	• Provide user-friendly access to quality transportation services	• Expand fixed route transit and paratransit services • Radio and TV traffic reports	• Multimodal pre-trip and en route traveler information services • Real-time response to changing demand • Personalized public transportation services • Enhanced fare card
Disconnected transportation modes	• Improve intermodality	• Construct intermodal connections	• Regional transportation management systems • Regional transportation information clearinghouse • Disseminate multimodal information pre-trip and en route
Budgetary constraints	• Use existing funding efficiently • Leverage new funding sources	• Existing funding authorizations and selection processes	• Public–private partnerships • Barter right-of-way • Advanced maintenance strategies • Restructure public support (subsidies of transportation modes) • Increased emphasis on fee-for-use services
Transportation following emergencies	• Improve disaster response plan	• Review and improve existing emergency plans	• Establish emergency response center • Internet with law enforcement, emergency units, traffic management, and transit
Crashes, injuries, and fatalities	• Improve safety	• Improve roadway geometry and sight distance • Grade-separate crossings • Driver training • Sobriety checkpoints • Install traffic signals • Reduce speed limits • Post warnings in problem areas	• Partially and fully automated vehicle control systems • Vehicle conditions monitoring • Driver condition monitoring • Advanced grade-crossing systems • Automated detection of adverse weather and road conditions, vehicle warning, and road view notification • Automated emergency notification

1 SOURCE: Smith, 1998

distribution of transportation demand for the purpose of reducing the impact of traffic or enhancing mobility options. Such actions can include offering commuters one or more alternative transportation modes and/or services, providing incentives to travel on these modes or at noncongested hours, providing opportunities to better link or "chain" trips together, and/or incorporating growth management or traffic impact policies into local development decisions.

Available evidence suggests that well-conceived and aggressively promoted demand reduction programs can decrease peak period traffic at many sites by as much as 10 to 15 percent. In fact, significantly higher demand reduction levels have been achieved at several employment sites. Demand reduction efforts, however, unless undertaken on a truly massive scale, can have only a local impact. They can relieve spot congestion—for example, at entrances and exits to large employment centers—but they cannot appreciably reduce traffic on freeways and major arterials. The only exception to this seems to be areawide road pricing schemes that at least from a modeling perspective indicate significant influence on travel demand.

Managing Land Use   One of the fundamental relationships in understanding transportation system performance is the linkage between land use and transportation. Put simply, trip-making patterns, volumes, and modal distributions are largely a function of the spatial distribution and use of land. Thus, at individual development sites, exercising control over the trip-generating characteristics of the land use (e.g., development density) can make the resulting demand consistent with the existing transportation infrastructure and the level of service desired.

Over the long run, the spatial distribution of land use can greatly influence regional travel patterns, and, in turn, this land use distribution can be influenced by the level of accessibility provided by the transportation system. Changing the economic equation for travel by equalizing subsidies for all modes could also affect location decisions. Avoiding future transportation problems therefore requires careful attention to zoning and land-use plans in coordination with the strategic provision and pricing of transportation services to influence where development occurs.

Even though these program components are listed separately, they are really complementary to one another. For example, a ride-sharing program (an effort to influence demand) can become more effective if some form of preferential treatment is provided en route (e.g., a high-occupancy vehicle lane) or at the destination (e.g., preferential parking), both changes to the transportation system. The effectiveness of the ride-sharing program could be even greater if employers were required to incorporate enhanced ride-sharing activities into the design and use of the site (a land-use/development decision). Similarly, linkages exist between different actions that often work at cross purposes. For example, development regulations that require high minimum numbers of parking spaces as a condition of development work against employer demand management programs that focus on reducing single-occupant vehicle use to that site.

An example of multimodal transportation planning is found in Maryland DOT's mobility planning studies that have adopted a multimodal perspective in identifying possible actions for solving transportation problems. These actions include transportation demand management such as parking pricing, transportation system management consisting primarily of traffic operations improvements, public transit and highway capacity expansion, high-occupancy vehicle treatments, nonmotorized transportation, and growth management and activity center land-use strategies [Maryland DOT, 1995]. The metropolitan planning organization (MPO) in Albany, New York, has adopted a strategy for better managing arterial road corridors that includes access management (driveway access and signal spacing guidelines), land-use planning and coordination, nonmotorized transportation, and use of traffic calming techniques to reduce vehicle intrusions into neighborhoods [CDTC, 1996]. In Denver, the prospective transportation strategy list is extensive and includes land-use and pricing approaches (Table 1.2). Such "transportation" strategies will likely continue to be characteristic of future transportation planning.

Table 1.2   Transportation strategies in Denver, CO

TDM Measures

Regional ride-share match assistance

Vanpool program

Guaranteed ride home

Parking management (pref. treatment and supply)

Parking management (on-street permits)

Alternative work schedules

Telecommuting

Employer trip-reduction support measures

Employer trip-reduction ordinances

Eliminate parking subsidies

Neighborhood support measures

Corridor/subarea support measures

Driving prohibitions

Truck restrictions

Restrict start/finish of major special events

Traffic Operations Improvements (Traditional)

At-grade intersection improvements

At-grade intersection replacement

Link traffic flow improvements—urban and rural

Traffic signal improvements

Conversion to one-way streets

Railroad grade crossings

Reversible traffic lanes

Improved traffic management during construction

High-Occupancy Vehicle Measures

HOV lanes

HOV bypass lanes or ramp meters

Direct HOV ramps to freeway HOV lanes

Public Transit Capital Improvements

Fixed exclusive facility

Bus lanes

Deregulate/eliminate barriers for private service

Bus-only ramps to freeway

Bus bypass lanes at ramp meters

Transit malls

Park-n-ride, transit center

Paratransit service

More/better amenities (shelters, newer buses)

Public Transit Operational

Additional service with current fleet

Transit schedule coordination

Signal priority for buses/rail vehicles

Transit promotion/marketing

Express bus service in mixed-use lanes

Service revision focusing on congestion relief (more service in congested corridors)

Nontraditional Modes

Bicycle paths/lanes

Pedestrian paths/malls

Grade separation for bicycles/pedestrians

Purchase/provide "free" bicycles

Bicycle/pedestrian encouragement measures

Table 1.2   Transportation strategies in Denver, CO (continued)

Transportation Pricing	*Incident Management*
Tolls (with peak-period pricing)	Rapid vehicle removal laws and clearance policies
HOV lane tolls for single-occupant vehicles	Freeway service patrols
Peak-period parking surcharges	Accident investigation site/policies
Fuel tax increases	Institutional arrangements for rapid response
Registration fee increase or vehicle surcharge	Detour route preplanning
Transit fare reductions/reduced fare zones	Rapid detection/response
VMT tax/emissions tax	*Intelligent Transportation Systems*
Increased assessment ratio for parking lots	Computerized signal control and enhancements
Growth Management/Activity Center Strategies	Freeway ramp metering
Land-use policies/regulations	Electronic toll collection
Site design standards	Real-time motorist information
Growth limitations/controls	Dynamic route guidance
Concurrency management; trip budgets	Real-time ride matching and dynamic bus routing
Incentives to encourage employees to live closer to work	Real-time transit information
Subsidized housing near transit facilities	Public transit operations management
	Port-of-entry automation
Access Management	Weigh-in-motion
Adopt access management plan for new roads	*General Purpose Improvements*
Retrofit existing facilities	Lane additions
Construct frontage roads—freeways and arterials	Auxiliary lanes
	Interchange reconstruction

SOURCE: As reported in Smith, 1999

1.3   A CHANGING SOCIETY AND ITS IMPACT ON URBAN TRANSPORTATION PLANNING

The previous discussion on decision-oriented planning implies a direct cause–effect relationship between transportation decisions and resulting system performance. In reality, travel behavior and travel patterns on transportation networks are affected by numerous factors outside the control of planners and government officials. For example, overall levels of travel reflect the state of the economy—when the economy is good, more travel occurs. Knowing how these factors have changed over time becomes an important point of departure for understanding how the urban transportation planning process has evolved. Although there are many factors that have had such influence, five in particular merit special attention: changes in (1) population characteristics, (2) the metropolitan economy, (3) societal concerns and issues, (4) transportation legislation/regulation, and (5) the technology of planning.

Population Characteristics   The defining demographic characteristics of almost all aspects of North American society post-World War II have been the tremendous growth in population and the movement of the "baby-boom generation" through the aging process. This huge increase in population resulted in demands for housing, employment, and government services. Rising household incomes created a market demand for personal transportation that was satisfied in the majority of cases by the automobile. Although most pronounced in the United States, a similar population "bump" occurred in Canada (the population surge occurred later than in the United States) [Foot, 1996]. The growth in population over the past 40 years has been accompanied by changes in other factors important to transportation planning and critical in explaining corresponding increases in urban travel. Figure 1.3 illustrates the interaction of key variables that help explain the dramatic increases in travel witnessed over the past several decades.[2] Some data from the United States that correspond to these relationships include

Population: Between 1960 and 1998, the population of the United States grew by 49 percent (to over 270 million). From 1970 to 1990, 86 percent of total U.S. population growth occurred in the suburbs. In 1960, approximately 52 percent of the population in metropolitan areas of over one million people lived in the county having the central city, and 48 percent lived in suburban counties; by 1990, the respective percentages were 42 percent and 58 percent.

Employment: The number of workers in the labor force almost doubled between 1950 and 1990. Women joining the labor force were a major reason for this significant increase. Almost 65 percent of the total growth in worker population occurred in the suburbs, 18 percent in the central city, and 17 percent in nonmetropolitan areas. The suburbs in 1990 had 42 percent of the nation's jobs (up from 37 percent in 1980). Almost 50 percent of the growth in jobs from 1980 to 1990 occurred in suburban locations, whereas central cities represented only 23 percent of the growth.

Households and Persons/Household: Between 1960 and 1998, the number of households grew by 94 percent (to just over 102 million). The largest share of the growth in households occurred in single person and single-parent households. The average household size also declined from 3.33 persons to 2.62 over the same time period.

Trips/Capita: Between 1983 and 1990, annual daily trips per capita increased by almost 7 percent to 3.13 trips per female and 3.04 trips per male.

Person Trip Length: The average distance for commuting each day in 1983 was approximately 8.5 miles/13.7 kilometers; in 1995 this number had risen to 11.6 miles/18.7 kilometers.

---

[2] Note that the transportation system measure that results from the relationships in Fig. 1.3 is vehicle miles traveled, not congestion. Congestion, which results when facility usage exceeds what users perceive to be a desirable level of service for a specific period of time, is a different measure. Indexes have been developed that provide some indication of changes in congestion in metropolitan areas (see, for example, [TTI, 1998; BTS, 1999]).

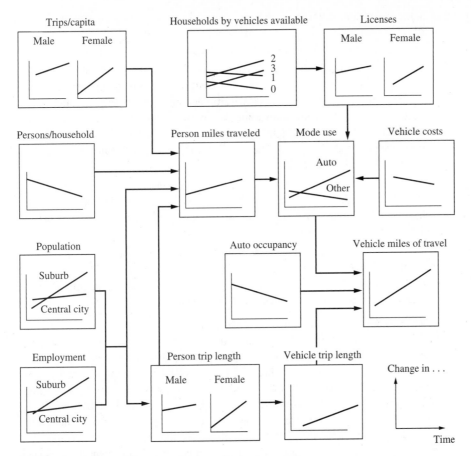

Figure 1.3 Factors influencing urban travel behavior over time
SOURCE: Meyer, 1998
NOTE: The relationships between these variables over the passage of time (denoted by the x axes) are represented as being linear. This is done only to illustrate basic relationships and their direction. There is clear evidence that such linearity is not the correct functional form for many of these relationships.

Vehicles/Household: Between 1969 and 1995, the average number of vehicles per household rose from 1.1 to 1.78, with only 8.1 percent of U.S. households—an estimated 8 million households—in 1995 not having a car available for a trip (one in five households had three or more cars) [Pickrell and Schimek, 1997]. During just one decade, 1980 to 1990, when population and number of households increased by less than 10 percent and 14 percent, respectively, the total vehicles available to households grew by over 17 percent. In 1995, every worker had on average 1.3 vehicles available for the work trip.

Licenses: Between 1969 and 1995, the number of licensed drivers in the United States increased by almost 60 percent (for women from 61 percent in 1969 to 85 percent in 1995; for men from 87 percent to 93 percent over the same time frame). For those currently aged 30 through 49, over 96 percent of the men and roughly 93 percent of the women are licensed, almost reaching driver saturation level for this population group.

Mode Use: Transit's share of all national travel has declined to about 2 percent, although in many urban markets transit retains a strong presence. This decline is associated with increased auto availability to those population groups most likely to use transit and the continued suburbanization of metropolitan regions. This results in land uses difficult to serve with traditional transit services. At the same time, the operating cost of the automobile declined in real dollar terms.

Auto Occupancy: Average vehicle occupancy (measured as person miles per vehicle mile) has declined from 1.9 in 1977 to 1.6 in 1990. In many urban highway corridors, the average automobile occupancy for work trips is below 1.1 persons per vehicle.

Vehicle Miles Traveled (VMT): From 1969 to 1995, the total number of vehicle miles traveled (number of vehicles multiplied by average annual miles driven) in the United States increased by almost 85 percent, reaching an estimated 2.2 trillion vehicle miles (3.6 trillion vehicle kilometers).

The trends and corresponding relationships shown in Fig. 1.3 occur in a political environment that can influence the outcome. As has been noted by several observers, government transportation policies do not always provide a "level playing field" [Holtzclaw, 1995; Pucher and Lefevre, 1996; Transportation Research Board, 1998; Vuchic, 1999]. In the United States, for example, numerous subsidies (direct or otherwise) have provided strong support for automobile use. Similarly, over the past 2 decades, substantial transit investment has occurred in the United States, often having little impact on mode share in a metropolitan area even though actual ridership levels have increased. In other countries, government policies such as high gas taxes and heavy subsidization of transit have resulted in very different urban travel patterns, although even in these cases, automobile use has increased dramatically in recent years.

Travel behavior can be partly explained by looking at the characteristics of who is traveling. Such things as traveler income, age, availability of an automobile, number of children, and occupation influence a person's propensity to travel, by what means this travel will occur, and the spatial distribution of travel in a transportation network. In particular, understanding the household characteristics of a community's population is important because transportation planners use the demographic and transportation characteristics of households to model travel behavior. Thus, changing characteristics of the typical household could lead to changing travel demands and, ultimately, to different mobility needs and types of transportation services in metropolitan areas (e.g., the transportation needs of an older population are very different in magnitude and type than of a younger population). Table 1.3 illustrates some of the changing characteristics of the population growth from 1960 to 1998. Clearly, the 1960s household of a mother, a father (as the head of household), and two children gave way to a much more diverse set of households in the 1990s.

Table 1.3   Selected changes in characteristics of U.S. population (1960 to 1998)

	1960	1970	1980	1990	1998	Compounded Annual Rate of Change
Average household size						
	3.33	3.14	2.94	2.63	2.62	−0.63%
Single- v. multiple-person households						
1 person	13%	17%	23%	25%	26%	1.8%
2+ persons	87%	83%	77%	75%	74%	−0.43%
Workers per household						
0	9%	12%	13%	13%		1.2%
1	53%	45%	33%	28%		−2.0%
2+	37%	43%	54%	59%		1.5%
Labor-force participation by gender						
Male	80%	77%	75%	74%	73%	−0.24%
Female	36%	41%	50%	57%	59%	1.3%
Percent of married women in labor force	32%	41%	58%	58%	62%	1.8%
Percent of married women in labor force having children under age 6	19%	30%	45%	59%	64%	3.2%
Distribution of population over age 65						
65–74	67%	62%	61%	58%	54%	−0.57%
75+	33%	38%	39%	42%	47%	0.93%

SOURCE: U.S. Census, http://census.gov

The relationships shown in Fig. 1.3 can help explain urban travel in the United States over the past 40 years. However, these relationships can also be used to predict what the future might hold for urban transportation (e.g., needs and system use), assuming one knows the characteristics of the future population and that the underlying behavioral relationships do not change. Such growth will likely be very different from the past. Examination of census data and of the data presented in Table 1.3 reveals that the following trends will likely have important impacts on transportation [Dunphy, 1997; Meyer, 1998]:

1.   Population and worker growth rates will slow significantly from previous decades (in comparison, the average annual rate of population increase between 1990 and 1995 was 1.1 percent; between 2040 and 2050 it is predicted to be 0.54

Something to Think About

The travel-influencing factors discussed in this chapter relate primarily to the United States. Other countries in the world are facing similar dramatic changes. Consider

- The annual growth rate from 1970 to 1992 in car ownership for low-income economies (e.g., China, India, Nigeria) was 9.4 percent, whereas the population grew annually on average for these countries 2.3 percent.

- From 1970 to 1993, the average growth rate in vehicle kilometers traveled was 2.7 percent in the United States; comparable rates for other countries include France, 3.2 percent; Great Britain, 3.8 percent; and Japan, 6.5 percent.

- Private car usage has increased in most world cities: in German urban areas, from 34 percent in 1972 to 47 percent in 1992; in Norwegian cities, from 32 percent in 1970 to 68 percent in 1990; in Manchester, England, from 32 percent in 1971 to 64 percent in 1991.

- Central city shares of metropolitan population have declined: Paris, from 32 percent in 1968 to 23 percent in 1990; Amsterdam, from 80 percent in 1970 to 66 percent in 1994; and Zurich, from 38 percent in 1970 to 29 percent in 1995.

---

percent). However, the growth rates will still add substantial new levels of population and workers, especially in high-growth metropolitan areas (population in the United States is predicted to reach 392 million by 2050).

2.   On a percentage basis, women labor rates will continue to increase while those for men will likely decrease.

3.   That portion of the urban population older than 65 years of age is predicted to increase at a proportionately much faster rate than before. This is especially true for those aged 75 and above.

4.   A substantial portion of metropolitan growth in many U.S. cities will likely occur because of immigration and subsequent increase in family size. For example, growth in the Hispanic population is expected to constitute more than 39 percent of the nation's population growth to the year 2010, 45 percent from 2010 to 2040, and 60 percent from 2030 to 2050. Under these predictions, by 2050, just less than 53 percent of the population will be non-Hispanic white, 16 percent will be African American, and 23 percent will be of Hispanic origin [U.S. Census, 1997].

5.   Household size and the number of workers per household will likely continue to decrease from previous levels but not at the high rates seen in previous decades (Table 1.3 shows indications of this declining growth rate).

6.   The importance of suburbs as locations of population and employment growth will probably continue at levels similar to those seen in previous decades, although several metropolitan areas are trying to guide this growth through growth management approaches.

7.   To all intents and purposes, a saturation level has been reached in the percentage of the driving age population having a driver's license. As these drivers get older, the number of elderly with a license will be much greater than in previous years.

8.   Vehicle ownership rates are predicted to slow down as compared to previous decades. The only likely exception will be for immigrants whose vehicle ownership will increase dramatically as their improved economic status enables them to own a vehicle.

These trends will have potentially significant impacts on the providers of transportation services such as transit agencies. Table 1.4 shows the assumed consequences to public transit of some selected changes in society [Cambridge Systematics, 1999]. In almost every case, the trend is likely to make the provision of transit, at least the traditional service of fixed bus and rail routes, more challenging.

Table 1.4   Implications of changes in society on provision of public transit services

Trend	Likely Implications
Movement Toward Service-based Economy	
• Growth in service jobs (73% from 1970 to 1990)	• Fast-growing market hard to serve with transit
• Large percentage of civilian jobs in service sector	• Traditional transit does not provide good service
• Retail trade will likely replace manufacturing as second largest employment category	• Traditional transit does not cover typical job hours
• Service sector job growth is dispersed	• Fast-growing market hard to serve with transit
• Service businesses tend to be smaller in size	• Less concentration of employees reduces transit efficiency
• Less pronounced effects in older urban areas	• Older urban areas can support traditional transit service
Increasing Income Disparities Between High- and Low-Income Households	
• Service sector requires highly paid "knowledge" worker and lower paid "support" worker	• Traditional low-income transit-dependent population will grow and require service
• Large portion of service workers are not well paid	• Service-oriented job access will be dispersed
• Wage gap will concentrate low-income workers in minority populations	• Inattention to transportation needs of low-income persons will result in public costs in health, welfare, etc.
Increasing Work Force Flexibility	
• One-quarter of workers have flexible hours	• Traditional transit service is of limited usefulness
• 90% of new jobs created monthly are part-time jobs	• Places and hours of employment change regularly
• 40% of women workers do not have day-shift job	• Access to jobs must be available over non-standard hours
• Increased flexible labor reflects economic hardship	• Ability to pay for transit is diminished for many

Table 1.4   Implications of changes in society on provision of public transit services (continued)

Trend	Likely Implications
More Women in Labor Force	
• Majority of women have paid employment	• Women's travel more likely to include trip chaining
• In 1990, more than 44% of new mothers returned to work before baby was 6 months old	• Changing family structure is reinforcing auto dependence
• Women work-force participation higher for minorities	• Income and ethnic characteristics may provide a counterbalance to gender in terms of transit's appeal
More Older People	
• Elderly are the fastest-growing population group	• Need for access, assistance, and travel options will increase
• More than 25% of population are over 60 years old, which will increase dramatically in 50 years	• Work trip-oriented services do not serve this market
• Large percentage of population are licensed drivers	• Reliance on personal vehicle for elderly taken for granted
• Elderly tend to remain in locales where they worked	• Personal services might not be easily accessed
• 2 of every 5 poor households are elderly	• Two key transit submarkets are converging
Growing Single-Person and Single-Parent Households	
• Household growth is outpacing population growth	• Increase in households increases auto ownership and use
• Single-person and single-parent household growth outpacing population growth	• These households take more trips than 2+ households
• Work and household demands on single-person household more severe than 2+ households	• These households less likely to use transit
Increasing Migration and Immigration	
• Migration from northeast/midwest to south and west	• Movement away from established transit cities
• Migration focused in metropolitan areas	• Important market for transit
• Immigration focused on south and west	• Movement to cities with less transit
• Immigrant workers tend to have low income	• Increasingly important transit market
• Immigrant employment patterns tend to be dispersed	• Typical travel patterns difficult to serve via transit

SOURCE: Cambridge Systematics, 1999

The previously listed factors, and their relationships as shown in Fig. 1.3, have contributed to three fundamental characteristics of the urban travel phenomenon in the United States. First, the number of vehicle miles traveled and the number of trips made have increased proportionately much faster than the individual factors that influence trip making, although with changing demographics, the rate of change in the future might not be as great (Fig. 1.4). Second, the automobile is the predominant mode of urban travel in North America and increasingly in the world [Ross and Dunning, 1997]. Third, the effectiveness of the transportation system is closely tied to household and employment location, transportation availability, and travel characteristics of the population. This latter point will especially hold true for the future.

The Metropolitan Economy   An important characteristic of today's urban economy is that it is defined primarily at a metropolitan scale and is often closely linked to larger state/provincial, national, and international markets [Barnes and Ledebur, 1998]. Unlike previous decades where the central city dominated a region's economy, the economic health of a region now depends on the flow of people, goods, and information within and around a metropolitan area. Given that a transportation system serves as the primary means of moving people and goods, the characteristics of a region's economic base, that is, the types and locations of industries and services, linkages to national and international markets, and corresponding occupations of a region's employees, become important factors in understanding metropolitan travel. For several decades, the shift from a heavy industry and manufacturing base to services has been the major trend characterizing metropolitan economic activity. More recently, changes in the use of information technology and the globalization of the marketplace have created new opportunities and challenges for the business community. In particular, five major economic trends have affected the nature of business transportation requirements:

1.   *Importance of trade and globalization of the economy* (share of trade as a percentage of U.S. gross domestic product [GDP] increased from 12.4 percent in 1970 to 24.8 percent in 1993).

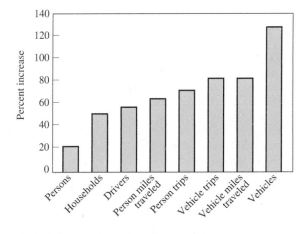

Figure 1.4   Percent increase in travel and corresponding factors (1969–1990)
 SOURCE: Federal Highway Administration, 1993

2.   *Growth of service industries* (from 1992 to 2005, of the top 10 expected fastest-growing U.S industries, three relate to computers, two to medical technology, two to child and elderly care, two to business services, and one to social services).

3.   *Restructuring of traditional manufacturing to increase competitiveness and emergence of high technology and knowledge-based industries* (from 1977 to 1992, the share of technology-based industries in the U.S. manufacturing industry rose from 35 percent to 42 percent).

4.   *Industrial location and demographic trends, including increased flexibility for businesses in their location decisions and an aging population* (illustrated by the dispersal of population and employment to suburban locations, as discussed in the previous section).

5.   *Reduced government roles and increased privatization* (government's share of GDP fell from 19 percent in 1977 to 16 percent in 1992; significant trends have occurred at the federal and state levels to "not stand in the way" of the new knowledge-based industries; the government has also initiated efforts to turn over infrastructure provision and management traditionally held in the public domain to private firms) [Berger International, 1999].

The changing metropolitan economy and the importance of transportation can be illustrated by what has happened over the past 25 years in the Los Angeles metropolitan area [SCAG, 2000]. With a gross regional product of nearly $500 billion, Los Angeles represents the 12th largest economy in the world. From 1970 to 1998, three million jobs were created, with a significant share of these jobs occurring in the region's service sectors. Foreign trade has also become a major sector of the region's economy, growing from $6.2 billion in 1972 to $186 billion in 1997. In addition, the Southern California Association of Governments (SCAG), the region's MPO, estimated that the region receives nearly $14 billion in spending from overseas tourists and travelers. Transportation investment is viewed as a critical ingredient of a regional strategy to expand the economic vitality of the region. As noted by SCAG,

> Economically vital and competitive regional economies must have modern and efficient infrastructures—roads, bridges, highways, rail, energy systems, water and sewer, telecommunications, and airport and air cargo facilities—that facilitate business expansion, relocation, trade, and investment. The most attractive regions for business expansion and investment during the 21st century will be those that give increasing attention to cutting-edge infrastructures that make business operations more efficient and responsive to international economic trends and allow businesses to operate more effectively in the global economy [SCAG, 2000].

The effect of such economic trends on urban transportation is dramatic. A service-oriented economy changes the focus of business users from needing a low-cost, high-capacity transportation system to one that is fast and reliable. An example of the latter is the emergence of just-in-time production processes that require the inputs to a manufacturing process to arrive at the production site "just-in-time" for assembly, thus reducing the need for costly inventory. A transportation system that allows production managers to know with certainty when these inputs will arrive is a critical factor for the success of this type of production process.

A service-oriented economy also results in one that, at least initially and in the absence of technologies that can substitute for transportation, experiences an increased number of service-oriented trips in a metropolitan region. The important caveat in this statement, however, is "in the absence of technologies that can substitute for transportation." With the advent of the Internet and the importance of knowledge transfer, the normal activities of society are becoming very different than in previous years. The potential impact of this transformation on transportation is yet unclear. The scholarly and popular literature refers to this transformation in a variety of ways: the development of virtual communities [Rheingold, 1998]; cyber cities [Mitchell, 1999]; cyberopolis [Roche, 1997]; infocities [Hall, 1996]; telecommunities [Warson, 1995]; and network cities [Batten, 1995], among others.

Some thoughts on likely impacts, in what is arguably the early stages of transition to an information society, include

> Virtual communities that networks bring together are often defined by common interests rather than by common location . . . the story of virtual communities so far, is urban history replayed in fast forward—but with computer resource use playing the part of land use, and network navigation systems standing in for streets and transportation systems . . . as more and more business and social interactions shift into cyberspace, we are finding that accessibility depends even less on propinquity, and community has come increasingly unglued from geography [Mitchell, 1999].

> The old demarcation between work and home will evaporate—in its place will be a shift in the location of work, a new role for cities, a new role for the home, and reshaped communities . . . In half a century's time, it may well seem extraordinary that millions of people once trooped from one building (their home) to another (their office) each morning, only to reverse the procedure each evening . . . All this might strike our grandchildren as bizarre [Cairncross, 1996].

> More telecommuting could change the need for centralized places of business, . . . may erode the retail sector . . . For the first two-thirds of the 20th century, cities had their fates tied to their traditional assets of physical capital. In the knowledge-based economy of the 21st century, cities will have their fates inextricably tied to their human capital assets [Clarke and Gaile, 1998].

> By increasing the flexibility of locational decisions, telecommunications help spatially to rearrange the distribution of work, retail, services, manufacturing, and leisure activities . . . The new city is likely to be characterized by the reduction in face-to-face communication in certain occupations, and a decreasing number of journeys to work with potential to relieve urban congestion [Graham and Marvin, 1997].

Preliminary studies on the impact of telecommunications on urban transportation have not provided answers to the questions of greatest concern to transportation planners: What impact will the information society have on where people live, work, and recreate? and subsequently, What will this mean to urban travel? [Mokhtarian, 1991; Pendyala et al., 1991; U.S. DOT, 1993; Turnbull et al., 1995; Gurstein, 1996; and Salomon and Mokhtarian, 1997.] In one of the few studies that has looked at the impact of the information society on urban form, Horan et al. [1996] suggest that telecommunication technologies reduce the importance of spatial concentration for some industries. However, these industries still need access to a diverse workforce, which suggests a need for an urban location. Salomon also argues that the technologies of transportation and telecommunications are complementary more than substitutable and concludes that cities will stay with us for a long time [Salomon, 1997]. As noted by Giuliano, however, there are few economic incentives for firms or households to locate in dense, high-cost centers, and the policies that would be needed to reduce private vehicle use have no political constituency [Giuliano, 1999]. So, whatever the defining forces of telecommunications on spatial form might be, they do not seem to be strongly constrained by countervailing forces.

There is little doubt, however, that the information system, sensor, and vehicle/network control technologies of intelligent transportation systems (ITS) will have an important role in the future of transportation system operations. Future transportation planning will likely be concerned as much with fiber optic and wireless communication connections of traffic management centers as it will with building highway or transit capacity. In addition, the application of these technologies to system operations could very well cause realignments in the traditional ways transportation services are provided. For example, private vendors might replace government agencies as the provider of technologically advanced transportation services.

In summary, the impact of the information society on urban transportation is unclear. Given the rapid change in information technologies, it is perhaps futile to predict likely impacts because of the unforeseen and unimagined directions these technologies will take. However, metropolitan areas and urban transportation over the next 20 years will likely be facing the early consequences of this transformation. It is therefore important for transportation planners to begin thinking about how information technologies could change urban form and urban transportation in their community.

Societal Concerns and Issues   Meeting the transportation needs of urban areas during the 1950s and 1960s generally meant providing the necessary highway capacity to accommodate an increasing demand for automobile travel. The planning process facilitated "rational" decision making by developing comprehensive plans. These comprehensive plans outlined, in some detail, a pattern for urban development 20 to 25 years into the future, a pattern that often did not materialize. Disenchantment with these comprehensive plans grew because they were perceived as not responding to the information needs of decision makers, and, in general, were considered unresponsive, myopic, and slow in meeting societal needs [Boyce et al., 1970; Hill, 1973; Altshuler, 1979].

By the late 1960s, citizen unrest over the disruption caused by the construction and operation of the large-scale facilities (most often highways) resulting from this process led to a re-examination of comprehensive planning. This dissatisfaction also raised serious questions about the underlying attitudes of the professionals responsible for planning, and generated debate over the implicit assumptions used in the analysis approach. In many cases, new planning studies were initiated to "open" the process to affected interests and other community groups [Gakenheimer, 1976; Pill, 1978]. Instead of developing plans primarily to accommodate auto use, the challenge for transportation planners became one of providing desired levels of mobility through any means, in a fashion complementary to the surrounding urban area. This led to important questions regarding the appropriate role for the automobile, related roles for transit and nonmotorized transportation, the linkage between transportation investment and urban development, and the effectiveness of policies relating to the pricing and financing of transportation infrastructure and services.

Making the urban transportation planning process more inclusive over the past 4 decades has resulted in a greater number of often conflicting issues being addressed by the process. An increasing societal awareness of the consequences of human action on the environment that started in the late 1960s has resulted in an urban transportation planning process concerned with the air, land, water, and ecological impacts of transportation facilities. The importance of economic growth and the increasing linkage to global markets provided new planning emphasis in the 1980s and 1990s on freight movement, intermodal connections, and public/private partnerships. The potentially varying impact of transportation accessibility and mobility on different groups led to a planning emphasis in the late 1990s on "equitable distribution" of investment resources and "environmental justice." Quite often, these different issues have had to be resolved in court. To this day, conflicts in societal values, as they play out in transportation decision making, are often reflected in legal challenges [Garrett and Wachs, 1996].

Table 1.5 presents a review of major planning conferences that have occurred in the United States over the past 40 years. As seen in this table, the professional interest in transportation planning has evolved primarily from one focused on the methods for best locating highways to a much broader consideration of the myriad factors that influence the type and scope of multimodal transportation investment decisions. Not surprisingly, the timing of many of these conferences closely corresponded to federal transportation legislation that mandated new directions for transportation planning. Thus, for example, the 1962 Hershey conference reflected the "3C" (comprehensive, continuing, and cooperative) planning process that surfaced from the 1962 Federal Aid Highway Act. The 1992 Charlotte conference followed the 1990 Clean Air Act Amendments that closely linked the national goal of improving air quality to transportation investment. Likewise, the 1992 Irvine conference followed the Intermodal Surface Transportation Efficiency Act (ISTEA) of 1991, which placed great emphasis on incorporating intermodal connectivity issues into planning and investment decision making.

An important aspect of this evolution has been a critical examination of the impacts of an automobile-based transportation system on society. Many of the reviews in the planning literature have been very critical of these impacts and of the government policies that have allowed them (see, for example, [Kunstler, 1993; McShane, 1994; Carlson et al., 1995; Pucher and Lefevre, 1996; Kay, 1997; Safdie, 1997; and Lewis, 1997]), although others have taken a more favorable view (see, for example, [Dunn, 1998]). At a more general level, societal concerns for environmental consequences of human action, combined with enhanced awareness of the role the automobile plays in many of these consequences, has led to policy attention on actions that provide more environmentally sensitive development. Known in general terms as sustainable development, actual state and local actions have usu-

Table 1.5   Transportation planning-related conferences 1957 to present

Conference	Major Issues
1957 Hartford, CT	Designing urban interstates to fit into an urban environment; importance of comprehensive land use plans and linkage to transportation plans
1958 Sagamore, NY	Extension of interstates into urban areas; linking highway investment to economic development; highway design characteristics; need for comprehensive focus in planning; benefit/cost evaluation strategies
1962 Hershey, PA	Conflict between highway, housing, and land use goals; desire for broader perspective in transportation planning
1965 Williamsburg, VA *Highways and Urban Development*	Cooperative planning among different groups; community values and goals; land use plan coordination with transportation planning; desire for more formalized transportation planning process
1971 Mt. Pocono, PA *Organization for Continuing Urban Transportation Planning*	Linkage between transportation investment and environment; community values and their incorporation into transportation planning; multimodal perspectives; citizen participation
1972 Williamsburg, VA *Urban Travel Demand Forecasting*	Improvements to travel forecasting procedures needed; better understanding of travel behavior and disaggregate urban travel demand forecasting
1975/1976 Airlie House, VA *Evaluation of Urban Transportation Alternatives*	Transportation improvement plans should be multimodal; major mass transportation projects should be planned and implemented in stages; focus should be given to improving operation of existing system; use of cost-effectiveness as criterion for project evaluation
1982 Airlie House, VA *Urban Transportation Planning in the 1980s*	Need for systematic urban transportation planning; more flexibility in planning process; streamline regulations; corridor perspectives; more responsibility to state and local officials
1982 Easton, MD *Travel Analysis Methods*	New travel analysis methods need to be implemented in practice; models should be better able to provide information to decision makers on consequences of actions
1987 Orlando, FL *National Conference on Transportation Planning Applications (held many times hereafter)*	Corridor and local-level scale of planning important; short-term planning becoming more important; microcomputer applications used at all scales of planning; new approaches to data collection being developed
1987 Smuggler's Notch, VT *Understanding the Highway Finance Evolution/Revolution*	Importance of private sector participation in transportation financing; need to include sound fiscal planning as part of the planning process; transportation linkage to important issues such as economic development and tourism
1988 Washington, D.C. *A Look Ahead: Year 2020*	Linkage between transportation investment and economic productivity; need to monitor demographic changes and impacts on travel; environmental impacts; institutional responsibilities; urban form and relationship to transportation investment; role of technology
1989 Williamsburg, VA *Transportation and Economic Development*	Methods needed to evaluate economic development impacts of transportation investment; primary benefits are travel time, cost, and accident savings; transportation as a necessary but not sufficient condition for development

Table 1.5 Transportation planning-related conferences 1957 to present (continued)

Conference	Major Issues
1989 Boston, MA *Statewide Transportation Planning*	Relating planning to decision making; importance of vision; system management; multimodal perspectives in evaluation; role of technology
1990 Irvine, CA *Transportation, Urban Form, and the Environment*	Importance of good data; dynamics of demographic and social changes; transportation and air quality; accessibility and its measurement; judging the effectiveness of the planning process; institutional arrangements and financial innovation
1992 Charlotte, NC *Moving Urban America*	Importance of partnerships to get things done; serving needs of customers and system users; mobility as a goal; social costs of transportation provision and use; importance of public involvement; transportation/land use connection; transportation and air quality; management systems in context of transportation planning; measuring quality-of-life indicators
1992 Irvine, CA *ISTEA and Intermodal Planning*	Focus on effectiveness of intermodal connections; partnerships; role of freight movement in transportation planning; stakeholder participation; performance-orientation in planning; institutional barriers
1992 Seattle, WA *Transportation Planning, Programming, and Finance*	Multimodal planning and programming; transportation and land use; consideration of freight in planning process; need for cooperation among many different groups; importance of demographics in travel characteristics; performance-oriented planning and evaluation
1995 Washington, D.C. *Institutional Aspects of Metropolitan Transportation Planning*	Roles and responsibilities of different actors in planning; importance of public involvement; improvements needed in analyzing transportation and air quality and intermodal planning issues
1996 Coeur d' Alene, ID *Statewide Transportation Planning*	Private sector role in transportation; system preservation as a goal of planning; financial constraint; performance-based planning; incorporating operations issues into planning; freight planning; system monitoring; multistate planning efforts
1998 Irvine, CA *Statewide Travel Forecasting*	Investment methods to provide support for decisions among modes and between capacity and operational improvements; methods need to be tied into asset management; performance measures; integration of economic activities into forecasting; need to test modes that do not exist today; transportation and land use connections
1999 Washington, D.C. *Refocusing Transportation Planning*	Environmental justice; environmental impacts of transportation investment; role of technology and operations in planning; growth management and importance of transportation investment; linking planning to decision making

SOURCE: Weiner, 1999; Meyer, 2000

ally come under the terms of smart growth, growth management, livable communities, and sustainable transportation. In most cases, the argument for sustainable development (at least as it relates to transportation) is that changes in travel behavior must occur in order to minimize transportation's impact on the environment [Banister and Button, 1993; Tolley, 1997; Roseland, 1998; Warren, 1998; Newman and Kenworthy, 1999]. These changes include encouraging travelers to switch from single-occupant vehicles to ride share, transit or nonmotorized means of transportation; developing more benign vehicle technology as it relates to the consumption of resources (e.g., better air pollution controls and use of alternative, nonpetroleum-based sources of energy) [Sperling, 1988]; and pricing the use of transportation to more accurately reflect the true costs to society of its use [Hart and Spivak, 1993; Hohmeyer et al., 1997].

Table 1.6 shows a comparison of transportation planning characteristics as practiced over the past 20 years and the characteristics of a future planning process more concerned with sustainability. Although the profession is a long way from a process focused on sustainability, it is likely that transportation planners will be asked increasingly to provide decision makers with the type of information shown in Table 1.6.

Table 1.6   Traditional transportation planning compared to sustainable development orientation

Characteristic	Traditional Process	Sustainable Development Oriented
Scale	• Regional and network level	• Local, state, national, and global perspective
Underlying "science"	• Traffic-flow theory • Network analysis • Travel behavior	• Ecology • Systems theory
Focus of planning and investment	• Accommodate travel demand • Promote economic development • Enhance system safety • Catch up to sprawl	• Efficient use/management of existing infrastructure • Provide transportation capacity where appropriate (from ecology perspective) • Redevelopment of development sites • Reduce demand for single-occupant vehicles • Reduce material consumption and throughput
Government economic policies	• Promote new development on new land • Economic policy focuses on productivity • Do not include secondary and cumulative impacts in policy analysis	• Promote reuse and infill development • Economic policy is fully integrated with environmental policy • Secondary and cumulative impacts are part of policy decision analysis
Time frame	• 15–20 years planning • 4–8 years for decision-maker interest (elections)	• Short (1 to 4 years) • Medium (4 to 12 years) • Long (12 to — years)
Focus of technical analysis	• Trip-making and system characteristics between origins and destinations • Air-quality conformity • Benefits defined in economic terms	• Relationships between transportation, ecosystem, land use, economic development, and community social health • Secondary and cumulative impacts

Table 1.6   Traditional transportation planning compared to sustainable development orientation (continued)

Characteristic	Traditional Process	Sustainable Development Oriented
Role of technology	• Promote individual mobility • Meet government-mandated performance thresholds to minimize negative impacts • Improve system operations	• Travel substitution and more options • Benign technology • Total life-cycle perspective to determine true costs • More efficient use of existing system
Land use	• Considered as a given based on zoning that accommodates autos • Land use and transportation planning separated	• Integral part of solutions set for providing mobility and sustainable community development • Infrastructure funding tied to sound land-use planning • Increased density and preservation of open space and natural resources
Pricing	• Subsidies to transportation users • True "costs" to society not reflected in price to travel	• Societal cost pricing including environmental cost accounting • Value, that is, transportation priced as utility
Types of issues	• Congestion • Mobility and accessibility • Environmental impact at macroscale • Economic development • Little concern for secondary and cumulative impacts • Social equity (increasingly)	• Global warming and greenhouse gases • Biodiversity and economic development • Community quality of life • Energy consumption • Social equity
Types of strategies	• System expansion/safety • Efficiency improvements • Traffic management • Demand management (from perspective of system operating more smoothly) • Intelligent transportation systems	• Maintenance of existing system • Traffic calming and urban design • Multimodal/intermodal • Transportation–land-use integration • Demand management (from perspective of reducing demand)/nonmotorized transportation • Education

NOTE: Characteristics for sustainable development-oriented process synthesized from Haq, 1997; Maser, 1997; Maser et al., 1998; Newman and Kenworthy, 1999

Transportation Legislation/Regulation   Government legislation and regulation has been one of the important driving forces behind changes in U.S. transportation planning over the past 40 years. The evolution of such legislation and regulation has been dramatic and mirrors to some extent the changing issues discussed in the previous section. Appendix A presents the major pieces of U.S. transportation-related legislation and regulation since 1962. Of note in this list is the inclusion of legislation that has significantly affected transportation planning, but which is not transportation legislation per se. Thus, for example, the Americans With Disabilities Act of 1990, the Clean Air Act Amendments of 1991, and the Executive Order on Environmental Justice of 1994 have provided important linkages between transportation planning and other policy concerns that have filtered down to the metropolitan transportation planning process. In addition, transportation legislation itself has provided important points of departure for new directions in transportation planning. Because of their role in defining the post-interstate transportation program, the following two federal transportation laws are perhaps most relevant to current transportation planning processes.

Intermodal Surface Transportation Efficiency Act (ISTEA) of 1991   The ISTEA was the first piece of federal transportation legislation intended to define the federally aided transportation program in the post-interstate era. Most of the federal transportation aid prior to ISTEA was allocated by program categories, with very little flexibility in using funds from one category to support projects in another. This legislation created a program called "The National Highway System," an approximate 161,000-mile (259,200-kilometer) system of highway routes serving major population centers, border crossings, military bases, ports, airports, and other major travel destinations. This system is now the primary network of the federally aided highway system in the United States. More importantly, however, ISTEA provided new flexibility that allowed funds to be used for different purposes. A program called the Surface Transportation Program (STP) was created that allowed federal funds to be used for such things as transit facilities and services, ride-share, bicycle, pedestrian, and historic preservation projects. Transportation planning was considered an important process for determining the most appropriate mix of projects to be funded with these dollars.

Other important planning components of ISTEA included the formal incorporation of intermodal (i.e., connections between modes of travel for people and freight) issues into transportation planning; the consideration of numerous planning factors (such as system preservation, energy conservation, and economic development) in the metropolitan and state transportation planning process; the development of a financial plan indicating the funding sources for proposed projects; the creation of information management systems to be used by decision makers for monitoring and prioritizing projects (although this requirement was later rescinded except for the case of metropolitan areas having air-quality problems); the reinforcement of the importance of public involvement in the planning process; and the linkage between achieving air-quality goals through the development of a transportation plan and program that conformed to the air-quality plan. And, for the first time, states were required to have a statewide transportation plan that was coordinated with the transportation plans of their metropolitan areas.

Transportation Equity Act for the 21st Century (TEA-21) of 1998   TEA-21 continued many of the planning provisions of ISTEA. Metropolitan and statewide transportation planning processes were required to consider transportation projects and strategies that (*http://www.fhwa.dot.gov/environment/tea21opt.htm*)

1. Support economic vitality by enabling global competitiveness, productivity, and efficiency.

2. Increase the safety and security of the transportation system for motorized and nonmotorized users.

3. Increase the accessibility and mobility options available to people and freight.

4. Protect and enhance the environment, promote energy conservation, and improve air quality.

5. Enhance the integration and connectivity of the transportation system, across and between modes, for people and freight.

6. Promote efficient system management and operation.

7. Emphasize the preservation of the existing transportation system.

TEA-21 also continued the emphasis on public involvement in the planning process and mandated that transit operators and freight suppliers be given the opportunity to comment on transportation plans and programs. In addition, TEA-21 regulations required that opportunities be provided for stakeholder participation in the planning process with special concerns given to minority and low-income populations. The law also sought better integration of the planning process with environmental process requirements (so-called environmental process streamlining).

Although national legislation and regulation play important roles in setting the context for transportation planning, many states/provinces and local governments have their own laws and regulations that often affect transportation decision making in more direct ways. This legal context for transportation planning and decision making can reflect state/provincial environmental laws (that are often more stringent or precise than national laws), regulations establishing institutional relationships for transportation project approval, zoning requirements mandating development approval thresholds for transportation access and system performance, and, more recently, growth management legislation that directs programs of investment in ways designed to minimize environmental and community impacts [Freilich, 1999; Weitz, 1999].

The Technology of Planning   The ability of planners to analyze and evaluate transportation systems directly depends on the tools and methods available to collect data, model system performance, analyze results, and communicate this information to those making decisions. Large-scale land-use/transport network modeling efforts in the 1960s relied heavily on mainframe computers and software packages designed to process large quantities of data. However, the rapid evolution in computer processing capability has had as significant an impact on transportation planning as it has on other aspects of society. Today, desktop computers allow planners to process massive data sets and to utilize more complex mathematical relationships that describe the dynamic behavior of urban travel flows and the interaction between the transportation system and land use [Southworth, 1995]. The personal computer and corresponding software have enabled a new generation of transportation planners to provide detailed predictions of urban travel phenomena at various scales of analysis. Instead of focusing on aggregate flows of vehicles on a coarsely defined network, future analysis could very well disaggregate trip activities and corresponding travel behavior down to the individual traveler.

Two major advancements in the technology of analysis over the past 5 years portend momentous changes in the way transportation planning will likely occur in the future. The first, the use of geographic information systems (GIS), is already commonplace in most planning agencies in North America and Europe. Such systems allow the user to connect different data layers in a database to answer spatial inquiries about different elements of transportation system behavior. For example, planners are now able to identify at a very fine scale the characteristics of households in traffic analysis zones. Before, such identification relied on zonal averages. Important questions concerning different impacts of transportation investment decisions can now be easily analyzed (e.g., what income groups will be negatively affected by the location of a highway, and how many households/businesses will be displaced given different highway alignments?). GIS systems, when combined with travel demand modeling software, allow planners to not only predict person movement on a network but also to assess the socioeconomic and environmental impacts of this movement. GIS systems, when combined with global positioning systems (GPS) and satellite imagery, allow planners to analyze remotely captured data within a spatial and temporal context for important environmental and community planning decisions. Hence, the "richness" of information produced by the urban transportation analysis process is much greater than ever before. As new technologies for data collection and analysis are deployed, urban transportation planning will be able to provide even more detailed information on the consequences of different courses of action.

The second advancement in planning technology relates to the continuing revolution in computer capabilities to analyze large amounts of data at very minute scales of analysis. A current focus of transportation modeling research is the development of microsimulation models that simulate the movement of vehicles in a network down to an individual unit. In the United States, one of the resulting analysis packages, called Transportation Analysis Simulation System (TRANSIMS), differs significantly from the current state-of-practice [Texas Transportation Institute, 1999]. TRANSIMS tracks the simulated movements of individuals, households, and vehicles. Travel is estimated on a second-by-second basis throughout the simulation period. An activity-based travel demand module estimates the number, characteristics, and locations of activities that drive individual trip making during the simulation period. The predicted network travel flows are based on a microsimulation of vehicle interactions that continuously computes operating status of all vehicles and engines on a second-by-second basis. Although it is too early to tell how widely this particular model will be used, the approach of using the enhanced power of computers to better analyze urban travel at very fine levels of detail is a trend that will likely continue for many years.

Summary   This section has presented a brief overview of the important contextual trends that have influenced the evolution of transportation planning. In general, the nature of urban transportation planning can be summarized as evolving [Meyer, 2000]

From Emphasis on large-scale, network-level capital improvements (and on the methods and data support of such decisions)

To Consideration of a wide variety of capital, operational, pricing, lifestyle, and land-use decisions (and on the corresponding technical support).

From Expansion of the transportation system

To System management and preservation.

From Focus on the *efficiency* of highway networks and corresponding levels of service (speed and travel time)

To Focus on the *effectiveness* of multimodal systems operation and broad performance measurement (e.g., equity, accessibility, and mobility).

From Perspective on how to get from point A to point B, focusing on transportation for transportation's sake

To Broader context of transportation's role in a community and in the global, national, state, and local economic market.

From Primary attention to passenger/person movement

To Commensurate attention to freight movement and productivity improvements.

From Vehicle and system technology viewed as a given for technical analysis

To Innovative technologies considered as possible strategies to improve system operation and influence travel behavior.

From Acceptance of land-use patterns as a given and not part of the solutions set

To Use of growth management tools in connection with corresponding transportation policies as a major strategy for dealing with mobility issues.

From Environmental impacts as a project-level mitigation issue

To Linkage between transportation decisions and a broader systems and sustainability framework of ecological and community health.

From Investment benefits and costs considered from the perspective of system users and operators

To Equitable distribution of societal benefits and costs within the concept of a community.

From Perspective on today's system operation (e.g., speed and level of service) as a means of evaluating models, to show that model outputs replicate existing system behavior

To Use of today's system operation for real-time control and development of historic files based on monitoring and measurement that can be used for traffic control strategies.

From Lowest first cost as a primary determinant in making design choices

To Life-cycle considerations, including benefits of reducing material requirements, secondary economic inputs, environmental consequences, and recycling/disposal.

From   What should the planning and/or transportation agency do to "solve" the transportation problem

To   What should *all of us* do to "solve" the transportation problem (e.g., partnerships)?

Something to Think About

And perhaps, as was described earlier, the revolution in the use of information systems and the dash toward the information society leads

From Transportation as the major backbone of a metropolitan area's economy

To Transportation as a secondary, maybe tertiary, support system for the human capital that has become the major driver of modern economic prosperity.

Many of these "new" characteristics of transportation planning are still unfolding and will likely be a major component of transportation planning over the next several decades. The challenge to transportation planners will be in making sure this transition is done in ways that preserve the substance of the planning process while at the same time reflecting the changing technological and market characteristics of society. These trends indicate that an image of transportation planning as a technical process focusing mainly on the transportation system is inappropriate. Many of the social and economic trends of the past decades also suggest that the underlying assumptions of the methodologies used in this technical process may no longer be valid. A new perspective on the transportation planning process and how it responds to the new demands of the decision-making process is needed.

1.4 CHAPTER SUMMARY

1. Several political, economic, and social trends have influenced the substance and form of transportation planning during the past 40 years. These trends have included changes in population and employment characteristics, the metropolitan economy, societal concerns and issues, legislation/regulations, and advances in the technology of planning.

2. One of the most important changes facing society today is the rapid advance in the use of telecommunications technologies in all facets of business and personal interaction. The effect of this transformation on urban form and transportation is not yet known but is something that will deserve a great deal of attention by transportation planners in coming years.

3. There are many transportation planning processes underway in an urban area at any given time, each defined at a different level of complexity and purpose. In general, the focus of transportation planning has in recent years changed from large-

scale infrastructure actions to managing the existing transportation system. The basic purpose of transportation planning, however, has not changed—to generate information useful to decision makers concerning the consequences of alternative transportation actions.

4.   A decision-oriented approach to urban transportation planning should focus on the information needs of the decision makers. These decision makers can include publicly elected officials, transportation agency managers, private sector managers, and corporate officials, although elected or appointed government officials are likely to play the biggest role in determining the characteristics of urban transportation systems. Before the information is ultimately supplied to decision makers, however, the planning process should identify the consequences of decision alternatives and relate the various alternatives to established policy or performance objectives.

5.   Urban transportation planning should include a long-range perspective, a short-range perspective, and a mechanism for linking the two. Long-range planning is a continuing activity that represents a statement of need and policy direction, thereby providing a foundation for short-range transportation decisions. Short-range planning should, in turn, consider the extent to which short-range decisions will change the future design and performance of the transportation system.

6.   The urban transportation planning process presented in this chapter encompasses a broad set of activities, starting with the development of a community vision and continuing through project implementation and operations monitoring.

7.   Opportunities for public involvement should be provided throughout the planning process. There are many different "publics" that are affected by transportation, and thus this public involvement program will likely consist of many different strategies and means of providing opportunities for input. This requires the direct interaction among all groups—especially the professional, the politician, and the public.

8.   The consequences of not making a decision should be examined closely. Avoiding or delaying actions that are necessary for the future of a metropolitan area could lead to significant and eventually more costly solutions later.

QUESTIONS

1.   For one of the factors shown in Fig. 1.3, identify the underlying reasons for the shown trend. What do you think will happen to this factor over the next 15 years? The next 25 years? How will this affect transportation planning?

2.   What impact do you think the information society will have on the use of the transportation system? Use the relationships shown in Fig. 1.3 to trace this impact to the resulting effect on vehicle miles traveled. Are there countereffects that will work against your hypothesized result?

3.   Table 1.6 presents possible characteristics of a transportation planning process oriented toward sustainable development. Compare and contrast one of these

characteristics to the current state of planning. Do you believe society is ready for such a change?

4. Search the Internet for the home page of a metropolitan planning organization (MPO). Examine the documents that are available on this site. How does this MPO define the purpose of transportation planning as evidenced by the information shown on the site? What are the key transportation-related issues this metropolitan area seems to be facing?

5. How do you define multimodal transportation planning? What type of data and analysis tools would you need to successfully undertake multimodal planning?

6. You have been hired by a coalition of local governments in a metropolitan area that in a recent census exceeded the population threshold for having a metropolitan transportation planning process. The local officials want you to explain to them what such a planning process is and what benefit they will gain from it. Write a memorandum not to exceed four pages that provides the desired information.

REFERENCES

Altshuler, A. 1979. The politics of urban transportation innovation. *Technology Review.* May.

Banister, D. 1994. *Transport planning.* London: E&FN Spon.

——— and K. Button. 1993. *Transport, the environment, and sustainable development.* London: E&FN Spon.

Barnes, W. and L. Ledebur. 1998. *The new regional economies.* Thousand Oaks, CA: Sage.

Batten, D. 1995. Network cities: Creative urban agglomerations for the 21st century. *Urban Studies* 32 (2): 313–327.

Berger International, Inc. 1999. *Economic trends and multimodal transportation requirements.* National Cooperative Highway Research Program Report 421. Washington, D.C.: National Academy Press.

Bhat, C. and F. Koppleman. 1993. A conceptual framework of individual activity program generation. *Transportation Research* 27A: 433–466.

Boulding, K. E. 1974. Reflections on planning: The value of uncertainty. *Technology Review.* Oct/Nov.

Boyce, D. E., N. D. Day, and C. MacDonald. 1970. *Metropolitan plan making.* Philadelphia: Regional Science Research Institute.

Bureau of Transportation Statistics. 1999. *Transportation statistics annual report, 1999.* Washington, D.C.: U.S. Department of Transportation.

Cairncross, F. 1996. *The death of distance, how the communications revolution will change our lives.* Cambridge, MA: Harvard Business School Press.

Cambridge Systematics. 1999. New paradigms for local public transportation organizations. *TCRP Report 53.* Washington, D.C.: National Academy Press.

——— et al. 1998. Multimodal corridor and capacity analysis manual. *NCHRP Report 399.* Washington D.C: National Academy Press.

Cambridge Systematics, M. Meyer et al. 1996. *A framework for performance-based planning.* Task 2 Report. National Cooperative Highway Research Program, Washington, D.C.: Transportation Research Board. Feb. 3.

Carlson, D. et al. 1995. *At road's end: Transportation and land use choices for communities.* Washington, D.C.: Island Press.

Clarke, S. and G. Gaile. 1998. *The work of cities.* Minneapolis, MN: University of Minnesota Press.

Dunn, J. 1998. *Driving forces, the automobile, its enemies, and the politics of mobility.* Washington, D.C.: The Brookings Institution.

Dunphy, R. 1997. *Moving beyond gridlock, traffic and development.* Washington, D.C.: ULI—Urban Land Institute.

Federal Highway Administration. 1993. *1990 national personal transportation survey.* Report FHWA-PL-94-010A. Nov.

———. 2000. ITS architecture and standards. *Federal Register.* Docket FHWA-99-5899. Washington, D.C. May 25.

Florida Department of Transportation. Undated. *Implementing the 2020 Florida transportation plan.* Tallahassee, FL.

Foot, D. K. 1996. *Boom bust & echo: How to profit from the coming demographic shift.* Toronto: MW&R.

Freilich. R. 1999. *From sprawl to smart growth, successful legal, planning, and environmental systems.* Chicago: American Bar Association.

Gakenheimer, R. 1976. *Transportation planning as response to controversy: The Boston case.* Cambridge, MA: MIT Press.

Gakenheimer, R. and M. D. Meyer. 1979. Urban transportation planning in transition: The sources and prospects of transportation system management. *Journal of the American Planning Association.* Jan.

Garrett, M. and M. Wachs. 1996. *Transportation planning on trial: The Clean Air Act and travel forecasting.* Thousand Oaks, CA: Sage.

Giuliano, G. 1999. Land use policy and transportation: Why we won't get there from here. In *Transportation Research Circular 492.* Washington, D.C.: National Academy Press. Aug.

Gordon, D. 1991. *Steering a new course: Transportation, energy and the environment.* Cambridge, MA: Union of Concerned Scientists.

Graham, S. and S. Marvin. 1997. *Telecommunications and the city, electronic spaces, urban places.* London: Routledge.

Gurstein, P. 1996. Planning for telework and home-based employment: Reconsidering the home–work separation. *Journal of Planning and Education Research* 15: 212–224.

Hall, P. 1996. *Cities of tomorrow.* Oxford, England: Blackwell.

Haq, G. 1997. *Towards sustainable transport planning.* Aldershot, England: Ashgate.

Hart, S. and A. Spivak. 1993. *The elephant in the bedroom: Automobile dependence and denial.* Pasadena, CA: New Paradigm Books.

Hill, M. 1973. *Planning for multiple objectives.* Monograph Series, no. 5. Amherst, MA: Regional Science Research Institute.

Hohmeyer, O., R. L. Ottinger, and K. Rennings (eds). 1997. *Social costs and sustainability.* Berlin: Verlag-Springer.

Holtzclaw, J. 1995. America's autos and trucks on welfare: A summary of subsidies. *Mobilizing the Region* 15. Feb. 3.

Horan, T., B. Chinitz, and D. Hackler. 1996. *Stalking the invisible revolution: The impact of information technology on human settlement patterns.* Los Angeles: Claremont Graduate University. July 29.

Kay, J. H. 1997. *Asphalt nation: How the automobile took over America and how we can take it back.* New York: Crown Publishers.

Kunstler, J. H. 1993. *The geography of nowhere: The rise and decline of America's man-made landscape.* New York: Simon and Schuster.

Lewis, T. 1997. *Divided highways, building the interstate highways, transforming American life.* New York: Viking Penguin Press.

Maser, C. 1997. *Sustainable community development.* Delray Beach, FL: St. Lucie Press.

Maser, C., R. Beaton, and K. Smith. 1998. *Setting the stage for sustainability: A citizen's handbook.* New York: Lewis.

McShane, C. 1994. *Down the asphalt path: The automobile and the American city.* New York: Columbia University Press.

Meyer, J. and J. Gomez-lbanez. 1981. *Autos, transit and cities.* Cambridge, MA: Harvard University Press.

Meyer, M. D. 1980. *Monitoring system performance: A foundation for TSM planning.* Special Report 190, Washington, D.C.: National Academy Press.

———. 1993. Jumpstarting the move toward multimodal planning. *Transportation Research Circular 406.* Washington, D.C.: National Academy Press. April.

———. 1995. *Alternative performance measures for transportation planning: Evolution toward multimodal planning.* Final Report FTA-GA-26-7000. Washington, D.C.: Federal Transit Administration.

———. 1998. *A toolbox for alleviating congestion and enhancing mobility.* Washington, D.C.: Institute of Transportation Engineers.

———. 1999. Statewide multimodal transportation planning. In J. Edwards (ed.) *Transportation planning handbook.* Washington, D.C.: Institute of Transportation Engineers.

———. 2000. *Refocusing transportation planning.* Special Report. Washington, D.C.: National Academy Press.

Mitchell, W. 1999. *Space, place, and the infobahn, city of bits.* Cambridge, MA: MIT Press.

Mokhtarian, P. 1991. Telecommuting and travel: State of the practice, state of the art. *Transportation* 18 (4): 319–342.

National Cooperative Highway Research Program. 1998. *Multimodal transportation: Development of a performance-based planning process.* Research Results Digest, Number 226. Washington, D.C.: National Academy Press. July.

Newman, P. and J. Kenworthy. 1999. *Sustainability and cities: Overcoming automobile dependence.* Washington, D.C.: Island Press.

Pendyala, R., K. Goulias, and R. Kitamura. 1991. Impact of telecommuting on spatial and temporal patterns of household travel. *Transportation* 18 (4): 383–410.

Pickrell, D. and P. Schimek. 1997. Trends in personal motor vehicle ownership and use: Evidence from the nationwide personal transportation survey. In *Office of highway policy information, Federal Highway Administration, searching for solutions, a policy discussion.* Series Number 17. Feb.

Pill, J. 1978. *Planning and politics: The metro Toronto transportation plan review.* Cambridge, MA: MIT Press.

Pucher, J. and C. Lefevre. 1996. *The urban transportation crisis in Europe and North America.* London: MacMillan.

Rheingold, H. 1998. Virtual communities. In F. Hesselbein (ed.) *The community of the future.* San Francisco: Jossey-Bass Publishers.

Roche, E. M. 1997. Cyberopolis: The cybernetic city faces the global economy. In Crahan and Vourvoulias (eds.) *The city and the world: New York's global future.* New York: Council on Foreign Relations.

Roseland, M. 1998. *Toward sustainable communities.* Gabriola Island, BC, Canada: New Society Publishers.

Ross, C. and A. Dunning. 1997. Land use transportation interaction: An examination of the 1995 NPTS data. In *Office of highway policy information, Federal Highway Administration, searching for solutions, a policy discussion.* Series Number 17. Feb.

Safdie, M. 1997. *The city after the automobile: An architect's vision.* New York: Basic Books.

Salomon, I. 1997. Telecommunications and the "death of distance": Some implications for transport and urban areas. In L. Day (ed.) *Urban design, telecommuting, and travel forecasting conference: Summary, recommendations, and compendium of papers.* Washington, D.C.: U.S. DOT. Nov.

Salomon, I. and P. Mokhtarian. 1997. Why don't you telecommute? *Access* 10 (Spring): 27–29. Berkeley, CA: University of California Transportation Center.

Smith S. 1998. *Integrating intelligent transportation systems within the transportation planning process, An interim handbook.* Report FHWA-SA-98-048. Washington D.C.: FHWA. Jan.

———. 1999. *Guidebook for transportation corridor studies: A process for effective decision making.* NCHRP Report 435. Washington, D.C.: National Academy Press.

Southern California Association of Governments. 2000. *New solutions for a new economic environment.* Los Angeles, CA.

Southworth, F. 1995. *A technical review of urban land use-transportation models as tools for evaluating vehicle travel reduction strategies.* Report ORNL-6881. Center for Transportation Analysis, Oak Ridge National Laboratory, Oak Ridge, TN. July.

Sperling. D. 1988. *New transportation fuels.* Berkeley, CA: University of California Press.

Texas Transportation Institute. 1998. *Urban roadway congestion annual report, 1998.* College Station, TX.

———. 1999. *Early Deployment of TRANSIMS, issue paper.* College Station, TX. June.

Tolley, R. (ed.). 1997. *The greening of urban transport.* New York: Wiley.

Transportation Research Board. 1998. *The costs of sprawl—revisited.* TCRP Report 39. Washington, D.C.: National Academy Press.

Turnbull, K., L. Higgins, D. Puckeu, and C. Lewis. 1995. *Potential of telecommuting for travel demand management.* Research Report 1446-1. Texas Transportation Institute, College Station, TX. Nov.

U.S. Census. 1997. *How we're changing, demographic state of the nation: 1997.* Series P23-193, Washington, D.C. March.

U.S. Department of Transportation. 1993. *Transportation implications of telecommuting.* Washington, D.C. April.

———. 1995. *National transportation system performance measurement conference.* Washington, D.C. Nov. 1–2.

Vuchic, V. 1999. *Transportation for livable cities.* New Brunswick, NJ: CUPR Press.

Wachs, M. 1985. *Ethics in planning.* New Brunswick, NJ: CUPR Press.

Warren, R. 1998. *The urban oasis: Guideways and greenways in the human environment.* New York: McGraw-Hill.

Warson, A. 1995. Building telecommunities. *Urban Land* (May): 37–39.

Weiner, Edward. 1999. *Urban transportation planning in the United States.* New York: Praeger.

Weitz, J. 1999. *Sprawl busting: State programs to guide growth.* Chicago: Planners Press.

2

Transportation Planning and Decision Making

2.0 INTRODUCTION

Because transportation is critical to the social, environmental, and economic health of every metropolitan area, decisions to change this system must be considered with knowledge of the likely impacts of proposed actions and of the consequences if no decision is made. The underlying premise of this book is that, in order to provide such knowledge effectively, the planning process and the related technical analysis should be consistent with the substance and form of transportation decision making. There are, however, many different ways of viewing the decision-making process. Understanding the nature of alternative decision-making processes and the needs and capabilities of those who participate in them[1] are thus prerequisites for the development of an effective transportation planning process. This chapter examines the characteristics of different decision-making models and develops a framework for transportation planning that reflects these characteristics.

2.1 INSTITUTIONAL FRAMEWORK FOR TRANSPORTATION DECISION MAKING

Given that many factors can influence how decisions occur and why certain choices are made, it is difficult to provide a general description of decision making that applies to all situations. However, a common characteristic of all transportation decision making is that it occurs within an institutional framework that is often similar from one metropolitan area to another. Such a framework consists of the *organizations* created to provide and manage transportation services, each having specific (and sometimes contradictory) mandates; the formal *process* of interaction among, and production of outputs from, these organizations that is often mandated by other levels of government; the informal personal and group dynamic *relationships* that make the process work (or slow it down if desired); the political, legal, and fiscal *constraints* that can either provide strong support for desired outcomes or that can

---

[1] An examination of the uncertainty associated with modeling and forecasting led to the observation, "even if predictions are possible and reliable, the use of the information is often suboptimal. Better user interfaces and easily understandable output might be a better improvement than more expensive modeling" [van Zuylen et al., 1999].

become nearly insurmountable barriers; and the positive or negative *roles* of specific individuals or groups.

This institutional environment is especially challenging when a proposal is considered to be new or innovative.[2] Long-standing institutional inertia or comfort with the status quo can make new initiatives difficult to implement unless external pressure is applied by a higher level of government or by citizen activists. Thus, one often hears that the difficulties in project implementation are not technical, but rather "institutional." What this usually means is individuals or organizations are unwilling to change the status quo or that the procedural/political/legal/financial prerogatives of the current institutional environment are hindering implementation. An example of institutional barriers is shown in Fig. 2.1. This figure shows the findings of a study that investigated the institutional issues associated with developing an intermodal planning process in response to the requirements of ISTEA. As noted, these institutional barriers were found in three major areas: organizational constraints, interjurisdictional issues, and resource limitations.

The dynamic nature of the urban transportation system and of the institutional environment within which it operates suggests that dealing with change will be a continuing characteristic of the transportation planning profession. Change in the policy context of transportation planning has certainly characterized the past 10 years in the United States. Beginning with ISTEA (1991), the planning process has had to respond to new demands. As noted by the Advisory Commission on Intergovernmental Relations (ACIR), the new planning processes introduced to the transportation field by ISTEA included "(1) a much more inclusive set of partners; (2) greatly expanded opportunities for public involvement; (3) financially constrained planning and programming; (4) a major increase in quantitative analysis;

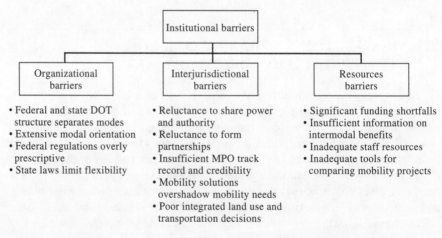

Figure 2.1   Institutional barriers to implementing an intermodal transportation planning process

SOURCE: Crain and Assocs., 1996

---

[2] The organizational literature on barriers to change is extensive, yet informative to this discussion. Readers are referred to the following citations: Jacobs [1994], Rogers [1995], and Kotter [1998].

(5) more attention to operating, maintaining, and managing transportation systems for peak performance (rather than concentrating mostly on new construction); and (6) formal power sharing among transportation agencies." Figure 2.2 shows the factors that were considered essential by one study to improve this MPO-directed intermodal planning process [Crain and Assocs., 1996]. Lockwood [1997] also described those characteristics of the transportation institutional environment that he felt would change the way transportation agencies do business in the future, such as a focus on customer service and innovative financing. He concluded that most transportation agencies were capable of achieving "mid-20th century missions but not well prepared for those of the 21st century" (see also [McDowell, 1988]).

Over the past 20 years, several important trends have strongly affected the institutional environment for transportation planning that will likely characterize the future as well. Changes have occurred in the composition of the institutional environment for planning and decision making, including an increased role for nongovernment groups. Also, the linkage of transportation planning to other public policy concerns has resulted in a corresponding broadening of the mix of projects considered in the planning process. An evolving financial structure for transportation investment has placed greater emphasis on nontraditional sources of revenues. Understanding the changes in each of these institutional contexts for transportation decision making becomes an important point of departure for subsequent discussions on the decision-making processes themselves.

Changing Composition of the Institutional Environment   When urban transportation studies first started in the 1950s, very few government organizations had sole responsibility for transportation planning. These first metropolitan transportation studies created new organizational structures that served as models for planning studies elsewhere. The rapid spread of this comprehensive planning approach is evidenced by the fact that in 1954 only 11 U.S. cities had comprehensive planning agencies, but by 1970, 276 metropolitan areas had transportation planning organizations, and by 2000, this number had reached 339 [ACIR, 1995].

Today, in a typical metropolitan area, a large number of organizations are involved with transportation planning and decision making. In the United States, federal regulations require each urbanized area over 50,000 population to have a metropolitan planning organization (MPO) responsible for transportation planning. The MPO has five core functions:

1. Establish and manage a level playing field for effective multimodal, intergovernmental decision making in the metropolitan area.

2. Develop, adopt, and update a long-range multimodal transportation performance plan for the metropolitan area that focuses on three types of performance: mobility and access for people and goods, system operation and preservation, and quality of life.

3. Develop and continuously pursue an appropriate analytical program to evaluate transportation alternatives and support metropolitan decision making, scaled to the size and complexity of the region and to the nature of its transportation issues and the realistically available options.

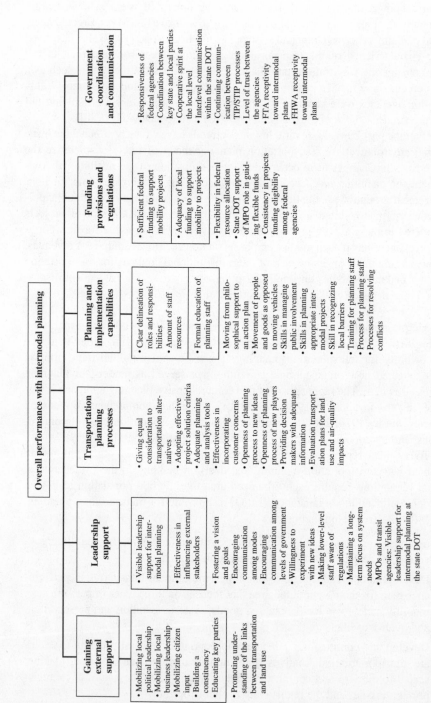

Figure 2.2   Factors for improving intermodal planning
SOURCE: Crain and Assocs., 1996

4. Develop and systematically pursue a multifaceted implementation program designed to reach all the metropolitan transportation plan goals, using a mix of spending, regulating, operating, management, and revenue enhancement tools.

5. Develop and pursue an inclusive and proactive public involvement program designed to give the general public and all the significantly affected subgroups access to and important roles in the four essential functions listed above [ACIR, 1997].

In many cases, MPOs also have responsibilities under state law. For example, California state law requires MPOs to allocate all nonfederal transportation funds in their region, while other states give MPOs limited growth management and land-use planning roles.

In addition to the MPO, most regions have regional and local transit providers; county or local transportation or public works agencies; city planning departments; social service agencies that provide transportation services; representatives from the state transportation and environmental protection agencies; special authorities such as parking, airport, port, or recreational districts with transportation responsibilities; enforcement agencies; and ride-sharing/commute options organizations, just to name a few. Elected and appointed officials are also important parts of the government component of the institutional environment. For example, almost half of the MPOs in the United States are regional councils of government, which consist primarily of local elected officials. Thus, these officials have important influence over the metropolitan transportation program.

In addition to government agencies, participants in the planning process can include a variety of private sector organizations and community groups. In recent years, transportation legislation and regulations have provided for a much more inclusive process that promotes participation by anyone who wants to be involved. Typical participants include environmental groups, community associations, bicycle/walking advocates, business associations, civic groups, freight providers, groups focused on specific issues (e.g., "Stop the Freeway!!"), government groups, developers, trade associations, representatives from disadvantaged populations, professional organizations, and tourism groups.

As one would expect with such a diverse set of actors involved with transportation, transportation planning and decision making are very complex.[3] The MPO is responsible for coordinating the participation of all these groups in the planning process. Because so many stakeholders can be part of this process (and usually align themselves on opposite sides of an issue), a challenge for transportation planning is to provide as much information as possible on the consequences of alternative decisions so that decision makers understand who gains or loses from the decision. In addition, with such a diverse set of interests potentially engaged in the process, transportation planning often becomes a means of educating groups on the underlying causes of transportation problems and on the likely results of actions to solve them.

---

[3] North American metropolitan areas are not alone in having complex institutional structures. A study of four world cities—Paris, New York, London, and Tokyo—observed that, in some cases, national governments provide a strong centralizing influence on the provision of transportation (Paris and Tokyo) and that in others, fragmentation of authority has led to a weakened planning function (New York and London) [London Research Centre, 1998].

Increasing Linkage to Other Policy Goals   Because transportation provides opportunities for social, economic, and community activities, policy makers have turned to the transportation sector as a means of achieving a variety of societal goals. In so doing, they have linked (but not often integrated) the transportation planning process to other planning and policy initiatives. A good example of this linkage is the relationship between transportation and air-quality planning. Figure 2.3 shows how transportation plans, programs, and projects must be shown to conform to the air-quality plan. This conformity process is in place in every U.S. metropolitan area that is not in attainment of national air-quality standards. From an institutional perspective, this linkage introduces a variety of organizations and groups into the transportation planning process, each having a mandate or agenda to make sure the transportation plan reflects their concerns. In addition, this linkage introduces process requirements as well as legal and financial constraints on the transportation investment program.

Similar policy linkages exist between transportation and growth management, welfare, energy, and other environmental policies. The net effect on the institutional environment for transportation is once again to broaden the number and variety of actors involved in the planning process. Unlike the changing composition of the institutional environment described in the previous section, these actors often have little interest in the other purposes served by a transportation system; that is, they are primarily focused on "solving" their problem. This can easily create an adversarial relationship unless approached carefully. For example, interaction between transportation agencies and environmental resource agencies occurs so often that the FHWA has developed guidelines on strategies to foster improved coordination (see [FHWA, 1996; Mickelson, 1998]).

TEA-21 contained language whose intent was to streamline the environmental analysis and project development process. As stated in the proposed regulation to implement this requirement, "a new approach to NEPA (National Environmental Policy Act) is needed, one that emphasizes strong environmental policy, collaborative problem-solving approaches involving all levels of government and the public early in the process, and integrated and streamlined coordination and decision making processes." Importantly, "the transportation planning process needs to be coordinated with the project development/NEPA process so that transportation planning decisions can alternately support the development of the individual projects which arise from the transportation plan" [FHWA and FTA, 2000]. Although the legislative intent was primarily to reduce the amount of time it takes a project to be developed, the effect of the regulation could very well be to more strongly link planning and environmental analysis.[4]

The Mix of Transportation Strategies   The mix of strategies considered in the transportation planning process as described in Sec. 1.2 leads to the participation in the planning process of a variety of groups. In many cases, this mix requires technical skills/knowledge that are very different from those needed to construct infrastruc-

---

[4]   As of the date of this book, the environmental and planning regulations were in the initial stages of being promulgated. Readers should check with the Federal Highway Administration or the Federal Transit Administration to obtain the most up-to-date regulations guiding the planning process.

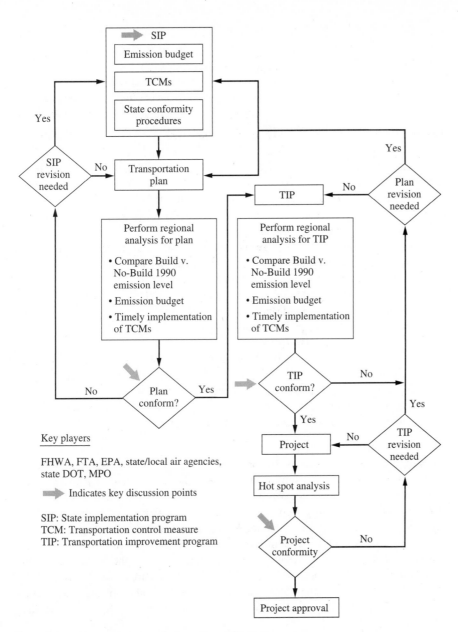

Figure 2.3   Transportation and air-quality conformity process
| SOURCE: U.S. DOT, 1997

ture. The institutional issues relating to the mix of transportation strategies can be found in three areas. First, many of these strategies are implemented at the metropolitan or subregional level. This most likely means that many jurisdictions and organizations have to be involved to coordinate strategy implementation. However, as noted by Cervero, this can be a challenge:

> [T]he diffusion of decision making throughout suburbia has more often than not hindered efforts to engage in meaningful cooperation on problems that transcend municipal boundaries . . . without a regional focus, individual suburban communities often lapse into a 'band-aid' style of problem solving—trying to put out fires by widening an intersection here and retiming signals there . . . [Cervero, 1986].

Downs [1992] agrees with these observations and suggested that "regional anti-congestion policies," such as road pricing, would most likely be adopted only if (1) there was a perception of a regional crisis; (2) there was a belief that regional remedies were essential to solving this crisis; and (3) credible regional institutions existed to implement the program (see also [Dunphy, 1997]).

The second institutional issue relates to the delivery of services that have not commonly been provided by public agencies. This is especially true for those ITS technologies that rely on private firms and companies and that have much faster implementation and upgrade cycles than traditional infrastructure. Organizational capability in providing these types of strategies is a serious question that must be answered when assessing implementation strategies, including whether the public sector should even be involved. Indeed, it is likely that new types of service provisions based on third-party providers or innovative public–private partnerships will evolve to provide these services.

Finally, given that the mix of strategies is so diverse, a large number of organizations, groups, and individuals will likely be involved in any major planning initiative in a region. This means that the planning process will have to provide participation opportunities for a diverse set of interests, each having different perspectives on the likely consequences of project/plan alternatives.

Evolving Financial Structure for Transportation Projects The transportation system represents a substantial investment on the part of the public sector. For example, the value of just the highway network in the United States is estimated to be $1.3 trillion, with state and local governments owning about 98 percent of this asset base. All levels of government in the United States spend about $140 billion annually on the nation's transportation system [BTS, 1999]. Mechanisms to finance the transportation system have been in place for a long time. Some of the first roads in colonial America, for example, were toll roads, as were the canals that followed in the early 1800s. Ever since the 1950s, the most important source of highway finance has been the motor fuel tax. However, there are several concerns with the stability of this funding source:

- Revenues from fuel taxes will fail to keep pace with inflation in that rates are usually fixed and not indexed to the rate of inflation.
- Indexing fuel taxes to the price of fuel can provide a roller-coaster effect, with the revenues increasing or decreasing depending on fuel price.
- Fuel-efficiency increases will reduce the revenue collected per mile of travel.
- The combined effects of inflation and increased fuel efficiency will be to lower the real yields of fuel taxes per mile, while costs per mile will not decrease.

- Petroleum-based fuels may become more scarce or more risky, accelerating a switch to nonpetroleum-based fuels.

- Taxation of alternative fuels can complicate the revenue-raising efforts of all levels of governments.

- Reliance on fuel taxes leaves the door open to proposals to subsidize alternative fuels by taxing them at a lower rate or not taxing them at all.

- The opportunity to use advanced vehicle identification technologies may render fuel taxes obsolete as a means of measuring vehicle use.

- The potential of electric or alternative-fueled vehicles raises the challenge of measuring and reporting fuel consumption by these vehicles [Reno and Stowers, 1995].

Importantly, federal transportation legislation in 1991 (ISTEA) required for the first time that transportation plans and programs be financially constrained, meaning that realistic funding sources had to be associated with all proposed transportation projects in the region. This was a major change from previous practice where most plans and programs included many more projects than could be reasonably funded. In response to fiscal pressures that constrained transportation funding, public officials and planners have sought alternative sources of funding from, in many cases, innovative and nontraditional sources.

Innovation in transportation financing has occurred primarily in three areas: (1) new revenue sources; (2) new roles for the public and private sectors that support tapping new resources and that include a greater role for the private sector in developing, financing, and even owning facilities; and (3) financing structures and techniques that leverage existing revenue sources and encourage private investment. Many innovative financing approaches are often discussed in the context of privatization or public/private partnerships. Public/private partnerships represent any mixture of public and private financial sponsorship that is different than the traditional public sector model of providing transportation infrastructure. In general, the following benefits have been attributed to such partnerships:

- Private sources of funds can extend public funds in supporting more transportation projects.

- Private investors can explore new and untested markets and initiate transportation projects where the government cannot.

- Private sector involvement introduces efficient opportunities for value capture and joint commercial development and may be more likely to take advantage of innovative pricing, marketing, and service strategies.

- Private provision of project construction could proceed much more quickly and efficiently than under public procurement regulations.

- Private sector participation places a premium on life-cycle cost reduction via innovations in design and construction methods and the installation of new technologies such as intelligent transportation systems (ITS) [FHWA, 1994].

Different combinations of financing strategies can be developed for a specific project and for the different partners involved. Figure 2.4 illustrates the different

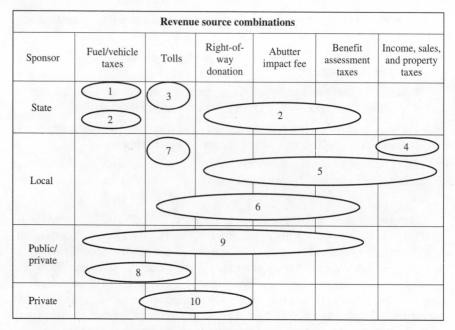

1. Conventional public highway ownership with fuel/vehicle taxes
2. Innovative public/beneficiary mix, especially transportation development corporations
3. Conventional public toll highway
4. Conventional public street/road ownership through general taxation
5. Road utility districts
6. Local jurisdiction toll entity
7. Public toll highway with target in local jurisdiction
8. State franchised public/private toll corporations
9. Full public/private partnerships
10. Privatization

Figure 2.4   Packaging of transportation revenue sources
| SOURCE: Lockwood, 1995

types of financing strategies that could be considered for project development [Lockwood, 1995]. Not only does innovative financing relate to capital costs, but similar concepts can be applied to operating costs as well [Horowitz and Thompson, 1994]. For example, Table 2.1 shows the variety of proposed funding sources to support transit operations in Chicago.

The many sources of transportation funding have, in some cases, made project development much more complex. New financing arrangements have faced questions on the appropriate role for the government and have introduced into the institutional environment groups and organizations having very different goals (Table 2.2). The need to produce realistic transportation plans and programs, however, necessarily requires new and creative approaches to transportation finance. This makes the transportation institutional environment more challenging.

Table 2.1   Sources of operating funding for transit line in Chicago

Who Pays	Funding Source
Riders	Farebox, bulk sales
Property owners	Property tax, special assessments, impact fees, joint development fees, mortgage recording tax, transfer tax
Business	Employment tax, business use tax, income tax, hotel/motel tax, rent
Vehicle owners	Gas tax, parking tax, vehicle registration fees, traffic fines
Public agencies (although ultimately, corporate and individual taxpayers)	City general fund, state appropriation, federal appropriation, transit agency funds
City residents	Sales tax, wage tax
System vendors	Advertising revenues, concession revenues, rent

SOURCE: Horowitz and Thompson, 1994

Table 2.2   Different perspectives in public/private partnerships

Attributes	Private Sector	Public Sector
Stewardship	Private investors	Public trust, safety, and welfare
Response mechanism	Proactive	Response to constituents
Work/assignment orientation	Outcome oriented	Process oriented
Funding	Investment	Budgets, taxes, fees
Usual type of service provided	Specialized, short-term, high-technology applications	Normal engineering and design; emergency and incident management
Decision-making functions	Centralized in headquarters	Moving toward decentralized
Major management functions	Innovates, designs, moving toward operation and maintenance	Directs, plans, operates, maintains, and regulates
Response to opportunities	Flexible	Standardized, regulated
Efficiency	Driven by competition	No competition, except within agencies
Business orientation	Profit seeking, quality improvement oriented	No profits, public service oriented
Production orientation	Sets own pace, progress oriented, tends to be exclusive	Consensus decision-making process, slower procurement, inclusive

SOURCE: As adapted from Hauser, 1999

2.2   AN EVOLVING PERSPECTIVE ON THE PLANNING AND DECISION-MAKING PROCESS

The transportation planning process that evolved out of early experiences with large-scale urban transportation studies in the mid-1950s and early 1960s, and that was then formalized in governmental directives on comprehensive transportation planning in the mid-1960s, was based on the premise of rational choice. That is, the adoption of transportation programs and projects would be determined through a sequence of choices arrived at in a rational manner with significant technical assistance from systems analysis, operations research, and computer models. The set of procedural steps followed by a decision maker in this rational model included [Dror, 1968]:

1. Understand the context for decision making by identifying and weighing societal values and goals.
2. Establish operational objectives for the specific problem area under consideration.
3. Identify all possible alternatives.
4. Evaluate all the consequences of each alternative.
5. Select the alternative whose probable consequences maximize the likelihood of achieving the specified goals.

This view of decision making was reflected in the structure of the community planning processes that began to evolve during the 1950s in response to U.S. federal requirements for the development of comprehensive community plans. These planning efforts first identified community goals and objectives, then examined in a comprehensive way the alternatives for community development by considering all geographic parts of the community and all functional elements that influenced physical development. The recommended alternative was the one that optimized some performance measure.

Within the context of this general trend toward comprehensive planning, transportation planners began to undertake large-scale urban transportation studies that closely reflected the characteristics of the community development planning process. These studies incorporated the assumption of a rational decision-making process into the overall purpose and methodology of transportation planning. The structure of this transportation planning process began with an articulation of policy or community goals, leading to an identification of transportation system problems. Once these problems were identified, alternative solutions were identified and evaluated, and a set of actions was recommended based on which alternatives returned the most benefit for the costs incurred. This process (shown in Fig. 2.5), with minor variations, has served as the basis of most transportation planning efforts for decades.

A good example of this process is provided by the 1962 transportation plan for Chicago [Chicago Area Transportation Study, 1962]. The transportation planning

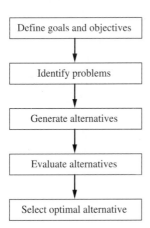

Figure 2.5   The rational approach toward transportation planning

process was described as consisting of "fact gathering, forecasting and plan making" (Fig. 2.6). The stated objective of planning was to provide a transportation system for the Chicago metropolitan area that reduced "travel frictions within the constraints of safety, economy, and the desirable development of land use." As stated in the plan, "it does not matter, basically, whether people in urban areas move by bus, automobile, suburban railroad, or elevated-subway train, as long as the main purpose is achieved."

The most important characteristic of the Chicago plan that reflects the rational nature of its approach was the selection of a single criterion of choice—lowest transportation cost. As described in the plan:

> The single objective chosen is to provide that transportation system for the region which will cost least to build and use over a period of thirty years. In other words, the target is to plan a system the sum of whose measurable costs for all travelers and taxpayers in the region will be at a minimum. . . . Total costs are defined here as construction and travel costs, the latter including time, accident, and other user costs [CATS, 1962].

Table 2.3 shows the recommended Chicago transportation plan in comparison to other alternatives. Note, as mentioned, the index of choice was the dollar cost per vehicle mile of each alternative. As was typical of the plans of this era, the only modes of transportation investigated were automobile and transit (see Fig. 2.6).

Much of the early effort by engineers and planners to structure the planning process focused on the analysis and evaluation techniques that could provide quantitative estimates of the benefits and costs of alternative transportation projects and system configurations. A large-scale transportation network and land-use modeling methodology was developed to provide transportation planners with a means of determining transportation system impacts. In the minds of many engineers and planners, the transportation planning process soon became synonymous with this

Table 2.3   Evaluation matrix for the transportation alternatives in the 1962 Chicago plan

Characteristics	Plan					
	A	**B**	**K**	**L–3***	**I**	**J**
Miles of proposed routes	288	327	466	520	681	968
Cost of completion in millions (after 1960 and including arterial streets)	$907	$1,274	$1,797	$2,007	$2,457	$3,180
Average weekday vehicle miles of assigned travel to arterial and express facilities for 1980 (in thousands)						
Arterials	45,036	41,963	34,380	33,149	31,531	24,245
Expressways	22,878	25,191	33,320	34,414	35,061	41,574
Total	67,914	67,154	67,700	67,563	66,592	65,819
Daily vehicle equivalent hours of travel (in thousands)	2,420	2,283	2,049	1,990	1,937	1,990
Estimated annual traffic fatalities	781	698	638	626	606	638
Estimated daily traffic accidents	504	450	378	359	346	416
Costs converted to cents per vehicle mile						
Travel (accident, time, and operating)	9.10	8.71	8.11	7.96	7.90	8.04
Interest/principal on construction	0.43	0.62	0.86	0.96	1.20	1.57
Total	9.53	9.33	8.97	8.92	9.10	9.61

*Recommended plan
SOURCE: CATS, 1962

modeling methodology.[5] Early research efforts aimed at improving transportation planning and decision making focused almost exclusively on developing quantitative evaluation methodologies, often ignoring the question of what type of decision-making process really existed and what type of information was needed for making

[5] In the United States, federal guidelines to local transportation planners on an appropriate structure for the transportation planning process encouraged this technical approach to planning. *Policy and Procedure Memorandum 50-9*, issued by the Federal Highway Administration in 1969, is a good example of such guidelines [Federal Highway Administration, 1969]. According to this memorandum, the urban transportation planning process was to be based on "the collection, analysis, and interpretation of pertinent data concerning existing conditions and historical growth; the establishment of community goals and objectives; and the forecasting of future urban development and future travel demands."

The comprehensive character of the planning process was found in the requirement

that the economic, population, and land use element be included; that estimates be made of the future demands for all modes of transportation both public and private for both persons and goods; that terminal and transfer facilities and traffic control systems be included in the inventories and analyses; and, that the entire area, within which the forces of development are interrelated and which is expected to be urbanized within the forecast period, be included.

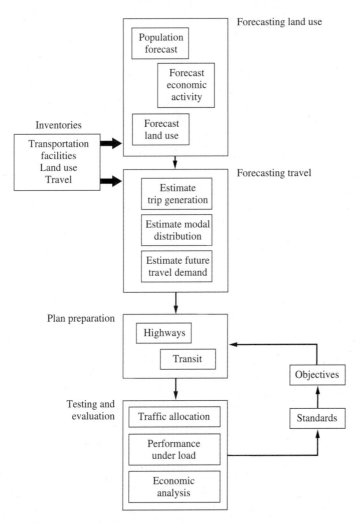

Figure 2.6   The planning process for Chicago's 1962 transportation plan
I SOURCE: CATS, 1962

decisions within this context [Bureau of Public Roads, 1966; Jessiman et al., 1967; Highway Research Board, 1970; Smith, 1971].

In examining the professional and societal context of transportation planning during this period, Altshuler argued that the "rational" nature of such a planning process could be found in the transportation profession's main assumptions, its self-image, its procedures, and its orientation toward key societal values [Altshuler, 1979]. Further, transportation planning was an activity "whose theories and preoccupations at any moment in time closely reflected the government programs and political moods which were dominant or which had been in the recent past." The characteristics of the rational transportation planning process, many of which created severe problems for it when political moods began to change, were identified by Altshuler [1979] as:

1.   Transportation planners assumed there was a general public consensus that the mission of transportation planning was to provide the most cost-effective means of expanding the highway network. Because of this consensus, little attention needed to be paid to those who opposed the transportation policies and programs designed to achieve this objective.

2.   This consensus on the purpose of transportation planning was expected to last a long time because it was based on long-term economic and cultural trends (e.g., the steady growth of motor vehicle travel) that were becoming permanent elements of society. The vision of a future heavily dependent upon the automobile for mobility was firmly ingrained into the professional attitudes of most planners and engineers.

3.   Transportation planners were confident that they could plan for the public without ever dealing directly with elected officials or affected citizens. The assumed public consensus discussed previously gave planners a clear mandate for their activities that, when combined with their expertise, left little doubt that the planners were the critical actors in providing the "best" solutions to the problems facing the community.

4.   The collective concern of the transportation profession was to perfect techniques that would demonstrate the "one best way" to solve given problems. As stated previously, much of the planning research undertaken at this time focused on improving travel-demand forecasting techniques so that planners could better estimate the impacts of alternative project specifications.

5.   Transportation planning was comprehensive in its scope and was structured to produce regional network plans to accommodate long-term forecasts of travel demand. The planning focus was thus "calibrated to the regional scale."

These characteristics of rational transportation planning present an image of a process centralized in a few agencies, dominated by technical procedures, and placing a high value on the provision of personal mobility. By the late 1960s, however, this image of transportation planning was being severely questioned in several cities by large numbers of people who protested the large-scale expansion of the highway network and the local disruption it created. A new image of planning, one based on public participation, development of consensus, and focusing on the amelioration of project impacts rather than accommodating demands for increased personal mobility, started to replace the previous style of planning. The test of good planning was not the ability to produce comprehensive plans consisting of large-scale projects, but rather the ability to produce modest plans focused on site-specific problems. New legislation and governmental guidelines that began to reflect these new characteristics of the planning process provided access to the planning process to those interested in implementing this new perspective (see Appendix A).[6]

---

6   In some situations, local planners adopted a more aggressive role in the planning process and became resources to local opposition groups, helping to transform their often vaguely stated aspirations to well-defined projects and programs. This style of planning, called advocacy planning, represented a complete break from the rationalist planning tradition and reinforced the participatory planning process described above [Goodman, 1971; Lupo et al., 1971; Kaplan, 1973; Catanese, 1974; National Wildlife Federation, 1977; Smith, 1979].

This evolving professional image of the planning process was being reinforced by studies and research that began examining the link between the rational planning model and decision making. Perhaps the best known critique of the rational planning model was made by Braybrooke and Lindblom [1970], who argued that comprehensive planning was doomed to fail because of the practical limits on rationality (i.e., a human's limited problem-solving capabilities, the lack of truly comprehensive information, the high cost of comprehensive analysis, and the difficulty in evaluating values or goals). The decision-making process was defined by these authors as being

1. Incremental or tending toward relatively small changes.

2. Remedial, in that decisions are made to move away from ills rather than toward goals.

3. Serial, in that problems are not solved at one time but are successively attacked.

4. Exploratory, in that goals are continually being redefined or newly discovered.

5. Fragmented or limited, in that only a limited number of alternatives rather than all possible alternatives are considered.

6. Disjointed, in that there are many dispersed "decision-points."

Other studies of planning and its relationship to the political decision-making process generated further questions about the validity of the planning structure and its technical rationality [Meyerson and Banfield, 1970; Altshuler, 1965; Perin, 1967; Rabinovitz, 1969; Friedman, 1971; Rondinelli, 1973; Greenberger et al., 1976].

Within the urban transportation sector, only a few studies were made during this period of the investment decision process. An even smaller number examined how the information generated from the evaluation stage of the planning process was used by decision makers in making their final choice. A 1974 study of the urban political and planning contexts of 12 North American and European cities concluded that the institutional fragmentation of responsibility and influence in the United States had created a structure of decision making that was unresponsive to the transportation needs of urban areas [Colcord, 1974]. A subsequent study by the Office of Technology Assessment (OTA) of the U.S. Congress examined the process by which U.S. metropolitan areas made decisions about the development and modernization of rail transit systems and reached similar conclusions:

> Because no single lead agency has the authority to set priorities among transit improvement projects, plans are developed to offer something for everyone. . . . The division of responsibilities for implementing transit and highway plans makes it difficult to utilize traffic management strategies. . . . Because regional organizations lack the power to control land use, transit cannot be used effectively as a tool for shaping regional development [OTA, 1976].

Studies of the politics of transportation reemphasized the importance attributed to political factors in transportation decision making by these earlier research efforts. Hamer, for example, examined the decision-making and planning processes for investments in rapid rail transit in five U.S. cities [Hamer, 1976]. In general, he found that the planning that supported the decision to construct a rail-rapid transit system was based on faulty projections of key variables and a premature imposition of constraints. Although not the focus of the study, Hamer also suggested that "an interesting hypothesis links the advocacy of rail-rapid transit to an unhealthy relationship between regional decision makers and various pressure groups. These include downtown property owners, contractors, and consultants who might be called on to build the recommended systems."

In addition to these studies, several others examined the importance of political conflict in transportation decision making and planning during this era [Gakenheimer, 1976; Banks, 1977; Steiner, 1978]. The general conclusion that can be drawn from all of these studies is that *the planning of transportation systems is as much a political process as it is a technical one.*

Beginning in the early 1970s, serious questions were raised about the relevance and use of the information generated by transportation planners. These questions were first raised with respect to the large-scale models used to predict land-use and transportation variables [Lee, 1973] and then later extended to the entire methodological approach [Barker and Roark, 1979; de Bettencourt et al., 1979; Schulz et al., 1979]. These critiques focused on the weak linkage between the results of planning and their use by decision makers. The conventional transportation planning process "had not involved the relevant decision makers nor had it adequately informed them on policy issues" [Lee, 1977]. Thus, part of the solution was to understand better the decision-making process and to tailor the planning methodology and analysis style to the information needs of the decision-making body.

To some, understanding the decision-making process meant understanding politics. The politics of transportation decision making included those who had influence in the decision-making process, the formation and composition of coalition groups, the important role played by key elected officials in the final choice, and the importance of funding and those who controlled it.[7] To others, understanding the decision-making process meant undertaking a detailed examination of the institutional environment in which it occurred. A good opportunity for examining this

---

[7] In the United States, federal transportation planning guidelines were changing to reflect these new characteristics of planning. An interesting illustration of this evolution was the changing definition of who was responsible for transportation investment decisions. Originally, state governments were considered the major party responsible for such decisions. The 1962 Federal-Aid Highway Act, however, stated that

> The Secretary shall not approve . . . any program for projects in any urban area of more than 50,000 population unless he finds that such projects are based on a continuing, comprehensive transportation planning process *carried on cooperatively by States and local communities* . . . [emphasis added].

This responsibility was redefined in the 1970 Federal-Aid Highway Act, which specified that

> No highway project may be constructed in any urban area of fifty thousand population or more *unless the responsible public officials of such urban area in which the project is located have been consulted* and their views considered with respect to the corridor, the location and the design of the project [emphasis added].

This act also established a new federal-aid urban highway system and required that the system components be

> . . . selected so as to best serve the goals and objectives of the community as determined by the *responsible local officials* of such urbanized areas [and the routes were to be] . . . *selected by the appropriate local officials and the State highway departments, in cooperation with each other* . . . [emphasis added].

The Federal-Aid Highway Act of 1973, however, amended this section and required that urban system projects be *selected* by appropriate local officials with the *concurrence* of the state highway department. This act also allowed local officials, for the first time, to substitute nonhighway transit projects for highway projects. By 1973, then, the role of local officials in some major federal transportation programs had gone from consulting to shared authority [U.S. Congress, 1977].

environment arose in 1975 when the U.S. DOT issued new planning regulations designed to change the focus of local transportation planning and decision making from a long-range perspective to one more concerned with operational improvements to the existing transportation system. Several studies of the response to this regulation were nearly unanimous in concluding that the institutional relationships existing at the local level were not conducive to the type of planning process envisioned in the regulations [System Design Concepts, Inc., 1977; Meyer, 1978; Jones and Sullivan, 1978]. Specifically, these studies identified the institutional gap between planning agencies and those responsible for project implementation as a critical barrier for successfully changing the planning process as outlined in the regulations. For example, Jones noted,

> The planning style embodied in the regulations does not match the negotiated character of implementation planning. Modal agencies—as opposed to metropolitan planning organizations—are staffed and organized to implement projects, not policy. Project outcomes are structured by funding availability, eligibility criteria, design standards, rules-of-thumb, and political give-and-take. They rarely reflect explicit policy objectives or policy tradeoffs at a regional or systemwide scale. They more typically reflect ad hoc responses to local pressures than the pursuit of system efficiencies [Jones, 1976].

Transportation decision making today is still largely defined by the political and institutional character of a region. For example, although the concept of multimodal transportation planning (that is, a consideration of all modes of transportation equally with unbiased evaluation criteria) has been promoted for many years, such an approach is not commonly found in practice [Meyer, 1993; Rutherford, 1994]. This is true primarily because of the organizationally based focus of service provision; in other words, highway agencies provide highways, transit agencies provide regional transit service, and no one has responsibility for general mobility. Even within one modal category, different jurisdictional responsibilities can create barriers. For example, a study of implementing regional traffic management operations strategies concluded that the major constraint was the fragmentation of responsibility among traffic management agencies both geographically and at different levels of government [Stearman, 1993].

Importantly, the structure of funding for transportation investment in the United States often contributes to this fragmentation by limiting the use of these funds. For example, federal transportation dollars coming from the Highway Trust Fund for many years had to be used for highways only.[8] This constraint was changed in a substantive way when ISTEA allowed a large portion of the highway funding to be "flexed" to nonroad building purposes. However, although such flexibility was provided at the federal level, many states still restricted gas tax revenues to highway investment, and perhaps because of this, initial experience with the use of highway funds for nonhighway purposes has been very limited.

---

[8] As noted by Soberman, in the case of Toronto, spending funds that are not your own results in biases in decision making. "Toronto's best transit decisions were taken directly by accountable agencies in response to locally determined needs when the money they were spending was their own; the poorest decisions were taken in response to funding incentives offered by agencies with no direct accountability, when money they were spending was someone else's" [Soberman, 1997].

In sum, an effective transportation planning process is one that provides opportunities for a wide-ranging examination of many alternatives. It also generates information on the consequences and implementation characteristics of these alternatives that is useful and understandable to those who must make the final decision. The role of the planner thus becomes one of planning *with* the interested public groups and officials rather than planning *for* a unitary general public, as in the rationalist tradition. As will be seen in the next section, however, the role and characteristics of a planning process depend very much on the type of decision-making approach that is assumed. There is no clear consensus on what type of approach really exists for metropolitan-level decision making.

2.3   CONCEPTUAL MODELS OF DECISION MAKING

One of the first difficulties in relating decision making to transportation planning is that different levels of decision making exist in organizations and governments, involving a wide variety of participants and often requiring different forms of information support. There are many differences involved, including (1) type, frequency, structure, and complexity of the decisions; (2) the characteristics, capabilities, and "needs" of the decision makers; and (3) the organizational and political context [Keen and Morton, 1978]. These differences make it very difficult to identify a single model of decision making that can be used to guide the development of a planning process. This chapter focuses on the characteristics of the decision-making process that are evident at higher managerial and political levels, given that most transportation planning activity is focused on influencing these decisions.

Five major conceptual models of decision making relevant to transportation planning emerge from past studies of the decision-making process. These models can be classified as (1) the rational actor approach; (2) the satisficing approach; (3) the incremental approach; (4) the organizational process approach; and (5) the political bargaining approach. These models are based on the principles and concepts of two major disciplines—political science and management science. Because they are such different disciplines, one must be careful in comparing the different models. For example, the incremental, rational, and satisficing models were developed on the basis of single-decision processes, while the organizational process and political bargaining models were developed to reflect the organizational and political settings within which decisions occurred. Even with these different backgrounds, however, the importance of these models to planners lies in their representation of common ways to understand decision making. As will be shown later, different characteristics of the decision-making process imply different strategies for providing information as input to that process. These models can thus provide a useful way to identify the important linkages between planning and decision making.

2.3.1   The Rational Actor Approach

As described earlier, this model traditionally assumes a rational, completely informed set of decision makers whose decision process is based on maximizing the attainment of a set of goals and objectives. Modified versions of the rational model have relaxed some of the more rigid requirements of complete information and have developed a model of rational decision making that recognizes the limitations of decision makers' capability to digest information [March and Simon, 1958; Allison and Zelikow, 1999]. The rational model has most often been used in a normative sense (i.e., as a model of what decision making should be). Indeed, much of the effort in operations research, management science, decision analysis, and systems analysis during the past several decades has adopted the "rational" logic of decision making. Thus, although there is little evidence to support the validity of the rational model as a descriptive tool, it might still prove valuable as a means of formulating an analysis framework and of "forcing" rationality onto the political process.

An example of the rational actor approach toward decision making is the environmental assessment process established in the United States. Although it is certainly a process that is very participatory and subject to public input, the final decision of the preferred alternative rests with the governor of the state. The supporting documentation is intended to examine the environmental consequences of each alternative and to identify possible mitigation strategies. The governor (that is, the rational actor) then "selects" the preferred alternative based on the information provided.

This model requires the most structured and data-intensive planning effort of the five models discussed here. The planning process would be structured to identify all feasible alternatives, compare these alternatives along some set of evaluation criteria, and rank them in order of preference with respect to defined goals and objectives. These tasks would require the development of analysis techniques to provide estimates of all impacts related to the evaluation criteria.

2.3.2   The Satisficing Approach

Critiques of the rational actor approach have focused on its requirement for comprehensive knowledge and the selection of the "optimal" alternative. For most observers of decision making, these requirements are rarely satisfied. In the satisficing model of decision making, decision makers choose alternatives that satisfy some minimum level of acceptability or that induce the least harm or disturbance while conveying some benefit. The search process in decision making is thus best described as satisficing:

> We cannot, within practicable computational limits, generate all the admissible alternatives and compare their relative merits. Nor can we recognize the best alternative, even if we are fortunate enough to generate it early, until we have seen all of them. We satisfice by looking for alternatives in such a way that we can generally find an acceptable one after only moderate search [Simon, 1969].

Even in this model, however, the underlying basis for decision making is rational choice, although this rationality is limited by the resources and ability of the decision maker to acquire and process information. This model of decision making has several important characteristics, including (1) alternatives and consequences of action are discovered sequentially through the search process; (2) each action deals with a restricted range of situations and consequences; (3) decision making is goal oriented and adaptive; and (4) decision makers will define a set of actions that can be implemented in recurrent situations [Simon, 1957; March and Simon, 1958].

A transportation example of the satisficing model is the selection of a particular alignment for an urban expressway that, while assumed to follow the rational actor approach, is in fact satisficing. The number of alignments that could be considered is theoretically infinite; however, only a select few are considered in the planning process. These alternatives and the one ultimately selected have to meet levels of performance in terms of cost, travel times, traffic capacity, and environmental impact. The final decision on an alignment, then, is a choice among relatively few alternatives based on the consideration of relatively few consequences of each. The first alternative alignment meeting the required levels of performance and surviving the process of public scrutiny will usually be the one selected, whether it is in fact the "optimal" alternative.

The type of planning that would support this approach toward decision making is one where acceptable levels of performance are identified and used to develop a feasible set of decision alternatives. Because the satisficing model is still based on rational choice, information on the impacts of alternatives must still be obtained, although the evaluation criteria can be limited to a small set that are most relevant to the decision makers. Attainment of specific goals and objectives still drives the planning process.

2.3.3   The Incremental Approach

This model of decision making argues that decisions are made on the basis of marginal or incremental differences in their consequences [Lindblom, 1959; Braybrooke and Lindblom, 1970]. This approach is different from the rational model in that it presents a limited strategic approach (in both the total number of alternatives considered and in the estimation of their consequences); is remedial in that it "problem solves" rather than proactively seeking to attain community goals; and assumes limited coordination and communication among key decision makers. The characteristics of the incrementalist model are

1. Rather than attempting a comprehensive survey and evaluation of all alternatives, the decision maker focuses on only those policies that differ incrementally from existing policies.

2. Only a relatively small number of policy alternatives are considered.

3. For each policy alternative, only a restricted number of "important" consequences are evaluated.

4. The problem confronting the decision maker is continually redefined. Incrementalism allows for countless "ends–means" and "means–ends" adjustments that, in effect, make the problem more manageable.

5. Thus, there is no one decision or "right" solution, but a "never ending series of attacks" on the issues at hand through serial analyses and evaluation.

6. As such, incremental decision making is described as remedial, geared more to the alleviation of present, concrete, social imperfections than to the promotion of future social goals [Lindblom, 1968].

The incremental approach to transportation decision making is evident in many instances. For example, as an urban corridor experiences increased congestion, traffic management policies can be implemented first to alleviate the perceived problem. Parking and stopping regulations may be implemented first to smooth traffic flow, followed by exclusive bus lanes years later, which may eventually be replaced by a rail facility when demand justifies such level of investment. This evolution of investment policies for a corridor reflects the tendency of decision makers to solve existing problems first with policies that take incremental steps from those in place.

A financial crisis experienced by Boston's transit system in the early 1980s is another example of incremental decision making. With the transit system on the verge of a shutdown because of insufficient operating funds, transit management, the unions, local governments, the state legislature, and the governor all blamed each other for the problems of the Massachusetts Bay Transportation Authority (MBTA). As a shutdown became imminent, local officials refused to allocate any more property tax revenues to the system, while both the legislature and the governor were reluctant to use more state tax revenues. Only after the transit system actually shut down for a day was a temporary solution formulated and a compromise reached, one which allowed the system to operate with emergency funds through the end of the year. A temporary solution was the only kind of response upon which decision makers could agree, while the more controversial issue of completely restructuring the MBTA or its financing was not debated.

The incremental approach raises serious questions about the appropriate role for planning given that decisions are made with limited information, time, and expertise; problem definitions usually vary between different levels of government; and the alternatives differ only slightly from existing policies and programs. At most, the purpose of planning in this decision-making model is to define those alternatives that differ marginally from the status quo and then provide information on the marginal differences among them.

2.3.4 The Organizational Process Approach

This model recognizes the fact that most individuals belong to organizations and that decision making is therefore influenced by the formal and informal structures of the organization, channels of communication, and standard operating procedures. The importance of this context is found in three areas [Allison and Zelikow, 1999]. First, given that governmental action is the output of organizations, decisions made by government leaders trigger organizational routines that guide implementation. Second, these organizational routines often define the "range of effective choice" open to government leaders. The alternatives considered by decision makers often come from agencies or organizational units whose own perception of the scope and severity of the problem heavily influences the set of alternatives presented to decision makers. Finally, policies and programs can be successful only to the extent that the organizations responsible for their implementation have the capability of carrying out their responsibility. For example, the physical resources available to a department of public works can be considered a constraint against which proposals for public works action can be measured.

The organizational process model is particularly well suited to transportation because transportation program and project implementation are guided by standard operating procedures. Highway construction, for example, must usually meet design standards of lane width, clearance, sight distance, and geometrics. In fact, much of the public opposition to expressway construction in the late 1960s can be related to highway design standards that required massive land acquisition. The standards that had been developed for construction of intercity highways were not appropriate in an urban setting and led to significant conflict between the agencies trying to apply the standards and the citizens affected by the projects.

The role of planning in this decision-making model is to provide the necessary information on alternatives to organization decision makers. Perhaps the most important impact on planning of this decision-making model is the significance that implementation has in the overall program and/or project development process. Because organizations are critical for program implementation, understanding the capabilities, skills, and resources of implementing organizations is important information for decision makers when choosing among program alternatives [Elmore, 1978; Nakamura and Smallwood, 1980; Meyer, 1982; Hatch, 1997].

2.3.5 The Political Bargaining Approach

This view of decision making recognizes that the large number of actors involved in a decision often have diverse goals, values, and interests, which creates conflict and a subsequent need for bargaining.[9] The important difference between this model and the rational actor approach is that the outcome of this bargaining process might not be "optimal" in a technical sense. An agreed-upon decision usually represents a compromise, with more controversial aspects potentially ignored or left for future discussion.

Some have argued that the bargaining nature of the decision-making process represents a degree of power sharing among diverse interests that often leads to stalemate [Altshuler and Curry, 1976]. As noted by Doig with respect to the evolution of the Port Authority of New York and New Jersey:

> The American political system is no friend to coherent planning. Fragmentation of power, within the federal system between government and private institutions, is deeply embedded in the constitutional system and in the political culture of the United States. On occasion, an institution is constructed to reach across that fractionalized system, but the jealous concerns of local and state governments, combined with the narrow interests of private corporations, tend to prevent the rhetoric of rational planning from being joined with the capacity to take effective action [Doig, 1993].

[9] For example, the size of policy boards for MPOs can be quite large—Los Angeles, 71 members; Philadelphia, 18 members; Washington, D.C., 33 members; Dallas–Ft. Worth, 223 members; and Atlanta, 39 members.

Others, however, have argued that the search for consensus (and thus the need for bargaining) is necessary for realizing the objectives of most political leaders and is thus the only effective strategy given the existing form of government [Schlesinger, 1973].

A good example of the political bargaining approach in transportation is the formation in 1999 of the Georgia Regional Transportation Authority (GRTA) in Atlanta, Georgia. A new governor, with pressure from the business community, decided that significant institutional change had to occur to solve the mobility problems in the Atlanta metropolitan area. He proposed sweeping changes that would not only take control of transportation investment decision making away from the state DOT and MPO but also would, for the first time, give a state agency veto power over local land-use decisions. In 1999, the state legislature provided much of what the governor wanted, but intense bargaining resulted in some revisions. These revisions included (1) changes to the veto provision (local governments could override); (2) GRTA could not look at airport issues (the responsibility of the city of Atlanta); and (3) the MPO and state DOT would still have primary responsibility for developing transportation plans, although GRTA must approve the transportation improvement program.

The role of planning in a political bargaining process is much broader than that for the other decision-making models. The planning process should be designed to provide as much information as possible on the alternatives being proposed by interest groups, which means that the analysis approach should be flexible enough to respond quickly to requests for information on alternatives that surface during negotiations. Also, the analysis should be sensitive to the issues likely to be raised by competing interests and incorporate as much information as possible in the evaluation results to clarify these issues.

2.3.6 Summary

As was illustrated by these five decision-making models, there are several ways to look at the decision process. All of these models relate in some respects to one another. For example, in each model, some form of rational behavior is assumed, although the rational actor, satisficing, and incremental models are clearly more dependent upon such an assumption than the other two. The organizational process model adds a sociological dimension to the perspective on decision making in that the results of decision making are really a product of the organizational context in which they occur. Finally, the political bargaining model introduces the political nature of decision making and the importance of power distribution.

The importance of these alternative perspectives is that they define, in many ways, the type and purpose of planning activities that would be most effective in each decision-making context (i.e., the matching of planning information with decision-making requirements). The challenge to the planner is to determine which characteristics of a decision-making process make one conceptual model or a combination of models—and the consequent planning process thereby implied—the most relevant description of the decision process. Table 2.4 provides a summary and comparison of the five models presented in this section.

Table 2.4   Comparison of decision-making models

Model of Decision Making	Rational Actor	Satisficing	Incremental	Organizational Process	Political Bargaining
Decision-making behavior assumed	Alternatives are selected to attain some set of predetermined goals and objectives in a utility-maximizing manner.	The first alternative to meet some minimal level of acceptability is selected.	Decision making is geared toward moving away from problems rather than toward the attainment of objectives. Decisions are made on the basis of the marginal differences in their consequences.	Decisions are highly influenced by organizational structures, channels of communication, and standard operating procedures (SOPs).	The decision process is pluralistic and is characterized by conflicts and bargaining.
Characteristics of the decision-making process assumed	All relevant alternatives are considered. Decision makers can attain a comprehensive knowledge of the impacts of each before making a decision. The evaluation criteria used can differentiate accurately among the choices considered. Alternatives can be ranked, and an "optimal" alternative can be selected.	It is impossible to generate all feasible alternatives and to compare them. Alternatives are sequentially discovered. Decision making is goal oriented but adaptive in nature. The underlying choice is rational but is constrained by available resources and the ability to acquire and process information.	Both the number of alternatives and consequences that can be identified are limited, meaning only a small number can be considered. There is limited coordination and communication among decision makers. Decision makers tend to focus efforts on policies differing marginally from those existing. There is no "right" solution, but a continual series of responses to problems. Problems are continually redefined to make them fit solutions. Actions are remedial in nature, addressing present problems, not future objectives.	Government action is the output of organizations. Organizational goals are important in the choice process, as members bargain to satisfy their own goals. Operating routines define the range of alternatives open to decision makers. Alternatives are initially proposed by organizational units with their own perceptions of problems. Selected policies can only be successful when the units chosen to implement them have the capacity to carry out the policy.	The large number of actors involved in decision making, with diverse goals, values, and interests, creates conflict and a need for bargaining. Outcomes of the process are not "optimal" but represent those aspects of a problem on which decision makers can agree. Controversial problems or issues tend to be ignored or put off for future discussion.

Table 2.4 Comparison of decision-making models (continued)

Model of Decision Making	Rational Actor	Satisficing	Incremental	Organizational Process	Political Bargaining
Implications for the planning process	The planning process is highly structured and data intensive. The process consists of the following steps: (1) identify all feasible alternatives; (2) compare them according to evaluation criteria; (3) rank order the alternatives with respect to defined goals; and (4) select the "optimal" alternative.	Because rational choice is still involved, provision of information on the impacts of each alternative is still crucial, but the set of evaluative criteria is limited to those relevant to decision makers. Planners should identify and employ the defined acceptable levels of policy performance to develop a feasible set of alternatives.	Because the alternatives selected differ only slightly from existing policies, planners have to define those alternatives that differ marginally from the status quo and provide decision makers with information on these marginal differences. Little if any information on the impacts of other alternatives is required.	Planners should understand the goals and objectives of the organizations involved so that specific types of information can be incorporated into any analysis. Understanding limits to implementation is important to both planners and decision makers when proposing and choosing among alternative projects or programs.	Planners should have a broader role. An analysis approach flexible enough to respond to the information needs related to alternatives arising from negotiation is needed. Issues likely to be raised by competing interests should be anticipated. Evaluation results must include as much information as possible to clarify these issues.

It is also important to realize that decision making is dynamic and not rigidly structured over time (see, for example, [Gifford et al., 1994]). Thus, in the same city but under different circumstances, alternative views of decision making might be appropriate. Once understood, these decision-making characteristics provide strong guidance on both the type of planning information that is most useful and the most appropriate structure for a planning process.

2.4   THE ELEMENTS OF DECISION MAKING: DEVELOPMENT OF A TRANSPORTATION PLANNING PROCESS

For purposes of this discussion, decision making will be viewed as having five major characteristics. Although when combined these characteristics most resemble the political bargaining model, several characteristics from the other models are also present. The major characteristics of the decision-making process include its being pluralistic, resource-allocative, consensus-seeking or constituency-building, problem-simplifying, and uncertainty-avoiding.

1.   *Pluralistic.* A basic tenet of the rational model of comprehensive planning is that there exists an identifiable public interest, most often defined through goals and objectives, that guide planning efforts. What has become increasingly clear over the past 40 years, however, is that many agencies and groups have interests in public issues and, through legislation and court intervention, have access to the decision-making process. Thus, many "publics" are involved in decision making, a reality that inevitably highlights conflicting objectives and leads to unquantifiable value judgments. Viewed from a slightly different perspective, centralized decision making and planning become most difficult in a metropolitan area that is politically and institutionally fragmented. One consequence of the pluralistic nature of decision making is that the number of public issues and problems brought before decision makers is so great and diverse that the time spent on one particular topic is often quite limited.

2.   *Resource-allocative.* Of all the decisions facing decision makers, perhaps the most difficult is equitably allocating financial and organizational resources. Fiscal austerity in state and local governments and decreasing revenues from higher levels of government create pressures on local officials to "solve" many societal problems with very limited resources. This necessarily leads to making trade-offs.

The resource-allocative purpose of decision making and the constraints imposed by a limited level of available resources focus much of the attention of decision makers (and those who would influence them) on the budget process. Programs and projects will not be implemented unless funding can be provided either from local sources (which automatically places them in competition with other community needs) or from sources external to the local government body, such as the private sector or higher levels of government.

3.   *Consensus-seeking or constituency-building.* The large number of groups involved in and trying to influence the decision-making process, combined with the

limited amount of resources available to satisfy their respective demands, inevitably leads to conflict. Resolution of conflict thus becomes one of the political necessities underlying the decision-making process, resulting in negotiated compromises, bargaining, and the formation of coalitions. The importance of resolving this conflict is underscored by the fact that elected officials (and those appointed by them) hold office for a set period, after which they must again win reelection or be reappointed. One way to avoid alienating any constituency powerful enough to create problems in a reelection effort is to seek a consensus on major decisions during the term in office. This is not to say that every actor in the decision process must agree to a decision or that a decision will not be made if significant opposition arises. The search for consensus is simply a decision maker's attempt to satisfy as many constituent groups as possible.

Development of a constituency is critically important for those projects or programs that will likely have widespread impacts or are perceived as being intrusive. For example, a study of the feasibility of implementing congestion pricing in the United States concluded

> The support of constituency groups willing to be advocates for congestion pricing is important because the American political system is organized in a way that makes change difficult. Proponents of change must be willing to advocate their cause successfully before state agencies, city and county councils, state legislative committees, and high-level government officials. Opponents of change, however, can stymie, if not block, progress by winning at any of these points [TRB, 1994].

Such an observation can often be made for any type of transportation project or program that is considered new or innovative. Figure 2.7 shows strategies recommended by state DOTs, MPOs, and transit agencies to improve intermodal transportation planning in the United States. Note that several of these strategies—building a constituency, educating key parties, inviting citizen input, and mobilizing business leadership—relate to constituency building.

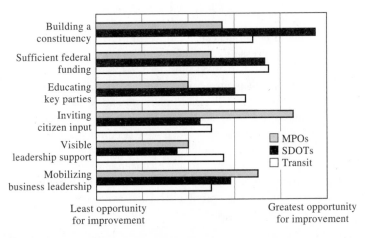

Figure 2.7   Strategies to improve intermodal transportation planning
SOURCE: Crain and Assoc. 1996

4. *Problem-simplifying.* Urban problems by their very nature are often extremely complex. An urban transportation problem can relate to physical patterns of development, land use, and property values; carry with it significant environmental impacts such as neighborhood disruption, noise, and air pollution; and can ultimately serve to alter human activities by changing levels of mobility or accessibility. Decision makers facing such a complex problem may find it very difficult to develop a coherent and effective policy response. To deal with this complexity, decision makers often try to develop separable, well-defined problem definitions that can be handled much more easily, and for which clearer assignments can be made, thus leading to results that are more easily monitored. Reducing the dimensionality of a problem, however, runs a risk of oversimplifying it to such an extent that the solution might address some of the symptoms without solving the root cause of the problem. Indeed, such a solution could actually exacerbate the original problem (through factors not taken into account) or even create new problems.

5. *Uncertainty-avoiding.* Most decisions are concerned with the future. However, because of the political nature of decision making and the need to show effective leadership, decision makers tend to shy away from options with uncertain results. One consequence of this aversion to risk is that the time horizon for many decisions is quite short, at most 2 to 3 years. Beyond this, political and economic forces can drastically change the context of a problem and the likely success of actions taken. This short time frame for decision-maker interest has often led to difficulties in obtaining decision-maker involvement in long-range planning efforts.

Another consequence of avoiding uncertainty is the nature of alternatives considered in the decision process. Except in situations where an individual or a group of individuals has initiated and strongly pushed a new concept, innovation or dramatic changes from previously accepted practice are rare. The uncertainty associated with the "newness" of innovative approaches tends to be sufficient to limit their chances for support, advocacy, or adoption.

These postulated characteristics of decision making—pluralistic, resource-allocative, consensus-seeking, problem-simplifying, and uncertainty-avoiding—suggest a decision process that is based on compromise, negotiation, and bargaining; a process that could be influenced by powerful interest groups; a process in which decision makers necessarily contend with many different issues; and a process oriented more toward perceived short-term issues of importance and crisis response than toward problems of uncertain dimensions likely to face a city 20 years hence. This is not to suggest that decision makers avoid participation in longer-range policy decisions or that some decision makers do not take strong leadership stances. There are many examples where such have occurred. However, the political nature of decision making, the limited time available for deliberation, and the tendency to avoid making choices in the face of uncertainty often lead to the type of decision process described earlier. Thus, the role for transportation planning and for the planner is very important—provide decision makers with the knowledge necessary to make informed decisions. As Wachs notes,

> By focusing on the roles that increased understanding, better information, and improved theories can play within a political framework for decision making, and by working to

understand the political mechanisms of public policy making, a great deal of progress can be made toward achieving a more effective, equitable, and efficient transportation system [Wachs, 1995].

Several key roles for a decision-oriented transportation planning process are discussed in the next section.

2.5 CHARACTERISTICS OF A DECISION-ORIENTED PLANNING PROCESS

The style of decision making described previously not only leads to the identification of the types of information desired by decision makers, but also identifies the type of information needed to overcome what some might consider deficiencies of this process. Within this context, a transportation planning process should address the following:

1. *Establish the future context.* Given the political tendency of decision makers to focus on the short term, it is essential that planners provide a good understanding of the future implications of decisions made today. This understanding might involve an examination of the opportunities forgone by making the decision, an identification of the degree to which a program and/or project diverges from long-range directions set by previous plans, and an analysis of the program and/or project impact on those aspects of community life that can occur only over a longer time frame, such as changes in land-use patterns. A long-range component of the planning process should provide a "framework whose role is constant but whose dimensions may shift; as a distant early-warning mechanism that seeks out both the opportunities and the limitations of the future and offers them as a background for the decisions that must be made today" [Schulz et al., 1979]. In addition, the future context implies having a vision of where a community wants to head, how transportation system performance fits into this vision, and what the implications would be of the no-decision alternative.

2. *Respond to different scales of analysis.* Many of the classical transportation planning efforts of the mid-1960s focused almost exclusively on large-scale, regional transportation system configurations. At this level of analysis, it was very difficult for political decision makers to obtain the information of most interest to them, that is, the benefits of specific improvements and an identification of those who will gain or lose. This, along with the fact that there are decision makers responsible for transportation actions at different jurisdictional levels, implies that the transportation planning process should have a capability to undertake analysis at several scales. Market segmentation, a process of analyzing travel markets from the perspective of travelers who exhibit similar behavior, is an example of how transportation planning must be able to respond to different scales of analysis. Examples of market segments could be transit dependent versus choice riders, commuters versus noncommuters, students, the elderly, special events patrons, and differentiation by geographic location [Elmore-Yalch, 1998].

3. *Expand scope of the problem definition.* As pointed out earlier, transportation affects many aspects of community life and can itself be affected by political, social, and economic factors. Transportation problems and their solutions are not as simple as some might define them. One purpose of planning is therefore to explore beyond the simply defined problem and examine the total system in which it occurs in order to identify both secondary and tertiary impacts of project implementation and alternative means of solution. For example, the shipment of goods in an urban area consists of many interrelated movements and decisions. Problems in urban goods movement could thus be caused by late arrivals at distribution terminals; loading delays caused by management, labor, or contractual work rules; travel delays resulting from inefficient vehicle routing or network congestion; and unloading delays because of congestion on local streets. In this particular case, one must look at the physical and organizational characteristics of the entire trip-making behavior and service delivery pattern to understand fully the range of potential problems (and solutions) related to urban goods movement.

This particular example of urban goods movement highlights an especially difficult problem in transportation planning that once again illustrates the need for a broadened scope of problem definition. Because goods movement is essentially a private sector activity, public planning efforts can run into problems of organizational responsibilities, inadequate information, conflict between different planning approaches for the public and private sectors, and disagreement over appropriate solutions. Dealing with urban goods movement solely from the perspective of the public sector can result in proposals that will not be implemented.

4. *Maintain flexibility in analysis.* Because of the dynamic nature of the political process, an effective planning process must possess a capability for a relatively quick analysis of the program and project alternatives that surface from political debate. The process should be capable of responding with credible information for a comparative evaluation among alternatives. Given the large number of possible alternatives that could be considered, this analysis procedure should also be used to screen initial alternatives to focus debate on those whose benefits and costs clearly dominate the others.

5. *Provide feedback and continuity over time.* The economic, political, fiscal, and social environment in which we live is constantly changing. With this change, new problems arise, and old problems reappear as old solutions are rendered ineffective by changed circumstances. Thus, an important attribute of a planning process is that it be continuous, that is, that it continually monitor environmental conditions and the corresponding response of the transportation system. This monitoring should then be used to update existing plans and programs so as to reflect evolving or changed economic, political, social, and fiscal realities. The short interest span of political decision makers on any one topic implies that planners could, in some ways, serve as the "memory" of the community as transportation issues continually resurface on the public agenda.

6. *Relate to the programming and budgeting process.* With limited resources available for improvements to the transportation system, decision-maker interest is often focused on the programming and budgeting process. The specifics of this process vary from one city or governmental structure to another. In some cases, political decision makers may be interested in specific projects because they are important politically. In other cases, decision makers may not be interested in specific projects, but rather in the overall policy they represent.

To provide this focus on implementation, one of the products of the planning process should include a continually updated plan that consists of a program of action staged over a multiyear period. Those actions programmed for implementation in the near term (i.e., within 1 to 3 years) should be outlined in detail, whereas those actions considered for implementation beyond this time frame can be less detailed. This is exactly the format of the required transportation improvement program (TIP) that must be produced by every MPO in the United States. By relating planning activities to a transportation program, it becomes much easier to see the impact of choosing one project over another or to see how desired results are often contingent upon alternative outcomes of earlier actions [Younger and O'Neill, 1998]. Relating the planning process to programming and budgeting can thus result in the development of an agenda for project implementation that appeals to the action orientation of decision makers.

7. *Provide opportunities for public involvement.* Given the pluralistic nature of decision making, it is important that opportunities be provided to all interested groups to become informed on the objectives and framework of the planning process, to exchange ideas and information with planners and decision makers, and to influence the final outcome of the planning process. As noted earlier, the "public" can include local and regional decision makers, professionals and technical staff who advise officials, organized interests such as business and civic groups, and the general public. Table 2.5 shows the different types of techniques that can be used as part of a public involvement program and the decision context for this information (see also [Wilson and Assocs., 1994]). A review of the planning processes for 70 MPOs found the following examples of good practice for public involvement: (1) comprehensive and easy-to-understand guidelines for participation; (2) hiring a public involvement expert on staff; (3) multiple, specialized advisory boards; (4) focused task forces for problem solving; (5) active consensus-building by citizen advisory committees; (6) broad and innovative notification of involvement opportunities; (7) proactive outreach; (8) two-way electronic communication facilities; (9) direct access to staff and policy board; and (10) evaluation and feedback to improve the process [ACIR, 1997]. Most importantly, successful public involvement is based on direct interaction (two-way learning) between citizens and planning professionals early in and throughout the planning process.

Table 2.5   Community involvement techniques, stratified by applicable stakeholders and involvement objectives

Group Types and Involvement Techniques	Provides General Information	Allows for Community Feedback	Helps Identify/ Resolve Impacts	Provides Key Input to Decisions
Community in General				
Resident panels	X	X	X	X
Business group meetings	X	X	X	
Neighborhood meetings	X	X	X	
Focus groups		X	X	X
Citizens advisory committees	X	X	X	X
Workshops	X	X		X
Newsletter inserts	X			
Documentation on Web	X			
Hot line	X	X		
Public area display		X		X
Opinion polls	X	X		
Community access TV	X	X	X	
Field tours	X	X	X	
Open house		X	X	
Planning		X		X
Community leadership	X		X	
Key interviews	X	X	X	
Speakers bureau	X	X		
Newcomer's packet	X			
Resources	X			
Elected Officials				
Interviews	X	X	X	
Information summaries	X			X
Presentations	X	X	X	X
Individual briefings	X	X	X	
Resolutions	X	X	X	X
Government cable TV	X	X		
Media				
Editorial board	X			
Info summaries/fact sheets	X			
Continuing education	X			
Fact-based editorials	X			
Public panel discussions	X	X		
Press briefings	X			
Resource Agencies				
Strategic meetings	X	X	X	X
Info summaries	X			
Letters		X	X	X
Interagency coordination procedures	X	X	X	X

SOURCE: Smith, 1999

2.6   DEVELOPMENT OF A DECISION-ORIENTED TRANSPORTATION PLANNING APPROACH

The four major stages of a decision-making process most often found in the literature include: problem identification and/or definition, debate and choice, implementation, and evaluation and feedback.

1. *Problem identification and/or definition.* Decisions are made and policies formulated in response to perceived differences between desired states of affairs and the decision maker's perception and/or interpretation of the actual situation. The political process is generally effective in identifying these differences, with the media, lobbyists, elected officials, and constituent groups all playing a role in identifying the problems that should be solved through government action. At other times, however, the political process can lag behind in problem understanding. Also within the political process, different levels of awareness and varied abilities to participate effectively can prevent timely identification of problems. For example, the public opposition to highway projects in many cities during the late 1960s reached dramatic levels because the political decision process had not yet understood the degree to which the highway program was opposed. There was little recourse left for opponents except to create public confrontations to put their issues on the public agenda. To some extent, the current public interest in growth management and concern for community quality of life has preceded the political will to face these issues.

   A critical issue in problem identification is the way in which the problem is perceived and thus defined. For many years, the urban transportation problem was perceived almost exclusively as one of highway congestion, and the response to this problem definition was straightforward—build or expand more highways. This definition has changed during the past decade so that problems associated with urban transportation now include, at the very least, the relationship between transportation and energy consumption, air quality, equity, safety, congestion, land-use impact, noise, and more efficient utilization of fiscal resources. The solutions to these problems are by no means obvious and often require policy action outside of the transportation sector.

2. *Debate and choice.* A decision is a choice among feasible alternatives. Because of limited resources and the need to set priorities, the selection of one alternative over another might occur in an atmosphere of conflict. Thus, a decision process can often consist of bargaining, the consideration of a limited number of alternatives, incremental adjustment to existing situations, and a search for consensus. The specific characteristics of the debate and choice process, however, are dependent upon many of the factors identified previously in the discussion on decision-making models. For the purposes of formulating a planning process, it is assumed here that individual decision makers act rationally within political and resource constraints. However, collective decision making (i.e., where several decision makers must agree to a course of action) places greater emphasis on politics and thus increases the possibility of choices being characterized more by compromise and incremental adjustments and less by optimality.

3. *Implementation.* The most interesting aspect of decision making to many observers of the process has been the manner in which the decision was reached. This was the key element in understanding the outcome of the decision process. Studies of several government programs, however, have concluded that the decision to adopt one policy, program, or project over others is just the beginning in understanding the outcome of a decision [Pressman and Wildavsky, 1979]. Implementing the final decision often implies a new type of politics and a new set of actors [Lloyd and Meyer, 1984; Marzotto et al., 2000]. In transportation, the political feasibility of project or policy implementation will depend on the degree to which individual choice in travel is restricted and where the "blame" for this inconvenience can be placed. Figure 2.8, for example, suggests that the more everyone is affected by a change and the greater probability the instigator of this change can be identified, the longer it will take to implement. In addition, this figure indicates that a strong leadership position by key participants in the process (e.g., a governor, mayor, or secretary of transportation) will tend to shorten implementation time, while the natural tendency of institutional inertia (i.e., the status quo) will act to lengthen it.

Within the transportation arena, the implementation of plans and projects cannot occur until specific projects are scheduled for implementation (called programming), a process requiring a determination of which projects should receive funding. Thus, the process of programming and budgeting transportation projects becomes a critical step in project implementation and in achieving the final objectives of the decision.

4. *Evaluation and feedback.* Because actions taken by governments do not always solve the problems for which they were designed, and given the implementation challenges discussed previously, monitoring is needed to correct any distortions that might occur during implementation. This monitoring also allows management to alter program implementation if desired, permitting an understanding not only of what occurred, but why. Monitoring can involve informal feedback mechanisms (e.g., constituent reactions and complaints, media reporting, or unoffi-

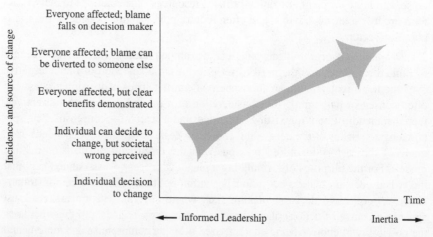

Figure 2.8   Implementation time for innovative programs/projects

cial agency reports) or more formal ones (e.g., use of management systems for performance monitoring).

A transportation planning process should provide not only that information which is of most interest to decision makers (e.g., cost, immediate impacts, and benefits), but also information that gives decision makers a more complete understanding of the important implications of their decision (e.g., opportunities forgone, long-run impacts, and equity issues). Figure 2.9 presents a general framework for a decision-oriented transportation planning process. Figure 2.10 shows how the components in this process relate to the four stages of decision making discussed previously. These planning components inform the decision-making process at each stage so decisions can be made with knowledge of likely consequences.

There are significant differences between the planning process shown and more traditional constructs. First, and perhaps most significantly, planning as shown encompasses a broad set of activities. To some, planning stops after the analysis and evaluation step, with program and/or project implementation and system monitoring (i.e., assessing how well the system is performing) occurring outside the planners' purview. The institutional structure for transportation system management in most metropolitan areas reinforces this view—there are planning agencies to develop plans and implementing agencies to build and operate facilities. However, planners need to be aware of the requirements of the entire process, from initial visioning to project implementation and system operation.

Second, the planning process begins with a vision of what a community desires for the future. The vision as portrayed in the figure reflects the interaction between desired states of prosperity, environmental quality, and social equity/community quality of life. (Note: This representation is often used to denote the important linkages in sustainable development.) This vision can consist of general statements of desired end-states or can be as specific as a defined land-use scenario. Because the vision is a "community" vision statement, this step in the planning process should provide numerous opportunities for public input.

Third, unlike previous depictions of a transportation planning process, performance measures are shown in Fig. 2.10 as a central concept. Performance measures, defined as indicators of transportation system effectiveness and efficiency, focus on the information of greatest concern to decision makers. This information could reflect concerns for system delays, travel-time reliability, average speed, and accident rates, all measures relating to system operations. Performance measures should also reflect the ultimate outcomes of transportation system performance, for example, the level of mobility for disadvantaged populations, pollutant levels from mobile sources, and economic development gains. Performance measures not only define data requirements and influence the development of analytical methods, but they become a critical way of providing feedback to the decision-making process on the results of previous decisions. The use of performance measures, however, becomes problematic if there is no agreement on the goals that are to be achieved. Many types of solutions can be considered in trying to meet the performance targets, and unless they are placed in the context of overall goals achievement, there is a strong possibility that conflicts over which strategies to implement could lead to a decision-making impasse.

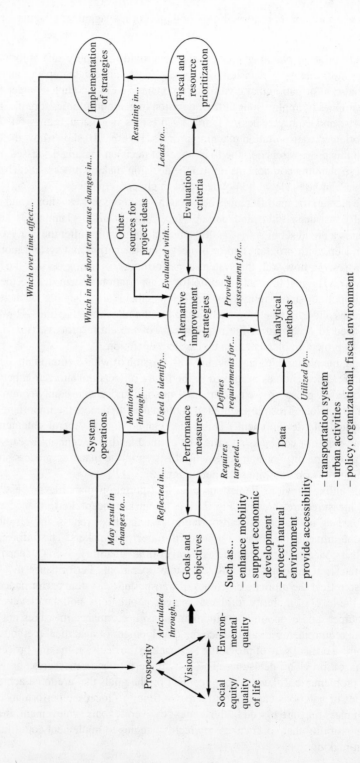

Figure 2.9 Approach to urban transportation planning
I SOURCE: Modified from Cambridge Systematics et al., 1996

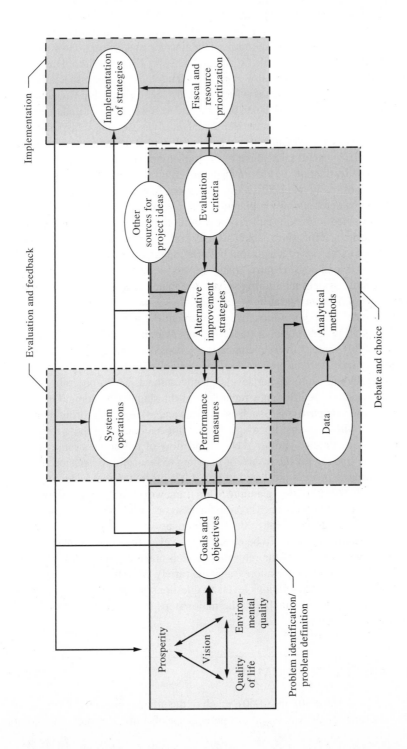

Figure 2.10   Linkage between transportation planning and stages of decision making

Fourth, different types of data are used for planning. Most urban areas already have a large database of transportation system characteristics and an inventory of land-use characteristics (although both are often outdated). This type of information has been the mainstay of transportation planning for many years. However, information relating to the political, organizational, and fiscal environment of transportation decision making is also important. This has become especially true in recent years in the United States with the requirement to develop financially constrained plans. This information can provide planners with a better understanding of the problems and opportunities for improvement in the existing transportation system. Information on the organizational and fiscal state of the transportation program can also provide a useful assessment of the implementation feasibility of alternative projects, an understanding of the organizational requirements of other agencies, and an awareness of likely competition for investment funds.

A final characteristic of the planning framework proposed here is the periodic feedback provided to the original vision definition, goals statement, and identification of performance measures. Analysis and evaluation are undertaken not only to assess the consequences of a decision, but also to better understand the definition of the problem, which may require changing this definition based on preliminary analysis results. System monitoring serves as a major source of information on the performance of the transportation system and, thus, is an important indicator of system deficiencies or opportunities for improvement [Meyer, 1980].

Although Fig. 2.9 portrays transportation planning as a sequential series of steps, frequently there is overlap among these steps. For example, the prioritization process may have to conform to deadlines established by local governments or process regulations from higher levels of government. Thus, program budgeting will likely occur at specific times regardless of the stage of planning. On the other hand, there are some activities, such as analysis and system monitoring, that should occur on a continuing basis. In any given planning agency, all of these steps will be occurring at one time or another. The overall flow of the process must keep pace with decision making for the results of planning to be useful to decision makers.

The planning process presented in Fig. 2.9 is offered to the reader as a general framework for transportation planning. This framework can be applied to establish a more detailed process for specific problem areas or for different planning contexts, for example, as a transportation system planning process, a subarea planning study, or for site-specific planning. To be effective, however, this process must reflect the needs and characteristics of the relevant decision-making process. This implies an approach to planning that explores a wide variety of actions and is capable of responding quickly and at various levels to questions posed by decision makers.

In the remainder of this book, the major steps in the transportation planning process will be discussed in detail.

2.7   CHAPTER SUMMARY

1.   During the late 1950s and 1960s, much of the urban transportation planning for North American cities was based on the concept of rational choice. The compre-

hensive plans that resulted from such planning often recommended system alternatives based on a single criterion of choice, usually lowest transportation cost. The characteristics of this type of transportation planning presented an image of a process centralized in a few agencies, dominated by technical procedures, and placing a high value on the provision of personal mobility. By the late 1960s and early 1970s, a new image of planning, based on public participation and development of consensus and focusing on the amelioration of project impacts, began to replace the previous style of planning in many cities. In some cases, planners provided support to community groups in their involvement in transportation planning, a concept called advocacy planning. Also beginning in the early 1970s, several studies of the politics of transportation decision making highlighted the political factors that characterized such a decision-making process, the formation and composition of coalition groups, the role played by elected officials in the final choice, and the importance of funding and who controlled it.

2.   The role and characteristics of a decision-oriented planning process depend very much on the type of decision-making approach assumed. There are five major conceptual models that have been used to explain the decision-making process—the rational actor approach, the satisficing approach, the incremental approach, the organizational process approach, and the political bargaining approach. The importance of these models to planners lies in their representation of decision making and in the strategies for providing information that each implies. The *rational actor* approach assumes a rational, completely informed set of decision makers whose criterion of decision is maximizing the attainment of an explicit set of goals and objectives. The *satisficing* approach, although still based on the concept of rational choice, suggests that decision makers choose alternatives that satisfy some minimum level of acceptability or induce the least harm or disturbance. The *incremental* approach argues that decision makers focus only on those policies that differ incrementally from existing policies. The *organizational process* approach places decision making within an organizational context and identifies the organizational characteristics that limit or constrain decision-maker choice. The *political bargaining* approach states that decisions result from bargaining and searches for consensus among the many participants in a decision process.

3.   The major characteristics of the decision-making process assumed in this book include its being pluralistic, resource-allocative, consensus-seeking, problem-simplifying, and uncertainty-avoiding. In response to these characteristics, a decision-oriented transportation planning process should perform the following tasks: establish the future context, respond to different scales of analysis, expand the scope of the problem definition, maintain flexibility in analysis, provide continuity over time, relate planning to programming and budgeting, and provide opportunities for public involvement.

4.   The approach to transportation planning described in this chapter consists of a process focused on system performance and providing feedback to earlier steps in the process. In addition, planning is directly linked to the vision established for a community and how this vision relates to stated goals and objectives.

5. Transportation decision making can be viewed as consisting of four major stages: (1) problem identification and/or definition; (2) debate and choice; (3) implementation; and (4) evaluation and feedback. Transportation planning needs to provide input into each stage of this process.

QUESTIONS

1. How did changing political moods and other historical events alter the five steps of the rational transportation planning process shown in Fig. 2.5?

2. For your urban area, review a transportation plan that was produced more than 10 years ago and compare it with a recent plan. What are the major differences between the two plans? What factors do you think contributed to these differences?

3. The different conceptual models of decision making suggest that there may be different roles appropriate for transportation planners under different circumstances.

   (a) Given your knowledge of decision making in your locality, what, in your opinion, is the most appropriate role for transportation planners in responding to the political process?

   (b) With recent trends toward austerity and other government initiatives to alter local government decision making, how would you expect decision making to change in the near future? How can transportation planners adapt to this change?

   (c) Advocacy planning was a big issue in the 1970s. Do you see a future for greater advocacy on the part of planners and "technocrats"? Is there any way that transportation planners can themselves influence the decision-making approach used in their situation?

4. For a transportation project or program with which you are familiar, analyze the decision-making approach used with the help of the models discussed in this chapter. Which model(s) best explains the process used? What factors make the application of any one model difficult? Develop a conceptual framework of the process undertaken.

5. The conceptual models of decision making discussed in this chapter assume an underlying basis of stability in the process. Assume that an urban area suddenly faces a crisis (e.g., a shutdown of transit service, or a fuel shortage). How will the decision-making process likely change in this situation? What impact will this change have on the role of the planner?

6. For each of the characteristics of decision making (pages 68–70), identify a political action you are familiar with that illustrates the characteristic (e.g., what recent political activity would support the notion that the political process is "pluralistic?"). Describe the actual or estimated effect of this characteristic on the policy or project outcome.

7. A leading transportation professor at a local university has proposed to metro-
    politan decision makers that tolls be put on all major highways in the metro-
    politan area. His argument is that modern ITS technologies allow every vehicle
    to be instrumented to measure the amount of driving that takes place on the
    freeway system. At the end of each month, a bill will be issued to the owner of
    the vehicle that represents the amount of highway use that occurred during the
    previous month. By so doing, motorists would be paying their fair share of the
    costs they incur on society. Discuss this proposal from an implementation per-
    spective using Fig. 2.8. Identify those groups who would likely oppose and
    those who would support this proposal. If you were asked to develop a strategy
    to get metropolitan decision makers to approve such a proposal, what would it
    consist of?

REFERENCES

Advisory Commission on Intergovernmental Relations. 1995. *MPO capacity:
    Improving the capacity of MPOs to help implement national transportation
    policies.* Report A-150. Washington, D.C., May.
_____. 1997. *Planning progress, Addressing ISTEA reauthorization in metropoli-
    tan planning areas.* Washington, D.C.: ACIR. Feb.
Allison, G. and P. Zelikow. 1999. *Essence of decision.* New York: Longman.
Altshuler. A. 1965. *The city planning process.* Ithica, NY: Cornell University Press.
_____. 1979. *The urban transportation system: Politics and policy innovation.*
    Cambridge, MA: MIT Press.
_____ and R. Curry. 1976. The changing environment of urban development pol-
    icy—shared power or shared impotence? *Urban Law Annual.* Winter.
Banks, J. 1977. *Political influence in transportation planning: The San Francisco
    Bay Area Metropolitan Transportation Commission's regional transportation
    plan.* Report UCB-ITS-DS-77-1. University of California. Jan.
Barker, W. and J. Roark. 1979. *The role of the urban transportation planner in pub-
    lic policy.* Informal paper series 11. Arlington, TX: North Central Texas Coun-
    cil of Governments.
Braybrooke, D. and C. Lindblom. 1970. *A strategy of decision.* New York: Free
    Press.
Bureau of Public Roads. 1966. *Modal split.* Washington, D.C.: U.S. Department of
    Commerce.
Bureau of Transportation Statistics. 1999. *Transportation Statistics Annual Report
    1999.* Washington, D.C.: U.S. DOT.
Cambridge Systematics, M. Meyer et al. 1996. *A framework for performance-based
    planning.* Task 2 Report. National Cooperative Highway Research Program.
    Washington, D.C.: Transportation Research Board. Feb. 3.
Capital District Transportation Committee. 1996. *Development of an arterial corri-
    dor management strategy for the Capital District.* Albany, NY: CDTC.

Catanese, P. 1974. *Planners and local politics, Impossible dreams.* Beverly Hills, CA: Sage.

Cervero, R. 1986. *Suburban gridlock.* New Brunswick, NJ: Center for Urban Policy Research, Rutgers University.

Chicago Area Transportation Study. 1962. *The 1962 transportation plan.* Chicago, IL.

Colcord, F. 1974. *Urban transportation decision making summary.* Report OST-TPI-76-02. Washington, D.C.: U.S. Department of Transportation. Sept.

Crain and Assocs. 1996. Institutional barriers to intermodal transportation policies and planning in metropolitan areas. *TCRP Report 14.* Washington, D.C.: National Academy Press.

Cyert, R. and J. March. 1963. *A behavioral theory of the firm.* Englewood Cliffs, NJ: Prentice-Hall.

de Bettencourt, J., M. Mandell, S. Polzin, V. Sauter, and J. Schofer. 1981. Making planning more responsive to its users: The concept of metaplanning. *Environment and Planning A.* Vol. 13.

Denver Regional Council of Governments. 1995. *Projects considered in congestion management systems.* Denver, CO.

Doig, J. 1993. Expertise, politics, and technological change: The search for mission at the Port of New York Authority. *Journal of the American Planning Association.* Vol. 59. No. 1. Winter.

Downs, A. 1992. *Stuck in traffic: Coping with peak-hour traffic congestion.* Washington, D.C.: The Brookings Institution.

Dror, Y. 1968. *Public policymaking reexamined.* San Francisco, CA: Chandler.

Dunphy, R. 1997. *Moving beyond gridlock: Traffic and development.* Washington, D.C.: Urban Land Institute.

Elmore, R. 1978. Organizational models of social program implementation. *Public Policy.* No. 2. Spring.

Elmore-Yalch, R. 1998. A handbook: Using market segmentation to increase transit ridership. *TCRP Report 36.* Washington, D.C.: National Academy Press.

Federal Highway Administration. 1969. Guidance on the transportation planning process. *Policy and Procedure Memorandum 50-9.* Washington, D.C.

————. 1994. *Summary of the Federal Highway Administration's symposium on overcoming barriers to public–private partnerships.* Report no. FHWA-PL-94-026. Washington, D.C. Sept.

————. 1996. *Interagency consultation: The key toward collaborative state and local decision making in the conformity process.* Washington, D.C. Oct.

————. 1998. *Transportation planning and ITS: Putting the pieces together.* Report HEP-20/4-98(10M)EW. Washington, D.C.

———— and Federal Transit Administration. 2000. NEPA and related procedures for transportation decision making, protection of public parks, wildlife and waterfowl refuges and historic sites. Notice of Proposed Rulemaking. *Federal Register.* May 25.

Friedman, J. 1971. The future of comprehensive planning: A critique. *Public Administration Review 31.* May/June.

Gakenheimer, R. 1976. *Transportation planning as a response to controversy: The Boston case.* Cambridge, MA: MIT Press.

Gifford, J., W. Mallett, and S. Talkington. 1994. Implementing Intermodal Surface Transportation Efficiency Act of 1991: Issues and early field data. *Transportation Research Record 1466.* Washington, D.C.: National Academy Press.

Goodman, R. 1971. *After the planners.* New York: Simon and Schuster.

Greenberger, M., M. Chenson, and B. Crissey. 1976. *Models in the policy process.* New York: Russell Sage.

Hamer, A. 1976. *The selling of rail rapid transit.* Lexington, MA: D.C. Heath.

Hatch, M. 1997. *Organization theory.* New York: Oxford University Press.

Hauser, E. 1999. Guidelines for developing and maintaining successful partnerships for multimodal transportation projects. *NCHRP Report 433.* Washington, D.C.: National Academy Press.

Highway Research Board. 1970. Transportation analysis: Past and prospects. *Highway Research Record 309.* Washington, D. C.: National Academy Press.

Horowitz, A. and N. Thompson. 1994. *Evaluation of intermodal passenger transfer facilities.* Final report. Federal Highway Administration. Sept.

Jacobs, R. 1994. Why common approaches to organizational change fall short. *Real time strategic change.* San Francisco: Berrett-Koehler.

Jessiman, W., D. Brand, A. Tumminia, and C. R. Brussee. 1967. A rational decision-making technique for transportation planning. *Highway Research Record 180.* Washington, D.C.: Highway Research Board.

Jones, D. 1976. *The politics of metropolitan transportation planning and programming—implications of transportation system management.* Report UCB-ITS-SR-76-11. Berkeley, CA: University of California. Nov.

_____ and E. Sullivan. 1978. TSM: Tinkering superficially at the margin? *Transportation Engineering Journal of ASCE.* Vol. 1. No. TE6. Nov.

Kaplan, M. 1973. *Urban planning in the 1960s: A design for irrelevancy.* Cambridge, MA: MIT Press.

Keen, P. and M. Morton. 1978. *Decision support systems: An organizational perspective.* Reading, MA: Addison-Wesley.

Kotter, J. 1998. Why transformation efforts fail. In Harvard Business Review *On Change.* Cambridge, MA: Harvard Business School Press.

Lee, D. 1973. Requiem for large-scale models. *Journal of the American Institute of Planners.* Vol. 39. No. 3. May.

_____ 1977. *Improving communication among researchers, professionals and policy makers in land use and transportation planning.* Report DOT-TPI-77-10-12. Washington, D.C.: U.S. Department of Transportation.

Lindblom, C. 1959. The science of muddling through. *Public Administration Review.* Vol. 19. Spring.

_____ 1968. *The policy-making process.* Englewood Cliffs, NJ: Prentice-Hall.

Lloyd, E. and M. Meyer. 1984. Strategies for overcoming opposition to project implementation. *Transportation Policy and Decision making.*

Lockwood, S. 1995. Public–private partnerships in U.S. highway finance: ISTEA and beyond. *Transportation Quarterly.* Washington, D.C.: Eno Transportation Foundation. Winter.

_____. 1997. Transportation infrastructure services in the 21st century. In TRB *Circular 450.* Washington, D.C.: National Academy Press.

London Research Centre. 1998. *The four world cities transport study.* London: The Stationery Office.

Lupo, A., F. Colcord, and E. Fowler. 1971. *Rites of way: The politics of transportation in Boston and the U.S. city.* Boston, MA: Little, Brown.

March, J. and H. Simon. 1958. *Organizations.* New York: Wiley.

Maryland Department of Transportation. 1995. *U.S. 301 corridor study: Actions studied.* Baltimore, MD.

Marzotto, T., V. Burnor, and G. Bonham. 2000. *The evolution of public policy: Cars and the environment.* London: Lynne Rienner.

McDowell, B. 1988. Transportation institutions in the year 2020. In TRB, A look ahead, year 2020. *Special Report 220.* Washington, D.C.: National Academy Press.

Meyer, M. 1978. Organizational response to a federal policy initiative in the public transportation sector: A study of implementation and compliance. Unpublished Ph.D. thesis. Department of Civil Engineering, MIT. June.

————. 1980. Monitoring system performance: A foundation for TSM planning. *Transportation Research Board Special Report 192.* Washington, D.C.: National Academy Press.

————. 1982. Public policy development process. *Transportation Research Record 837.* Washington, D.C.: National Academy Press.

————. 1993. Jumpstarting the start toward multimodal planning. *Transportation Research Board Special Report 237.* Washington, D.C.: National Academy Press.

Meyerson, M. and E. Banfield. 1970. *Politics, planning and the public interest.* New York: Free Press.

Mickelson, R. 1998. Transportation development process. *NCHRP Synthesis 267.* Washington, D.C.: National Academy Press.

Nakamura, R. and F. Smallwood. 1980. *The politics of policy implementation.* New York: St. Martins.

National Wildlife Federation. 1977. *The end of the road: A citizen's guide to transportation problem solving.* Washington, D.C.

Office of Technology Assessment. 1976. *An assessment of community planning for mass transit.* Washington, D.C.: United States Congress. Feb.

Perin, C. 1967. The noiseless secession from the comprehensive plan. *Journal of the American Institute of Planners.* Sept.

Pressman, J. and A. Wildavsky. 1979. *Implementation.* Berkeley, CA: University of California Press.

Rabinovitz., F. 1969. *City politics and planning.* Chicago: Atherton.

Reno, A. and J. Stowers. 1995. Alternatives to motor fuel taxes for financing surface transportation improvements. *NCHRP Report 377.* Transportation Research Board. Washington, D.C.

Rogers, E. 1995. *Diffusion of innovations.* 4th edition. New York: Free Press.

Rondinelli, D. 1973. Urban planning as policy analysis. *Journal of the American Institute of Planners.* Vol. 39. Jan.

Rutherford, S. 1994. Multimodal evaluation of passenger transportation. *NCHRP Synthesis 201.* Washington, D.C.: National Academy Press.

Schlesinger, J. 1973. *Systems analysis and the political processes.* RAND Corporation.

Schulz, D., J. Schofer, and N. Pedersen. 1979. An evolving image of long-range transportation planning. *Traffic Quarterly.* Vol. 3B. No. 3. July.

Simon, H. 1957. A behavioral model of rational choice. In H. Simon *Models of man.* New York: Wiley.

_____. 1969. *Sciences of the artificial.* Cambridge, MA: MIT Press.

Smith, H. 1979. *The citizen's guide to planning.* Washington, D.C.: American Planning Association.

Smith, S. 1998. *Integrating ITS within the planning process: An interim handbook.* Report FHWA-SA-98-048. Washington, D.C.: Federal Highway Administration. Jan.

_____. 1999. Guidebook for transportation corridor studies: A process for effective decision-making. *NCHRP Report 435.* Washington, D.C.: National Academy Press.

Smith, W. 1971. Rational location of a highway corridor: A probabilistic approach. *Highway Research Record 348.* Washington, D.C.: Highway Research Board.

Soberman, R. 1997. Rethinking urban transportation: Lessons from Toronto. *Transportation Research Record 1606.* Washington, D.C.: National Academy Press.

Stearman, B. 1993. *Institutional impediments to metro traffic management coordination, Task 5—final report.* Cambridge, MA: Volpe Transportation Center. Sept. 13.

Steiner, H. 1978. *Conflict in urban transportation.* Lexington, MA: D.C. Heath.

System Design Concepts, Inc. 1977. *Operating multi-modal urban transportation systems.* Washington, D.C.: U.S. Department of Transportation.

Transportation Research Board. 1994. Curbing gridlock: Peak-period fees to relieve traffic congestion. vol. 1. *Special Report 242.* Washington D.C.: National Academy Press.

_____. 1997. *The future highway transportation system and society.* Washington D.C.: National Academy Press.

U.S. Congress. 1977. *Urban system study: Report of the U.S. secretary of transportation to the U.S. Congress.* Washington, D.C.: Committee on Public Works and Transportation.

U.S. Department of Transportation. 1997. *Transportation conformity: A basic guide for state and local officials.* Report FHWA-PD-97-035. Washington, D.C.

Van Zuylen, H., M. van Geenhuizen, and P. Nijkamp. 1999. Unpredictability in traffic and transport decision making. *Transportation Research Record 1685.* Washington, D.C.: National Academy Press.

Wachs, M. 1995. The political context of transportation policy. In S. Hanson (ed.) *The geography of urban transportation.* 2d ed. New York: Guilford Press.

Wilson, F. and Assocs. 1994. Public outreach handbook for departments of transportation. *NCHRP Report 364.* Washington, D.C.: National Academy Press.

Younger, K. and C. O'Neill. 1998. Making the connection: The transportation improvement program and the long-range plan. *Transportation Research Record 1617.* Washington, D.C.: National Academy Press.

3

Urban Travel and Transportation System Characteristics: A Systems Perspective

3.0 INTRODUCTION

At the level of the individual traveler or goods movement, transportation is a trip from an origin to a destination taken primarily to accomplish some purpose. At the level of a metropolitan region, transportation is the aggregate of thousands or, in many cases, millions of individual trip-making decisions. These decisions result in vehicle and passenger trips during specific time periods (e.g., peak-hour travel flows). A transportation system consists of the facilities and services that allow these travel movements to occur. The characteristics of these travel flows and of the facilities and services that enable them are basic to an understanding of transportation. In fact, it is the relationship among travel patterns (as determined by individual trip-making behavior); transportation facilities (as shaped by the transportation planning and decision-making processes); and the economic, social, and environmental context of a region that forms the basis of most transportation analysis and policy decisions.

This book views urban transportation from a systems perspective. Chapter 3 begins by describing what a systems perspective means to transportation planning. The chapter then identifies the important urban transportation system impacts that are often the focus of planning activities. The final section presents the important characteristics of travel in a metropolitan area. Transportation planners should be familiar with each of these characteristics as they relate to their metropolitan area because they define the substance and scope of transportation problems, as well as provide useful indications of possible solutions.

3.1 TRANSPORTATION FROM A SYSTEMS PERSPECTIVE

Systems analysis techniques have been used for many years to understand and predict the performance of civil infrastructure systems. These studies have used models to examine both the individual performance of specific system components as well as of the system as a whole (see, for example, [Manheim, 1979; and Jewell, 1986]). More recently, a systems perspective on infrastructure planning and assessment has received increasing attention from researchers and national organizations (see, for example, [Revelle et al., 1997]). This attention coincided with the development of many transportation policies whose intent was to provide transportation systems that were described as "balanced," "integrated," "coordinated," "intermodal," "seamless," and "multimodal." No matter what descriptor was used in combination with the term "transportation," the common element of all the phrases was "system." A systems perspective on transportation is thus an appropriate segue to an examination of the salient characteristics of transportation and urban travel. But first, what is a system?

Interestingly, many different disciplines—ecology, biology, systems engineering, management science, planning, sociology, etc.—have developed definitions of a system that, although focusing on different phenomena, have common characteristics. A definition of a system that includes elements recognizable to all of these disciplines and that will serve as a point of departure for this book, is as follows:

> *A system is a group of interdependent and interrelated components that form a complex and unified whole intended to serve some purpose through the performance of its interacting parts.*

As will be seen in the following sections, a transportation system fits this definition well. In particular, there are several commonly accepted characteristics of systems that are useful in understanding a systems perspective of transportation. Figure 3.1 shows in very simple terms the concepts that are described in more detail in the following sections.

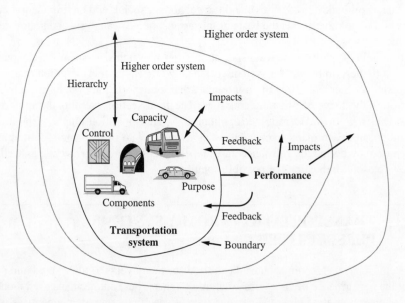

Figure 3.1   Important variables for a systems perspective of transportation

3.1.1   System Hierarchy

A common theme in most disciplinary perspectives is that every system is a part of another system. This perspective leads to a focus on the linkages and relationships among these systems and on the overall importance of one to another. A system hierarchy thus provides order and function to the operation of the individual components in the context of more global system goals. How this system hierarchy is defined affects how one views problems and conducts planning.

Functionally, the transportation system is just one of many systems that allow urban areas to exist (Fig. 3.2). Others might include water supply and distribution, electric power grids, sanitary and waste facilities, education facilities, public safety services, and telecommunications infrastructure. Relevant planning questions at this level in the hierarchy include: How does the transportation system interrelate with these other systems to provide livable communities? What is the cause and effect of relative investment in transportation as compared to other infrastructure systems as it relates to community development? What are the demands placed on transportation systems by changes in the urban system?

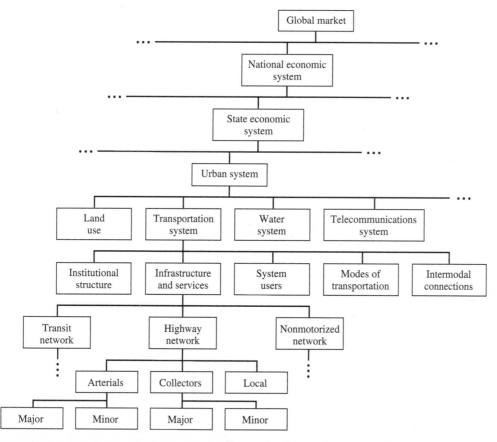

Figure 3.2   Transportation in a systems hierarchy

Urban systems, however, are just one part of larger ecological, community/land use, and economic systems (i.e., looking vertically in Fig. 3.2). For these larger systems, the hierarchical influence can range from the community level to regions, states/provinces, nations, and to the world (e.g., a metropolitan area's role in the global market or contribution to global warming). Relevant planning questions from this perspective might include: How does the transportation system as it operates in an urban system affect larger economies and environments? What goals can be established for transportation systems that will achieve desirable effects at higher levels? How is the cause–effect relationship between transportation system performance and other systems' performance modified as one gets to higher levels of the hierarchy?

Finally, a transportation system itself consists of a hierarchy of subsystems, components, and relationships. Starting at its lowest level, this hierarchy consists of individual vehicles or travelers, facilities and terminals, and modal networks/services, all leading to an integrated transportation system with effective connections among all of the other system elements (e.g., transit services, pedestrian facilities, airports, ports, and bicycle paths). But even within each subsystem, a hierarchy can be established to define function and form. For example, it is quite common to establish a highway functional classification scheme that categorizes roads based on the function they serve in the highway network—those that provide mobility are called arterials, and those that primarily provide access to land are classified as local roads. An example of how transportation system components can be considered in a hierarchical structure is found in the 1999 transportation plan for Portland, Oregon [Portland METRO, 1999].

Regional Motor Vehicle Functional Classification

Principal arterials should provide an integrated system that is continuous throughout the urbanized area; serve the central city, regional centers, industrial areas, and intermodal facilities; connect freight routes; and provide direct service from each entry point to each exit point.

Major arterials should serve as primary links to the principal arterial system and are intended to provide general mobility within the region. Motor vehicle trips between central city, regional centers, industrial areas, and intermodal facilities will occur on these routes. The principal and major arterial networks should comprise 5 to 10 percent of the road network, but carry between 40 to 65 percent of the vehicle miles traveled.

Minor arterials complement and support the principal and major arterial network but are primarily oriented toward motor vehicle travel at the community level connecting town centers, corridors, main streets, and neighborhoods. They serve shorter trips and thus balance a mobility function with that of accessibility to land. The minor arterial network should comprise 15 to 25 percent of the road network, but carry between 65 to 80 percent of total vehicle miles traveled.

Collectors operate at the community level to provide local connections to minor and major arterials. They carry fewer vehicles with reduced speed, and serve as freight access routes. The collector network should comprise 5 to 10 percent of the road network and carry between 5 to 10 percent of the total vehicle miles traveled.

Local streets, which provide for local circulation and access, are the primary connections between neighborhoods and give access to adjacent land uses. The local street network should comprise 65 to 80 percent of the road network, but carry 10 to 30 percent of the total vehicle miles traveled.

Public Transportation Service Hierarchy

The *regional transit network* provides fast and frequent transit service to the central city, regional centers, industrial areas, and intermodal facilities. The network consists of five major service types operating at frequencies of 15 minutes or less—light rail transit, commuter rail, rapid bus, frequent bus, regional bus, and street cars.

A *community transit network* is focused more on accessibility and service coverage; thus, speed is not the major concern. Service types will include community bus, minibus, paratransit and park-and-ride.

Interurban public transportation services provide transit opportunities for intercity movements. Included in this service are passenger rail, intercity bus, and passenger intermodal terminals or transfer points.

Transit service for special needs populations provides service to students, the elderly, economically disadvantaged, mobility impaired, and others with special needs.

Regional Freight System Functional Classification

Main roadway route	Road connectors	Main railroad lines
Branch railroad lines	Marine facility	Railroad facility
Air cargo facility	Distribution facility	Truck terminal
Intermodal railyard		

Regional Bicycle Functional Classification

Regional access bikeways focus on accessibility to and within the central city, regional centers, and some larger town centers. These bikeways have higher volumes, and travel time is important.

Regional corridor bikeways provide point-to-point connectivity between the central city, regional centers, and larger town centers. They are of longer distance and have high speeds.

Community connector bikeways connect small town centers, main streets, rail stations, industrial areas, and regional attractions.

Multiuse paths with bicycle transportation function connect work sites, schools, transit stations, stores, and other work/recreational/shop destinations. These paths are shared with walkers, skaters, joggers, etc.

Regional Pedestrian System Functional Classification

Pedestrian districts are areas of high or potentially high pedestrian activity where a walkable environment is desired. Such areas include the central city, regional and town centers, and light rail stations.

Transit/mixed-use corridors are located along transit lines and will be the subject of increased development densities. Substantial pedestrian activity can be expected in retail areas, schools, and parks.

Multiuse paths with pedestrian transportation function are paved, off-street regional facilities that accommodate pedestrian and bicycle travel and are designed to meet the requirements of the Americans With Disabilities Act.

One of the major purposes of functional classification and service schemes is to provide guidance on the level and type of designs that are appropriate for each facility. Often, technical manuals are used to provide such guidance (see, for example, [AASHTO, 1994]). However, a functional classification approach to system definition and hierarchy can create impediments to multimodal transportation planning if the definition is too modal or facility oriented. Transportation funding programs where funds can be spent only on specific modal networks or facility types instead of being used for the "best" transportation solution is an example of such an impediment.

An alternative approach to system definition has been to define travel markets or market segments that focus on the travelers or goods movers instead of which modal network is being used. Total flows or movement of people and goods are most important in such an approach, and the transportation system is characterized by functionality, that is, the purpose it serves. For example, a study of freight movement in eastern Washington state utilized Standardized Industrial Classification (SIC) codes to characterize commodity flows. The transportation system was viewed in this study simply as a means of transporting these commodities to market [Transmanagement et al., 1998].

Planning Applications of System Hierarchy

• Transportation can be viewed as one system that relates to and is part of many other systems. This perspective leads to important planning questions reflecting the interaction among (a) transportation and the other systems that help an urban area function, and (b) transportation and higher-level systems, for example, ecological or economic systems.

• Transportation systems themselves consist of hierarchies or classification schemes that relate to the respective role of different system components, for example, a road functional classification system or transit service categories.

3.1.2 System Purpose

Every natural and engineered system serves some purpose within the overall system hierarchy. In some cases, this purpose can be fairly self-contained and focused on a small niche in the hierarchy. In other cases, a particular system could be an important component in the effective functioning of other systems, and thus have a high level of interdependence. A transportation system can be viewed in both ways. One perspective, focusing primarily on the transportation function itself, identifies the purpose of a transportation system as providing opportunities for mobility and accessibility, defined as follows [Meyer, 1995]:

> Mobility: *The ability and knowledge to travel from one location to another in a reasonable amount of time and for acceptable costs.*

> Accessibility: *The means by which an individual can accomplish some economic or social activity through access to that activity.*

A broader definition of system purpose would establish the linkage between transportation system performance and other systems such as the economy, environment, or community. In this perspective, transportation system performance becomes an "enabler" of other system activities as well as an important cause of negative impacts. Thus, for example, transportation enables economic activity to occur or community interaction to happen, while at the same time, it could be a major source of air pollutants. The evolution of transportation planning over the past 40 years has broadened the definition of system purpose to include these important linkages as considerations in planning activities and in decision making. In the United States, for example, TEA-21 identified several issues that had to be considered by the planning process, including economic vitality, safeguarding the environment, energy conservation, and quality of life. The Portland transportation plan referenced earlier gives a good example of the diverse set of purposes that the transportation system is supposed to serve. According to this plan, the region's transportation policies should [Portland METRO, 1999]

1. Protect the economic health and livability of the region.

2. Improve the safety of the transportation system.

3. Provide a transportation system that is efficient and cost-effective, investing limited resources wisely.

4. Provide access to more and better choices for travel in the region and serve special access needs for all people, including youth, the elderly, and disabled.

5. Provide adequate levels of mobility for people and goods within the region.

6. Protect air and water quality and promote energy conservation.

7. Provide transportation facilities that support a balance of jobs and housing.

8. Limit dependence on any single mode of travel and increase the use of transit, bicycling, walking, and carpooling and vanpooling.

9. Provide for the movement of people and goods through an interconnected system of highway, air, marine, and rail systems, including passenger and freight intermodal facilities and air and water terminals.

10. Integrate land use, automobile, bicycle, pedestrian, freight, and public transportation needs in regional and local street designs.

11. Use transportation demand management and system management strategies.

12. Limit the impact of urban travel on rural land through the use of green corridors.

The process of defining the purpose(s) of the transportation system occurs as part of the visioning process and in defining goals and objectives. An overall vision of a community includes (either explicitly or implicitly) a sense of how the transportation system fits in. The very nature of goals and objectives as most often defined in transportation studies reflects different perceptions of what the transportation system should be achieving. Thus, goals set often includes statements ranging from desirable network performance levels (a perspective of interest to engineers and planners) to the positive impact of system performance on an urban economy (a perspective of interest to business leaders and elected officials). The proactive involvement of the public and stakeholders early in the process results in a diverse set of transportation system purposes being included in the vision for the region.

Planning Applications of System Purpose

• The purpose(s) of a transportation system can be defined in a variety of ways and often reflects the perspective of those involved in the planning process. The definition could be narrowly targeted on the specific functioning of the system or more broadly reflect the enabling influence of transportation on other systems.

• Defining the transportation system purpose during the planning process most often occurs in the vision statement and in the statement of goals and objectives.

3.1.3   System Boundary

Given that systems are parts of other systems, analysts often establish a boundary that permits them to focus on the key relationships within and across this boundary. The application of this concept in transportation planning is seen in the study boundaries that are established at the beginning of each planning effort to define the focus of analysis and the domain for decisions. Clearly, the scale of the problem will influence the size of the system boundary. For example, an investigation of traffic impacts at a new development site will encompass surrounding transportation facilities and services and perhaps key locations in the road network some distance away that will be affected by the new trips being generated. At the other end of the scale, the boundary of a metropolitanwide planning study could include numerous counties and, on occasion, often more than one state, for example, the metropolitan areas of Washington, D.C., St. Louis, Chicago, and New York City.

The level of analysis within the boundary is usually quite detailed, whereas the activities outside the boundary are represented in very general ways. For example, passenger flows across a study boundary that originate from or are destined to the "outside world" are represented simply as external nodes. Those trips internal to the boundary are considered internal-to-internal trips; those that cross the boundary are considered external-to-internal or vice versa depending on the direction; and those that pass through are external-to-external trips. The boundary itself can be used as a cordon line to judge whether predicted passenger flows match measured flows, thus acting as a quality control for the validity of the model used to predict such flows (Fig. 3.3).

A key challenge to transportation planners is establishing the appropriate definition of the system boundary, especially as it relates to the economic and environmental contexts of transportation and in defining impacts on other systems. For example, the Mid-Ohio Regional Planning Commission of Columbus, Ohio, conducted a study of intermodal freight transportation in the region that included analysis of economic conditions in southeast Asia, the origin of the goods that might use such facilities for a port of entry [Mid-Ohio Regional Planning Commission, 1994]. In this case, the study boundary did not stop at jurisdictional lines and, in fact, crossed national boundaries. In the environmental arena, transportation systems can have impacts that extend far beyond the areas adjacent to facilities and often cannot

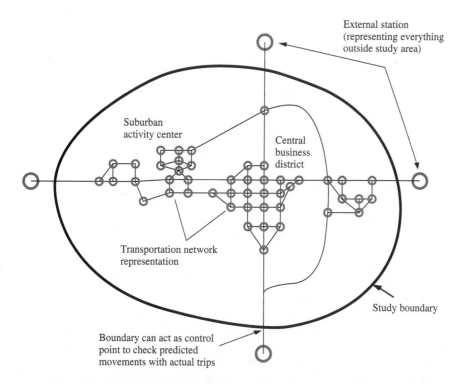

Figure 3.3   System boundary as defined for a study

be addressed well by transportation planners. Ecosystems are intricately complex systems; an impact upstream could have disastrous effects miles downstream. Pollutants originating from motor vehicles in an urban area could be transported to different regions of a country. In addition, the fact that many of the most serious impacts (especially on ecosystems and social relationships) often occur many years later provides a temporal aspect of boundary definition that becomes a significant challenge to planners.

Planning Applications of System Boundary

- One of the first steps in any planning effort is to establish the boundaries of the system being analyzed. These boundaries will vary in relation to the problem definition, decision domain, and scale of analysis.

- The level of technical analysis within the system boundary is much more detailed than that associated with the area outside, which is often represented very simply. The boundary itself can become an important part of the analysis by examining the inputs and outputs that cross it.

- Because a system boundary becomes a key determining factor in the breadth and depth of the attention focused on system operation and impacts, this boundary must be initially defined broadly enough (or allowed flexibility to change later on) to cover the spatial and temporal interrelationships associated with system effects.

3.1.4   System Components

Systems consist of individual parts or components that, when interconnected, allow the system to perform. A transportation system consists of many different components, all of which have to function together for the system to work successfully. For the purposes of this book, five major system components are presented as the key elements for a basic understanding of transportation systems. These components can be best illustrated by considering the characteristics of a hypothetical trip from an origin to a destination. A traveler leaving home walks to a bus stop and uses a local bus service to reach a suburban subway station (known as a trip collection process); proceeds through the station to the subway platform (a transfer process); rides the subway to a downtown station (a line-haul process); and walks to a place of employment (a distribution process). One can similarly view a home-to-work trip by car as consisting of similar segments, with the local street network providing the trip collection process; a freeway providing the line-haul capability; a parking lot in the central business district serving as a transfer point; and walking serving as before, the distribution function.

The first critical component of this trip is the *system user*. As noted in Chap. 1, the characteristics of individual travelers strongly influence the type and propensity of trip making. Thus, transportation planners devote considerable attention to understanding the characteristics of system users that influence travel behavior. Important in this understanding is the motivation for choosing one means of transportation or mode over another. The second major component of the transportation system is thus the *mode* or means of transportation used, which in modern society includes information technologies. Much of the technical analysis in the transportation planning process focuses on estimating the level of usage for different transportation modes given the performance characteristics of each available mode and the characteristics of the individual user. Although many trips in major central cities consist simply of walking, most often a trip will consist of at least two modes of transportation—walking plus some other mode. The modes of transportation available today for urban transportation are common to most metropolitan areas, although they exhibit different performance characteristics and can be applied in different ways to serve different purposes (Fig. 3.4). For example, as noted in the hypothetical trip discussed earlier, each part of the trip utilized a different transportation mode best suited for the trip function being served.

Infrastructure, the third component of a transportation system, provides the modal networks, facilities, and services necessary for mobility in metropolitan areas. The operational performance (i.e., the ease of travel, the quality of service provided, and service reliability) of the facilities that permit travel is therefore an important consideration in maintaining acceptable levels of mobility and accessibility. Historically, this has been the component of the transportation system that has received the most attention in the planning process. Developing the networks and facilities to accommodate transportation demand has relied on sophisticated modeling

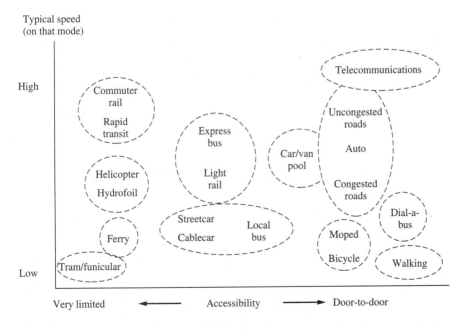

Figure 3.4   Typical urban transportation modes

approaches to determine where capacity additions were needed. In recent years, transportation planning has provided more focus on the operations and management of this infrastructure, rather than physically expanding it. In addition, planning has begun to focus on changing the demand itself through demand management techniques and pricing, instead of simply accommodating what future trips were expected.

Of particular importance to the overall effectiveness of transportation system performance is the degree of system connectivity. This means the ease with which an individual trip (passenger or goods movement) can occur from an origin to a destination with acceptable levels of system performance, regardless of the modes used. Thus, the *intermodal connections*—for example, transfer points, terminals, and stations where movements occur between modes—are critical components of an effective multimodal transportation system.

Finally, although not explicitly described in the hypothetical trip, there are many *stakeholders* that are affected by transportation. Employers expect employees to arrive on time; producers expect goods to be delivered reliably; government services rely on the ability to reach the public; retail stores depend on customer accessibility; and social/cultural organizations that foster a sense of place in an urban area exist because of the human interactions made possible by transportation. Stakeholders can also include those segments of a community negatively affected by transportation system impacts, for example, air pollution, noise, community disruption, and so on.

Planning Applications of System Components

- Multimodal transportation systems consist of numerous components whose interaction becomes a key factor for system effectiveness. These components include system users; transportation modes; infrastructure such as networks, facilities, and services; intermodal connections; and stakeholders. The challenge to transportation planning is to provide the coordination and foresight for all these components to work together effectively.

3.1.5   System Performance

The definition of a system offered previously links the success or effectiveness of a system to its performance. In a very simple system, this performance can be directly related to the level and quality of the outputs being produced. In a more complex system, which the transportation system certainly is, performance extends to how these outputs ultimately affect society and the environment. Thus, a simple perspective on system performance that focuses on immediate transportation *outputs* might include such things as the number of vehicle miles traveled, the level of congestion and thus average travel delay, the degree of on-time delivery for goods movement, schedule adherence for transit service, and the number/severity of transportation accidents.

The term *level of service* is often used in transportation planning to describe these types of measures. Level of service is a measure that describes performance conditions in terms of operational characteristics of interest to users, for example, speed and travel time, freedom to maneuver, and comfort and convenience. Typically, such measures are considered from the best performance (level of service A) to the worst (level of service F). An example of level of service as applied to different transportation facilities/services is shown in Tables 3.1 to 3.4. As illustrated, level of service is directly related to the experience of the user, for example, the amount of space available for movement or the delay experienced by the traveler. Much of the technical analysis in transportation planning focuses on the estimation of these types of measures. For example, network models for predicting travel flows use performance functions on each link of the network to influence the level of demand that will be assigned to that link, given link performance.

Table 3.1   Level of service for transit, space per passenger

Level of Service	Bus		Rail		Comments
	m²/pass (ft²/pass)	Pass/seat	m²/pass (ft²/pass)	Pass/seat	
A	>1.20 (>12.9)	0.00–050	>1.85 (>19.9)	0.00–0.50	No passenger need sit next to another
B	0.80–1.19 (8.6–12.9)	0.51–0.75	1.30–1.85 (14.0–19.9)	0.51–0.75	Passengers can choose where to sit
C	0.60–0.79 (6.5–8.5)	0.76–1.00	0.95–1.29 (10.3–13.9)	0.76–1.00	All passengers can sit
D	0.50–0.59 (5.4–6.4)	1.01–1.25	0.50–0.94 (5.4–10.1)	1.01–2.00	Comfortable standee load for design
E	0.40–0.49 (4.3–5.3)	1.26–1.50	0.30–0.49 (3.3–5.3)	2.01–3.00	Maximum schedule load
F	<0.40 (<4.3)	>1.50	<0.30 (<3.2)	>3.00	Crush loads

NOTE: English units shown in parentheses below metric values
SOURCE: TRB, 1999

A more complex perspective on system performance might include the impact of transportation on economic development, environmental quality, societal equity, and sustainable community development, that is, the eventual *outcomes* of system performance. A broader definition of system performance should also reflect the impressions and concerns of the customer or user of the system. In addition, the distributional effects of mobility and accessibility changes on different population groups should be part of measuring system performance [Meyer, 1995].

Table 3.2   Level of service for signalized intersections

LOS	Delay or Time Spent Waiting in Queue (secs/veh)
A	≤10
B	10 < but ≤ 20
C	20 < but ≤ 35
D	< 35 but ≤ 55
E	< 55 but ≤ 80
F	>80

SOURCE: TRB, 2000

Table 3.3   Level of service for stairways

Level of Service	Average Pedestrian Space in m²/ped (ft²/ped)	Unit Width Flow in ped/m/min (ped/ft/min)	Description
A	>1.9 (>20)	<16.4 (<5)	Sufficient area to freely select speed and to pass slower-moving pedestrians
B	1.4–1.9 (15–20)	16.4–23.0 (5–7)	Sufficient area to freely select speed with some difficulty in passing slower-moving pedestrians
C	0.9–1.4 (10–15)	23.0–32.8 (7–10)	Speeds slightly restricted due to inability to pass slower-moving pedestrians
D	0.7–0.9 (7–10)	32.8–42.6 (10–13)	Speeds restricted due to inability to pass slower-moving pedestrians; reverse flows cause significant conflicts
E	0.4–0.7 (4–7)	42.6–55.8 (13–17)	Speeds of all pedestrians reduced. Intermittent stoppages likely to occur; reverse flows cause serious conflicts
F	<0.4 (<4)	Variable to 55.8 (17)	Complete breakdown in traffic flow with many stoppages; forward progress dependent on slowest-moving pedestrians

SOURCE: TRB, 2000

Table 3.4   Level of service for uninterrupted bicycle facilities

LOS	Hindrance (%)
A	≤10
B	>10–20
C	>20–40
D	>40–70
E	>70–100
F	100

SOURCE: TRB, 2000

Figure 2.9 illustrated the important role of system performance in the transportation planning process. Not only does system performance reflect the vision that guides the planning process, but it acts as a means of monitoring system changes over time and influences the type and scope of planning data and analysis tools. From the perspective of a decision-oriented planning process, incorporating measures of system performance into the planning process also provides accountability for previous investment decisions.

Planning Applications of System Performance

• System performance is an important consideration guiding the definition of problems and opportunities that become the focus of planning efforts.

• System performance measures should be defined to provide information to the decision-making process, thus implying that such measures should be broadly defined not only as outputs, but also as the ultimate impacts or outcomes on society.

3.1.6   System Capacity

A system can handle only a finite number of inputs in the process of producing desired outputs. This limit is called *system capacity*. Natural and engineered systems have such capacity limits. In recent years, for example, scientists have been attempting to identify and assess the capacity of the ecological system to handle the ever-increasing impacts of an urban society. Similarly, engineered systems, such as water distribution, solid waste disposal, and electric power distribution, are limited by the physical capacity of the individual components and by the relative capacity of these components as part of the interconnected system.

Determining the capacity of transportation networks has been the focus of much research over the past 50 years. Indeed, two of the most important references in a transportation library are the manuals that present this information [TRB, 1999; TRB, 2000]. For both roads and transit, the capacity of a *facility* is defined as the number of units passing a given point during a given time period. For example, the theoretical design capacity of a transit line is defined as the number of people per hour that can be carried by the service under specified operating conditions. (Note: The TRB *Transit Capacity and Quality of Service Manual* also defines an *achievable capacity* as the maximum number of passengers that can be carried given *variability* in passenger loading and vehicle operations.) For roads, capacity is defined as the maximum hourly rate at which persons or vehicles can reasonably be expected to traverse a point or uniform section of a lane or roadway during a given time period under prevailing roadway, traffic, and control conditions. For example, the theoretical capacity of a freeway with ideal traffic and geometric conditions is 2,400 passenger cars per lane per hour. With the increasing interest on transportation policies that emphasize people movement rather than vehicle movement, road capacity can also be considered from a people-moving perspective by multiplying the average vehicle occupancy of different vehicle types by the number of these vehicles, and summing over all vehicle types. For both road and transit facilities, important relationships exist that link achievable capacity to key factors. Importantly, control strategies such as intelligent transportation system (ITS) technologies or transit vehicle guidance systems can influence achievable capacity. Table 3.5, for example, shows those factors that can affect the capacity of a road or transit facility. The capacity manuals mentioned previously provide equations that use these factors to adjust capacity levels based on the effect each has on passenger or vehicle flow.

Like transportation networks, the capacity of *terminals* can be become an important factor in the overall performance of a transportation system. Transportation terminals serve several functions: (1) loading and unloading of passengers or freight onto vehicles; (2) storage of passengers or freight from the time of arrival to time of departure; (3) documentation of movement (e.g., ticketing or billing); (4) vehicle storage and maintenance; and (5) concentration of passengers or freight into groups of economical size for movement [Morlok, 1978]. In the case of passenger terminals, achievable capacity can be defined as that volume that yields waiting times of an acceptable magnitude. This capacity can be measured for the many different types of terminals that exist in an urban area. In the case of freight terminals, relationships can also be developed to determine terminal capacity based on the equipment or labor needed to load or unload a shipment.

The capacity of a *system* directly depends on the capacity of the individual parts. Often, the system capacity will reflect the capacity of the weakest link in the network. No matter what the theoretical capacity at a particular location, the defining capacity for the entire route or facility will be at that location where the frequency of vehicle movements is determined by the maximum constraints facing vehicle flow. For example, a multilane freeway can handle thousands of vehicles per hour. But if a ramp entrance area each morning backs up traffic because of the merging of vehicles into the main traffic stream, the freeway will carry only the number of vehicles that can make it through this congested location. This is why some metropolitan areas have implemented ramp-metering systems that control the rate of vehicle entry onto the mainline freeway, thus minimizing congestion. Similar bottlenecks can occur at intersections, transit stations, freight terminals, and passenger loading bays. At a modal network level, where multiple paths and routes can be used by travelers, the identification of such bottlenecks becomes a complex undertaking and relies on sophisticated network models. At the systems level, where individual travel movements depend not only on network capacity, but also on the ability of travelers to substitute travel modes and transfer across networks (through intermodal connections), the analysis challenge becomes even more complex.

Table 3.5   Factors affecting road and transit capacity

Transit Capacity	Road Capacity
Vehicle Characteristics	***Roadway Factors***
Allowable number of vehicles per transit unit	Number of lanes
Vehicle dimensions	Type of facility and surrounding land
Seating configuration and capacity	Lane widths
Number, location, and width of doors	Shoulder widths and lateral clearances
Number and height of steps	Design speed
Maximum speed	Horizontal and vertical alignments
Acceleration and deceleration rates	Turn lanes at intersections
Type of door actuation control	
Right-of-Way Characteristics	***Traffic Conditions***
Cross-section design (i.e., number of lanes/tracks)	Vehicle type and percentage in traffic
Degree of separation from other traffic	Lane and directional distribution of trucks/buses
Intersection design	
Horizontal and vertical alignment	
Stop Characteristics	***Control Conditions***
Spacing (frequency) and duration	Type of intersection control
Design (online or offline)	Signal phasing
Platform height	Signal cycle length and green time
Number and length of loading positions	Linkage to nearby control measures
Method of fare collection	Parking controls
Type of fare	Turn restrictions
Common or separate areas for boarding/alighting	Lane-use controls
Passenger accessibility to stops	One-way street routings
Operating Characteristics	
Intercity versus suburban operations at terminals	
Layover and schedule adjustment practices	
Time losses to obtain clock headways/driver relief	
Regularity of arrivals at given stop	
Passenger Traffic Characteristics	
Passenger concentrations and distribution at stops	
Peaking of ridership	
Street Traffic Characteristics	
Volume and nature of other traffic	
Cross traffic at intersections if at-grade	
Method of Vehicle Control	
Automatic or by driver/train operator	
Policy spacing between vehicles	

SOURCE: TRB, 1999; TRB, 2000

Planning Applications of System Capacity

- Transportation system capacity (person flow or vehicle flow per hour) is an important consideration in transportation planning and is the focus of much technical analysis. Transportation policies aimed at reducing the use of single-occupant vehicles has led to increasing use of person flow-rate capacity measures.

- The capacity of individual facilities or services can be affected by a variety of factors, most often incorporated into the design of the facility or service itself.

- Network and system capacity is directly related to the capacity of the "weakest" link in the network. Thus, much planning attention is given to improving the efficiency of flow through, or enhancing the capacity of, these bottlenecks.

3.1.7   System Control

The interaction of the many different components of a system usually requires some means of system coordination. In a natural system, this guidance can be provided by a central intelligence or through instinctive responses to environmental stimuli. In a transportation system, control can be viewed from many different perspectives. A technical/operational perspective would look at such things as traffic management centers, coordinated traffic signal strategies, transit guidance technologies, and traffic control devices. Intelligent transportation system (ITS) technologies have the potential for providing new levels of system control at the metropolitan level by monitoring system performance and feeding this information back to system users and service providers.

Another perspective on system control, however, focuses on the institutional structure for transportation planning, operations, management, and decision making. The provision and maintenance of an urban road network and public transportation services have been, in most cases, considered the responsibility of government. Ever since "the urban transportation problem" emerged as a major public issue in the 1950s, the number of organizations established at all levels of government to deal with urban transportation has greatly increased. The proliferation of these organizations created the need for an institutional structure at the metropolitan and/or local level that would encourage coordinated transportation planning.

In the United States, this coordination has been carried out through the creation of formal organizational relationships and roles that rely on a bargaining and negotiating process to resolve conflicts and formulate policy [Marzotto, Burner, and Bonham, 2000]. In many cities, memoranda of understanding among representatives of the major transportation agencies in the region define the responsibilities of each agency and the formal mechanisms for adopting regional transportation policies. However, fragmentation of jurisdictional responsibilities, conflicting organizational mandates, the fact that those agencies responsible for project implementation are often not those doing the planning, and the inherently political nature of many transportation investment decisions have created a complex institutional structure at the metropolitan and/or local level that often hinders coordinated transportation action.

In comparison to the United States, numerous Canadian urban areas have metropolitan or regional governments with authority over regionwide services and facilities to coordinate the planning and implementation of transportation projects. The introduction of a metropolitan or regional "tier" of government has in many cases served to centralize transportation planning and decision-making activities and to increase the efficiency of transportation service provision. On the other hand, the retention of municipal jurisdictions with their own transportation-related functions has, in some cases, increased the amount of fragmentation and conflict over transportation policies between local and regional governments. Such conflicts, when combined with provincial government involvement in most urban transportation projects, have kept the debate over coordinated transportation planning alive in Canada as well as in the United States.

In both the U.S. and Canadian cases, the institutional structure for transportation has usually been considered from the public sector perspective. That is, the major providers of urban transportation, and the most important actors in the planning process, have traditionally been public agencies and officials. However, the opportunities for transportation in most urban areas include a variety of services, many of which are provided by private sector groups. Many employers are actively involved with employee ride-sharing programs; land developers are concerned about transportation access to their sites; private sector groups such as taxi companies, bus firms, and school bus operators can provide substantial transportation services; and business groups (such as chambers of commerce) can influence the policies and the planning process of government agencies. In addition, the urban goods movement industry consists of numerous firms, each providing goods delivery throughout a metropolitan area.

The institutional structure for transportation planning also includes many groups not associated with public agencies. These groups can include community groups, special interest groups (such as transit rider associations), "good government" groups, business organizations, and so on. These groups, as well as individual citizens, play an important role in the transportation planning process.

The institutional foundation of an urban transportation system is determined by the political, social, and historical characteristics of the locality. It is therefore difficult to describe, in general terms, the major transportation actors found in most urban areas. Not only do local agencies, public officials, private sector groups, and the general public have a role, but federal and state agencies, as the source of funds and planning process requirements, also influence metropolitan and/or local transportation decisions. In Canada, while the federal government does not play a major role in urban transportation, the presence of a metropolitan and/or regional body between local and provincial governments makes the decision process just as complex.

Planning Applications of System Control

- The transportation system is usually planned, designed, built, operated, and maintained by organizations and individuals with different objectives, mandates, constituencies, and problem definitions.

- Changes to the urban transportation system can include a wide variety of infrastructure and service actions that are applied at different geographic scales by the public and private sectors to improve mobility and the urban environment.

- Increasingly, private sector groups and organizations are playing a major role in providing transportation services and funding. The effective cooperation of public and private sector entities thus becomes an important criterion for successful transportation planning.

3.1.8   System Feedback

The dynamic nature of system behavior is often the result of environmental responses to system outputs and the feedback to individual components. In a system strongly directed from a central authority, a change in system behavior will be dictated by this authority. In a transportation system, which was defined earlier as consisting partly of (perhaps millions of) system users, it is highly unlikely that a central authority could dictate system response. There are two primary feedback mechanisms that relate to the components of the transportation system defined earlier—monitoring of system performance by the agencies that operate facilities and networks, and the influence of travel costs on user behavior as determined by the market.

The approach to urban transportation planning shown in Fig. 2.9 has an important feedback loop that relates the implementation of transportation strategies to their impact on community vision, goals/objectives, and on the operational characteristics of the system itself. Performance measures can be an important means of monitoring system performance (broadly defined) and subsequently of identifying strategies aimed at improving this performance. Many states and metropolitan areas have developed computer-based management systems that are used to provide a systematic monitoring of this linkage between system performance and desired performance measures.

The other important feedback mechanism in the transportation system is the market, that is, the general interaction between the demand for transportation services and the supply of these services. This interaction results in a "price" for the good being consumed. As the price of a good decreases, economic market theory states that demand for that good will increase. Thus, as the price of transportation decreases (e.g., new facilities constructed that shorten travel time and thus reduce the "price" of travel), more transportation will be consumed, that is, more trips will be made. From a policy perspective, the opposite relationship is gaining interest,

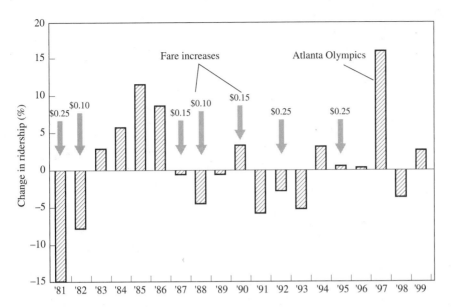

Figure 3.5   Consumer response to fare increases in Atlanta
| SOURCE: Metropolitan Atlanta Rapid Transit Authority

that is, increase the price of travel through a variety of means and thus decrease the number of auto trips. This cause-and-effect relationship also applies to transit service, although the same effect on transit trip making is *not* the desired outcome (Fig. 3.5). Known as fare elasticity, the transit industry standard is that a 10 percent increase in fare will result in an approximate 4 percent ridership loss [APTA, 1994].

Several studies have suggested that the most effective way of influencing travel behavior is through the use of pricing [ECONorthwest, 1995; Anderson and Mohring, 1996; and Roth, 1996]. A study by the Transportation Research Board concluded that peak-period fees averaging $0.06 to $0.09 per kilometer ($0.10 to $0.15 per mile) or $2 to $3 per daily trip would reduce peak-period travel on major roads by 10 to 15 percent [TRB, 1994]. Likely consumer responses would include diversion to other routes and time periods, different destinations, linking of trips, and changes in trip frequency. The types of strategies that can be used to increase the price of auto use and the resulting likely change in travel behavior are shown in Table 3.6. Over the longer run, changes in the costs of travel could affect urban form. For example, most studies have shown that the significant increases in highway capacity during the 1960s and 1970s that had the effect of reducing the cost of automobile travel led to a decentralizing effect on urban development [TRB, 1995b].

Market response is also an important feedback loop to the movement of goods. For example, a 1992 study by the Port Authority of New York and New Jersey showed that from 1985 to 1991 the distribution of truck trips heading easterly through the Hudson River tunnels and over the bridges into New York City shifted significantly to off-peak hours [Cambridge Systematics, 1992]. This change occurred because the trucking companies were avoiding the time "cost" associated with delays

Table 3.6 Traveler response to changes in travel cost

Measure Category	Overall Intent	Major Variations	Emissions-Related Travel Effect
Parking pricing	Remove/reduce subsidized parking privileges	Employee parking fees Rates to favor short-term parking over long-term parking General parking fees	Shift work trips to alternate modes, work arrangements (telecommuting, compressed work weeks), or off-peak periods, leading to reduced peak-period vehicle trips and vehicle miles traveled Shift travel to alternate modes and/or destinations, leading to reduced vehicle trips and vehicle miles traveled (VMT)
Modal subsidies	Improve attractiveness of nonsingle occupant vehicle modes by reducing their cost	Parking "cash out" HOV subsidies Nonmotorized mode subsidies	Increase use of subsidized non-SOV modes, leading to reduced vehicle trips and VMT, particularly in peak period
Pump charges (including their relative, VMT fees)	Raise cost of fuel to reflect actual costs to construct or maintain infrastructure	Per gallon fuel prices reflecting different cost allocations	Raise overall cost of driving for all travelers, causing shift in modes, reduced trip lengths, and more grouping of trips, leading to reduced vehicle trips and VMT
Emissions fees	Impose charges on vehicle owners proportional to vehicle emissions production	Age-based registration fees (periodic fee indexed to vehicle age and VMT) Emissions fee (VMT × measured emissions rate)	Reduce ownership and use of older, higher-emitting vehicles Reduce ownership and use of higher-emitting vehicles and encourage better vehicle maintenance
Roadway pricing	Impose direct charges for use of transportation facilities	Full-time facility tolls/fees Congestion pricing: variable prices linked to congestion levels of time of day	Shift travel to alternate modes, routes, or destinations, leading to reduced vehicle trips and VMT Shift travel to alternative modes, routes, destinations, and time of day, leading to reduced vehicle trips and VMT; also improved flow/speeds

SOURCE: U.S. EPA, 1998b

due to congestion. Given that the transportation cost of goods delivery is part of the cost of the good (and thus passed on to the consumer), reducing this cost becomes an important motivation to those competing to provide goods and services. Similar to responses in personal transportation, those responsible for urban goods movement could respond to changes in transportation cost by choosing different routes, modes (truck, rail, or truck/rail combination), or times of delivery, or by providing more efficient delivery tours. In the long run, increasing transportation costs could influence the technology of goods transportation (e.g., larger or combination trailer trucks or use of intelligent transportation system technologies) and the location of terminals and/or distribution centers.

Much of the technical analysis in transportation planning is based on economic market principles. Network models incorporate cost functions that represent the cost of travel on individual links in the network. As more trips are predicted to be carried on these links, the resulting congestion raises the cost of travel, and additional trips seek lower cost paths through the network. At equilibrium, no user can improve his/her cost by changing paths, or at the systems level, by changing mode. The market characteristics of the transportation system as determined by the aggregation of the individual choices of hundreds of thousands of users becomes critical feedback that influences travel demand and thus system performance.

Planning Applications of System Feedback

• The most important feedback mechanism to the users of a transportation system is the "cost" of travel, determined in the market by the interaction of supply and demand.

• Changing transportation costs as perceived by users can influence in the short term route choice, time of travel, trip frequency, and linking of trips. Over the long term, these changes could influence urban form through residential and business location decisions, auto ownership, and the technology of transportation.

• Increasing attention has been given to the use of road pricing strategies as a means of reducing congestion (i.e., as a feedback mechanism to system users). Although such strategies are often difficult to implement from a political perspective, they will likely become part of the strategy "mix" that planners will be investigating in the future.

• Transportation service providers have increasingly used more sophisticated management systems and advanced technologies to provide surveillance and control of transportation networks. These types of systems and technologies will increasingly be considered essential to effective transportation system management. Thus, they will be part of the strategies likely to be considered during the planning process.

3.2 TRANSPORTATION SYSTEM IMPACTS

One of the most important characteristics of the transportation system concept shown in Fig. 3.1 is the impact of the transportation system on other systems. During the 1970s, the physical impacts of the construction and operation of transportation facilities received increasing attention from public officials. In part, this attention was caused by a growing awareness of the important contribution of transportation system performance to environmental quality. Automobile travel in many urban areas is a primary source of many air pollutants. In terms of energy consumption, total auto travel accounts for approximately 25 percent of U.S. oil consumption, with about two-thirds of this consumption occurring in urban areas [Davis, 1999]. It is therefore not surprising that public officials, when adopting policies designed to improve environmental quality, have repeatedly turned to changes in the transportation system as one method of achieving these objectives. The following sections will describe those system impacts of most concern to transportation planners. Chapter 8 will provide more detail on the methods and tools used to analyze these impacts.

3.2.1 Context

The transportation system can impact the natural environment and urban systems in a variety of ways. As shown in Table 3.7, these impacts can range from the physical effects on a community's ecology to the social and cultural effects on surrounding neighborhoods. Some impacts, such as changes in air quality, noise levels, energy consumption, and the displacement of residential and commercial buildings,

Table 3.7   Transportation system impacts of concern to transportation planners

Natural System Impacts	*Social and Cultural Impacts*
Terrestrial ecology (habitats and animals)	Historic and archaeological
Aquatic ecology (habitats and animals)	Displacement of people
	Community cohesion
Physical Impacts	Resource consumption
Air quality	Land use
Noise	Aesthetics
Vibration	Infrastructure effects
Water quality	Accessibility of facilities, services, and jobs
Hazardous wastes	Environmental justice
Storm water	Employment, income, and business activity
Energy consumption	
Erosion and sedimentation	
Farmland conversion	

can be measured and thus incorporated into plan and project development in a quantifiable way. Other impacts are not easily measured but are still important considerations for plan or project selection. These include such things as aesthetics, social and cultural impacts, and psychological impacts on those forced to relocate.

Several characteristics of these impacts are especially important to transportation planners because they not only guide the environmental analysis associated with plan and project development, but they have also become an important basis over the past 20 years for legal challenges to agency actions. The first important distinction is that impacts can originate with the *construction* of the transportation facilities themselves, while others relate to the *use* of the system. For example, the construction of a new highway could have physical impacts on the adjacent ecological system, possibly including impacts on wetlands, wildlife, air quality (construction dust), and soil erosion. The subsequent use of this new highway could result in continuing impacts relating to air quality (motor vehicle emissions), noise, water quality, and neighborhood disruption. Both types of impacts should be considered as part of an environmental analysis.

Second, impacts can be directly linked to an action both locationally and temporally, called *direct impacts,* or they might occur much later in time or farther removed in distance, called *indirect impacts* [Canter, 1996]. Indirect impacts are particularly relevant to transportation projects and plans because they include transportation-related changes in land use, population density, and urban growth that often occur years after the change in the transportation system.

A third characteristic that is related to indirect effects is the concept of *cumulative impact.* The additive effects of the incremental impacts of individual actions in the past, present, or reasonably foreseeable future are called a cumulative impact if they collectively result in a significant impact. For example, a series of road projects might each have minor impacts on an aquifer that is the source of drinking water for a community, but taken together, they could seriously cause the quality of water to deteriorate to an unusable level.

Fourth, in much of the world, requirements to conduct an environmental analysis are linked to a determination of the *significance* of likely impacts (see, for example, [World Bank, 1989; European Communities, 1985]). In the United States and Canada, the determination of significance is a precursor to a decision on how much more detailed analysis is required. In the United States, for example, impact significance reflects its *context* (e.g., the affected region, population groups impacted, and short- and long-term consequences) and its *intensity* or severity (e.g., public health effects, uniqueness, and magnitude). The expectation of a significance level could lead to one of three determinations:

Class I: Develop an environmental impact statement (EIS): Actions whose significant impacts on the environment are anticipated and must be analyzed.

Class II: Seek a categorical exclusion: Actions whose impacts individually or cumulatively do not have a significant environmental impact, and thus do not warrant an analysis. In transportation, these actions include bicycle paths, pedestrian walkways, safety projects, noise barriers, landscaping, ride sharing, utility installations, and so on.

Class III: Conduct an environmental assessment: Actions in which the significance of environmental impact is unknown, but there is a likelihood that such an impact will occur.

Finally, the *scope* of the proposed action very much influences the types of impacts that will be of concern to an analysis. Thus, for example, intersection improvement projects could have impacts relating to noise, local air quality, local economic activities, safety, and possibly community and wetland impacts, depending on the circumstances. Major investment projects or regional plans, on the other hand, would likely be concerned about regional air quality, community equity, urban form, public finance, and land consumption.

Those interested in more detail on the legal, procedural, and methodological approaches to impact assessment are referred to [Kreske, 1996; Canter, 1996; Marriott, 1997; and Bregman, 1999]. Readers interested in knowing more about the environmental impacts associated with transportation are referred to *<http://www.fhwa.dot.gov/environment/guidebook/contents/htm>*.

3.2.2   Natural System Impacts

The state-of-the-practice today for analyzing natural system impacts is to focus on the individual components of the environment and the direct relationship between transportation system performance/facility construction and the components' function in the ecosystem. For example, the construction of any transportation facility requires land that, in its natural state, determines the physical and ecological characteristics of a region. Physical attributes include soil and topography and involve such physical processes as erosion and sedimentation. Biological attributes involve local plant and wildlife communities and the various interrelated processes comprising the ecological system of the area. These become the focus of the environmental impact analysis.

Potential negative impacts on biological systems near proposed transportation facilities have been a major focus of environmental studies concerned with the natural system. Plants and animals interact in complex ecosystems, even in urban areas. The construction of a transportation facility can disrupt the equilibrium established between the components of these ecosystems. Consideration must therefore be given to the loss of or injury to important organisms both directly (e.g., through destruction of vegetation) or indirectly (e.g., through contamination of water supplies).

Ecosystem management, at least as considered in transportation planning, is a relatively new concept (see, for example, [Herbstritt and Marble, 1996]). The U.S. DOT in 1995 developed guidance to states and urban areas as to how ecosystem considerations should be incorporated into planning. As noted by Garrett and Bank [1995]:

The ecosystem approach is characterized as a method for sustaining or restoring natural systems and their functions and values. It is goal driven, and is based on a collaboratively developed vision of desired future conditions that integrates ecological, economic, and social factors.

Ecosystems can be defined very broadly to include almost every natural process that supports life. For example, Cairns [1996] states that

> ecosystem services are those functions of ecosystems that society deems beneficial, including the maintenance of atmospheric gas balance, flood control, carbon storage, capture of solar energy, and subsequent production of food and fiber, maintenance of water quality, and maintenance of a genetic library that provides the raw material for improved foods, materials and drugs.

The art and science of environmental analysis has yet to reach the level where such functions can be readily analyzed in the context of a transportation study. However, it seems likely that one day such issues will be an important foundation for environmental impact analysis [Wachs, 1999].

The following questions have been developed by the U.S. DOT to serve as the basis for any investigation of transportation-related impacts on the ecosystem [U.S. Department of Transportation, 1979a]:

1. Are threatened or endangered organisms found in the project corridor?

2. Are organisms found in the corridor that are valuable for commercial, recreational, or ecological reasons?

3. Are nuisance organisms (such as rodents and insects) present that might migrate or proliferate because of project-induced habitat changes?

4. Are organisms present that should be retained for aesthetic purposes?

5. Will ecosystem food webs be damaged, endangering important or protected organisms?

6. Will the diversity of important biological communities be lessened, promoting less ecosystem stability and ability to withstand stress?

7. Will ecosystem productivity be lessened, reducing the system's ability to support important or protected organisms?

Inherent in these questions is the concern for the function or role that the particular organism/biological community/habitat plays in the larger ecosystem. Indeed, this issue is often the major focus of such environmental investigations (Table 3.8). As noted previously, it is important not only to consider the ecological impacts resulting from the actual construction of the facility, but also those relating to its subsequent operation and maintenance (e.g., the runoff of salt used to remove ice and snow from roadways).

3.2.3   Physical Impacts

The effects of transportation systems and facilities on air quality, noise, storm-water runoff, energy consumption, and erosion are often the most visible to the public. Not surprisingly, these impact categories have been the target of the earliest environmental legislation. Referred to as physical impacts, four in particular are often found in transportation planning studies—air quality, noise, energy consumption, and water quality.

Table 3.8 Example of the role of "function" in environmental analysis: Functions and effects of wetlands

Function	Effects	Societal Value	Indicator
Hydrologic			
Short-term surface water storage	Reduced downstream flood peaks	Reduced damage from floodwaters	Presence of floodplain along river corridor
Long-term surface water storage	Maintenance of base flows, seasonal flow distribution	Maintenance of fish habitat	Topographic relief of floodplain
Maintenance of high water table	Maintenance of hydrophytic community	Maintenance of biodiversity	Presence of hydrophytes
Biogeochemical			
Transformation, cycling of elements	Maintenance of nutrient stocks within wetlands	Wood production	Tree growth
Retention, removal of dissolved substances	Reduces transport of nutrients downstream	Maintenance of water quality	Nutrient outflow lower than inflow
Accumulation of peat	Retention of nutrients, metals, other substances	Maintenance of water quality	Increase in depth of peat
Accumulation of inorganic sediments	Retention of sediments, some nutrients	Maintenance of water quality	Increase in depth of sediment
Habitat and Food-Web Support			
Maintenance of characteristic plant communities	Food, nesting, cover for animals	Support for furbearers, waterfowl	Mature wetland vegetation
Maintenance of characteristic energy flow	Support for populations of vertebrates	Maintenance of biodiversity	High diversity of vertebrates

1 SOURCE: National Research Council, 1995

Air Quality   The urban transportation system is a major source of air pollution in many metropolitan areas. Vehicle emissions can account for close to 74 percent of pollutants in some cities (Table 3.9). The most important pollutants emitted from motor vehicles include the following.

Table 3.9   Highway vehicle emissions as a percentage of total emissions in selected North American metropolitan areas, 1998

Area	NO_x	VOC	CO
Atlanta	49	47	74
Boston	45	32	67
Chicago	32	27	59
Cincinnati	25	36	66
Cleveland	39	35	64
Dallas	50	47	68
Detroit	32	54	73
Houston	21	29	57
Los Angeles	47	47	63
Milwaukee	43	29	67
Minneapolis	38	30	63
New York	43	29	61
Philadelphia	41	33	65
Pittsburgh	40	37	72
St. Louis	37	36	72
San Diego	49	45	56
San Francisco	49	44	60
Seattle	55	38	63
Washington, D.C.	40	36	56

Carbon Monoxide (CO)   CO emissions are caused by incomplete fuel combustion within the engine. The most abundant of motor vehicle pollutants, CO is a colorless, odorless gas that interferes with oxygen transfer in the bloodstream. Because carbon monoxide concentrations are found most often near congested roadway facilities in metropolitan areas, CO is considered to be more of a localized air pollution problem (often referred to as hot spots). Advances in vehicle technology over the past 10 years have resulted in significant reductions in CO.

Hydrocarbon (HC) or Volatile Organic Compounds (VOCs)   Hydrocarbon emissions are the result of unburned fuel. Hydrocarbons are present in automobile exhaust and are also emitted as gasoline vapors from the fuel tank. The most important impact of nonmethane HC emissions is that they play a dominant role in the formation of photochemical oxidants (smog).

Oxides of Nitrogen   Nitric oxide $(NO)_x$ is produced in high-temperature combustion processes. The toxic potential of NO_x lies in its ability to oxidize and produce nitrogen dioxide (NO_2), another major component in the formation of photochemical oxidants. By itself, NO_2 can irritate the eyes, nose, and lungs.

Sulfur Oxides   Although generated primarily by nontransportation sources, sulfur is found in small amounts in gasoline. During combustion, this sulfur can be converted into sulfates; combine with other pollutants and moisture; and irritate the eyes, nose, throat, and lungs.

Particulates   Particulate matter is a general term used for the mixture of solid particles and liquid droplets in the air. The automobile is not a major source of particulate matter in the atmosphere, although some unburned carbon can be emitted by engines. The U.S. EPA has established health standards relating to particles of specific size— particles of aerodynamic diameter less than or equal to 10 micrometers (PM_{10}) and less than or equal to 2.5 micrometers $(PM_{2.5})$. Particulates can cause breathing difficulties by themselves or can promote the toxic effects of other pollutants.

Ozone   Ground-level ozone is formed through the interaction of NO_2 and volatile organic compounds (nonmethane hydrocarbons being the major contributor) in the presence of sunlight. Ozone, the principal ingredient of smog, has been linked to respiratory infections, decreases in lung functions, and aggravation of asthma. Recent studies have also identified ozone's negative impacts on vegetation. Because ozone formation requires favorable meteorological conditions, year-to-year trends in urban ozone concentrations can be quite variable. Even with such variation, the EPA estimates that urban ozone concentrations have decreased about 1 percent per year since 1987 [U.S. EPA, 1998a].

Air Toxins   Some air pollutants can potentially be quite harmful to human and ecosystem health. The U.S. government has identified 188 pollutants, called air toxins, which are considered hazardous; 60 percent are considered carcinogenic and 30 percent have some evidence of reproductive or developmental effects. Mobile (i.e., transportation) sources contributed approximately 21 percent of the emissions of the 166 air toxins monitored by the U.S. EPA in urban areas. On-road gasoline vehicles emit approximately 76 percent of these emissions. Of the 33 most dangerous toxins, mobile sources emitted almost 40 percent of the amounts estimated in 1990 [BTS, 1999].

All of the above pollutants can be harmful to human and ecosystem health when found in high concentrations. Accordingly, the U.S. EPA has developed National Ambient Air Quality Standards (NAAQS) to protect against adverse health effects (called primary standards) and against welfare effects such as damage to crops and vegetation (called secondary standards). Examples of such standards are shown in Table 3.10.

When any of these standards are not achieved, a metropolitan area is considered to be in nonattainment of the NAAQS and is subject to a variety of policy and technical requirements, depending on how severe the problem is. As of 1999, there

Table 3.10   Ambient air quality standards, United States

Pollutant	Primary (Health Related)		Secondary (Welfare Related)
	Type of Average	**Standard**	
CO	8 hour	9 ppm	No secondary standard
	1 hour	35 ppm	
NO_2	Max quarterly average	1.5 $\mu g/m^3$	Same as primary standard
O_3[a]	8 hour	0.08 ppm for 3-year average of annual 4th highest daily max 8-hour concentration	Same as primary standard
	Max 1-hour average	0.12 ppm	
PM_{10}	Annual	50 $\mu g/m^3$	Same as primary standard
	24 hour	150 $\mu g/m^3$	
$PM_{2.5}$[a]	Annual	15 $\mu g/m^3$	
	24 hour	65 $\mu g/m^3$	

[a]NOTE: As of the date of this writing, the O_3 8-hour and $PM_{2.5}$ standards are in litigation

were 119 nonattainment areas in the United States, having a population of approximately 120 million people (see <http://www.epa.gov./air/nonattn/html>).

Six strategies have been used to reduce the amount of pollutant emissions coming from on-road motor vehicles: (1) reducing the emissions from new vehicles that displace older, high-emitting vehicles; (2) accelerating vehicle fleet turnover to get new vehicles into the fleet more quickly; (3) reducing emissions from in-use vehicles through such strategies as vehicle inspection and maintenance strategies; (4) reducing travel demand to reduce vehicle activity; (5) improving traffic flow to reduce emission rates; and (6) use of low-polluting, alternative fuels [Guensler, 2000]. Strategies to reduce travel demand and improve traffic flow are referred to as *transportation control measures* and are the subject of many transportation planning activities in nonattainment areas. By far, the most dramatic improvement in motor vehicle emissions, however, has come from government-mandated vehicle emission standards. Tailpipe emission standards for HC and NO_x have reduced the emissions coming out of the vehicle tailpipe by 98 percent and 90 percent, respectively, since 1967. The EPA has proposed new standards for model year 2004 vehicles that will result in cars that emit less than 1 percent of the HC and NO_x emissions of their 1960s counterparts. In addition, new hybrid electric and fuel cell vehicles, characterized as "clean" vehicles because of much-reduced vehicle emissions, will likely enter the consumer market in increasing numbers over the next decade. However, even with these new technologies, the growth in vehicle miles traveled and the time it takes for the vehicle fleet to be replaced suggests that air-quality concerns will be an important issue for some time.

In recent years, concern over global climate change has focused attention on another set of pollutants—carbon dioxide (CO_2), methane, and nitrous oxides—called greenhouse gases. The United States accounts for approximately 23 percent of global energy-related carbon emissions, with the transportation sector constituting about 32 percent of this amount [U.S. DOT, 1998]. Importantly, the CO_2 emissions from the U.S. transportation sector have increased steadily on a percentage basis—9.5 percent from 1990 to 1997, and in absolute amount, 41 million metric tons—over this period. The Kyoto Protocol, an international agreement to set binding targets for countries to reduce greenhouse gas emissions, has become the framework for international efforts to deal with global warming. Although the United States has signed the Protocol, as of the date of this book, the Senate has yet to approve it. If such approval happens, the United States is committed to a 7 percent reduction below 1990 levels in all greenhouse gases averaged over the 2008 to 2012 period. In the United States, strategies to achieve such goals have included efforts to develop motor vehicles with reduced emissions and increased fuel economy, switching to fuels having lower carbon content, and reducing vehicle travel.

In summary, much progress has been made in reducing pollutant emissions in North America. However, as noted before, given increasing levels of travel and the lengthy time it takes for new, low-polluting vehicles to enter and penetrate the market, it is likely that air quality will remain a concern in many metropolitan areas.

Noise   The production of noise (and vibration) is one of the most apparent physical impacts of a transportation facility's operation. Exposure to high levels of noise over an extended period can have detrimental effects on the physical and mental health of human beings. In the case of urban transportation, however, noise levels are usually not high enough to actually cause physical harm. As shown in Fig. 3.6, noise levels vary for different types of transportation facilities and operations. As a rule of thumb, noise levels have to exceed 85 decibels on a fairly continual basis to cause hearing damage.

The most common unit of noise measurement is the decibel (dB), a logarithmic measure of sound pressure, filtered with an A-scale to be more sensitive to human hearing (dBA). A doubling of noise at its source produces a 3-dBA increase in the sound pressure level, which is barely perceptible to the human ear (an increase of 10 dBA in noise level will cause the noise to be perceived as being about twice as loud). For example, if a highway carrying 500 vehicles per hour produces a noise level of 50 dBA for an observer a specific distance away, an increase in traffic volume to 1000 vehicles per hour (under identical operating conditions) would produce a noise level of 53 dBA, a change an individual would hardly notice. A volume increase to 5000 vehicles per hour, a tenfold increase in traffic volume, would increase noise level to about 60 dBA, a doubling of the sound level for the observer.

Another important characteristic of noise in terms of human hearing is that sound intensity decreases with the square of the distance from a point source. In the case of a transportation facility, the noise level will decrease either 3 or 4.5 dBA (depending on the characteristics of the surface between the noise source and receptor) each time the distance from the facility is doubled.

The FHWA has established noise impact criteria for different land uses close to highways (Table 3.11) and the Federal Transit Administration (FTA) has similar criteria for train operations (Table 3.12). Models are used to predict noise levels of traffic volumes (under assumed levels of service) at sensitive receptor sites adjacent to

Type and location	Noise level (dBA)	Individual reaction
Rocket engine	180	Pain threshold
Motorcycle at a few feet	110	Deafening
Loud auto horn at 10 ft	100	
Lawn mower	98	Vocal effort
Freight train	95	
Philadelphia rail car (underground)	93-98	Loud and very annoying
Station platform	82-95	
Inside cars		
Large truck at 50 ft	90	
Busy city street		
Toronto subway car		
Station platform	84	
Inside cars	78	Annoying
Philadelphia trolley car (above ground)		
Station platform	80-85	
Inside cars	65-75	
Highway traffic at 50 ft	70	Telephone difficult to use
Light car traffic at 50 ft	60	Intrusive
Normal breathing	10	Barely audible

Figure 3.6   Transportation noise in urban areas
SOURCE: Central Mortgage and Housing Corp., 1977; Kreske, 1996; Marriot, 1997

Table 3.11   FHWA exterior noise abatement criteria (dBA)

Activity Category	Maximum 1-hour L_{eq}
A: Land where serenity and quiet are of extraordinary importance	57 dBA
B: Picnic area, playgrounds, residences, motels, schools, churches, libraries, and hospitals	67 dBA (52 dBA indoors)
C: Developed lands	72 dBA

SOURCE: U.S. DOT 1992

the road or track. If these levels substantially exceed current levels or violate noise standards, some form of mitigation will be necessary.[1] Many cities also have their own standards to regulate the production of noise during construction activities. Airports, with their own particular set of noise problems, have taken significant steps to mitigate impacts, including relocating nearby residences, retrofitting houses

---

[1] For an interesting discussion of noise impacts on natural habitats, see [Barrett, 1996].

Table 3.12   Criteria for maximum noise from train operations (dBA)

Community Area	Single-Family Dwellings	Multifamily Dwellings	Commercial Buildings
Low-density residential	70	70	75
Average residential	70	70	80
High-density residential	70	75	80
Commercial	75	80	85
Industrial	80	80	85

Building Type	Maximum Pass-by Noise Level (dBA)
Amphitheaters	65
"Quiet" outdoor recreation areas	70
Concert halls, TV studios	70
Churches, hospitals, schools, theaters, museums, libraries	70–75

SOURCE: U.S. DOT, 1989, 1995

and businesses with noise insulation, changing take-off and landing paths, and limiting nighttime operations.

Several studies have estimated the economic cost of noise by looking at the value of property in situations of different noise exposure. The currently accepted relationship is that residential property values decrease approximately 0.4 percent per decibel above a daily average of 55 dBA [Nelson, 1978]. Halig and Cohen [1996] used this approximation and estimated the cost of noise to be $34.61 per decibel per housing unit (1993 dollars).

In general, then, the production of noise by a transportation facility can be one of the most perceptible impacts of its operation. Transportation-related noise becomes a special concern in those instances where adjacent land uses are particularly sensitive to noise disruption, for example, hospitals and schools. Measures that can be taken to reduce this impact range from changes in traffic operation (e.g., reduce speed limits) to noise shielding (e.g., construct noise barriers).

Energy Consumption   Motor vehicles, the largest single consumers of petroleum in the United States and Canada, represent a significant portion of both countries' total energy consumption. Beginning in the mid-1970s with gasoline shortages caused by foreign embargoes on petroleum exports to the United States, concern for transportation energy consumption has been an important transportation policy issue. Recent trends indicate that the United States is becoming even more dependent on imported oil. In terms of energy value, the volume of imported oil and petroleum products for the first time exceeded domestic U.S. production in 1994. In 1973, transportation accounted for 51 percent of the U.S. consumption of petroleum products; by the mid-1990s this use had reached 67 percent of the total consumption. Indeed, in the United States, transportation is the only sector that consumed more petroleum in 2000 than it did in 1973 [BTS, 1999]. The following data illustrate the important position of the U.S. transportation sector in energy usage [Davis, 1999]:

- United States' share of world oil consumption 25.6 percent; share of world oil produced 9.7 percent (1996).
- Transportation oil use as percentage of U.S. oil production: 196 percent (1998).
- Net imports as a percentage of oil consumption: 51 percent (1998).
- Transportation share of oil consumption: 66 percent (1998).
- Transportation share of total energy consumption: 27.7 percent (1998).
- Transportation energy use by mode (1997):

    Auto and light-duty trucks: 63 percent.

    Heavy trucks and buses: 18 percent.

    Transit: 1 percent.

    Other (air, rail, pipeline): 18 percent.

Table 3.13 shows the relative energy use of passenger and freight modes of transportation [Howes and Fainberg, 1991; Gordon, 1991; Greene, 1996]. As can be seen, the automobile (especially the luxury auto) is one of the most energy-inefficient transportation modes of those most often found in an urban area. The freight energy efficiency figures are particularly important in that the energy efficiency of trucks has been declining over the past decade, while at the same time, the number of truck miles traveled has increased significantly.

Similar to national policy on air quality, the thrust of government action aimed at reducing petroleum consumption has been to focus on changes to vehicle technology (in this case, per vehicle fuel consumption) by establishing fuel economy standards. Automobiles are far more fuel efficient in 2000 than they were in the 1970s. On-road passenger car fuel economy increased about 60 percent between 1975 and 1994, from 13.5 miles per gallon (mpg) to 21.5 mpg. Average fuel economy for new passenger cars ranged from 27.9 to 28.8 mpg (in 1999). However, more recent years have seen a leveling off of fuel efficiency as less-fuel-efficient cars (increased vehicle weight and power) have entered the fleet [BTS, 1999].

In addition to fuel economy standards, policies have been adopted to encourage the use of alternative-fuel vehicles. For example, in the United States, the Alternative Motor Fuels Act of 1987 provided incentives for the purchase of alternative-fuel vehicles, and the 1992 Energy Policy Act set a national goal of displacing 30 percent of U.S. petroleum motor fuel use with nonpetroleum fuels by 2010. The 1992 Act also required vehicle fleets (e.g., transit vehicles) to use alternative fuels by target dates. In combination with the Clean Air Act clean fuel requirements, which encouraged the development of nonpetroleum-based, low-polluting fuels (e.g., reformulated gasoline, oxygenated gasoline, and gasohol), U.S. national policy has created an increasing market for nonpetroleum-based fuel. In 1992, for example, 230 million gasoline-equivalent gallons of alternative fuels and 1.875 billion gasoline-equivalent gallons of replacement fuels/oxygenates were consumed in the United States (out of 134.231 trillion). By 1998, these numbers had risen to 335 million and 3.938 billion, respectively [BTS, 1999]. The most popular alternative

Table 3.13   Energy consumption of passenger and freight transportation modes

Mode	Passengers	Vehicle Miles Per Gallon	Passenger (pax) Miles Per Gallon	Fuel Energy BTU/Gal	Energy Use BTU/Pax-Miles Per Gallon
Bicycle	1	50 kcal/mile	650 (equiv)		200
Walking	1	70 kcal/mile	450 (equiv)		300
Auto—fuel efficient	4	100	400	125,000	300
Bus—intercity	45	5	225	138,700	600
Auto-high econ	4	50	200	125,000	600
Maglev vehicle	140				800
10-car commuter train	800		160	138,700	900
10-car subway train	1,000		150	138,700	900
Auto	4		120	125,000	1,000
Bus—local	35	30	105	138,700	1,300
4-car intercity train	200	3	80	138,700	1,700
A300 jet plane	267		80	135,000	1,700
Motorcycle	1	60	60	125,000	2,100
Auto—luxury	4	12	48	125,000	2,600
747 jet plane	360		36	135,000	3,800
Light plane—2 seat	2	15	30	120,200	4,000
Executive jet plane	8	2	16	135,000	8,400
Concorde SST	110		13	135,000	10,400
Snowmobile	1	12	12	125,000	10,400
Ocean liner	2,000		10	150,000	15,000

Mode	Mileage (ton-miles/gal)	Energy consumption BTU/Ton-Mile			
Oil pipelines	275	450			
Railroads	185	670			
Waterways	182	680			
Truck	44	2,800			
Airplane	3	42,000			

SOURCE: Howes and Fainberg, 1991

fuel in 1998 was liquified petroleum gas (about 274,000 vehicles in 1998), followed by compressed natural gas (about 96,000 vehicles in 1998). There were only 6,000 electric vehicles in the United States in 1998, although this number is expected to increase rapidly once emerging battery technology provides the capability of producing the equivalent energy as liquid-fuel-powered vehicles. The most promising approach in the near future is likely to be hybrid vehicles, which combine both electric and fuel propulsion in one vehicle.

Finally, it is important to note that many of the actions transportation planners examine as part of the planning process will have energy-consumption-reducing benefits. Decreasing the use of the single-occupant vehicle, the target of many transit, demand management and telecommunications strategies, will generally have positive impacts on reducing overall energy use.

Water Quality   One of the most important environmental problems in urban areas is the degradation of water resources. And given that many transportation facilities have impervious surfaces (e.g., roads and parking lots), water runoff and drainage are a significant contributor to this problem. A much larger transportation-related issue is the inappropriate disposal of motor fuel oil. One estimate suggests that the volume of used motor oil improperly dumped in sewers, drains, and soil annually is 15 to 20 times greater than the 10 million gallons of crude oil spilled into Prince William Sound by the *Exxon Valdez* [BTS, 1996].

Table 3.14 shows the constituents and sources of highway runoff that might be found in a typical metropolitan area. The most common contaminants of highway runoff are heavy metals, inorganic salts, aromatic hydrocarbons, and suspended solids. However, because the amount and nature of contaminants are site-specific, it is difficult to generalize the extent of the problem. One study of 31 sites in 11 states showed that urban sites produced much higher concentrations than rural sites [FHWA, 1990], while another study found that metal and sodium concentrations in topsoil next to highways could have serious effects on ecosystem processes [FHWA, 1987].

Table 3.14   Constituents and sources of highway runoff

Constituent	Primary Sources
Particulates	Pavement wear, vehicles, atmosphere, maintenance
Nitrogen, phosphorus	Atmosphere, roadside fertilizer application
Lead	Leaded gasoline (auto exhaust), tire wear (lead oxide filler material), lubricating oil and grease, bearing wear
Zinc	Tire wear (filler material), motor oil (stabilizing additive), grease
Iron	Autobody rust, steel highway structures (e.g., guard rails), moving engine parts
Copper	Metal plating, bearing and bushing wear, moving engine parts, brake-lining wear, fungicides
Cadmium	Tire wear (filler material), insecticide application
Chromium	Metal plating, moving engine parts, brake-lining wear
Nickel	Diesel fuel and gasoline (exhaust), lubricating oil, metal plating, bushing wear, brake-lining wear, asphalt paving
Manganese	Moving engine parts
Bromide	Exhaust
Cyanide	Anticake compound (ferric ferrocyanide, Prussian Blue or sodium ferro-cyanide, Yellow Prussiate of Soda) used to keep deicing salt granular
Sodium, calcium	Deicing salts, grease
Chlorine	Deicing salts
Sulfate	Roadway beds, fuel, deicing salts
Petroleum	Spills, leaks or blow-by of motor lubricants, antifreeze and hydraulic fluids, asphalt surface leachate
Polychlorinated biphenyls, (PCBs), pesticides	Spraying of highway rights-of-way, background atmospheric deposition, PCB catalyst in synthetic tires
Pathogenic bacteria	Soil, litter, bird droppings, and trucks hauling livestock and stockyard waste
Rubber	Tire wear

SOURCE: BTS, 1996

The impacts of highway runoff on receiving waters can be defined in order of severity [FHWA, 1985]:

1. Any measurable increase in pollutant concentration or load compared to background levels.

2. An increase in sediment pollutant concentration, demonstrated bioaccumulation of toxic materials, and/or subtle changes in biological communities.

3. An increase in pollutant concentrations sufficient to exceed acute or chronic water-quality criteria established by the U.S. EPA.

4. An increase in pollutant concentrations sufficient to cause contravention of state water-quality standards.

5. Dramatic, highly visible impacts such as fish kills, water-supply taste problems, shellfish water closures, and/or severe alteration of the aquatic biological community.

Surface water and groundwater are the most susceptible to runoff contaminants. Although groundwater has the advantage of layers of soil that can filter contaminants as they seep through the ground, once contaminated, they are much more difficult to clean up. For those communities using salt or deicing chemicals on the roads during the winter months, special care must be taken to avoid harmful chemicals making their way to adjacent water supplies. Strategies that can be used to minimize the effects of pavement runoff include infiltration strategies to remove harmful substances; detention and retention practices to capture and retain suspended solids and other contaminants; and filtering systems that capture large particles and suspended constituents [FHWA, 1999].

A likely direction of transportation-related environmental analysis in the future will be to investigate combined transportation actions in the context of urban watersheds. Such an approach might include [Shrouds, 1996]

1. Protecting areas that provide water-quality benefits (e.g., wetlands and aquatic water systems) and protecting areas susceptible to erosion (e.g., unstable soils and landslide areas).

2. Developing erosion and sediment control strategies at the planning and design stages to be implemented during construction, operation, and maintenance.

3. Ensuring proper use, storage, and disposal of toxic substances at construction sites.

4. Identifying watershed pollution-reduction opportunities.

5. Promoting the use of vegetation methods to control erosion and to reduce pollution loadings.

6. Performing water-quality monitoring to assess pollution load reduction and changes in water quality.

Those interested in a more detailed discussion of highway runoff are directed to [FHWA, 1985].

Planning Applications of Natural System Impacts

- Transportation system impacts of the ecosystem have become an important consideration in transportation planning. These impact categories have legal and regulatory requirements associated with their consideration in project development, and, in some cases, plan development.

- These impacts are increasingly being viewed from an ecosystem perspective, that is, an interrelated set of natural processes supporting life. The linkages among the construction, operation, and maintenance of transportation facilities and the natural environment must often be considered from a broader perspective of the spatial and temporal linkages that characterize such processes. Thus, for example, watersheds or air basins become the boundaries of impact analyses.

- For several types of impacts, national standards have been established that relate to the health effects of high incidence exposures. These standards have become important targets for transportation and other measures to help achieve.

- National policies aimed at minimizing these impacts have adopted a variety of approaches, including mandating changes in technology (e.g., fuel economy and pollutant emission standards and financial incentives to use alternative fuel vehicles) and encouraging transportation actions that reduce vehicle use. These latter actions are often a focus of regional transportation plans.

3.2.4 Social and Cultural Impacts

The impacts of transportation systems on a community reflect not only their influence on land development but also the potential disruption to community relationships. The characteristics of the dominant transportation technology of an era defines where people and businesses locate [Muller, 1995]. Thus, for example, some of the basic theories of city development rely heavily on transportation access as a critical variable. Location theory, as espoused by Von Thunen [1826], Isard [1956], Alonso [1964], and Wingo [1972], argues for a direct causality between accessibility as provided by the transportation system and the value (and thus the use) of land. Business location theory suggests that businesses will tend to congregate at locations where transportation costs are low [Deakin, 1991]. Many of the most prominent land-use and transportation models in use today rely on measures of accessibility that connect predicted residential locations with economic activities (Chap. 6 provides a more detailed description of these models). There is thus a strong theoretical and empirical basis for understanding the linkages between the functioning of an urban system and transportation system performance. (For an interesting discussion on the impact of changes in transportation technology on urban systems, see [Ausubel and Herman, 1988].)

This section focuses on three of the most prominent of these impacts—land use, economic activity, and community/social/cultural impacts. Experience indicates that these three impacts, along with safety and mobility, are the primary motivation behind many regional transportation plans.

Land Use and Urban Form   The linkage between land use and transportation is a fundamental relationship in the study of transportation. Put simply, the trip-making characteristics of a region—spatial travel patterns and modal distributions—are largely a function of how land is organized and used. Likewise, the pattern of land use is influenced by the level of accessibility provided by the transportation system. For example, in recent years, the automobile (and thus the highway network) has been viewed by many as a primary cause of urban decentralization and sprawl. This view is best summarized by Muller [1995], who states,

> The maturing freeway system was the primary force that turned the metropolis inside out after 1970, because it eliminated the regionwide centrality advantage of the central city's CBD. Now *any* location on that expressway network could easily be reached by motor vehicle, and intraurban accessibility swiftly became an all-but-ubiquitous spatial good.

A TRB study that examined the impacts of highways on metropolitan development concluded that

1. Early highway capacity expansions, such as construction of interstate highways, dramatically reduced travel costs and increased access to undeveloped land. Lower land costs enticed households and firms to move to areas on the urban fringe that had improved accessibility.

2. Highway capacity expansions interacted with population growth, rising personal income, increased automobile ownership, decreased cost of transportation, and land-use policies to channel the location of growth within metropolitan areas.

3. Additions to the highway system made at the same time a metropolitan area was growing influenced the location of residential and employment development because the corridor where the investments were made became more attractive for development.

4. Additions to highway capacity that reduced the cost of travel supported sprawl when other conditions also supported dispersed development. The effect was greatest when access to large tracts of rural land on the urban fringe was improved [TRB, 1995b].

This conclusion—that major highway investment coincided with many other trends and factors to foster decentralization and that, by itself, highway investment was not the sole influence on urban decentralization—is supported by several other studies [Altshuler et al., 1979; Meyer and Gomez Ibanez, 1981; Jackson, 1985; Linneman and Summers, 1993; Parsons et al., 1996b; Boarnet and Haughwout, 2000].

Similar studies have looked at the land-use impact of transit investment (primarily of rail investment, little if any on bus transit). In the 1880s and early 1900s, urban rail lines providing access to the then urban fringe were the major cause for rapid city decentralization, which was heralded in that time as a way of escaping the ills of the central city [Warner, 1962; Middleton, 1967; Fogelson, 1967]. Investment in urban rail systems since the 1960s has had the effect of continuing to support urban decentralization trends [Webber, 1976; Donnelly, 1982]. However, such investment has also been shown to reinforce the economic viability of the downtown [Arrington, 1995; Parsons et al., 1996a]. For example, for the most-studied new North American rail system—the BART system in the San Francisco Bay area—Cervero concluded that 20 years after opening, its most significant impact on land use was the strengthening of the central business district [Cervero, 1995].

Perhaps the most important land-use impact of rail investment is its attractiveness to potential development near stations if market factors are conducive and if government policies are used to encourage such development. Because of the increased value of land that proximity to transportation accessibility brings (although this effect is mitigated in some cases by the quality of service provided and the desirability of the neighborhood), the resulting combination of numerous individual development decisions has caused a clustering effect near many urban rail stations [Voith, 1970; Damm, 1980; Rice Center, 1987; Landis, 1995; Nelson and McCleskey, 1992]. For example, a sample of real-estate transactions in Washington, D.C., from 1969 to 1976 showed that (1) the distance of a parcel from a subway station appeared to be a determinant of the variation in values of properties; (2) the value of a parcel tended to increase as a nearby station's opening date approached; and (3) the availability of parking positively influenced retail property values [Lerman et al., 1977]. With respect to location decisions, Cheslow and Olsson [1975] concluded that, apart from house and neighborhood-related attributes such as lot size, quality of schools, extent of crime, and housing cost, household location decisions were also influenced by accessibility to workplaces, shopping and business centers, recreation opportunities, and health facilities. In essence then, each household must make a trade-off between housing price, transportation costs, and these other attributes when deciding upon a housing location.

The land use–transportation relationship is shown in Fig. 3.7. The land development pattern for metropolitan areas is the starting point for the relationships shown in this figure. Such patterns influence business and household location decisions. The subsequent land uses (e.g., shops, schools, recreational facilities, and employment sites) influence the activity schedules of daily trip making, which influences overall activity patterns. The activity schedules create new travel demands and, consequently, a need for transportation services, whether in the form of new infrastructure or more efficient operation of existing facilities. The demand for transportation in conjunction with land-use patterns can influence auto ownership. For example, in a suburb without transit service, auto ownership is likely to be quite high. Improvements to transportation networks make the land more accessible for additional development to occur. Increased accessibility and improved land values, in turn, influence the location decisions of individuals and firms, once again spurring new land development and starting this cycle again, until an equilibrium is reached or until some other external factor intervenes.

An important dimension of the relationships shown in Fig. 3.7 is the time it takes for the cycle to complete itself. In the short run, the predominant influence is that of land use on the performance of the transportation system. For example, the

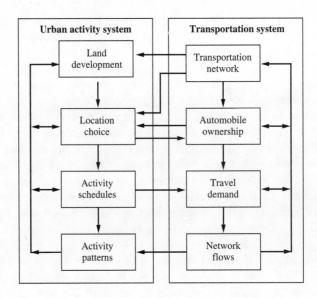

Figure 3.7   Urban activity and transportation systems interaction

impact of a new shopping center on the surrounding street system could well require major street and/or transit improvements. Similarly, the opening of a large office building in the center city could significantly tax the ability of a nearby transit station to handle the peak-hour rush.

In the long run, the provision of transportation infrastructure and the introduction of new technologies will influence urban form because of the improved accessibility that results. In the 1800s, when urban travel was mainly by foot and horseback, city structures were highly centralized with dense concentrations of commerce and industry located in the urban core. The introduction of more advanced transportation technology (urban rail for passengers and motor truck for goods movement) allowed the urban area to expand and decentralize. Taking place over decades, the influence of transportation technology on urban form has resulted in the North American metropolitan areas of today (see Table 3.15 for an example of recent geographic dispersal).

Table 3.16 summarizes many of the studies that have examined land use and transportation relationships. These studies have ranged from site-specific studies on the impact of a transportation facility (e.g., a subway station or road) on surrounding land uses, to regional studies of the impact of changes in transportation accessibility on metropolitan spatial patterns. One of the interesting characteristics of this research record is that the evidence is not at all clear, given the high level of access currently provided by urban transportation systems (in particular road networks), whether incremental improvements in this access can significantly affect metropolitan patterns of urban development. For a more extensive discussion of this issue, see [Giuliano, 1995; Burchell et al., 1998].

Of greatest interest to transportation planners over the next several decades will be the question of whether a proactive land-use policy (i.e., where government takes

Table 3.15   Population increases compared to increases in developed area, 1970 to 1990

Metropolitan Area	Population Growth	Increase in Developed Area
Chicago	4%	46%
Los Angeles	45	30
New York	8	65
Cleveland	−6	31
St. Louis	35	355

| SOURCE: Diamond and Noonan, 1996

a lead role in influencing land-use decisions) can be combined with transportation investment decisions to provide a more "desirable" urban form. Several authors have argued that transit investment can play such a role:

> In the long run, if lasting and effective transportation improvements that act as a permanent, positive force for livability are to be achieved, then they must take place within the context of an overall land-use policy designed to further the preservation and revitalization of dense, lively town centers, as well as the creation of new nodes near public transportation [Project for Public Spaces, 1997].

> Transit can influence urban form, and the design of the urban environment can encourage transit use, and reduce dependence on the automobile [Parsons et al., 1996a].

> Promoting transit use through coordinated land use and transportation planning is a sine qua non for permanent avoidance of aggravated urban transportation problems in the future, and for the creation of a livable city [Vuchic, 1999].

Others have suggested that the key to the land use/transportation strategy is on the land-use side. Some of the policies and planning tools include [Deakin, 1991; Ewing, 1997]

1. Urban limit lines and urban development reserves designed to produce compact development in areas where urban services are already available or are scheduled.

2. Mandatory consistency between local land-use plans and local and regional transportation plans.

3. Requirements for the provision of adequate public facilities concurrent with development, or attainment of minimum level-of-service standards.

4. Mandatory balancing of job growth with housing development, priced and located to match the needs and incomes of the work force.

5. Minimum as well as maximum development densities and floor area ratios to ensure adequate development for transit to work.

6. Incentives and bonuses for desired land uses and for developments that provide desired transportation and land-use amenities.

7. Site design planning emphasizing pedestrian access and transit serviceability.

Table 3.16   Studies of land use and transportation impacts

Author	Date	Facility	Conclusions
Adkins	1959	Expressway	Value of land closest to expressway increased on the order of 300 to 600%; land farther from expressway experienced smaller increases
Allen and Mudge	1974	Heavy rail	Major impact of investment was to transfer land development from one part of the region to another
Arrington	1995	Light rail	Investment has reinforced downtown Portland's (OR) role as central business district for region
Ashley and Bernard	1965	Highway	Major development at interchanges caused by many factors, including market condition and financing arrangements
Cambridge Systematics	1994	Transit	Mixed land use and accessibility of services result in an approximate 3.3 to 3.5% increase in transit ridership
Cervero	1993 1995	Rapid transit	BART's most significant effect has been to strengthen development of San Francisco's downtown; commercial real-estate values rose faster for property adjacent to rail stations than for those farther away
Davies	1976	Subway	Increased population density occurred near subway stations as compared to rest of region
Donnelly	1982	Rail transit	Land-use planning in affected communities responded to new rail transit and will likely shape future growth; development focused within 0.25 mile of transit station
Dornbusch et al.	1978	Rail transit	BART system has been responsible for some redirection of office building activity
Downing	1973	Bus routes	Property values higher in bus corridor depending on accessibility to services
Forkenbrock and Foster	1990	Highway	Suggests that many highways did not cause development to occur but were built where development was going to happen
Frank and Pivo	1995	Transit	Positive relationship between land use/population/employment density and nonauto trip making
Gannon and Dear	1972	Rapid transit	Lindenwald transit line improved attractiveness of downtown Philadelphia, but study did not examine diverted development
Gauthier	1970	Railroads	Improvement in transportation may help some parts of region while harming others
Green and James	1993	Rapid transit	Areas with transit access consistently grew more quickly than areas without transit accessibility
Hansen et al.	1993	Highways	Statistically significant short-term relationships found between highway expansion and residential construction in corridor
Knight and Trygg	1977	Rapid transit	No evidence that any rapid-transit improvements by themselves have led to net new urban economic or population growth
Landis	1995	Rapid transit	Property values near rail stations in California found to be higher than comparable nonstation sites
Leinberger	1996	Highway	Interstate highway program had important impact on urban form, but effect was delayed for development dynamics to realize advantages of increased accessibility
Lerman et al.	1977	Rapid transit	Land closer to subway station experienced greater increase in property value
Lewis-Workman and Brod	1997	Rapid transit	Measured the neighborhood benefits of rail transit accessibility; nontransit user benefits may account for up to 50% of property value premium for near-station properties

Table 3.16 Studies of land use and transportation impacts (continued)

Author	Date	Facility	Conclusions
Lichter and Fuguitt	1980	Highway	Population and employment growth correlates to existence of interstate highways in metropolitan area
Metropolitan Transportation Commission	1979	Rapid transit	Development within BART transit corridors measurably different than that in non-BART corridors. System may have prevented further central city decline
Meyer, Kain, and Wohl	1965	Transit	Availability of transit did not appear to be sufficient condition to guarantee continued growth of a central business district
Mohring	1961	Highway	Increase in land value near highway balanced by relative decreases elsewhere
Muller	1995	Highway	Urban freeways major force in decentralization of urban areas
Northern Virginia Planning District	1993	Commuter rail	43% of new home buyers in commuter rail service area influenced by presence of commuter rail service
Parsons et al.	1996a	Rapid transit	Transit lines have strongly supported development of employment centers in inner suburbs; importance of supporting policies
Parsons et al.	1996b	Highway	Interstate highway program in conjunction with other public policies and economic trends caused urban decentralization
Payne-Maxie and Blaney-Dyett	1980	Beltways	No strong evidence to suggest that beltways improve a metropolitan area's competitive advantage; difference in housing patterns between beltway and nonbeltway cities not statistically significant
Pushkarev and Zupan	1977	Rapid transit	Transit usage requires minimal residential densities and connection to major employment center
Spengler	1930	Subway	New transit lines tend to shift value, not create new value; areas already developed do not show marked increase in land value when new transit lines are opened
Transportation Research Board	1995	Highway	Highway capacity increases influenced urban form, but only as one element of many other factors
U.S. EPA	1975	Highway	Value of land near new or improved highways increases significantly; value of single-family dwelling, on average, not affected as much
Voith	1970	Rapid transit	Proximity to rail stations in Philadelphia raised property values
Warner	1962	Streetcar	Streetcar expansion in the late 1800s led to expansion of cities along rail corridors, creating first suburbs

SOURCE: Lerman, 1977; TRB, 1995; Porter, 1997

Growth management (or also known as "smart growth") is the general policy umbrella for these strategies. The goal of growth management is to integrate land-use and transportation policies, resulting in growth that is more sustainable (see Table 1.6) and that better utilizes existing infrastructure [Urban Land Institute, 1998]. However, even with a concerted policy effort, the results of growth management efforts might be slow to materialize.

In summary, land-use development and the provision of transportation services and infrastructure are, in large part, functions of one another. In the past, when urban areas were experiencing demand for rapid expansion and growth, the location of transportation facilities (and the accessibility they afforded) provided a strong influence to growth. In more recent times, it has become apparent that with the high level of accessibility already available through existing transportation systems, the use of transportation investment *by itself* to influence land use is likely to produce minimal results. Thus, one of the key challenges to transportation officials and planners who desire to have a significant impact on land use is the development of a package of policies, incentives, and investment strategies that together can reinforce the movement to a desired development pattern [Moore and Thorsnes, 1994; Carlson and Billen, 1996; Dunphy, 1997].

With the amount of attention that is given to the development-influencing nature of transportation investment in many planning studies, some additional conclusions seem appropriate. First, transportation is only one factor that influences development decisions and, hence, land-use impacts. In general, land development impacts near transportation facilities are caused by many economic, governmental, and social factors including (1) regional demand for new development; (2) availability of developable land; (3) complementary local government actions (e.g., zoning and land-use plans); (4) appropriate adjacent land uses; and (5) attractive sites for development. Public officials can, and have, used tax incentives, financial support, and technical assistance programs to influence developers' decisions as to the location of proposed projects. The use of government influence in this way can indeed affect the land-use configuration at a particular site and perhaps more broadly if applied consistently.[2]

Second, transportation planning and investment should be coordinated with land-use planning. In Canada and in many other countries, transportation is regarded as a land use and is fully integrated with general urban planning. Although this is generally not the case in the United States, such coordination is essential for the development of a transportation investment program that meets the development needs of an urban area.

Third, the development of land in market-oriented economies is mainly the responsibility of private entrepreneurs and developers. Thus, the feasibility of new

---

[2] Because of concentrated development surrounding its stations, the Toronto subway system is one of those pointed to by many as an example of how such investment can influence development. In reality, the following factors contributed to this phenomenon:

1. The entire region and particularly the central business district was experiencing rapid growth and economic expansion when the subway line was built.

2. The original subway lines followed already heavily used surface streetcar routes. The subway thus increased accessibility to areas where demand for additional residential and commercial floor space was high.

3. The metropolitan government was able to provide capital support for both land acquisition and construction. This resulted in a great deal of publicly owned surplus land available for sale back to developers.

4. The policy-making powers of the metropolitan government gave it substantial control over both its own and other properties adjacent to subway lines. Metropolitan Toronto was able to coordinate the land-use policies of the different municipalities where the subway lines ran.

5. The transit commission sold air rights, that is, the right to build over the subway right-of-way, to developers.

6. There was a great deal of cooperation between the public and private sectors in developing land adjacent to subway stations, particularly in the form of direct access to the subway stations.

development and the use of transportation investment to encourage such development need to be evaluated at least in part from a private investment perspective, with an understanding of the factors that can influence development decisions [Witherspoon, 1979; U.S. Department of Housing and Urban Development, 1979; U.S. Conference of Mayors, 1980]. Such coordinated transportation and private investment planning characterizes many joint development projects found throughout North America [Cervero et al., 1991]. In such projects, private sector interests play key roles in the implementation of the development proposal and in defining the desirable characteristics of the complementary transportation investment.

Fourth, the land-use impact of transportation investment tends to involve changes in the spatial distribution of activities within a metropolitan area, both in terms of the location of new development and changing property values. A major investment in transportation infrastructure seldom creates a net increase of development within a metropolitan area. Regional transit investments have a tendency to reinforce already existing decentralization trends, while still having the potential of strengthening the downtown core [Cervero et al., 1998]. Experience in San Francisco; Washington, D.C.; Toronto; and Baltimore shows this combined effect.

Economic Impacts   In addition to land-use impacts, the impact of transportation investment on the regional or community economy is one of the most important concerns to elected decision makers and stakeholders. This impact can be defined in many different ways, for example, jobs, overall economic health, the influx of federal funds to the region, and industrial productivity (see, for example, [Apogee et al., 1998] and [Baniser and Berechman et al., 2000). Transportation represents a huge economic resource to a region. In 1997, transportation-related goods and services contributed $905 billion to the U.S. gross domestic product, a total of 11 percent of the GDP. At the individual level, transportation expenditures represent an important economic cost as well. The average American household spent approximately $6,400 on transportation in 1996—of which 44 percent went for vehicle purchase; 17 percent for gasoline and oil; 32 percent for other vehicle expenses such as insurance; and 7 percent for purchased transportation service such as taxi.[3] This expenditure represented about 11 percent of the average household budget, as compared to 24 percent for housing, 14 percent for health, 12 percent for food, and 7 percent for education [BTS, 1999].

The definition of economic benefits associated with a transportation investment differs depending on the specific objectives outlined for a decision. For example, a TRB study on the economic impacts of transit investment identified three types of economic benefits that might result from such investment [Cambridge Systematics, 1998]:

1. *Generative* impacts produce net economic growth and benefits in a region, such as travel-time savings, increased regional employment and income, improved environmental quality, and increased job accessibility. This is the only type of impact that results in a net economic gain to society at large.

---

[3] If one is interested in the total transportation-related expenses for a household, including the freight costs relating to the delivery of goods and food, the average household expenditure is much higher. Kulash, for example, estimated an average household spent $12,000 on transportation in 1996, of which $4,000 was for freight bills associated with the price of purchased goods [Kulash, 1999].

2. *Redistributive* impacts account for locational shifts in economic activity within a region such that land development, employment, and, therefore, income occur in a transit corridor or around a transit stop, rather than being dispersed throughout a region.

3. *Transfer* impacts involve the conveyance or transfer of moneys from one entity to another; for example, the employment stimulated by the construction and operation of a transit system financed through public funds, joint development income, and property tax income from development are redistributed to a particular transit corridor.

Another study looked at the potential for transportation capital investment to be a catalyst in producing productivity gains, economic growth, and improved regional competitiveness [Lewis, 1991]. Table 3.17 shows the potential effectiveness of transportation policies in achieving alternative economic objectives. The study concluded that "transportation infrastructure investment is substantially more effective in promoting net productivity growth than it is in stimulating regional economic gains." Simply stated, the reduction of trip delays, vehicle operating costs, and accident costs have a very positive impact on regional economic welfare in areas economically strong. Economically weak areas would not benefit as much from transportation investment as they might from an educated and trained labor force or competitive market advantages. Interestingly, the study also concludes that although transportation investment can promote local job growth, job creation is usually at the expense of job growth elsewhere in the region or state. Thus, transportation investment promotes growth through productivity gains rather than net increases in the rate of employment. This conclusion is reinforced by research that asked business managers to identify the "costs" of congestion or travel delay to their businesses. The results are shown in Table 3.18. The benefits of reducing congestion are perceived by business managers as relating to enhanced economic productivity associated with reduced travel times and enhanced quality of life [Grenzeback and Warner, 1995].

One of the more unusual studies that illustrates the economic benefits of an efficient road system looked at the cost of delay due to earthquake-induced road closures [Wesemann et al., 1996]. The 1994 Northridge earthquake in southern California caused several major freeways to be closed. The cost of delay to the two million person-trips affected by this closure was estimated to be $1.6 million per day. In addition, 30 percent of the Los Angeles area shipping firms indicated that their operations were severely affected, causing an average 8 percent increase in shipping costs as a result of freeway closures.

Transit's role in enhancing economic development has also been the subject of several studies. Typical estimates of such economic impacts for different scenarios include [Cambridge Systematics, 1999]

City	*Scenario*	*Economic Impacts*
Philadelphia	System shutdown	175,000 loss in jobs $10.1 billion in lost annual personal income $16.3 billion in lost annual business sales $632 million loss in state/local revenues

(continues)

City	*Scenario*	*Economic Impacts*
Chicago	Restore system to good repair	41,209 gain in jobs $4.6 billion in business sales in 2020
New York	System disinvestment	319,800 loss in jobs (2016) $18.9 billion loss in business sales (2016)
Los Angeles	System investment with rail/bus improvements	131,200–261,700 increase in jobs (2020) $8.9–16 billion increase in personal income
Dayton, OH	Immediate shutdown	$3.8 million loss in direct/indirect spending 985 loss in direct/indirect jobs
Commuter rail services in U.S.	Value of current services	420,000 increase in jobs (1986–1996) $3.5 billion increase in tax revenue $300–450 million time savings to trucks $247–865 annual time/fuel savings to riders

Table 3.17   Overview of potential effectiveness of transportation policies in achieving alternative economic objectives

	Potential Effectiveness Relative to Nontransportation Policies	**Potential Effectiveness Relative to Status Quo or Base Case**
Distributional Objectives		
Employment	Low	Low/moderate
Personal income	Low	Low/moderate
Regional output	Low	Low/moderate
Sectoral output	Low/moderate	Moderate/high
Growth Objective		
Productivity	Moderate/high	High
Output	Moderate/high	High
Welfare/living standards	High	High

SOURCE: Lewis, 1991

A study commissioned by the American Public Transit Association (APTA) concluded that the economic impacts associated with transit investments are likely to be greatest where the following conditions exist [Cambridge Systematics, 1999]:

1. Moderate to high congestion exists in corridors designated for transit investment.

2. Opportunities for highway capacity improvements expansion are limited.

3. Good access is provided to significant land-use activity at the destination end and residences at the origin end.

4. Public policies exist that support and abet transit usage, such as zoning and land-use policies, transit-supportive parking policies, and employer-based commute options.

Table 3.18   Impact of urban congestion on business

	Client Travel to Obtain Goods and Services	Commuting	Delivering Goods and Services	Receiving Goods and Services
Direct Traveler Impacts	Not applicable	Increased travel time Increased vehicle operating costs Change in travel hour Change in trip frequency	Not applicable	Not applicable
Indirect Traveler Impacts	Increased stress and aggravation Decreased quality of life Change in destination	Increased stress and aggravation Decreased quality of life Change in residence Change in destination	Increased stress and aggravation Increased pressure to work harder Decreased quality of life	Not applicable
First-Order Business Impacts	Low sales	Recruitment and retention problems Tardiness or stress concerns Alternative work schedule complication Trip reduction requirements	Increased staff and vehicles Increased inventory New branch locations	Higher prices for goods and services Disruptions to operations
Second-Order Business Impacts	Change in prices or profits on sales to final consumers Change in land use Decline in business growth Relocation of business Decline in local spending Loss in business economies of scale			

SOURCE: Grenzeback and Warner, 1995

5. The transit service is competitive with the highway alternative in terms of time and cost. Systems operating on dedicated rights-of-way, for example, are likely to be more time competitive with auto times as compared to bus systems running in mixed traffic.

In summary, the perceived positive impact of transportation investment on economic development is one of the most important considerations to transportation decision makers. Given the importance of metropolitan areas as the economic engines of national economies (see Sec. 1.2 and [Armstrong, 1999]), providing an efficient, urban transportation system is a necessary precursor for a healthy economy. As previously noted, however, the major economic benefit for well-established urban economies is likely to be in productivity gains rather than growth in jobs. In most cases, transportation investment by itself is not likely to cause significant change in economic activity. Supportive market forces and public policies are also important to spur economic development.

Planning Applications of Land Use and Urban Form

- The relationship among land use, economic activity, and transportation investment is one of the most important to decision makers and to the planning process. Accordingly, considerable effort is undertaken to inventory current land use and economic development in a region and analyze likely future patterns of both.

- For both land use and economic activity, transportation investment rarely creates new economic activity or land use, but rather acts to redistribute activity that would likely occur anyway. Through the use of public policy and financial incentives, this redistribution can be guided to provide public benefits such as increased transit ridership.

- Recent interest in using land-use policies to provide more efficient utilization of transportation and other infrastructure will likely continue for many years. Growth management will be an overriding policy framework for transportation investment decisions in many metropolitan areas.

Community/Social/Cultural Impacts   Not only is transportation an integral component of a region's economy and land-use system, it is also a defining factor in community quality of life. The ability of people to interact, to enjoy the cultural and recreational benefits of living in an urban area, and to have access to the multitude of public and private services afforded by the clustering of urban activities are primary benefits of living in a "community." As noted previously, transportation enables other activities to occur and provides accessibility to employment, cultural, and recreational opportunities. Thus, disparities in transportation system performance could cause inequities to those living in a metropolitan region. Given a transportation system's purpose of providing mobility and accessibility, the important question is, mobility and accessibility for whom? In addition, because the construction of a transportation facility can disrupt a community just as it can an ecosystem, the consideration of social and cultural impacts also includes the displacement of people, the potentially negative effects on historical and cultural resources, and the degree to which community cohesion is affected.

This section focuses on two important social/cultural impact categories: community impacts and equity/environmental justice.

Community Impacts   Community impacts refer primarily to the effects of transportation investment on neighborhoods or groups of neighborhoods. The broader definition of "community," which could include an entire region, will be discussed in the next section with regard to equitable distribution of costs and benefits. Indeed, one of the first steps in conducting a community impact analysis is defining the study boundary. Defining this boundary could itself be quite a challenge given that "a community is defined in part by the behavior patterns which individuals or groups of individuals hold in common . . . and is defined by shared perceptions or attitudes, typically expressed through individuals' identification with, commitment to, and attitude toward a particular area" [U.S. DOT, 1996]. Because of the types of impacts associated with this definition of a community, a very proactive approach toward public involvement in the planning process is needed.

The displacement of people or businesses due to project implementation is of particular concern. Associated impacts can relate to such things as [Voorhees, 1979]

1. *Family and social ties* could be weakened or broken by geographic separation or relocation.

2. *Attitudes and behavior* could change as part of a psychological response to separation from neighbors or familiar surroundings.

3. *Disruption of neighborhood patterns* could occur due to relocation of large numbers of households. "Community cohesion" could weaken or dissolve in both the abandoned and new neighborhoods.

4. *Business viability* could be affected by immediate financial losses, as well as the loss of steady customers.

Special consideration should be given to those groups that are especially sensitive to changes in community structure, such as the elderly, low-income families, minorities, and longtime residents. These groups often experience the most negative and long-term effects of displacement [Fried, 1966]. Impacts on neighborhoods and on these groups could reach far beyond the construction period. As noted in a highway design manual produced by the U.S. DOT [1980]:

A new transportation facility may induce extensive business or industrial development and population growth, straining local services and facilities. Moreover, if the needs, desires, and social values of the incoming residents are at variance with the current residents, conflicts may arise and a new social structure may evolve.

Table 3.19 shows the types of impacts that should be part of community impact analysis. Guidelines for assessing the community impact of transportation facilities have identified several attributes of community character and structure that can be directly measured [Kaplan et al., 1972; Manheim et al., 1975; U.S. DOT, 1979a; Burdge, 1987; U.S. DOT, 1996]. More sophisticated and robust models and databases developed over the past 15 years have provided planners with a strong foundation for analyzing community impacts. For example, geographic information systems (GISs) can be a powerful tool for assessing the impacts of transportation investment on different population groups and on different geographic areas. It seems likely that a detailed assessment of social impacts and of the actions needed to mitigate negative effects will become an ever more important effort in the planning process in future years.

Table 3.19   Questions for community impact analysis

Social and Psychological Aspects

Will the project cause redistribution of the population or an influx of population?

How will the project affect interaction among persons and groups?

How will it change social relationships and patterns?

Will certain people be separated or set apart from others?

Will the project cause a change in social values?

What is the perceived impact on quality of life?

Physical Aspects

Is a wall or barrier effect created (such as from noise walls or fencing)?

Will noise or vibration increase?

Will dust or odor increase?

Will there be a shadowing effect on property?

Visual Environment

Will the community's aesthetic character be changed?

Is the design of the project compatible with community goals?

Has aesthetics surfaced as a community concern?

Land Use

Will there be loss of farmland?

Does it open new areas for development?

Will it induce changes in land use and density? What changes might be expected?

Is the project consistent with local land-use plans and zoning?

Economic Conditions

Will the proposed action encourage businesses to move to the area, relocate to other locations within the area close by, or move out of the area?

What is the economic impact on both the region and individual communities?

How is the local economy affected by the construction activities?

Are there both positive (jobs generated) and negative (detours and loss of access) impacts?

Will the proposed action alter business visibility to traffic-based businesses?

How will visibility and access changes alter business activity?

What is the effect on the tax base (from taxable property removed from base, changes in property values, changes in business activity)?

What is the likely effect on property values caused by relocations or change in land use?

Mobility and Access

How does the project affect nonmotorist access to businesses, public services, schools, and other facilities?

Does the project impede or enhance access between residences and community facilities and businesses? Does it shift traffic?

How does the project affect access to public transportation?

How does the project affect short- and long-term vehicular access to businesses, public services, and other facilities? Does it affect parking availability?

Table 3.19   Questions for community impact analysis (continued)

Provision of Public Services

Will the proposed action lead to or help alleviate overcrowding of public facilities (i.e., schools and recreational facilities)? Will it lead to or help alleviate underuse?

How will it affect the ability to provide adequate services?

Will the project result in relocation or displacement of public facilities or community centers (e.g., places of worship)?

Safety

Will the proposed action increase or decrease the likelihood of accidents for nonmotorists?

Will the proposed action increase or decrease crime?

Will there be changes in emergency response time (fire, police, and emergency medical)?

Displacement

What are the effects on the neighborhood from which people move and into which people are relocated?

How many residences will be displaced? What types—multiunit, single family, others?

Are there residents with special needs (disabled, minority, elderly) being displaced?

How many businesses and farms will be displaced? What types? Do they have unique characteristics, such as specialty products or a unique customer base?

Are there available sites to accommodate those displaced?

SOURCE: U.S. DOT, 1996

Mobility-Limited Populations/Equity/Environmental Justice   Because the transportation system plays such an important role in a region's economic and social well-being, it seems inevitable that change to this system will benefit some, while possibly hurting others. As noted by Hodge [1995],

> Certain places will be advantaged by investment in transportation, and others will be disadvantaged . . . because of the strong correspondence between social class and location in most U.S. cities, the impacts of transportation investments are strongly connected to social class.

The assessment of the geographic and population-related distributional impacts of transportation system investment and performance is called *equity analysis*. Three broad categories of inequity in transportation have been identified by Bullard and Johnson [1997]: (1) procedural inequity relating to the fairness and openness of the transportation decision-making process; (2) geographic inequity relating to the spatial distribution of benefits and costs; and (3) social inequity relating to the disproportionate impacts or burdens of transportation system outcomes borne by minority and low-income populations. These latter inequities, that is, the incidence of environmental impacts on minority and low-income communities, has led to a concern for *environmental justice*. The transportation planning process must carefully examine both issues—equity and environmental justice—in the ongoing effort to identify, assess, and mitigate the negative effects of transportation system performance.

Transportation systems in most urban areas provide unprecedented levels of mobility for urban residents. However, in most cases, this mobility is directly related to the ability of individuals to own and drive a car.[4] In those situations where individuals are too old or infirm to drive a car, or too poor to own one, the high level of mobility provided by the automobile is of little use. Yet the spatial distribution of urban activities, heavily influenced by the development of metropolitan highway networks, generally requires some form of vehicle for transportation. Four population groups in particular often have limited mobility options: the elderly, those with disabilities, low-income households, and minority populations. Although past studies have examined the mobility challenges of each as a separate group, there is a great deal of overlap in their respective travel behavior and mobility needs. For example, a study in Portland, Oregon, found that 60 percent of those disabled and 40 percent of the nonhandicapped elderly also lived in low-income households [Crain and Courington, 1977]. A nationwide survey in the United States found about 53 percent of those requiring special transportation services to be elderly [Ellis et al., 1977]. As noted by Revis [1976],

> The transportation disadvantaged (1) have no car or cannot drive because they are poor or physically unable to, (2) they often live in areas that are poorly served by public transportation, (3) for many, particularly the elderly and handicapped, the available transportation services do not meet their need for personalized door-to-door or door-through-door service, and (4) they may be confronted by serious physical design features of available transportation that create problems of orientation and maneuverability and that frequently discourage the transportation disadvantaged from making any trips.

Tables 3.20 and 3.21 and Fig. 3.8 present data on who is traveling by what means in the United States. Studies in both the United States and Canada have shown that the makeup of the elderly is quite diverse, resulting in travel patterns and

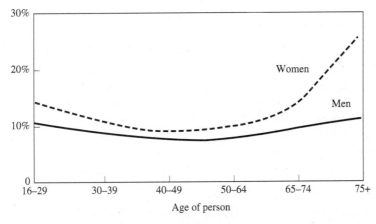

Figure 3.8   Percent of Americans in households with no vehicles
SOURCE: Spain, 1999

---

[4] Even in those cases where low-income households own automobiles, the tax structure for motor vehicle fees can be regressive. See Dill et al., 1999.

Table 3.20   Travel characteristics by age, sex, and household income

Age and Daily Trip Characteristics		Males Household Income			Females Household Income		
		<$15,000	$15,000–$59,999	$60,000+	<$15,000	$15,000–$59,999	$60,000+
16–29	Ave. no. trips	3.73	4.26	4.40	3.74	4.28	4.64
	Ave. no. person miles	25.6	37.4	37.8	23.6	32.3	36.7
	Ave. no. veh. trips	2.08	3.09	3.21	2.09	2.72	2.88
	Ave. no. veh. miles	17.1	28.1	29.5	13.8	20.8	23.5
30–39	Ave. no. trips	3.69	4.33	4.27	4.12	4.67	4.67
	Ave. no. person miles	26.4	39.7	42.2	25.4	33.6	37.0
	Ave. no. veh. trips	2.57	3.60	3.55	2.41	3.40	3.52
	Ave. no. veh. miles	20.0	33.8	36.9	14.6	24.1	27.4
40–49	Ave. no. trips	3.63	4.36	4.55	3.64	4.59	4.84
	Ave. no. person miles	28.9	39.5	42.2	23.9	32.2	37.1
	Ave. no. veh. trips	2.49	3.72	3.95	2.20	3.54	3.58
	Ave. no. veh. miles	21.1	34.7	38.3	15.2	24.6	26.4
50–64	Ave. no. trips	3.40	4.13	4.42	2.86	3.72	3.84
	Ave. no. person miles	22.3	36.0	39.0	16.6	25.7	28.5
	Ave. no. veh. trips	2.42	3.51	3.88	1.63	2.47	2.63
	Ave. no. veh. miles	16.9	30.6	34.5	8.7	15.7	18.2
65–74	Ave. no. trips	2.97	4.20	4.08	2.71	3.56	3.75
	Ave. no. person miles	18.5	28.3	30.2	15.2	22.2	26.3
	Ave. no. veh. trips	2.16	3.50	3.38	1.54	2.12	2.13
	Ave. no. veh. miles	12.9	23.8	25.6	8.7	11.9	13.7
75+	Ave. no. trips	2.55	3.17	3.56	1.93	2.60	2.08
	Ave. no. person miles	15.4	20.4	21.6	8.8	14.4	11.9
	Ave. no. veh. trips	1.62	2.44	2.68	0.89	1.41	1.24
	Ave. no. veh. miles	9.9	15.1	17.6	3.6	7.0	7.3
Column average	Ave. no. trips	3.44	4.23	4.39	3.15	4.19	4.49
	Ave. no. person miles	23.7	37.7	40.0	18.8	37.7	34.6
	Ave. no. veh. trips	2.24	3.42	3.64	1.77	2.89	3.16
	Ave. no. veh. miles	16.8	30.3	34.5	10.6	20.0	23.9
For U.S.	Ave. no. trips		4.02				
	Ave. no. person miles		31.4				
	Ave. no. veh. trips		3.03				
	Ave. no. veh. miles		24.2				

SOURCE: Spain, 1999

needs that differ significantly from one another [Wachs, 1979; Wolfe and Miller, 1982]. However, as is shown in Table 3.20, trip making by the elderly is clearly much less frequent than that by other adults. The data in this table indicate the income effect as well; that is, the low-income elderly travel much less and for shorter distances than those with higher incomes. Perhaps a more significant measure of the mobility challenge facing the elderly is lack of a vehicle for trip making,

Table 3.21   Mode split by income

		Household Income				
Mode	<$15,000	$15,000–$29,000	$30,000–$49,999	$50,000–$79,999	$80,000 and over	All
Total auto	75.9%	87.0%	90.1%	91.2%	91.1%	88.3%
2+ occupants	40.7	43.6	46.0	47.4	48.1	45.5
Only driver	35.1	43.3	44.0	43.7	42.9	42.4
Total transit	6.8	2.5	1.5	1.2	1.2	2.1
Bus/LRT	5.6	1.7	1.0	0.5	0.5	1.4
Heavy rail	0.9	0.6	0.4	0.4	0.5	0.5
Commuter rail	0.3	0.2	0.2	0.2	0.2	0.2
School bus	1.8	1.8	1.9	1.7	1.5	1.8
Taxicab	0.5	0.2	0.1	0.2	0.3	0.2
Bicycle	1.6	1.1	0.7	0.9	0.5	0.9
Walk	12.8	7.1	5.4	4.2	5.0	6.2
Other	0.6	0.4	0.5	0.5	0.5	0.5
All	100	100	100	100	100	100

SOURCE: Pucher et al. 1998

especially for elderly women (see Fig. 3.8). Two observations concerning these results, however, merit special attention. First, a 65-year-old and a 25-year-old having similar socioeconomic and auto-ownership levels would generally not have the same trip-making behavior. Second, the mobility problems for the future elderly might be different from those experienced by the elderly today. For example, the fastest-growing population cohort having drivers licenses as compared to previous years is those over 65 years of age.

Much of the concern with equity in transportation during recent years has related to the mobility opportunities and costs of travel for specific income groups. Low-income households often do not have access to automobiles and thus must rely on public transportation or other means for their mobility. For example, close to 33 percent of total transit ridership in the United States comes from households with incomes less $15,000 [Pucher et al., 1998]. This income group also accounts for 22 percent of the taxi and 21 percent of the pedestrian travel in the United States.

Figures 3.9 and 3.10 illustrate the typical mobility challenge facing low-income households, in this case for the Albany, New York, metropolitan area. Over 50 percent of the households in the region's central cities have income less than $25,000 (1990 dollars). The central cities also have more zero-car households than the rest of the region combined and the largest percentage of disabled, close to 5 percent of the central city population. The distribution of building permits in the region, an indication of where new retail, employment, and residential opportunities will exist in the future, showed substantial new development in the suburban locations, thus suggesting continuing mobility challenges for the low-income population [Capital District Transportation Committee, 1995].

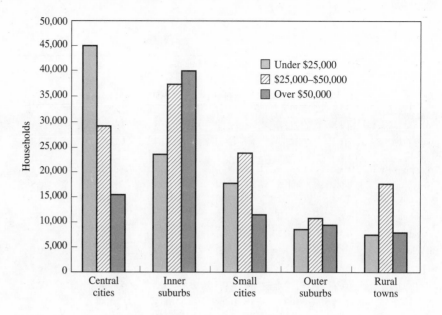

Figure 3.9   Residence of low-income population, Albany, New York, 1990

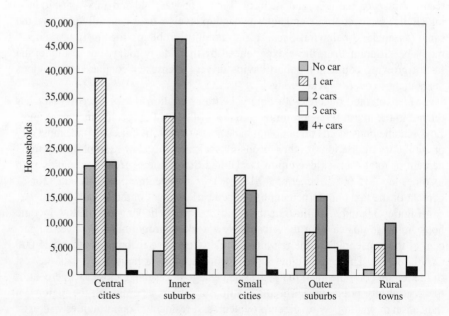

Figure 3.10   Household auto ownership in Albany, New York, 1990
SOURCE: CDTC, 1995

To a large extent then, the mobility of many lower-income urban residents depends on government-provided transit service. Indeed, one of the key justifications of transit subsidization has been the premise that low-income populations are primary beneficiaries of the public subsidy. Several studies, however, have concluded that the costs for such benefits occur at a much higher incidence among low-income households [Altshuler et al., 1979; Cervero et al., 1980; Pucher and Hirschman, 1981]. An interesting study of the distribution of fare payments by income class was undertaken by Pucher et al. [1981]. Their results of a national sample are shown in Table 3.22.

A study of the incidence of all transportation funding sources for southern California showed a similar relationship. Table 3.23 shows that those in the bottom 20 percent income group spent 3.35 percent of their income to support transportation, while those in the upper quintile spent 1.36 percent [EDF, 1994].

Table 3.22 Distribution of fare payments by income class

Income Class	Fare Payments as % of Total Fare Revenues	Fare Payments as % of Money Income in Each Class
<$6,000	22.8%	1.30%
$6,000–$10,000	17.0	0.63
$10,000–$15,000	19.4	0.38
$15,000–$20,000	14.4	0.20
$20,000–$25,000	10.8	0.20
>$25,000	15.5	0.08

SOURCE: Pucher et al., 1981

Table 3.23 Distribution by income quintile of transportation taxes and fees in southern California, 1991 (percent of per capita median income)

	Quintile 1	Quintile 2	Quintile 3	Quintile 4	Quintile 5	Ratio 1:5
Federal gas tax	0.62%	0.56%	0.42%	0.35%	0.22%	2.9
State gas tax	0.73	0.65	0.49	0.42	0.25	2.9
State sales tax on gas	0.35	0.31	0.24	0.20	0.12	2.9
State motor vehicle license fees	0.52	0.40	0.41	0.49	0.40	1.3
State registration and other fees	0.34	0.26	0.20	0.16	0.10	3.5
Retail sales tax	0.80	0.58	0.48	0.42	0.28	2.9
All trans. taxes	3.35%	2.75%	2.24%	2.03%	1.36%	2.5

SOURCE: EDF, 1994

Perhaps the most contentious example of the incidence of transportation-funding burden on low-income and minority populations occurred in Los Angeles in the mid-1990s. A significant fare increase (needed to pay for subway, light rail, and commuter rail operations) was viewed by many as a "disproportionate and irreparable impact on the metropolitan area's minority communities and bus-riding public of whom 80 percent are Latino, African-American, Asia-Pacific Islander, and Native American" and thus a violation of Title VI of the Civil Rights Act [Mann, 1997]. After two years of litigation, the issue was settled out of court with the transit agency reducing the price of transit passes and agreeing to increase bus service at a substantial cost.

In recent years in the United States, mobility for low-income households has also become an important element of national welfare policy. The Personal Responsibility and Work Opportunity Reconciliation Act of 1996 provided time limits for welfare recipients to find jobs (the so-called "welfare-to-work" program). However, two key issues have repeatedly been pointed to by welfare recipients as barriers— child care and transportation. In larger cities, where the bulk of the welfare and low-income families live, transit service exists but often does not provide good service to suburban employment sites where the demand for low-skill, entry-level workers is strongest [Loveless, 1999]. Given that almost one-third of all households having less than $15,000 income do not own a car and that just 48 percent own only one, access to job opportunities can be a tremendous barrier to welfare recipients seeking a job [Pucher et al., 1998].

Equity issues as they relate to mobility disadvantaged populations can concern finance, distribution of benefits, and service provision. However, in 1978 another key equity issue surfaced when contaminated soil was buried in a predominantly poor and African-American North Carolina county. The national outcry over this incident, and a subsequent study by the U.S. General Accounting Office (GAO) showing that three of four hazardous landfill sites in the U.S. South were located in African-American communities, gave rise to the environmental justice movement.

In 1994, a presidential Executive Order, *Federal Actions to Address Environmental Justice in Minority Populations and Low-Income Populations,* laid out the general framework for federal agencies (and more importantly for transportation, to federally aided programs) to assure that disproportionately high and adverse human health and environmental effects on minority and low-income populations did not occur. In response, the U.S. DOT, FTA, and FHWA issued their own guidance on how environmental justice was to be included in transportation decision making. (See <*http://www.fhwa.gov/legsregs/directives/orders/6640_23htm*>.) Although it is too soon to tell how transportation planning will be affected by such concerns, it seems likely that at a minimum, the planning process will have to reach out to low-income and minority communities for their input, collect data on issues relating to environmental justice, and document how environmental-justice-related data are incorporated into decision making. For a good overview of environmental justice as it relates to transportation, see [Kennedy, 1999].

In general then, the mobility problems of the low-income, elderly, disabled, and minority populations merit special attention in the transportation planning process. As will be discussed in Chap. 8, the equitable distribution of costs and benefits is an issue that consistently surfaces in the decision-making process and, consequently, in the evaluation phase of urban transportation planning.

Planning Applications of Community/Social/Cultural Impacts

• Transportation systems provide tremendous benefits to those who can afford to use the system. Assessment of the equitable distribution of benefits and costs is an important part of transportation planning.

• Community impact analysis has become an important component of transportation planning. Community impacts of transportation tend to affect elderly, low-income, and disabled groups to a greater extent than most other groups in urban areas. Guidelines for assessing the community impacts of transportation facilities have been developed to help ensure that the actions required to reduce the negative effects of a new facility are planned for early in project development.

• Geographic information systems provide a powerful tool for analyzing the distribution of benefits and costs and will likely be one of the most important technical approaches in these types of studies.

3.3 CHARACTERISTICS OF URBAN TRAVEL

Previous sections have used a systems perspective to describe the general characteristics and impacts of transportation. However, there are some specific characteristics of transportation systems relating to their use, that is, urban travel characteristics, that are important for understanding much of the technical analysis in the planning process and the types of strategies considered. Six characteristics of this trip-making behavior merit special attention: trip purpose, temporal distribution of trip making, spatial distribution of urban travel, selection of the mode used, transportation safety, and the cost of making the trip. Each of these characteristics will be found in some form in the analysis tools or models used to analyze urban travel.

3.3.1 Trip Purpose

Traditionally, transportation planners have modeled passenger trips by the purpose they serve, classified by the trip origin. Common trip types include

Work trips: Trips made to a person's place of employment such as a factory, a store, or an office.

Shopping trips: Trips made to a retail establishment regardless of the size or type of purchase. Trips made to a store "just to look" are shopping trips even though no purchase is made.

Social or recreation trips: Cultural trips made to recreational or entertainment facilities (e.g., church, civic meetings, concerts, sporting events). Travel to social activities (parties, visiting friends) would be included.

Business trips: Trips made in the course of performing a normal day's work. The origin of such trips is often the place of employment.

School trips: Trips made by students to an institution of learning.

Because trips are defined as one-way movements, another purpose called "home" is often appended to the trip purpose and is used to classify trips into one of five categories: home-based work, home-based shop, home-based school, home-based other, and nonhome based. For smaller urban areas, the home-based work, home-based other, and nonhome-based trip categories are usually used.

Table 3.24 shows the frequency of trip purpose of daily urban passenger travel in the San Francisco Bay Area. The home as an anchor for trips accounts for over 75 percent of the trips. Home-based work trips account for about 27 percent of all trips but 43 percent of all person-miles traveled. Nonhome-based trips are the largest trip purpose share at 28 percent but represent only 22 percent of the person-miles traveled. This distribution of urban passenger trips by purpose in the Bay Area is fairly typical of most major urban areas. Also note in this table that the school trip is further disaggregated to grade school, high school, and college.

One of the important changes in urban travel behavior in recent years has been the increase in multipurpose trip making, also referred to as trip chaining. Most likely due to higher family incomes and more women in the work force, the trip to and from work is now often the opportunity to accomplish other tasks, such as to drop off or pick up children, shop, or take care of personal business [Nishii et al., 1988; Strathman and Dueker, 1995; Levinson and Kumar, 1995; McGuckin and

Table 3.24   Trip purpose characteristics in San Francisco Bay Area

Trip Purpose	Person Trips	Average Distance (miles)	Average Time (mins)	Total Distance (miles)
Home-based work				
First income quartile	534,639	9.47	18.8	5,063,031
Second income quartile	1,124,801	11.43	21.3	12,856,475
Third income quartile	1,620,069	12.91	23.2	20,915,091
Fourth income quartile	1,284,902	13.31	23.9	17,102,046
Total home-based work	4,564,411	12.25	22.4	55,936,643
Home-based shop/other	4,259,935	5.73	11.8	24,409,428
Home-based social/recreation	1,910,361	7.39	13.7	14,117,568
Home-based school				
Home-based grade school	842,871	3.2	8.1	2,697,187
Home-based high school	345,542	4.41	10.0	1,523,840
Home-based college	438,063	8.81	17.9	3,859,335
Total home-based school	1,626,476	4.97	11.2	8,083,586
Nonhome based	4,716,990	6.11	12.8	28,820,809
Total trips, all purposes	17,078,173	7.69	15.1	131,368,034

SOURCE: Purvis, 1998

Murakami, 1998]. Current research in travel demand forecasting is developing an approach that models the entire activity associated with trip making instead of the traditional approach of modeling each trip purpose. Known as activity-based travel demand modeling, this approach is based on modeling several concepts shown in Fig. 3.11. The trip-based model is the approach most used today in transportation planning studies. The tour-based approach models the combination of trips taken with the trip anchors of home and work. Examples include home-to-work, work-to-home, work-to-work (midday trips) and home-to-home. The daily schedule approach ties together the tours taken during the entire day and comes closest to what the traveler actually did.

Figures 3.12 to 3.14 show data from the 1995 National Personal Transportation Survey (NPTS) that indicate different travel behavior for men and women. As shown, there is a significant difference in trip activity, most likely relating to women taking the primary responsibility for transporting children and other household tasks (for a good overview of women's travel issues see [FHWA, 1998]). Chapter 5 will discuss the modeling implications of this phenomenon in greater detail.

Urban goods movement can also be classified by the purpose of the trip. Trip purpose, as related to goods movement, reflects the type of activity being performed by the truck driver. Between 60 and 80 percent of the truck trips in *central business districts* (CBDs) are for goods delivery, between 10 and 20 percent are for goods

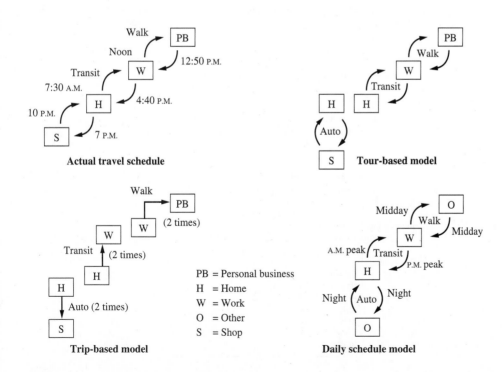

Figure 3.11   Three approaches for modeling trips

SOURCE: Bowman and Ben-Akiva, 1997

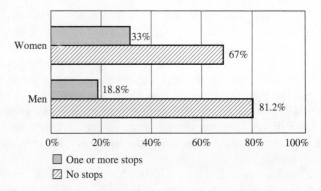

Figure 3.12   Percent of men and women who stop (home-to-work trips)
ǀ SOURCE: McGuckin and Murakami, 1998

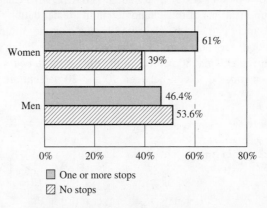

Figure 3.13   Percent of men and women to stop (work-to-home trips)
ǀ SOURCE: McGuckin and Murakami, 1998

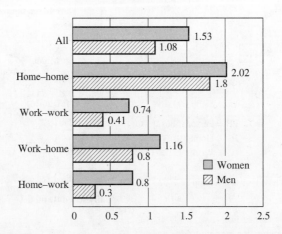

Figure 3.14   Mean number of stops by tour type
ǀ SOURCE: McGuckin and Murakami, 1998

pickup, and between 10 and 20 percent are for service [Wegmann et al., 1995]. Across *metropolitan areas,* truck trip purposes exhibit greater diversity, reflecting the wider variety of activities outside CBDs. In a sample of 11 areas, goods pickup and delivery (including mail) represented 46 percent of the truck trips, whereas home based and personal use was 28 percent, construction 5 percent, maintenance and repair 8 percent, and business use 7 percent, with the remaining 5 percent in other uses [Levinson, 1982].

An important trip characteristic associated with both passenger and goods movement is that the length of urban trips, because of the spatial location of origins and destinations, will vary by trip purpose. This was seen in Table 3.24, for example, where the commute trip had the longest average trip length, which is not surprising given that this is the least discretionary of all trip purposes (although the vacation and recreational trip purpose often has the longest trip distance when it is included in the trip mix). For trucks, trip lengths will vary significantly by trip purpose and by truck type. For example, in Chicago, the average trip length for the smallest trucks (less than 8,000 pounds) was 11.1 miles; for light trucks (8,000 to 28,000 pounds) 9.6 miles; for medium trucks (28,000 to 64,000 pounds) 10.4 miles; and for heavy trucks (more than 64,000 pounds) 24.9 miles [Wegmann et al., 1995].

Travel times associated with these trips will also vary by trip purpose because of the spatial differences in land-use locations and because of the different times of day when trips occur (e.g., congestion during the peak hours might cause even a short-distance trip to take a long time). A review of the 1990 Census data for the United States indicated that the average commute trip in metropolitan areas was 23.2 minutes, with 68 percent of commute trips taking less than 30 minutes (roughly 7.5 percent take more than 60 minutes in areas over one million population) [Pisarski, 1996].

Figure 3.15 shows that average vehicle occupancy also varies by trip purpose. This information is important for those interested in fostering increased ride sharing in a metropolitan area.

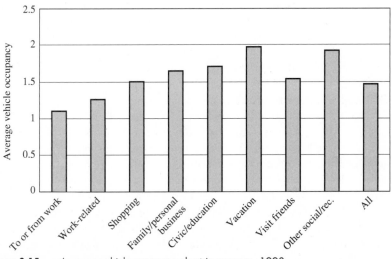

Figure 3.15   Average vehicle occupancy by trip purpose, 1990
SOURCE: Pisarski, 1996

From a transportation planning perspective, an analysis of urban travel by trip purpose is important because it can be used to deduce demand patterns of urban travel. For example, information on the number of home-to-work trips originating in one area and destined for another can be used to derive levels of demand on transportation facilities/services linking these two locations. Such information is also useful for determining the magnitude of problems that occur when the needs of two or more trip types conflict. For example, the allocation of limited downtown parking space between commuters who park all day and shoppers who park for only a few hours is an important policy issue in many urban areas. With regard to goods movement, information on trip purpose can be used to assess major truck routes and analyze terminal or truck stop congestion.

3.3.2   Temporal Distribution of Trip Making

Given that transportation is the primary means by which people participate in urban activities that by their very nature occur at different times of the day, it is not surprising that the profile of urban travel over the day shows temporal variation. The best example of this phenomenon is the peak periods or rush hours of 6:00 to 9:00 A.M. and 4:00 to 7:00 P.M. when many activities such as work and school either begin or end. According to the 1995 National Personal Transportation Survey (NPTS), 37 percent of trips for all purposes start during these two periods [McGuckin and Murakami, 1998].

Historically, most North American urban transportation systems have exhibited a pronounced "double peaking" of travel. Work-related travel in the morning and evening produced larger travel demands during the peak periods than during off-peak periods. Although work-related travel is still more peaked than other trip purposes, overall travel on an urban transportation system today tends to be more evenly distributed throughout the day because nonwork trip making has increased significantly over the past 20 years (Fig. 3.16). For example, nationally, between 1977 and 1995, the most growth in daily travel per person occurred in trips for family and personal business, which more than doubled. In addition, trips for social and recreational purposes increased by 51 percent, whereas work trips increased by 33 percent. However, the peaking phenomenon is still characteristic of transit service, which carries its primary ridership during the morning and evening peaks.

Peak periods for truck traffic tend not to correspond with peak passenger traffic. In fact, many urban goods providers try to avoid congested locations and times. Thus, in many cities, truck activity is most intense right after the morning passenger peak and just before the evening rush hour. In other cases, such as intermodal terminals, peaking characteristics reflect more the operations of the terminal and the need to transfer goods to waiting transportation services. At the metropolitan level, the majority of truck trips occur in midday between the hours of 10:00 A.M. and 3:00 P.M.

The peaking of travel demand in specific time periods results in congested highway facilities and transit services. Congestion is simply a condition of any transportation facility in which use of the facility is so great that there are delays for the users of that facility. Usually this happens when traffic approaches facility capacity. Urban highway peak demand is just less than twice the average hourly

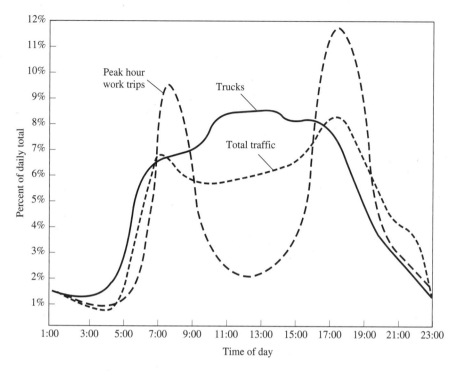

Figure 3.16   Typical distribution of trips during the weekday

demand, while for public transit services, the peak-hour demand can be three to four times the average. This pronounced peaking characteristic of transit ridership creates a challenge to transit operators who must provide sufficient capacity to handle peak demands, but which is not needed during off-peak periods.

In the 1960s, the solution to congestion often involved expanding the capacity of the highway and transit network to accommodate increased demand during peak hours. Due to a variety of factors relating to environmental concerns, community preservation, and inadequate funding, other strategies have now become part of the possible solutions to the congestion problem. One approach to handling the peaking of passenger travel demand involves arranging work hours so that large groups of employees arrive and leave work at different times in order to reduce or "flatten" the peak. These so-called "variable work-hour programs" have proved successful at spreading employee arrival times, which, at least for the specific employment site, can have significant congestion-reduction benefits. A more widespread impact of such programs would depend on the coordinated adoption of these programs on a regional or corridor basis.

Another approach for dealing with the peaking phenomenon is to increase the person-carrying capacity of facilities and services. By moving more people in fewer vehicles, the transportation system can handle more person trips when the demand is the greatest. Actions such as preferential treatment of high-occupancy vehicles, ride-sharing programs, and alternative transit operating strategies have been used in many cities seeking such a result. Another strategy is to reduce the demand by encouraging employees to telecommunicate.

3.3.3 Spatial Distribution of Trip Making

Each trip begins at an origin and ends at a destination located at specific geographic points in an urban area (or at a study boundary point if leaving or entering the urban area). Thus, the spatial distribution of travel in an urban area is directly related to the patterns of land use and the network configuration of the transportation system. The road network connects many more places than does a transit network. Thus, transit travel tends to be more spatially restricted. In North America, it is largely CBD oriented, although much attention has been paid in recent years to ways of providing transit services within and to suburban markets.

Figure 3.17 shows the flow of vehicle travel in Charlotte, North Carolina. This figure is very typical for North American metropolitan areas. As can be seen, there is a predominant flow of traffic into and through the central city. Importantly, however, one can see in this figure the existence of major activity centers throughout the region that serve as major attractors of trips. These suburban activity centers, usually located near major freeways, often become mini-downtowns, experiencing the same issues of mobility and congestion that the central city faces.

The spatial distribution of trip making requires transportation planners to represent or model travel flows on networks that reflect the movement of people and goods throughout the region. This is done in several ways. First, the study area is usually divided into *traffic analysis zones* that represent the geographic locations of trip origins and destinations. Each zone is characterized by land-use and population demographic characteristics that allow planners to estimate the number of trips likely to be produced and attracted to that zone. An *origin–destination matrix,* or *trip table,* is developed that forecasts the number of trips coming from each zone destined to all others. The trips are assigned to a *transportation network* that is spatially defined to represent the actual transportation system. Links in this network, for example, represent roads or transit facilities. The zonal system, trip table, and network representation allow planners to estimate trip movements throughout the region.

Understanding the spatial distribution of trip making becomes an important element of urban transportation planning in that it indicates where transportation problems are likely to occur in an urban area, the level to which the existing transportation system satisfies this need, and those areas where action must be taken to improve system performance. As will be seen in later chapters, the spatial nature of urban travel is incorporated into every aspect of the planning approach.

3.3.4 Modal Distribution of Trip Making

As noted in Sec. 3.1.4, modes of transportation are basic components of a transportation system. The proportion of trips made in an urban area by different travel

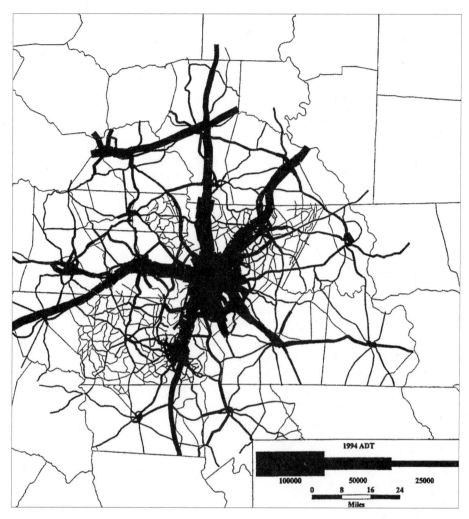

Figure 3.17   Vehicle flows in Charlotte, North Carolina, 1994
⏐ SOURCE: Hartgen et al., 1996

modes (e.g., transit, auto, bicycle, walking) varies markedly from city to city and from country to country. However, in the United States and Canada, the privately owned vehicle is the dominant travel mode (Fig. 3.18). At the national level, the percentage of trip making made by public transit is very low, accounting for just over 5 percent of total work trips in the United States in 1990. The aggregate numbers can be misleading, however, because the transit share of the modal split in large metropolitan areas can be dramatically higher, especially in central business districts. Disaggregating these national numbers by type of urban area gives different numbers— transit carries about 11.5 percent of commute trips in central cities and 8.7 percent

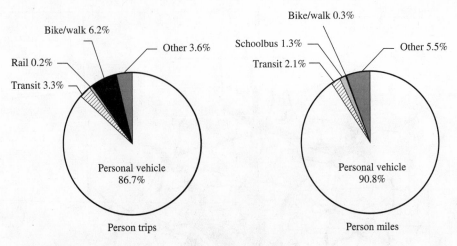

Figure 3.18   Person trips and person miles by mode
I SOURCE: Pisarski, 1996

of work trips in metropolitan areas over one million population. Transit share of commute trips will differ by city. For example, the transit mode share for commutes in New York City is 26 percent; Chicago and Washington, D.C., 14 percent; Atlanta and Los Angeles, 5 percent; and Kansas City, 3 percent. These percentages will also increase substantially for travel during the peak hours. For example, public transit accounts for about 90 percent of the peak-hour trips entering or leaving the New York City central business district, 82 percent of such trips in Chicago, 49 percent in Boston, 44 percent in Ottawa, and 37 percent in Los Angeles.

Several factors influence an individual's choice of travel mode. One of the most important factors is the difference in trip time between modes for particular trips. For example, in those cases where an automobile is available, an actual or perceived transit trip time that is longer than that for driving would probably result in the trip being made by car (all other factors being equal). A desire to make transit more competitive with the automobile has led to numerous projects that provide preferential treatment to high-occupancy vehicles (HOVs), thus resulting in travel-time savings. Other important factors include mode availability, such as auto ownership or accessibility to transit (Fig. 3.19), and differences in actual costs, perceived costs, comfort, or convenience, such as availability of parking close to the destination. Modal distribution can also be related to factors such as occupation, income, age, and other socioeconomic characteristics (Table 3.25).

Much of public policy in urban transportation during the past several years has focused on shifting the modal patterns of trip making. This has taken many forms. Preferential lanes on freeways for high-occupancy vehicles (i.e., car pools, vans, and buses) are designed to make these modes more attractive by reducing trip times. Parking policies that restrict access, increase parking prices, and reduce overall parking supply are meant to reduce the attractiveness of the automobile as a commuting mode. Actions to improve transit service are undertaken to increase its attractiveness as a mode of travel.

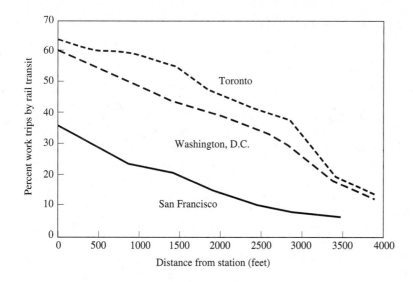

Figure 3.19   Market share related to walking distance
SOURCE: Weyrich and Lind, 1999

Table 3.25   Socioeconomic characteristics indicating a high propensity for transit use, 1990 Census

Characteristic	Transit Share (%)
Central city dweller	11.5
Renter	9.5
Household with no vehicles	39.1
Women	6.0
Young	6.4
Older (75+)	6.4
High income	
$75,000–$99,000	6.6
Over $100,000	5.8
One-worker household	6.1
Four-worker household	7.5
Female worker, living alone in central city	16.3
Black	14.8
Asian	11.0
Hispanic	8.8
All commuters	5.1

SOURCE: Pisarski, 1996

3.3.5   Transportation Safety

A primary goal of the urban traveler is to arrive at the destination safely. Over the past 3 decades, transportation fatality rates have declined as measured in relationship to system usage. Figure 3.20 shows fatality rates (fatalities per 100 million vehicle miles) for different modes. Note the different magnitudes of these rates associated with the characteristics of the modes. Occupants of passenger cars and light-duty trucks have much higher fatality rates than occupants of large trucks. Motorcyclists have fatality rates an order of magnitude greater than the other modes. The differences are clearly related to the greater size and mass of the larger vehicles. The declining fatality rates from 1975 are most likely due to safety belts, air bags, child safety seats, and motorcycle helmet use, in addition to police enforcement of tougher drunk-driving laws.

Because of population concentration, metropolitan areas are particularly susceptible to higher rates of pedestrian and bicycle crashes. In 1997, for example, 6,100 pedestrians and bicyclists in the United States were killed in crashes with motor vehicles [BTS, 1999]. Because of the significant disadvantage to pedestrians and bicyclists when colliding with motor vehicles, it is not surprising that they represent a disproportionate amount of the fatalities. For example, in 1997, pedestrian/motor vehicle crashes accounted for only 2 percent of the total injuries nationally, but 13 percent of the fatalities. Six percent of pedestrians involved in injury

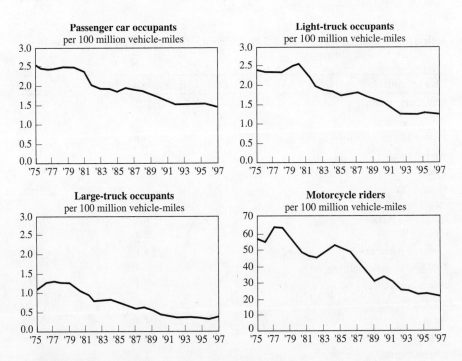

Figure 3.20   Fatality rates for selected modes, 1975–1997
I SOURCE: BTS, 1999

crashes were killed and 24 percent had incapacitating injuries; the respective percentages for motor vehicle occupants were 1 and 12 percent. Importantly, children and the elderly are disproportionately represented in crash statistics. Children 5 to 15 years old accounted for 16 percent of the U.S. population in 1997 but 29 percent of the pedestrian injuries and 9 percent of the pedestrian fatalities during this year. Adults 65 years and older, which represent 13 percent of the population, accounted for 8 percent of the injuries and 22 percent of the fatalities.

Tables 3.26 and 3.27 show the frequency of crash types for pedestrians and bicyclists. As can be seen, the most frequent type of crash occurs where the geometric design of the transportation system causes or allows the different travel flows to intersect. Especially in urban areas, this suggests that special care is necessary when designing intersections, pedestrian walkways, and bicycle paths to make sure that these crossings are safe.

Table 3.26   Pedestrian crash types

Crash Type	Percentage of Crashes
Crossing at intersections	32
Crossing at midblock	26
Not in road (e.g., parking or near curb)	9
Walking along road/crossing expressway	8
Backing vehicle	7
Working or playing in road	3
Other	15
Total	100

SOURCE: BTS, 1999

Table 3.27   Bicycle–motor vehicle crash types

Crash type	Percentage of Crashes
Crossing paths	57
Parallel paths	
Motorist turned into path of bicyclist	12
Motorist overtaking bicyclist	9
Bicyclist turned into path of motorist	7
Bicyclist overtaking motorist	3
Operator on wrong side of road	3
Operator or motorist loss of control	2
Other circumstances	7
Total	100

SOURCE: BTS, 1999

3.3.6   Travel Costs

Costs are incurred whenever a trip is made. This cost of travel is often defined and perceived differently by users, stakeholders, and system providers. For example, if asked, most travelers would identify just the out-of-pocket costs associated with travel—fuel costs, parking fees, tolls, or fares. Some might even consider the associated costs of vehicle purchase, maintenance, and insurance. (See *<http://www. apta.com/statstr/sauto/drivcost.htm>* for the most up-to-date information on automobile driving costs.) A broader perspective on the costs of travel, however, would include many other considerations. In recent years, a great deal of attention has been paid to the "social cost" of motor vehicle use [Apogee, 1994; Greene et al., 1997; Gomez-Ibanez, 1997]. Social costs extend traditional travel costs to include the cost of transportation services and goods that are not priced directly to the user but are incorporated into the price of the nontransportation good; public sector costs to build, maintain, and operate the transportation system; and the nonmonetary costs to society of transportation-related impacts such as health effects of air and water pollution (known as externalities), pain and suffering from accidents, and travel time [Delucchi, 1997].

An example of this approach toward cost accounting is shown in Fig. 3.21. In this example from Seattle, *direct costs* are out-of-pocket expenditures directly for transportation—the costs of cars, insurance, building and maintaining roads, paying police officers to provide traffic enforcement, operating transit service, transportation planning, and so on. These costs are paid by private individuals and businesses as well as by governments. Indirect costs include costs that occur as a result of

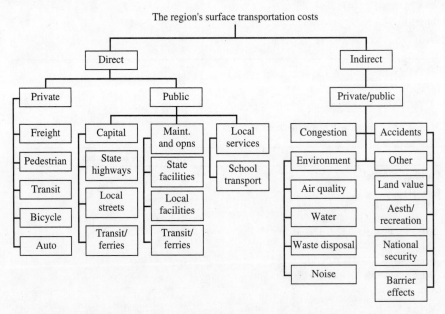

Figure 3.21   Transportation cost accounting in Seattle
| SOURCE: PSRC, 1996

transportation—congestion, accidents, air pollution, water impacts, solid-waste disposal, and noise. The Seattle report estimated that direct annual private costs totaled $18.3 billion, direct public costs were $1.7 billion, and indirect costs were conservatively estimated at $1.1 billion. The study concluded

- The Seattle region spent a "staggering" amount of money on transportation, about 25 percent of the region's personal income.

- Government expenditures on roads, transit, and ferries represented only 8 percent of all transportation expenditures.

- Even without considering gas taxes, licensing, and excise taxes, auto-related expenditures by individuals and businesses represented 62 percent of all expenditures.

- Less than 3 percent of the region's direct public and private transportation expenditures go toward transit, ferries, and nonmotorized transportation [Puget Sound Regional Council, 1996].

National estimates of social costs of motor vehicle use vary widely, with most variation explained by differing underlying assumptions. For example, the estimated underpayment by auto users, or what is referred to as net subsidy, ranges from 3 to 53 cents per passenger mile [Gomez-Ibanez, 1997]. In a similar vein, Delucchi [1997] estimated that motor vehicle users cost society more than they explicitly paid for transportation goods and services in a range from $1.7 to $2.8 trillion annually. Although one can argue over the assumptions used and the resulting magnitudes of the net social cost to society of motor vehicle use, it seems clear from all the studies so far that indeed motor vehicle users do not pay the total costs to society incurred by motor vehicle use. This is an important public policy issue that will continue to receive attention for many years and is discussed in greater detail in Chap. 8.

3.4   CHAPTER SUMMARY

Several of the sections presented in this chapter ended with observations on how the concepts discussed relate to transportation planning. These observations provide the summary for this chapter.

1.   Transportation can be viewed as one system that relates to and is part of many other systems. This perspective leads to important planning questions reflecting the interaction among transportation and the other systems that help an urban area function, and between transportation and higher-level systems, for example, ecological or economic systems. Transportation systems themselves consist of hierarchies or classification schemes that relate to the respective role of different system components, for example, a road functional classification system or transit service categories.

2.   The purpose(s) of a transportation system can be defined in a variety of ways and often reflects the perspective of those involved in the planning process. The definition could be narrowly targeted on the specific functioning of the system or more

broadly reflect the enabling influence of transportation on other systems. Defining the transportation system purpose during the planning process most often occurs in the vision statement and in the statement of goals and objectives.

3.   One of the first steps in any planning effort is to establish the boundaries of the system being analyzed. These boundaries will vary in relation to the problem definition, decision domain, and scale of analysis. The level of technical analysis within the system boundary is much more detailed than that associated with the area outside, which is often represented very simply. The boundary itself can become an important part of the analysis by examining the inputs and outputs that cross it. Because a system boundary becomes a key determining factor in the breadth and depth of the attention focused on system operation and impacts, this boundary must be initially defined broadly enough (or allowed flexibility to change later on) to cover the spatial and temporal interrelationships associated with system *effects*.

4.   Transportation systems consist of numerous components whose interaction becomes a key factor for system effectiveness. These components include system users; transportation modes; networks, facilities, and services; intermodal connections; and stakeholders. A multimodal transportation system consists of a variety of interconnected modes that provide effective transportation service from origins to destinations.

5.   System performance is an important consideration guiding the definition of problems and opportunities that become the focus of planning efforts. System performance measures should be defined to provide information to the decision-making process, thus implying that such measures should be broadly defined not only as outputs, but also as the ultimate impacts or outcomes on society.

6.   Transportation system capacity (person flow or vehicle flow per hour) is an important consideration in transportation planning and is the focus of much technical analysis. Transportation policies aimed at reducing the use of single-occupant vehicles have led to increasing use of person flow-rate capacity measures. The capacity of individual facilities or services can be affected by a variety of factors, most often incorporated into the design of the facility or service itself. Network and system capacity is directly related to the capacity of the "weakest" link in the network. Thus, much planning attention is often paid to improving the efficiency of flow or enhancing the capacity of these bottlenecks.

7.   The transportation system is usually planned, designed, built, operated, and maintained by organizations and individuals with different objectives, mandates, constituencies, and problem definitions. Changes to the urban transportation system can include a wide variety of infrastructure and service actions, applied at different geographic scales by the public and private sectors, to improve mobility and the urban environment. Increasingly, private sector groups and organizations are playing a major role in providing transportation services and funding. The effective cooperation of public and private sector entities thus becomes an important criterion for successful transportation planning.

8.   The most important feedback mechanism to the users of a transportation system is the *cost* of travel, determined in the market by the equilibration of supply and

demand. Changing transportation costs as perceived by users can influence, in the short run, route choice, time of travel, trip frequency, and linking of trips. Over the long term, these changes could influence urban form through residential and business location decisions, auto ownership, and the technology of transportation. Increasing attention has been given to the use of road-pricing strategies as a means of reducing congestion (i.e., as a feedback mechanism to system users). Although such strategies are often difficult to implement from a political perspective, they will likely become part of the strategy "mix" that planners will be investigating in the future.

9. Transportation service providers have increasingly used more sophisticated management systems and advanced technologies to provide surveillance and control of transportation networks. These types of systems and technologies will increasingly be considered essential to effective transportation system management. Thus, they will be part of the strategies likely to be considered during the planning process.

10. Transportation system impacts of the ecosystem have become an important consideration in transportation planning. These impact categories have legal and regulatory requirements associated with their consideration in project development, and, in some cases, plan development. These impacts are increasingly being viewed from an ecosystem perspective, that is, an interrelated set of natural processes supporting life. The linkages between the construction, operation, and maintenance of transportation facilities and the natural environment must often be considered from the broader perspective of the spatial and temporal linkages that characterize such processes. Thus, for example, watersheds or air basins become the boundaries of impact analyses. National policies aimed at minimizing these impacts have adopted a variety of approaches, including mandating changes in technology (e.g., fuel economy and pollutant emission standards, and financial incentives to use alternative-fuel vehicles) and encouraging transportation actions that reduce vehicle use. These latter actions are often a focus of regional transportation plans.

11. The relationship among land use, economic activity, and transportation investment is one of the most important to decision makers and to the planning process. Accordingly, considerable effort is undertaken to inventory current land-use and economic development in a region and analyze likely future patterns of both. For both land use and economic activity, transportation investment rarely creates new economic activity or land use, but rather acts to redistribute activity that would likely occur anyway. Through the use of public policy and financial incentives, this redistribution can be guided to provide public benefits such as increased transit ridership. Recent interest in using land-use policies to provide more efficient utilization of transportation and other infrastructure will likely continue for many years. Growth management will be an overriding policy framework for transportation investment decisions in many metropolitan areas.

12. Transportation systems provide tremendous benefits to those who can afford to use the system. Assessment of the equitable distribution of benefits and costs is an important part of transportation planning. Community impact analysis has become

an important component of transportation planning. Community impacts of transportation tend to affect elderly, low-income, and ethnic groups to a greater extent than most other groups in urban areas. Guidelines for assessing the community impacts of transportation facilities have been developed to help ensure that the actions required to reduce the negative effects of a new facility are planned for early in project development. Geographic information systems provide a powerful tool for analyzing the distribution of benefits and costs and will likely be one of the most important technical approaches in these types of studies.

13. A knowledge of the characteristics of urban travel patterns is required by transportation planners so that the needs for transportation facilities can be matched to actual planning and decision-making activities. *Trip purpose* determines the origins and destinations involved and, consequently, the length of urban trips. The *temporal distribution* of urban travel is the primary cause of congestion on transportation facilities at certain times of the day. The *spatial distribution* of urban travel is directly related to the configuration of the transportation system. The *modal distribution* of urban trips is related to both the trip purpose and the temporal distribution of travel and is generally determined by the costs (actual and perceived) of using different modes for a given trip. These characteristics of urban travel can differ substantially from one city to the next, but there are nevertheless similar trends exhibited in most urban areas.

14. Transportation safety and traveler costs are important considerations in transportation planning. Crash rates have declined in recent years in the United States due to better vehicle designs and laws; however, safety is still one of the most important goals for transportation investment. Travel costs are not only important to the individual decision on which mode to choose, but they become critical evaluation criteria for investment strategies when the definition is broadened to include societal costs.

QUESTIONS

1. For your metropolitan area, what are the different types of system hierarchies used to describe the transportation system? Give examples of transportation facilities or services that illustrate each level in the hierarchy.

2. Describe in detail the relationship between the transportation system and the ecosystem. Do this at the local, metropolitan, and national scales.

3. Transportation system purpose relates to different community goals and objectives. Identify three community goals and discuss how transportation system performance can support or hurt the achievement of these goals.

4. Performance-based planning makes an important distinction between outputs and outcomes. Identify five outcomes of transportation investment, and provide examples of corresponding outputs that will lead to these outcomes. How can such information be used by the decision-making process?

5. Summarize the interrelationships between trip purpose, temporal distribution, mode choice, and cost in urban travel. How do these interrelationships contribute to the "problems" associated with urban transportation?

6. Find the latest available travel data for your urban area (or one with which you are familiar) from census statistics or from a recent transportation planning study. Analyze the data on urban travel by trip purpose and/or modal split. Use your familiarity with the characteristics of the urban area under consideration to explain any significant patterns or any differences from the patterns outlined in this chapter.

7. In the land use–transportation interaction process conceptualized in Fig. 3.7, at what point is it possible for public agencies to intervene and influence the process? Specifically, what policies do you think would be the most effective in guiding the process, given the economic, institutional, and political realities in North American urban areas, and given the results of the transportation and land-use studies outlined in Table 3.16?

8. Many urban transportation planners accept the premise that, in general, lower-income persons tend to live in inner-city areas and tend to make shorter trips on public transit, more often, during all periods of the day. Conversely, it is assumed that middle- and upper-income persons live in suburban areas and tend to use transit less frequently, for longer suburb-to-CBD commute trips during the morning and evening rush hours.

   (a) In terms of equity and income distribution, what implications does this premise have for (i) distance-based transit fares and (ii) peak and off-peak transit fares? Use the data provided in Tables 3.20 to 3.25 to support your arguments.

   (b) How accurate is this premise? What trends and considerations might reduce its validity?

9. Table 3.6 presents several policies that can be implemented to influence travel behavior. Could such policies be implemented in your community? What are the likely barriers to such implementation? What arguments would you use to support their implementation?

10. Explain the difference between direct, indirect, and cumulative impacts. Give examples of each. What type of data would you need to analyze each impact?

11. Choose one of the transportation system impacts listed in Table 3.7. Do the following:

    (a) Describe the legislative and regulatory history that influenced how this impact is considered in transportation planning.

    (b) Identify the types of data that would likely be collected in the analysis of this impact.

    (c) Describe the methods and analysis tools that would be used to determine the scope and magnitude of this impact in a typical project.

    (d) Outline strategies that could be implemented to mitigate the negative effects of this impact.

REFERENCES

Adkins, W. 1959. Land value impacts of expressways in Dallas, Houston, and San Antonio, Texas. *Highway Research Bulletin 227*. Washington, D.C.: National Academy Press.

Allen, B. and R. Mudge. 1974. *The impact of rapid transit on urban development: The case of the Philadelphia-Lindenwold high speed line.* Rand Corporation Paper P-5246. Washington, D.C.

Alonso, W. 1964. *Location and land use: Toward a general theory of land rents.* Cambridge, MA: Joint Center for Urban Studies, Harvard University.

Altshuler, A., J. Womack, and J. Pucher. 1979. *The urban transportation system: Politics and policy innovation.* Cambridge, MA: MIT Press.

American Association of State Highway and Transportation Officials. 1994. *A policy on geometric design of highways and streets.* Washington, D.C.: AASHTO.

American Public Transit Association. 1994. *Fare elasticity and its application to forecasting transit demand.* Washington, D.C.

Anderson, D. and H. Mohring. 1996. *Congestion costs and congestion pricing for the Twin Cities.* Report No. MN/RC-96/32. St. Paul, MN: Minnesota DOT.

Apogee Research, Inc. 1994. *The costs of transportation: Final report.* Report prepared for the Conservation Law Foundation. Washington, D.C. March.

_____ and Greenhorne and O'Mara. 1998. Research on the relationship between economic development and transportation investment. *NCHRP Report 418.* Washington, D.C.: National Academy Press.

Armstrong, R. 1994. Impacts of commuter rail service as reflected in single-family residential property values. *Transportation Research Record 1466.* Washington, D.C.: National Academy Press.

_____. 1999. Economic and social relevance of central cities in the nation's 12 largest urban regions. *Conference Proceedings 18.* Washington, D.C.: National Academy Press.

Arrington, G. B. 1995. *Beyond the field of dreams: Light rail and growth in Portland.* Portland, OR: Tri-Met.

Ashley, R. N. and W. F. Bernard. 1965. Interchange development along 180 miles of I-94. *Highway Research Record 96.* Washington, D.C.: National Academy Press.

Ausubel, J. H. and R. Herman (eds.). 1988. *Cities and their vital systems: Infrastructure, past, present, and future.* Washington, D.C.: National Academy Press.

Banister, D. and J. Berechman. 2000. *Transport investment and economic development.* London. UCL Press.

Barrett, D. 1996. Traffic-noise impact study for Least Bello Vireo Habitat along California state route 83. *Transportation Research Record 1559.* Washington, D.C.: National Academy Press.

Berger, L. International. 1999. Economic trends and multimodal transportation requirements. *NCHRP Report 421.* Washington, D.C.: National Academy Press.

Boarnet, M. and A. Haughwout. 2000. *Do highways matter? Evidence and policy implications of highways' influence on metropolitan development.* Discussion paper. Washington, D.C.: The Brookings Institute.

Bowman, J. and M. Ben-Akiva. 1997. Activity-based travel forecasting. In Texas Transportation Institute. *Activity-Based Travel Forecasting Conference, June 3–5, 1996.* Report DOT-T-97-17. Washington, D.C.: U.S. DOT.

Bregman, J. 1999. *Environmental impact statements.* 2d ed. Boca Raton, FL: Lewis.

Bullard, R. and G. Johnson. 1997. *Just transportation: Dismantling race and class barriers to mobility.* Gabriola Island, BC: New Society Publishers.

Burchell, R. et al. 1998. The costs of sprawl—revisited. *TCRP Report 39.* Washington, D.C.: National Academy Press.

Burdge, R. 1987. The social impact assessment model and the planning process. *Environmental Impact Assessment Review* 7: 141–150.

Bureau of Transportation Statistics. 1996. *Transportation statistics annual report 1996: Transportation and the environment.* Washington, D.C.: U.S. Department of Transportation.

_____. 1999. *Transportation statistics annual report.* Washington, D.C.: U.S. Department of Transportation.

Cairns, J. 1996. Determining the balance between technological and ecosystem services. In P. Schulze (ed.) *Engineering within ecological constraints.* Washington, D.C.: National Academy Press.

Cambridge Systematics, Inc. 1992. *Interstate goods movement—trends and issues.* Prepared for the Interstate Transportation Division, New York: Port Authority of New York and New Jersey.

_____. 1994. *The effects of land use and travel demand strategies on commuting behavior.* Washington, D.C.: Federal Highway Administration.

_____. 1998. *Economic impact analysis of transit investments: Guidebook for practitioners.* NCHRP Report 35. Washington, D.C.: National Academy Press.

_____. 1999. *Public transportation and the nation's economy: A quantitative analysis of public transportation's economic impact.* Prepared for the America Public Transit Association. Washington, D.C. Oct.

Cameron, M. 1997. Transportation efficiency and equity in southern California: Are they compatible? In Bullard, R. and G. Johnson (eds.) *Just transportation: Dismantling race and class barriers to mobility.* Gabriola Island, BC: New Society Publishers.

Canter, L. 1996. *Environmental impact assessment.* New York: McGraw-Hill.

Capital District Transportation Committee. 1995. *Community quality of life: Measurement, trends, and transportation strategies.* Albany, NY. Aug.

Carlson, D. and D. Billen. 1996. *Transportation corridor management: Are we linking transportation and land use yet?* Institute for Public Policy and Management. Seattle, WA: University of Washington.

Central Mortgage and Housing Corporation. 1977. *Road and rail noise effects on housing.* Ottawa, Canada.

Cervero, R. et al. 1980. *Efficiency and equity implications of alternative transit fare policies.* Final Report. U.S. Department of Transportation Report No. DOT-1-80-32. Sept.

_____.1993. *Ridership impacts of transit focused development.* Berkeley, CA: University of California.

_____. 1995. *BART @ 20: Land use and development impacts*. Berkeley, CA: University of California. July.

_____. 1998. *The transit metropolis, A global inquiry*. Washington, D.C.: Island Press.

Cerrero, R., P. Hall, and J. Landis. 1991. *Transit joint development in the United States*. Washington, D.C.: Federal Transit Administration. Sept.

Chen, H, A. Rufolo, and K. Dueker. 1998. Measuring the impact of light rail systems on single-family home values: A hedonic approach with geographic information systems application. *Transportation Research Record 1617*. Washington, D.C.: National Academy Press.

Cheslow, M. and M. Olsson. 1975. *Transportation and metropolitan development*. Washington, D.C.: Urban Institute. Oct.

Comsis Corp. 1996. *Incorporating feedback in travel forecasting: Methods, pitfalls, and common concerns*. Report DOT-T-96-14. Washington, D.C.: Federal Highway Administration.

Crain, J. and W. Courington. 1977. *Incidence rates and travel characteristics of the transportation handicapped in Portland, Oregon*. Transportation Systems Center Report No. UMTA-OR-06-004-77-1. April.

Damm, D. et al. 1980. Response of urban real estate values in anticipation of the Washington Metro. *Journal of Transport Economics and Policy*. Sept.

Davies. G. W. 1976. The effect of a subway on the spatial distribution of population. *Journal of Transport Economics and Policy*. Vol. 10. No. 2. May: 126–136.

Davis, S. 1999. *Transportation energy data book*. Edition 19. Report ORNL-6958. Oak Ridge, TN: Oak Ridge National Laboratory. Sept.

Deakin, E. 1991. Jobs, housing and transportation: Theory and evidence on interactions between land use and transportation. In Transportation, urban form, and the environment. *Special Report 231*. Washington, D.C.: National Academy Press.

Delucchi, M. 1997. The annualized social costs of motor-vehicle use in the U.S., based on 1990–1991 data: Summary of theory, data, methods and results. In D. Greene, D. Jones, and M. Delucchi (eds.) *The full costs and benefits of transportation: Contributions to theory, method, and measurement*. New York: Springer.

Diamond, H. and P. Noonan. 1996. *Land use in America*. Washington, D.C.: Island Press.

Dill, J., T. Goldman, and M. Wachs. 1999. California license fees: Incidence and equity. *Journal of Transportation and Statistics*. Vol. 2. No. 2. Washington, D.C.: Bureau of Transportation Statistics, U.S. DOT. Dec.

Donnelly, P. 1982. *Rail transit impact studies: Atlanta, Washington, San Diego*. U.S. Department of Transportation Report DOT-1-83-3. March.

Dornbusch, D. et al. 1978. *Land use and urban development impacts of BART: Final Report*. Washington, D.C.: U.S. Department of Transportation. Oct.

Downing, P. 1973. Factors affecting commercial land values: An empirical study of Milwaukee, Wisconsin. *Land Economics*. Feb.

Dunphy, R. 1997. *Moving beyond gridlock: Traffic and development*. Washington, D.C.: Urban Land Institute.

ECONorthwest. 1995. *An introduction to congestion pricing for Oregon policy makers*. Portland, OR: Oregon DOT. March.

Ellis, R. H. et al. 1977. *Potential approaches for improving the mobility of different market segments of the transportation handicapped.* Washington, D.C.: Peat, Marwick, Mitchell, and Co.

Environmental Defense Fund. 1994. *Efficiency and fairness on the road: Strategies for unsnarling traffic in southern California.* Washington, D.C.

European Communities. 1985. Council directive of 27, June, 1985 on the assessment of the effects of certain public and private projects on the environment. *Official Journal of the European Communities.* No. L 175. Brussels. Sept 7.

Ewing, R. 1997. *Transportation and land use innovations.* Chicago, IL: American Planning Association.

Federal Highway Administration. 1985. Series of five technical reports. *Effects of highway runoff on receiving waters.* Reports FHWA-/RD/-84-062;063;064;065 and 066. Washington, D.C.

_____. 1987. *Sources and mitigation of highway runoff pollutants.* Technical Summary Report No. FHWA/RD/84/060. Washington, D.C.

_____. 1990. *Pollutant loadings and impacts from highway stormwater runoff.* Vol. 1. Report No. FHWA-RD-88-006, Washington, D.C.

_____. 1999. Is highway runoff a serious problem? *<http://www.tfhrc.gov/hnr20/runoff/runoff.htm>*

Fogelson, R. 1967. *The fragmented metropolis: Los Angeles from 1850 to 1930.* Cambridge, MA: Harvard University Press.

Forkenbrock, D. and N. Foster. 1990. Economic benefits of a corridor highway investment. *Transportation Research.* Vol. 24A, No. 4.

Frank, L. and G. Pivo. 1995. Impacts of mixed use and density on utilization of three modes of travel: Single occupant vehicle, transit, and walking. *Transportation Research Record 1466.* Washington, D.C.: National Academy Press.

Fried, M. 1966. Grieving for a lost home: Psychological costs of relocation. In James Wilson (ed.) *Urban renewal.* Cambridge, MA: MIT Press.

Fruin, J. 1981. Elements of terminal design and planning. *Proceedings of the conference on multi-modal terminal planning and design.* Irvine, CA: University of California. March.

Gannon, C. and M. Dear. 1972. *The impact of rail rapid transit systems on commercial office development: The case of the Philadelphia-Lindenwold high-speed line.* Philadelphia, PA: University of Pennsylvania. June.

Garrett, P. and F. Bank. 1995. The ecosystems approach and transportation development. Paper presented to AASHTO Standing Committee on the Environment. Washington, D.C. Oct. 30. (as found in *<http://www.fhwa.gov>*)

Gauthier, H. L. 1970. Geography, transportation and regional development. *Economic Geography.* Oct.

Giuliano, G. 1995. Land use impact of transportation investments: Highway and transit. In S. Handy (ed.) *The geography of urban transportation.* 2d ed. New York: Guilford Press.

_____. 1999. Land use policy and transportation: Why we won't get there from here. *Transportation Research Circular 494.* Washington, D.C.: National Academy Press, Aug.

Gomez-Ibanez, J. 1997. Estimating whether transport users pay their way: The state of the art. In D. Greene, D. Jones and M. Delucchi (eds.) *The full costs and benefits of transportation: Contributions to theory, method, and measurement.* New York: Springer.

Gordon, D. 1991. *Steering a new course: Transportation, energy and the environment.* Washington, D.C.: Island Press.

Green, R. and D. James. 1993. *Rail transit station area development: Small area modeling in Washington, D.C.* Armonk, NY: M.E. Sharpe.

Greene, D. 1996. *Transportation and energy.* Washington, D.C.: Eno Foundation.

_____, D. Jones, and M. Delucchi. 1997. *The full costs and benefits of transportation: Contributions to theory, method, and measurement.* New York: Springer.

Grenzeback, L. and M. Warner. 1995. Congestion impacts on business and strategies to mitigate them. *Research Results Digest. No. 202.* Washington, D.C.: National Research Council. Sept.

Guensler, R. 2000. Motor vehicle emissions. In *Macmillan Encyclopedia of Energy.* Farmington Hills, MI: Macmillan.

Halig, D. and H. Cohen. 1996. Residential noise damage costs caused by motor vehicles. *Transportation Research Record 1559.* Washington, D.C.: National Academy Press.

Hansen, M., D. Gillen, A. Dobbins, Y. Huang, and M. Puvathingal. 1993. *The air quality impacts of urban highway capacity expansion: Traffic generation and land use change.* UCB-ITS-RR-93-5. Berkeley, CA: University of California. April.

Hartgen, D., W. McCoy, and W. Walcott. 1996. Incremental regionalism: Staged approach to development of regional transportation organizations. *Transportation Research Record 1552.* Washington, D.C.: National Academy Press.

Herbstritt, R. and A. Marble. 1996. Current state of biodiversity impact analysis in state transportation agencies. *Transportation Research Record 1559.* Washington, D.C.: National Academy Press.

Hodge, D. 1995. My fair share: Equity issues in urban transportation. In S. Handy (ed.) *The geography of urban transportation.* 2d ed. New York: Guilford Press.

Howes, R. and A. Fainberg (eds.) 1991. *The energy sourcebook: A guide to technology, resources, and policy.* New York: American Institute of Physics.

Isard, W. 1956. *Location and the space-economy.* Cambridge, MA: MIT Press.

Jackson, K. 1985. *Crabgrass frontier: The suburbanization of the United States.* New York: Oxford University Press.

Jewell, T. 1986. *A systems approach to civil engineering planning and design.* New York: Harper & Row.

Kageson, P. 1993. *Getting the prices right: A European scheme for making transport pay its true costs.* Report T&E 93/6. Brussels: European Federation for Transport and Environment. May.

Kaplan, M., H. Gans, and M. Kahn. 1972. *Social characteristics of neighborhoods as indicators of the effects of highway improvements.* Report No. DOT/FH 11-7789. Washington, D.C.

Kennedy, L. 1999. Environmental justice. *Conference Proceedings 16.* Washington, D.C.: National Academy Press.

Knight, R. L. and L. L. Trygg. 1977. Evidence of land use impacts of rapid transit systems. *Transportation* 6: 231–247.

Kreske, D. 1996. *Environmental impact statements: A practical guide for agencies, citizens, and consultants.* New York: Wiley.

Kulash, D. 1999. Transportation and society. In J. Edwards (ed.) *Transportation planning handbook.* Washington, D.C.: Institute of Transportation Engineers.

Landis, J. et al. 1995. *Rail transit investments, real estate values, and land use change: A comparative analysis of five California rail transit systems.* Berkeley, CA: University of California. July.

Leinberger, C. 1996. Metropolitan development trends in the late 1990s: Social and environmental implications. In H. Diamond and P. Noonan (eds.) *Land use in America.* Washington, D.C.: Island Press.

Lerman, S. R. et al. 1977. *The effect of the Washington metro on urban property values.* Center for Transportation Studies Report 77-18. Cambridge, MA: MIT. Nov.

Levinson, H. 1982. Urban travel characteristics. In W. Homberger (ed.). *Transportation and traffic engineering handbook.* 2d ed. Englewood Cliffs, NJ: Prentice Hall.

Levinson, D. and A. Kumar. 1995. Activity, travel, and the allocation of time. *APA Journal.* Chicago, IL: American Planning Association. Autumn.

Lewis, D. 1981. Providing private cars to severely disabled people. In Transportation Systems Center *Transportation for the elderly and handicapped: Programs and problems 2.* Feb.

_____. 1991. Primer on transportation, productivity and economic development. *NCHRP Report 342.* Washington, D.C.: National Academy Press.

Lewis-Workman, S. and D. Brod. 1997. Measuring the neighborhood benefits of rail transit accessibility. *Transportation Research Record 1576.* Washington, D.C.: National Academy Press.

Lichter, D. and G. Fuguitt. 1980 Demographic response to transportation innovation: The case of Interstate highway. *Social Forces.* Vol. 59, No. 2.

Linneman, P. and A. Summers. 1993. Patterns and processes of employment and population decentralization in the United States, 1970–87. In A. Summers, P. Chesire, and L. Senn (eds.) *Urban change in the United States and western Europe: Comparative analysis and policy.* Washington, D.C.: Urban Institute.

Litman, T. 1994. *Transportation cost analysis: Techniques, estimates, and implications.* Victoria, BC, Canada. Dec. 3.

Loveless, S. 1999. Access to jobs: The intersection of transportation, social and economic development policies—a challenge for transportation planning in the 21st century. *Conference Proceedings 16.* Washington, D.C.: National Academy Press.

MacKenzie, J., R. Dower, and D. Chen. 1992. *The going rate: What it really costs to drive.* New York: World Resources Institute.

Manheim, M. 1979. *Fundamentals of transportation systems analysis, volume 1: Basic concepts.* Cambridge, MA: MIT Press.

_____ et al. 1975. Transportation decision making: A guide to social and environmental considerations. *National Cooperative Highway Research Program Report 156.* Washington, D.C.: National Academy Press.

Mann, E. 1997. Confronting transit racism in Los Angeles. In R. Bullard and G. Johnson (eds.) *Just transportation: Dismantling race and class barriers to mobility.* Gabriola Island, BC: New Society Publishers.

Marchese, A. 1999. Assessment of environmental justice issues in Atlanta: Proposed work plan. Washington, D.C.: FHWA. Letter sent to Atlanta Regional Commission, June 24, 1999.

Marriott, B. 1997. *Environmental impact assessment: A practical guide.* New York: McGraw-Hill.

Marzotto, T., V. Burner, and G. Bonham. 2000. *The evolution of public policy: Cars and the environment.* Boulder, CO: Lynne Rienner Publishers.

McGuckin, N. and E. Murakami. 1998. *Examining trip-chaining behavior: A comparison of travel by men and women.* Washington, D.C.: Federal Highway Administration.

Metropolitan Transportation Commission. 1979. *Land use and urban development impacts of BART.* Report HUD 0001682 and DOT/P-30/79/09. April.

Meyer, J., J. Kain, and M. Wohl. 1965. *The urban transportation problem.* Cambridge, MA: Harvard University Press.

Meyer, J. and J. Gomez-Ibanez. 1981. *Autos, transit, and cities.* Cambridge, MA: Harvard University Press.

Meyer, M. 1991. *Development impacts of transit investment.* Report prepared for Houston Metro. Houston, TX.

_____. 1995. *Alternative performance measures for transportation planning: Evolution toward multimodal planning.* Report No. FTA-GA-26-7000. Washington, D.C.: Federal Transit Administration. Dec. 31.

Middleton, W. 1967. *The time of the trolley.* Milwaukee, WI: Kalmbach Publishing.

Mid-Ohio Regional Planning Commission. 1994. *Transportation infrastructure improvement study for central Ohio inland port program.* Columbus, OH.

Miller, P. and J. Moffet. 1993. *The price of mobility: Uncovering the hidden cost of transportation.* New York: National Resources Defense Council.

Mohring, H. 1961. Land values and the measurement of highway benefits. *Journal of Political Economy* 79: 236–249.

Moore, T. and P. Thorsnes. 1994. *The transportation/land use connection.* Chicago, IL: American Planning Association.

Morlok. E. 1978. *Introduction to transportation engineering and planning.* New York: McGraw-Hill.

Muller, P. 1995. Transportation and urban form: Stages in the spatial evolution of the American metropolis. In S. Handy (ed.) *The geography of urban transportation.* 2d ed. New York: Guilford Press.

National Research Council. 1995. *Wetlands: Characteristics and boundaries.* Washington, D.C.: National Academy Press.

Nelson, A. and S. McCleskey. 1992. Improving the effects of elevated transit stations on neighborhoods. *Transportation Research Record 1266.* Washington, D.C.: National Academy Press.

Nelson, P. 1978. *Economic analysis of transportation noise abatement.* Cambridge, MA: Ballinger.

Nishii, K., K. Kondo, and R. Kitamura. 1988. Empirical analysis of trip chaining behavior. *Transportation Research Record 1203*. Washington, D.C.: National Academy Press.

Northern Virginia Planning District Commission. 1993. *Impact assessment of the Virginia Railway express commuter rail line on land use development patterns in Northern Virginia*. Report DOT-T-95-18. Washington, D.C.: Federal Transit Administration. Dec.

Ogden, K. 1997. *Urban goods movement: A guide to policy and planning*. Aldershot, England: Ashgate Publishing.

Parsons, Brinckerhoff, Quade, and Douglas, Inc. 1996a. *Transit and urban form*. Vols. 1 and 2. TCRP Report 16. Washington, D.C.: National Academy Press.

_____. 1996b. *Consequences of the interstate highway system for transit: Summary of findings*. TCRP Report 42. Washington, D.C.: National Academy Press.

Payne-Maxie Consultants and Blaney-Dyett. 1980. *The land use and urban development impacts of beltways*. Final report DOT-05-90079. Washington, D.C.: U.S. Dept. of Transportation.

Pisarski, A. 1996. *Commuting in America: Part II*. Washington, D.C.: Eno Foundation.

Porter, D. 1997. *Transit-focused development. Synthesis of transit practice 20*. Washington, D.C.: National Academy Press.

Portland METRO. 1999. *Regional transportation plan*. Portland, OR.

Project for Public Spaces. 1997. *The role of transit in creating livable metropolitan communities*. TCRP Report 22. Washington, D.C.: National Academy Press.

Pucher, J. and I. Hirschman. 1981. Distribution of the tax burden of transit subsidies in the United States. *Public Policy*. Vol. 29. No. 3. Summer.

_____, T. Evans, and J. Wenger. 1998. Socioeconomics of urban travel: Evidence from the 1995 NPTS. *Transportation Quarterly*. Vol. 52. No. 3. Washington, D.C.: Eno Foundation.

Puget Sound Regional Council. 1996. *The costs of transportation: Expenditures on surface transportation in the central Puget Sound region for 1995*. Seattle, WA.

Purvis, C. 1998. *Travel forecasting assumptions '98 summary. 1998 update of regional transportation plan*. Oakland, CA: Metropolitan Transportation Commission. Aug.

Pushkarev, B. and J. Zupan. 1977. *Public transportation and land use policy*. Bloomington, IN: Indiana University Press.

Revelle, C., E. Whitlach, and J. Wright. 1997. *Civil and environmental engineering systems*. Upper Saddle River, NJ: Prentice-Hall.

Revis, J. 1976. Transportation requirements for the handicapped, elderly, and economically disadvantaged. *National Cooperative Highway Research Program Synthesis 39*. Washington, D.C.: National Academy Press.

Rice Center for Urban Mobility. 1987. *Assessment of changes in property values in transit areas*. Houston, TX. Rice University.

Roth, G. 1996. *Roads in a market economy*. Aldershot, England: Ashgate Publishing.

Shrouds, J. 1996. Guidance on developing water quality action plans. *Memorandum to regional administrators*. Federal Highway Administration. Environmental Analysis Division. May 28.

Spain, D. 1999. Societal trends: The aging baby boom and women's increased independence. In U.S. DOT/FHWA *Searching for solutions: Nationwide personal transportation survey symposium. Oct. 29–31, 1997.* No. 17, Washington, D.C. Feb.

Spengler, E. H. 1930. *Land values in New York in relation to transit facilities.* New York: Columbia University Press.

————. 1995. Expanding metropolitan highways: Implications for air quality and energy use. *Special Report 245.* Washington, D.C.: National Academy Press.

Strathman, J. and K. Dueker. 1995. Understanding trip chaining. *1990 NPTS special reports on trip and vehicle attributes.* Report FHWA-PL-95-033. Washington, D.C.: Federal Highway Administration.

Transmanagement, Inc., M. Coogan, and M. Meyer. 1998. Innovative practices for multimodal transportation planning for freight and passengers. *NCHRP Report 404.* Washington, D.C.: National Academy Press.

Transportation Research Board. 1994. *Curbing gridlock: Peak period fees to relieve traffic congestion.* Vol. 1. Washington, D.C.: National Academy Press.

————. 1995a. Expanding metropolitan highways: Implications for air quality and energy use. *Special Report 245.* Washington, D.C.: National Academy Press.

————. 1995b. An evaluation of the relationships between transit and urban form. *TCRP Research Results Digest No. 7.* Washington, D.C.: National Academy Press. June.

————. 1999. *Transit capacity and quality of service manual.* Washington, D.C.: National Academy Press.

————. 2000. *Highway capacity manual.* Washington, D.C.: National Academy Press.

Urban Land Institute. 1979. *Joint development: Making the real estate-transit connection.* U.S. DOT Report UMTA-DC-06-0183-79-1. June.

————. 1998. *Smart growth: Myth and fact.* Washington, D.C.

U.S. Conference of Mayors. 1980. *Transportation and urban development: A review of federal programs and strategies that promote the coordination of public transportation and private investment.* Report DOT-P-30-80-33. Washington, D.C. Oct.

U.S. Department of Housing and Urban Development. 1979. *The private development process: A guidebook for local government.* Report No. HUD-PDR-353-2. Washington, D.C.: Feb.

U.S. Department of Transportation. 1979a. *Guidelines for assessing the environmental impact of public mass transportation projects.* Report No. DOT-P-79-00-001. April.

————. 1979b. *Joint development in the U.S.* Report No. TX-11-0006. Washington, D.C. June.

————. 1980a. *The land use and urban development impacts of beltways.* Report DOT-P-30-80-40. Oct.

————. 1980b. *Design of urban streets.* FHWA Technology-Sharing Report 80-204. Jan.

————.1996. *Community impact assessment.* Report HEP-30/8-96(10M)P. Washington, D.C. Sept.

_____ and Federal Highway Administration. 1995. *Highway traffic noise analysis and abatement policy and guidance.* Office of Environment and Planning. Washington, D.C. June.

_____ and Federal Highway Administration. 1998. *Transportation and global climate change: A review and analysis of the literature.* Report DOT-T-97-03. Washington, D.C. June.

_____, Federal Highway Administration, California Department of Transportation, and San Bernadino Associated Governments. 1992. *Draft environmental impact statement: Widening of I-215 between I-10 and SR 30 in San Bernadino, CA and supporting technical reports.* San Bernadino, CA.

_____, Urban Mass Transportation Administration, and City and County of Honolulu. 1989. *Alternatives analysis and draft environmental impact statement: Honolulu rapid transit development project.* Technical report on noise and vibration. Washington, D.C.

U.S. Environmental Protection Agency. 1975. *Secondary impacts of transportation and wastewater investments: Review and bibliography.* Washington, D.C.: Office of Research and Development. Jan.

_____.1997. *National air quality and emissions trends report.* Report EPA-454/R-98-016. Washington, D.C.: Dec. <*http://www.epa.gov/oar/aqtrnd971*>

_____. 1998a. *National air quality and emissions trends report.* Report EPA-454/R-97-013. Washington, D.C. Jan.

_____. 1998b. *Technical methods for analyzing pricing measures to reduce transportation emissions.* Report EPA-231-R-98-006.

Voith, R. 1970. Changing capitalization of CBD-oriented transportation systems: Evidence from Philadelphia 1970–1988. *Journal of Urban Economics.* 33: 361–376.

Von Thunen, J. 1826. *Der isolierte staat in beziehung auf landwertschaft und nationalekonomie.* Hamburg.

Voorhees, Alan M. and Assocs. 1979. *Guidelines for assessing the environmental impact of public mass transportation projects: Social impacts.* Report DOT-OS-80042. Washington, D.C.: Office of the Secretary. April.

Vuchic, V. 1999. *Transportation for livable cities.* New Brunswick, NJ: CUPR Press.

Wachs, M. 1979. *Transportation for the elderly: Changing lifestyles, changing needs.* Berkeley, CA: University of California Press.

_____. 1999. Linkages between transportation planning and the environment. *Conference Proceedings 16.* Washington, D.C.: National Academy Press.

Warner, S. 1962. *Streetcar suburbs.* Cambridge, MA: Harvard University Press.

Webber, M. 1976. The BART experience: What have we learned? *The Public Interest.* No. 45. Fall.

Wegmann, F., A. Chatterjee, M. Lipinski, B. Jennings, and R. McGinnis. 1995. *Characteristics of urban freight systems.* Report DOT-T-96-22. Washington, D.C.: U.S. Department of Transportation. Dec.

Wesemann, L., T. Hamilton, S. Tabaie and G. Bare. 1996. Cost-of-delay studies for freeway closures caused by Northridge earthquake. *Transportation Research Record 1559.* Washington, D.C.: National Academy Press.

Wehrich, P. and W. Lind. 1999. *Does transit work? A conservative reappraisal.* Washington, D.C.: Free Congress Research and Education Foundation.

Wingo. L. 1972. *Cities and space: The future use of urban land.* Baltimore, MD: Johns Hopkins Press.

Winston, C. 1997. U.S. industry adjustment to economic deregulation. Unpublished paper. Washington, D.C.: The Brookings Institution. March.

_____, T. Corsi, C. Grimm, and C. Evans. 1990. *The economic effects of surface freight deregulation.* Washington, D.C.: The Brookings Institution.

Witherspoon, R. 1979. Transit and urban economic development: How cities could use transit as a development tool: Why they don't: What to do about it. *Transit Journal.* Spring.

Wolfe, R. and E. Miller. 1982. *Transportation and seniors in Ontario.* Toronto, Canada: University of Toronto-York University Joint Program in Transportation. March.

World Bank. 1989. Operative directive 4.00. Annex A3. Environmental screening. Washington, D.C.: Sept.

4

Data Management and Use In Decision Making

4.0 INTRODUCTION

Transportation planners have long placed great emphasis on the importance of quality data for understanding the economic and societal context of the transportation system and for monitoring the performance and condition of this system. Data have also proved necessary for the more technical aspects of calibrating and applying travel forecasting models and for evaluating the effectiveness and impacts of proposed changes. This chapter will examine the types of data used in transportation studies and the techniques for collecting it. In addition, this chapter will describe how data analyses can provide input into the decision-making process.

4.1 THE TRANSPORTATION PLANNING DATABASE

Prior to the mid-1960s, travel data were collected for two basic reasons: (1) to determine the origin and destination of highway users entering urban areas; and (2) to estimate the total number of vehicles using urban highways. In the first case, data on the origin and destination of trips were used to solve site-specific problems, such as the location of a new bridge or a highway bypass of a city. In the second case, automatic or manual traffic counters were used to determine vehicle volumes on congested highways so that the need for highway expansion or highway control could be clearly established. Today, most state, provincial, and local transportation agencies conduct systematic programs of vehicle counting on selected highway segments, a legacy of these earlier efforts [AASHTO, 1992; U.S. DOT, 1995].

Beginning in 1944, when the U.S. Congress first made funds available for highway projects in urban areas, several state highway departments, in cooperation with the U.S. Bureau of the Census, developed a new approach toward data collection—interviewing individuals in a randomly selected sample of households. This procedure was found to provide good estimates of the total travel occurring within an urban area [Highway Research Board, 1944]. The information collected in these home interviews usually included the following: the type of housing structure, the number of vehicles available, the number of persons in the household, the household income category, and a description of the trips (i.e., origin, destination, trip purpose, trip times, and travel modes) made by household members 5 years or older. These initial household surveys led to important transportation-related data collection through the decennial census that today serves as the foundation for much of the socioeconomic and demographic data used in transportation planning.

In addition to data on travel behavior and demographics, these early transportation studies collected a large amount of data on land-use and transportation network characteristics. The data on the transportation network, called an inventory, identified roadway locations, roadway length, pavement width, speeds, parking restrictions, and, in the case of transit, vehicle headways, number of seats per vehicle, and overall line capacity. Collecting and processing these data consumed a substantial portion of the planning study budget, in some cases more than 50 to 60 percent of the total funds available.

Today, because of the variety of actions considered by transportation planners and the wide-ranging goals that transportation planning often serves, an even larger amount of data is collected by numerous agencies in a metropolitan area [TRB, 1997c]. An example of the range of data collected as part of transportation planning is found in a study that assessed the evaluation criteria used in multimodal transportation planning studies over the past 10 years [Rutherford, 1994]. Measures of transportation system performance, mobility, accessibility, system coordination, land use, freight movements, socioeconomic impacts, environmental and energy impacts, safety and security, equity, costs and cost-effectiveness, financial arrangements, and institutional capability were identified as being important for effective planning. Whereas much of the data collection in a metropolitan area occurs in response to planning requirements (e.g., the identification and analysis of government-specified impacts for environmental assessment), unique issues associated with a transportation plan, program, or project could cause data other than these to be collected. To be responsive to the decision-making process, transportation planning must thus have the flexibility and capability to produce the information (and thus the collection of data) needed to make informed decisions.

For purposes of this discussion, the transportation planning database will be divided into three major components. The first, and by far the most extensive, is the data collected on the transportation system itself, called an inventory. Inventories include such things as the physical condition of transportation facilities and the characteristics of traveler or freight movement. The second component is the data associated with developing a community vision and a corresponding set of goals and objectives. These data include such things as public attitudes, economic and fiscal trends, and quality-of-life indicators. The third component of the planning database relates to the monitoring of system performance. The first two data components act as inputs into the process of developing a system plan; the last component provides feedback on how well the system is performing, particularly after the adopted plan is implemented.

Before discussing specific data collection techniques, it is important to examine two important aspects of data collection that are found in many transportation planning efforts: classification schemes for data collection/interpretation and the use of sampling procedures.

4.1.1   Classification Schemes for Data Collection

Transportation planners often group data by geographic location or type of transportation facility. For example, every metropolitan area is divided into analysis units, often called traffic analysis zones, which form the basis for the analysis of travel movements within, into, and out of the region (Fig. 4.1). These zones are defined with several criteria in mind [Baass, 1981]:

1. Achieving homogeneous socioeconomic characteristics for each zone's population.

2. Minimizing the number of intrazonal trips.

3. Recognizing physical, political, jurisdictional, and historical boundaries.

4. Generating only connected zones and avoiding zones that are completely contained within another zone.

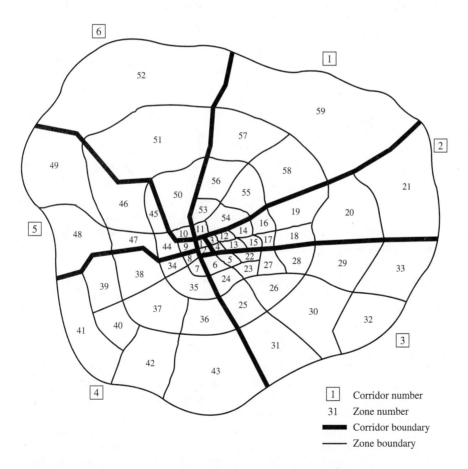

Figure 4.1   Typical zonal system for metropolitan transportation planning

5. Devising a zonal system in which the number of households, population, area, or trips generated and attracted are nearly equal in each zone.

6. Basing zonal boundaries on census zones.

This last criterion is especially important because both the U.S. and Canadian agencies responsible for census data conduct a census every 10 years. This information is used to update the socioeconomic data for the transportation planners' own zonal systems. (For a discussion of transportation planning uses of the census, see [TRB, 1997b] or visit *http://www.bts.gov.*) As shown in Fig. 4.1, the traffic analysis zones can also be grouped into larger analysis units reflecting corridors of travel or sectors of the urban area called districts. Data collected at the individual zone level can be aggregated to these larger geographical units for analyses undertaken at this level and for calibration/validation of travel forecasting models.

Corridor and/or subarea studies have become a major methodological approach in metropolitan transportation planning [McLeod, 1996; Smith, 1999]. These studies focus on well-defined travel markets for which very specific strategies can be considered, often at a very fine level of detail. Not only does this focus allow transportation planners to get a better handle on the types of problems that exist in a corridor, but it is often easier to motivate public involvement because participants are often more familiar with the specific locations or issues being discussed. In addition, multimodal trade-offs can be more easily determined at this scale of analysis as compared to a regional perspective. Figure 4.2 shows the corridor/subarea approach used in the San Francisco Bay Area [MTC, 1998].

Another scheme important for data collection is the functional classification of the different components of the transportation system. As noted in Sec. 3.1.1, transportation modal networks can be classified in a variety of ways, and data collection often reflects such schemes. Thus, one might expect more frequent and accurate data collection on principal arterial roads that carry 50 to 60 percent of the trips than on local streets. Similarly, high ridership transit lines would likely be monitored more often than those having low ridership (although one would certainly want to know why ridership was low).

In sum, transportation data in metropolitan areas are usually structured on the basis of analysis units and facility classification schemes. Because these schemes have been used for many years, periodic updates of the data can be used to monitor changes in socioeconomic and system characteristics over time.

4.1.2   Sampling Methods in Data Collection

Collecting and processing data can be difficult and costly. Because the basic unit of data collected for transportation planning is usually an individual household or a single-person trip (or a trip tour), it is too costly to develop a database consisting of data collected from every household or trip movement in a metropolitan area. Therefore, methods have been developed to make reliable inferences about the characteristics found in a carefully selected sample of households. The critical issue in this approach is how to select a sample that is representative of the characteristics of the entire population.

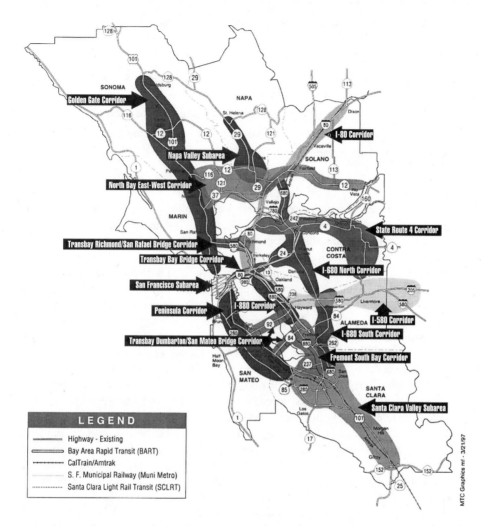

Figure 4.2 Planning corridors in the San Francisco Bay Area
I SOURCE: MTC, 1999

Cochran, in a classic text on survey methodology, outlined the steps in planning and executing a survey [Cochran, 1977]:

1. Establish a clear statement of the survey objectives.
2. Define the population to be sampled and the target groups to be focused on.
3. Identify the specific data that are relevant for the purpose of the survey.
4. Specify the degree of precision required from the survey results (i.e., how much error can be tolerated in the results).
5. Determine the methods to be used in obtaining survey results.

6. Divide the population into sampling units and list the units from which the sample will be drawn.

7. Select the sampling procedure and the sample size.

8. Pretest the survey and field methods to ensure that the procedures are workable and the survey is understandable.

9. Establish a good supervisory structure for managing the survey.

10. Determine the procedures for analyzing and summarizing the data.

11. Store the data and analysis results for future reference.

Step 7 of this survey methodology—selecting the sampling procedures and determining the size of the sample—merits special attention. In the former case, four major types of sampling procedures are commonly used: simple random sampling, sequential sampling, stratified random sampling, and cluster sampling.

Simple random sampling selects units out of a population such that each population unit has an equal chance of being drawn. The units in the population are assigned numbers from 1 to N, and numbers between 1 and N are drawn from a random number table or from a computer program especially designed to produce such numbers. The specific units in the population that correspond to these random numbers become the randomly drawn sample.

Sequential sampling draws a sample from every nth element in the population. This procedure is based on the assumption that the target population has been listed in random order.

Stratified random sampling divides the population of N units into subpopulations of $N_1, N_2, \ldots, N_L$ units, according to differences in some defining characteristic such as household income. Random samples are then taken within each stratum. For example, a household activity and travel behavior survey in Portland, Oregon, used 10 strata: residents in four counties, park-n-ride users, and five subgroups in the largest and most urban county. These five subgroups were residents in the following areas: (1) urban, good pedestrian environment, land-use mix, transit availability; (2) urban, bad pedestrian environment, transit availability; (3) urban, good pedestrian environment, transit availability; (4) light rail corridor; and (5) other residents of the county [Cambridge Systematics, 1996a].

Cluster sampling involves grouping sampling units, usually on a spatial or geographical basis (e.g., grouping households on the basis of neighborhood blocks). Clusters are then selected at random for the sample. Although cluster sampling is cheaper to undertake than other procedures, the statistical analysis associated with such sampling tends to be more difficult.

Each of these sampling strategies is designed to reduce the bias in the results that might come from measurement error, unrepresentative sample selection, high nonresponse rates, or varying question interpretations. In each case, the cost of survey sampling is different and must be balanced against the degree of accuracy required [Zimkowski et al., 1997].

The second important component of the surveying strategy is selecting the sample size (a typical methodology for making such a determination is found in Appendix B). Although the accuracy of population estimates will increase with the size of the sample, there is always a point at which the additional accuracy is not worth the associated costs of data collection and analysis. In transportation planning, determining an appropriate sample size can be complicated because data are seldom collected for strictly one purpose. For example, the same database could be used both to validate existing planning statistics (e.g., trip generation rates) and to calibrate travel forecasting models. Different levels of accuracy and a different sample size may thus be required, depending on the purposes for which the data will be used. The challenge to the transportation planner becomes one of first deciding how the collected data will or could be used and then determining the size of the sample needed to provide the required level of accuracy for the most sensitive of the data uses. Table 4.1 shows the sample sizes and costs for recent household surveys. As can be seen, sample sizes can vary substantially.

Table 4.1 Household survey sample sizes and costs

Metropolitan Planning Organization	Year	Method	Type of Diary	Sample Size	Approximate Cost
Albuquerque	1992	Mail	Travel/1 day	2,000	$130,000
Atlanta	1991	Phone	Travel/1 day	2,400	$225,000
Baltimore	1993	Phone	Travel/1 day	2,700	$400,000
Boston	1991	Mail	Activity/1 day	3,800	$360,000
Buffalo	1993	Mail	Travel/1 day	2,700	$180,000
Chicago	1990	Mail	Travel/1 day	19,314	$750,000
Dallas/Ft.Worth	1996	Phone/mail	Activity/1 day	6,000	$750,000
Los Angeles	1991	Phone	Activity/1 day	16,086	$1,300,000
Milwaukee	1991	Phone/home	Travel/ 1 day	17,500	$1,200,000
Pittsburgh	1990	Mail	Travel/1 day	450	$33,000
Salt Lake City	1993	Mail	Activity/1 day	3,082	$300,000
San Francisco	1990	Phone	Travel/1,3,5 days	10,900	$900,000
Tucson	1993	Phone	Travel/1 day	1,913	$215,000
Washington, D.C.	1994	Phone	Travel/1 day	4,800	$585,000

SOURCE: Cambridge Systematics, 1996a

For those interested in further reading on sample size determination as it relates to transportation data collection, see, for example, [Ferlis, 1980; Ferlis et al., 1981; Multisystems, 1985; Cambridge Systematics, 1996c].

4.2 TRANSPORTATION SYSTEM AND USER DATA

Monitoring the condition and performance of the transportation system has been the primary source of transportation planning data for many years. However, data on travelers, first collected in a systematic way in the late 1940s, has since become an additional input into effective transportation planning. And in the past 10 years, increased attention has also been given to the characteristics of freight movement and to the particular problems related to collecting data on the movement of goods where this information could be proprietary. The following discussion on data collection begins with a focus on the transportation system itself.

4.2.1   Data on Highway Network Performance and Condition

Traffic counts, obtained either by observation or through automatic equipment, are used for a variety of purposes—validating survey data, calibrating travel demand models, establishing traffic growth trends, assessing the transportation impact of large traffic generators, and determining the environmental impact of transportation facilities [Pedersen and Samdahl, 1982]. The major types of counts occurring in a metropolitan area include the following:

Continuous counts: Vehicle counts taken 365 days a year, almost always with permanent counting equipment often linked electronically to a central data processing location. The data from these counts are used to establish travel trends and growth rates for similarly classified roads.

Control or seasonal counts: Traffic counts taken from 2 to 12 times a year for short time periods (1 to 14 days), primarily used to factor traffic counts of short duration to an estimate of average annual daily traffic (AADT).

Coverage counts: Short duration counts (usually 6 to 8 hours) that provide traffic characteristics at specific locations.

Cordon counts: Counts undertaken to obtain estimates of the total number of movements occurring within the study area defined by a cordon boundary and the number of movements entering or leaving. For highways, automatic traffic recorders can be used to determine the total number of vehicles crossing the cordon line.

Screen-line counts: Imaginary lines or boundaries located in a transportation network to evaluate the completeness and accuracy of estimated trips within a study area by comparing trip volumes that are predicted to cross this boundary with those that actually do. Because the validity of a screen-line analysis depends on identifying as many of the movements that cross the screen line as possible, the screen line itself should be chosen very carefully. Often, barriers to traffic flow (e.g., rivers or railroad tracks) are used because they "funnel" movements through a small number of crossings, making those movements easier to count.

Vehicle classification: Identification of different types of vehicles found in the traffic stream at specific locations. This is important data not only for planning but also for road design, such as determining pavement thickness. Different vehicle classes have been established for this purpose, including motorcycle; passenger car or other two-axle, four-tire vehicles; and various combinations of axles/tires/trailer units for trucks. Recent research has shown the potential of using video imagery to recognize such classifications rather than the traditional approach of using induction loops in the pavement (which measure number of axles) or manual observations [Miller et al., 1993; Kyte et al., 1993; CUTR, 1996].

Truck weight data: Determination of vehicle weight most often collected at weigh-in-motion stations located on high volume, interurban roads. A road's deterioration relates not only to the number and types of vehicles using it but also to the weight of these vehicles. Knowing the frequency of use for each weight class of heavy vehicle on a region's road network is thus important information.

The traffic data collection methods discussed focus on determining the volume of traffic at specific locations or the characteristics of these vehicles. Other data collection methods are targeted at obtaining system performance data such as average travel speed. An example of one of these methods is the collection of spot speed data, which estimates—either by observation or with detection devices such as radar or inductive loops—the time it takes a sample of vehicles to traverse a measured distance. Another approach for determining average speed is called the moving vehicle method, which entails a driver matching the average speed of a traffic stream by passing as many vehicles as pass him. For both cases, new system surveillance technologies are providing better opportunities for collecting this type of data more efficiently.

As shown in Table 4.2, the type of traffic data collected in a typical metropolitan area is used for a variety of purposes, including planning, design, cost allocation, and safety analysis. In addition, the U.S. government requires states to collect and submit data on a sample of road segments stratified by functional classification. This data, part of the Highway Performance Monitoring System (HPMS), is used by the U.S. DOT to estimate the current condition of the national highway system and to forecast needs. A more recent model, the Highway Economic Requirements System (HERS), uses the HPMS database for investment forecasts based on a benefit/cost analysis. (For a critique of this model, see [U.S. GAO, 2000]). HPMS data items include data on existing traffic volumes and speeds and data related to the road's capacity, physical condition, and geometric characteristics. Most urban areas have data similar to this in a database called a road inventory, although many do not have as many data items as required to be part of the HPMS submittal. Special inventory surveys might also be needed to obtain data on other transportation-related activities. For example, inventories of urban parking capacity (supply, cost, and utilization data) have been undertaken in many urban areas. Such inventory data are important for both travel forecasting models and for assessing the impact of changes to the transportation system.

4.2.2   Data on Transit Network Performance and Condition

Similar to a road inventory, a transit inventory includes a variety of data—the number of vehicles in the fleet, number of seats available, measures of route utilization, speeds, and, in some cases, on–off passenger counts at specific transit stops (Table

Table 4.2   Use of traffic data for decision making in highway management

Highway Management Phase	Traffic Counting	Vehicle Classification	Truck Weighing
Engineering	Highway geometry	Pavement design	Structural design
Engineering economy	Benefit of highway improvements	Cost of vehicle operation	Benefit of truck climbing lane
Finance	Estimates of road revenue	Highway cost allocation	Weight distance taxes
Legislation	Selection of state highway routes	Speed limits and oversize vehicle policy	Permit policy for overweight vehicles
Planning	Location and design of highway systems	Forecasts of travel by vehicle type	Resurfacing forecasts
Safety	Design of traffic control systems and accident rates	Safety conflicts due to vehicle mix and accident rates	Posting of bridges for load limits
Statistics	Average daily traffic	Travel by vehicle type	Weight distance traveled
Private sector	Location of service areas	Marketing keyed to particular vehicle types	Trends in freight movement

SOURCE: U.S. DOT, 1995

4.3). In addition to this data, transit operations and system conditions data are typically collected on a periodic basis by transit agencies. The strategy for collecting operations data consists of two phases: (1) determining the "base" conditions of each route in the system and (2) periodic monitoring to detect changes which may have occurred [Multisystems, 1985]. The major types of transit data collection techniques excluding ridership surveys, which are discussed in a following section, include the following:

Ride check: Recording the number of passengers boarding and alighting at each stop and the bus arrival time at selected points. The individual recording the data is on the vehicle.

Point check: Estimating the number of passengers on the vehicle and recording vehicle arrival time. *Peak-load* count is taken at the peak-load point. *Multiple* point checks include several points along a route. The individual recording the data is located on the street.

Boarding count: Recording the total number of passengers boarding, most often by fare category.

Fare box reading: Recording of a fare box reading at selected route locations, most often done electronically.

Revenue count: Counting revenue in a fare box by bus, most often done electronically.

Transfer count: Counting the number of riders transferring by collecting the number of transfer tickets on each bus or through an automated fare system.

Table 4.3   Typical data items in a transit inventory

- List of transit companies and/or operating agencies
- Total number and type of transit vehicles
- Transit routes by type of service
- Total number of miles of routes by type and company
- Route number, description, and terminal-to-terminal mileage
- Location of transfer points, terminals, and parking facilities
- Location of stops
- Hours of operation
- Headway by hour of day
- Running time by route segment by hour of day
- Average turn-around time by time period
- Total annual and weekday vehicle miles and hours
- Fare structure
- Total annual and average weekday costs
- Accidents by type and location

Similar to the federally required HPMS data, the U.S. DOT requires each transit property to collect data on operating revenues, operating expenses, capital revenues and costs, organization/management costs, amount of service (e.g., vehicle revenue-hours), level of ridership (passenger trips and passenger miles), and service characteristics (service reliability and safety). This database is called the U.S. National Transit Database (also referred to as Section 15 data). Importantly, portions of the federal aid provided to transit properties are tied to several of the measures in this database. Transit properties also often use this data to compare their system's performance characteristics to those of peer systems.

Applying new technology for monitoring vehicle movements and for processing fare payments (and thus ridership information) has been one of the important advances in transit data collection in the past 10 years. Automated vehicle location (AVL) technology, which keeps track of vehicle position via GPS technology, permits transit planners to determine route performance characteristics and to identify where improvements might be necessary to reduce delay [Levy and Lawrence, 1991; Okunieff, 1997]. Microchip technology in "smart" cards allows transit planners to follow the travel patterns of transit riders, whereas automatic passenger count (APC) technology (pressure-sensitive mats, infrared sensors, or optical imaging) can be used to estimate boardings and alightings [TRB, 1997a; Boyle, 1998; Furth, 2000]. The types of decisions that such ridership data can inform include those relating to scheduling and operations planning (e.g., number of vehicles to put in service, schedule adjustments, and type of service); long-range planning and design (e.g., systems planning, priorities for expansion, design criteria, and station renovations); and financial planning and resource allocation (e.g., system revenue forecasts, market segment analysis, cost allocation to member communities, and prediction of revenue impacts of changes in service) [Pratt, 1991].

In addition to highway and transit counts, many metropolitan areas also conduct studies on pedestrian and bicycle movements. Classification schemes and sampling techniques similar to those used in highway and transit data collection are used to determine the volume of movement and system performance for these modes as well.

4.2.3   Evolution Toward Asset Management

As mentioned earlier, most metropolitan areas collect data on the inventory and physical condition of the facilities and vehicles that constitute the transportation system. Given the significant level of resources that has been spent on the transportation system, preserving this asset base is an important policy goal. In the United States, the HPMS database and data from the National Bridge Inventory are used to monitor the condition of the infrastructure. For example, in 1997, 67 percent of the bridges in urban areas were considered to be in good condition, 21 percent were functionally obsolete, and 12 percent were structurally deficient. Similarly, 76 percent of the pavement on urban roads was considered to be fair, good, or very good condition; 24 percent was poor or mediocre [BTS, 1999].

Transportation asset management is a decision-making process for making cost-effective decisions about the design, construction, maintenance, rehabilitation, retrofit, replacement, and abandonment of (transportation) assets, with the objective of maintaining or improving the value of these assets over time. Infrastructure management systems, most usually designed as computer-assisted decision support systems, are based on the inventory and physical condition data of transportation facilities. Most systems contain data on the causes and indicators of deterioration. These data are used to develop deterioration relationships to estimate the future system conditions under various demands or loads. The more sophisticated systems also employ priority assessment methods for ranking or optimizing projects under budgetary constraints. In addition to these elements, integrated asset management systems include procedures for system valuation and analysis of investment trade-offs for multiple modes of transportation infrastructure.

Key elements of management systems are as follows:

- *Inventory data*, which describe the more permanent aspects of facilities, for example, location.
- *Attribute data*, which describe the more changeable aspects of facilities, for example, condition.
- *Performance prediction models*, which predict the future condition of facilities by linking causes of deterioration, for example, roadway demand, with indicators of deterioration, for example, pavement distress.
- *Alternative selection methods*, which are used to identify deficiencies and select improvements for facilities or facility components.

- *Priority assessment methods*, which are used to prioritize projects in the context of budgetary constraints, based on such criteria as the benefit/cost ratio and life-cycle costing.

- *Validation procedures*, which are used to fine-tune the decision support system with respect to the actual facilities being managed [Grant and Lemer, 1993; McNeil, 1996; FHWA and AASHTO, 1997; Hudson, Haas, and Uddin, 1997; FHWA, 1999].

Current examples of asset management include pavement management systems for roadway networks, bridge management systems [Hudson and Hudson, 1994; TRB, 1994], and public transit facilities and equipment management systems for transit [Parsons et al., 1995; Kish and Meyer, 1996]. The concept of an asset management system for a metropolitan-level transportation system includes the total asset base and not just one component. Given the significant levels of investment in transportation systems and the decay associated with aging infrastructure, it is very likely that asset management will become a key aspect in managing metropolitan transportation systems in future years.

4.2.4 User Characteristics—Passenger

Because travel behavior predictions are partly based on the socioeconomic and demographic characteristics of those traveling, identifying traveler characteristics has become an important focus of data collection activities [Mahmassani et al., 1993]. Such information is becoming even more important because of the additional policy demands being placed on the transportation planning process. For example, Portland METRO, the MPO in Portland, Oregon, conducted a survey in 1995 for reasons that reflect these new demands:

- Oregon state law, in addition to federal air-quality policy, required a monitoring of vehicle miles traveled per capita.

- Population demographics and household structures had changed significantly.

- Travel demand forecasting was evolving to include daily trip activity, incorporating many shorter trips (walk and bike) that previous surveys had ignored. In addition, trip chaining as a transportation phenomenon needed data to support its modeling.

- New policies such as congestion pricing and urban design policies required surveys of likely traveler responses to these initiatives in order to model them.

- Air-quality analysis required household data on patterns of vehicle ownership [Cambridge Systematics, 1996a].

Figure 4.3 shows the travel survey that was used in Portland to collect data on the activities associated with individual travel.

The most common way of collecting trip information is with travel surveys. Travel surveys can collect many different kinds of data relating to such things as traveler and/or household characteristics, behavioral data about trips taken, and

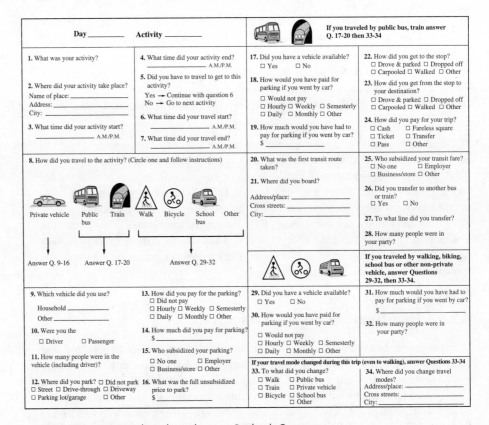

Figure 4.3 Activity-based travel survey, Portland, Oregon

attitudinal information. The most commonly used surveys in urban transportation planning focus on data collected in or at (1) households; (2) workplace or special trip generator sites; (3) visitor or tourist centers; (4) vehicle intercept and external stations; and (5) on transit lines [Cambridge Systematics, 1996c]. Two special types of surveys—panel and stated preference surveys—are used to collect travel behavior data. One of the challenges in survey methodology is determining which of these surveys is most appropriate for the desired data. Figure 4.4 shows the surveys most commonly used to collect the types of data shown. Several of these surveys will be discussed in the following paragraphs.

Household Travel Behavior Surveys While surveys that contact the household directly are probably the most expensive of all the data collection efforts listed previously, they provide the most detail. As shown in Table 4.4 for the case of Portland, Oregon, numerous person, household, and trip characteristics can be obtained from a household survey [Sen et al., 1998]. Three basic techniques can be used to collect these data—a personal home interview, a telephone interview, and a mail-back survey.

Survey data	Household travel/activity	Commercial vehicle (fleet manager surveys)	Workplace/hotel (centrally distributed surveys)	Transit on-board	Vehicle intercept/external station	Commercial vehicle (intercept surveys)	Workplace/hotel (intercept surveys)	Special generator	Parking
Socioeconomic/demographic data on travelers and/or their households	●	○	●	●	●	◑	●	●	●
Revealed preference travel data – travel diaries (multiple trips)	●	●	◐	○	○	○	○	○	○
– activity diaries	●	◑	◐	○	○	○	○	○	○
– on a specific trip	●	◑	●	●	●	●	●	◑	●
Attitudinal/perception data (rating, rankings, allocation of points)	●	●	●	◑	◑	●	●	◑	●
Knowledge data	●	●	●	◑	◑	◑	◑	◑	◑
Opinion data/open-ended questions	●	●	●	◑	◑	◑	◑	◑	◑
State response travel data (stated preference trade-off analysis)	◑	◐	◐	◐	◑	◑	◐	◐	◐
Longitudinal (panel) data	◑	◐	◐	○	○	○	○	○	○

NOTES:

● Data are commonly collected with this type of travel survey
◑ Data are sometimes collected with this type of travel survey
◐ Data could feasibly be collected with this type of survey but generally are not
○ Data are not collected with this survey method

Figure 4.4   Survey techniques and data collected

1 SOURCE: Cambridge Systematics et al., 1995

Table 4.4   Data items collected in the 1994 Portland household activity and travel
behavior survey

Household Data Elements

- Address
- Activity dates
- Household size and names
- Household structure type
- Household income
- Number of phone lines
- Number of cellular or car phones
- Presence/absence of household members or visitors
- Tenure at current address
- Zip code of previous address
- Own or rent
- Number of vehicles
- Shared phone lines
- Transportation disability

Person Data Elements

- Gender
- English proficiency
- Employment status
- Age
- Household language
- Drivers-license status
- Student status

If employed

- Occupation
- Industry
- Work at home
- Pay for parking?
- Parking cost
- Tenure at current job
- Address of primary job
- Zip code of secondary place of work
- Primary employer offers subsided
- Parking or transit?
- Number of days traveled by specific modes
- Zip code of previous employer

Activity Diary Data Elements/Questions

- What was the activity?
- Where did it take place?
- When did activity start?
- Did you have a vehicle available?
- Parking costs, if any
- How long did it take?
- Were you already there?
- How did you get there?
- Number in party
- Bus trip information (e.g. route, transfer)

Vehicle-Form Data Elements

- Vehicle year, make, model, type
- Year purchased
- Fuel type
- Vehicle ownership
- Purchased as replacement or add-on
- Odometer reading at end of 2d day

If student

- Name of school
- Number of days traveled by mode

SOURCE: Cambridge Systematics, 1996a

The *personal home interview* was the technique most often used in early transportation studies when new databases were being created. The ability of the interviewer to explain questions, a longer time per interview (compared to other techniques), and higher response rates because of personal interaction made the home interview a valuable technique in developing an extensive database. However, the home interview is a particularly time-consuming and expensive technique. From a methodological perspective, the possibility of biased results because of interviewer actions and statements is also a cause of concern. Even with these limitations, the home interview is often the best way of getting the most complete information.

Given the generally prohibitive costs of personal home interviews, planning agencies have turned to *telephone surveys* as a substitute. This approach has the advantages over personal household visits of requiring a shorter time to complete each interview; of using fewer people to administer a survey; of being able to closely supervise telephone interviews; and of having the ability to easily recontact those interviewed. The primary disadvantage is that those contacted can easily refuse to participate. To maintain the validity of the random sample, personal interviews should be ideally undertaken at households that could not be reached by telephone.

Several steps can be taken to reduce the potential for bias in telephone surveys [U.S. Department of Transportation, 1973; Dilman, 1978]:

1. A telephone book should not be used for sample selection because those not having a telephone or those with unlisted numbers would be missed (10 to over 50 percent of the households in an urban area). Other means are therefore required to choose household samples. For example, a semirandom dialing technique has been used in some instances in which telephone numbers are sampled randomly from a telephone book. The number is then increased by one (e.g., 924–3521 becomes 924–3522). This increases the chance of reaching both valid numbers as well as unlisted numbers.

2. The households for which no telephone number is available could be visited in person. A common way of choosing households is to use a document called a "reversed listing," which lists all households by street address (regardless of having a telephone) and provides telephone numbers for those listed in the telephone directory. Such a listing is available through most real-estate agencies.

3. Each household to be contacted should be sent a preinterview letter explaining the purpose and procedure of the survey.

4. In those cases where a household has a large number of trips to report, it might be necessary to send an interviewer to ensure that all the information is obtained.

A third technique, called a *mail-back survey,* sends a survey with return postage to a sample of households. The trade-off in this case is between the much-reduced cost of data collection and the potentially low response rate. Several actions have been shown to increase response rates, including mailing a second questionnaire or reminder to those households not responding to the initial request within a specified time period (e.g., 1 or 2 weeks), pretesting the questionnaire to avoid misleading or confusing questions, and using a personally signed cover letter (see, for example, [Cambridge Systematics, 1996c]).

If resources permit, a telephone or mail-back travel survey should be supplemented with a small stratified in-person home interview survey to validate the results from the larger random sample. This survey can be used to obtain additional information for specific household types (e.g., low-income families) or trip purposes, both of which may be poorly represented in the random sample survey. The size of this stratified sample depends on the resources available, the needs of the planners, and the degree to which the initial random sample is considered representative of the population.

The level of effort associated with obtaining a valid survey return is illustrated by the steps taken in Portland, Oregon, as part of a travel behavior survey (Table 4.5). This survey collected data from households through the use of a travel diary covering a 2-day period. These steps were taken for each survey participant.

Table 4.5   Tasks for conducting a travel behavior survey

Action	Household Contact	Timing	Data Collection Mode	Objectives
Advance letter	Yes	2 weeks prior	Mail	Promote survey and provide background information
Recruitment call	Yes	10 days prior	Phone	Get cooperation
				Confirm address
				Collect household/person data
Place diary packet	Yes	9 days prior	Mail	Record vehicle information
				Activity checklist
Reminder call	Yes	3 days prior	Phone	Ensure participation
Diary day 1	No		Paper and pencil	Record activities
Diary day 2	No		Paper and pencil	Record activities
Data collection	Yes	1 day after	Phone	Retrieve activity and travel data

SOURCE: Cambridge Systematics, 1996b

Workplace and Special Generator   Special site surveys are similar in nature to household surveys. These surveys are used to obtain information from employees at their places of employment and from shoppers and users of recreational facilities. The purpose of these surveys can range from identifying the characteristics of specific trip types to determining the feasibility of new services to these sites. This type of survey, which can be conducted in person or by distributing questionnaires, has been particularly effective in surveying target groups such as the elderly. In these cases, natural gathering points such as elderly housing complexes, social centers, church groups, and medical centers provide effective survey distribution locations.

Both household travel behavior and special site surveys can be useful data collection techniques. Such surveys provide timely data on socioeconomic characteristics and travel behavior in an urban area, data that are extremely important for calibrating and using travel forecasting models. However, the collection of these data tends to be both time-consuming and expensive.

Vehicle Intercept/External Station Surveys   These surveys obtain data on the patterns of movement of persons and goods in a study area. For transportation planning purposes, several types of trips are of interest—those internal to a planning study area, those made into or out of an area, and those that pass through. Three major techniques are used for collecting this type of data. A *roadside interview* involves stopping cars and commercial vehicles at interview stations (e.g., cordon or screen-line points) and asking the driver questions on origin, destination, trip purpose, route used, and intermediate stops made. Because it is impractical to stop all traffic at interview stations, a sample selection procedure is employed. Although the response rate can be quite high, traffic backups could be caused by drivers waiting to be interviewed (and thus police presence is essential to manage traffic).

The *roadside handout or postcard survey* consists of stopping vehicles and distributing mail-back postcards at roadside stations. The postcard requests information on the trip being made at the time of contact; thus, the origins and destinations of trips made on a specific date are obtained. This method is less expensive than the previous one but often results in a much lower response rate.

In a *license-plate survey,* roadside observers note the last four digits of license plates as cars pass their station. The route of a vehicle can thus be traced by its successive appearance at a series of recording stations [Crabtree and Krause, 1982]. A variation of this approach is to videotape license plates and match numbers with registration databases [CUTR, 1996]. A survey is then mailed to the owner's address requesting information about the observed trip. Because the license-plate survey is the only vehicle intercept technique not requiring vehicles to stop, it is the preferred method for high-volume locations. The major disadvantages are that nonresponse rates can be high and vehicles are often driven by individuals other than the owner. Thus, the possibility exists that the vehicle owner and not the actual traveler will be contacted.

Another option, albeit more expensive, is to obtain addresses from the motor vehicle registry and mail surveys to a randomly selected set of households in the study area. As with the household travel survey, the response rate for this strategy can be quite low.

The results of these surveys are expanded to represent a 100 percent estimate of the travel characteristics of the population being surveyed [Stopher and Stecher, 1993]. It is therefore important to know what portion of total trips the sample represents. The estimated results from the expanded sample should also be checked with other data and traffic counts. When validated, the data can be used to construct origin and destination tables that show trip patterns among the zones in the metropolitan area. The relative costs of the survey techniques used to determine origin/destination patterns are shown in Fig. 4.5.

Panel Surveys   All of the travel surveys described previously collect data from different travelers at one point in time. This is called a cross-sectional survey. How-

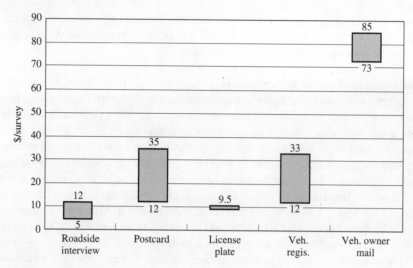

Figure 4.5   Cost ranges for origin–destination surveys
| SOURCE: CUTR, 1996

ever, it is important to know how travelers change their behavior over time in response to specific changes in the transportation system (e.g., a new transit line) or how they respond to changes in the general characteristics of system performance (e.g., increasing levels of congestion). In order to obtain this type of data, longitudinal or panel surveys follow the travel behavior of the same group of people over a long period of time, often 2 to 3 years. This is most often done by having participants complete a travel diary that records daily travel activities for a specific period of time, such as 1 or 2 days or weeks. Because a longer commitment time is required to participate in this effort, many participants will end up dropping out of the survey because of personal reasons. Thus, the validity of this approach depends on how representative the replacement households are that take the place of those that drop out or relocate. The best example of this approach is found in Seattle where a longitudinal panel first started in 1989. Each panel member was asked to complete a travel diary each year that contained information on trips taken during a 2-day period. Of the original sample of 1,713 households, 54 percent were still participating 4 years later [Cambridge Systematics, 1996b].

Stated Preference Surveys   Much of the data collected for transportation planning reflect actual travel behavior, that is, trip characteristics, origins/destinations, and purposes for trips that have already been made. This data is referred to as *revealed preference data*. However, transportation planning is often asked to assess the likely responses or consequences of policy actions that have not yet been taken or of mobility opportunities that do not yet exist. For example, new policy initiatives in urban design could influence household location decisions, or new technologies such as telecommuting or high-speed trains could change people's travel habits [Yen et al., 1994]. Without data on how individuals have actually responded to these changes, planners must rely on traveler reactions to hypothetical situations. The approach most commonly used for this is *stated preference surveys.*

Stated preference surveys ask such questions as, "Given the characteristics of the following options, which one would you choose?" or "Given the level of attributes of these alternatives, rank them in order of preference" or "Given the following characteristics of these options, how would you rate each alternative?" [Cambridge Systematics, 1996c]. Because respondents are being asked to react to alternatives with which they have no experience, the design of the stated preference survey is especially crucial. For example, experience has shown that at least three characteristics for each alternative should be available for relative comparison and that these characteristics should appear realistic to the respondent. For more information on stated preference surveys, see [Polak and Jones, 1995].

Transit Onboard Surveys   Onboard surveys are used to collect data on transit riders, trip characteristics, and route performance from the perspective of the customer. Given that onboard surveys are usually targeted at riders who are on the vehicle or in a terminal, the experimental design for this method is extremely important and often results in fairly labor-intensive activities. Several methods can be used to conduct the survey, ranging from drivers handing out and collecting questionnaires, to paid surveyors conducting face-to-face interviews. The most typical approach is for surveyors to hand out and collect self-administered questionnaires or to encourage respondents to mail them back. Typical response rates for the mail-back surveys have been between 20 and 30 percent of all boardings.

The design of onboard surveys must consider several other aspects of transit service. The socioeconomic and ethnic characteristics of transit riders require a survey that can be easily understood and returned. (For example, one should expect some difficulties with non-English-speaking riders.) The survey must also account for the changing character of transit ridership and service during the day. In addition, short trips might be underrepresented given the limited amount of time for filling out the form. The type of data collected with transit onboard surveys is shown in Table 4.6.

Table 4.6   Common data elements for transit onboard surveys

Boarding and alighting bus-stop location	Trip origin
Arrival and departure times	Trip destination
Activity or trip purpose	Access and egress mode
Bus routes used for trip	Fare payment type
Trip frequency	Auto ownership
Auto availability	Traveler age
Gender	Occupation
Race, ethnicity, or nationality	Household size
Household income	

SOURCE: Cambridge Systematics, 1996c

4.2.5   User Characteristics—Freight

The movement of goods represents a significant proportion of regional trip making. According to the U.S. Bureau of the Census, about 95 percent of all trips taken by trucks are less than 200 miles in length. Truck travel thus seems to be predominantly regional or local, and accordingly should be an important issue in metropolitan transportation planning. The freight data needs for transportation planning are shown in Table 4.7.

Table 4.7   Freight data needs for transportation planning

Function	Data Needs	Support for Planning
Congestion management	• Truck hours of travel • Average speed or travel rate • Added truck-hours due to delays • Truck transport cost per truck-mile • Added cost due to congestion • Transport time reliability • Types of trucks and commodities delayed • Energy consumption (truck-mile or ton-mile) • Emissions rates (truck-mile or ton-mile)	• Understand impact of congestion on goods movement • Understand contribution of trucks to urban congestion and air-quality problems
Intermodal access	• Volumes of trucks entering or exiting facility • Variability in demand for facility services • Congestion-related delays on access roads • Queuing delays related to capacity of facility • Accident rates on access roads • Travel time contours around facility • Number of people living/working within x miles of facility	• Identify land-side access improvement needs
Truck route designation and maintenance	• Truck traffic volumes • Origin/destination patterns • Truck size and weight data	• Identify high-volume truck routes and corridors • Assess pavement damage and replacement needs
Safety mitigation	• Accident rates • Rail-grade crossings • Low-clearance bridges • Steep grades	• Identify safety hazards and develop mitigation strategies
Economic development	• Truck volumes • Commodity movements • Origin/destination patterns • Shipping costs	• Assess economic benefits and costs of freight transportation investment projects

SOURCE: Cambridge Systematics and Meyer, 2000

Primary sources of freight data are similar to those used for passenger data. Truck surveys can be handed out at roadside stations and/or targeted at freight terminal operators or port authorities. A survey in Washington State, for example, interviewed 30,000 truck drivers at 28 weigh stations, ports of entry, and border crossings. Ninety individuals on any given day conducted the 3-minute interview, including noting vehicle characteristics. This process resulted in a 95 percent response rate.

Telephone interviews of shippers and carriers are another primary source of data that yields a high response rate [Lau, 1995]. A commercial vehicle survey in Phoenix, Arizona, consisted of an initial telephone contact followed by a mail-back questionnaire that included a 1-day travel diary. Data on approximately 3,400 truck trips made by 606 commercial vehicles were collected in this manner. See Table 4.8 for other examples of commercial vehicle surveys.

Table 4.8   Recent commercial vehicle surveys

City	Year	Number of Responses	Survey Method
Amarillo, TX	1990	444	Travel log, survey
Atlanta, GA	1996	1,000	Travel log, mail back
El Paso, TX	1994	188	Telephone
Harrisburg, PA	1992	240	Interview (at weigh station)
Houston, TX	1994	900	Mail and phone
Milwaukee, WI	1992	2,500	Mail
Philadelphia, PA	1991	2,500	Interview, mail back
San Francisco, CA	1991	10,200	2,200 mail back
			8,000 roadside interview

SOURCE: Cambridge Systematics, 1996b

MPO freight advisory committees are one of the more innovative primary sources of freight data. As part of the normal organizational structure of the MPO, these advisory committees provide policy input into the planning process representing those issues of most concern to the freight sector. These committees have also helped in collecting the data needed for effective planning. An example of this approach can be found in Columbus, Ohio, where representatives from rail lines, airlines, private businesses, and port authorities actively participated in a feasibility study for an inland port in that community [Mid-Ohio Regional Planning Commission, 1994]. Local freight or business associations can serve the same purpose.

However, the type of freight data of most interest to planners, such as commodity types, vehicle configuration and weight, trip origins and destinations, route taken (including intermodal transfers), and shipper/carrier characteristics, is often difficult to obtain. In addition, the data collected by government agencies are often available at such an aggregate level (e.g., zip codes) that their use in planning is problematic. Freight companies also view this data as important business information reflecting a company's customer base and overall productivity and thus are hesitant to release it. Accordingly, planners have often had to rely on secondary-source data. Such data are collected by government agencies and by companies that specialize in the data as a commercial product. The secondary sources shown in Table 4.9 could be of use in a regional transportation planning process. For a good overview of all sources of freight data, see [Capelle, 1999].

Table 4.9   Secondary sources for freight data

			Data on:		
Source	Vehicles/ Drivers	Shipment	Commodity	Origin/ Destination	Facilities
Commodity flow survey (census)		•	•	•	
HPMS (DOT)	•				•
Less-than-truck-load commodity and market survey (ATA)		•	•	•	
Nationwide truck activity and commodity survey (census)	•	•	•	•	
North American truck survey (AAR)	•		•	•	
Truck inventory/use survey (census)	•		•		

NOTE: ATA: American Trucking Associations
          AAR: American Association of Railroads
SOURCE: Cambridge Systematics et al., 1995; Cambridge Systematics and Meyer, 2000

4.2.6   Changing Technology and Data Collection

One of the important trends in data collection over the past decade has been the use of new technologies for collecting data. In some cases, the primary focus of these technologies has been data collection itself, for example, handheld computers for data input, infrared sensors to determine vehicle occupancy, satellite imaging to assess land-use changes, bar codes that identify the commodity being transported, and video imaging to identify vehicle types [Karimi et al., 2000]. In other cases, data collection is an ancillary use of the technology. For example, surveillance cameras to detect accidents, global positioning systems, and in-vehicle computer control of engine operations could each be used as sources of data on system operations even though they are used primarily for another purpose. New data collection technology will, in particular, have an important positive effect on the collection of data for person trips rather than vehicle trips.

The use of personal computers and GPS location technology provides useful capabilities for improving the accuracy and efficiency of individual trip reporting. For example, a study in Lexington, Kentucky, using GPS to monitor vehicle movement as compared to self-reported trip information showed that people tend to report trip start times to the nearest quarter or half hour and that trips over 5 miles are often rounded off to the nearest 5-mile increment [Wagner, 1997]. Thus, there is significant overstatement in travel time and distance in self-reported studies. It is likely that use of technologies such as GPS, handheld computers, and the Internet will be much more commonplace in future years (for overviews of the new technology of personal trip data collection, see [Sarasua and Meyer, 1996; Kalfs et al., 1997; Quiroga and Bullock, 1999; Doherty et al., 1999; Wolf, 2000]).

Much of the data collected with these new technologies is the same as that collected by traditional means. However, the data can now be collected continuously and at a great level of detail. For example, the various uses and benefits of data generated from intelligent transportation systems (ITS) technologies are explained in the following list:

- The continuous nature of most data generated by ITS removes sampling bias from estimates and allows the study of variability.
- The variability of ITS data provides the opportunity to analyze nonrecurring congestion issues, causes, and solutions.
- The detailed data needed to meet emerging requirements and for input to new modeling procedures can be provided by ITS.
- Use of data generated by ITS for multiple purposes is a way to stimulate the support of other stakeholders for ITS initiatives.
- Promoting the use of archived data for multiple purposes complements the initiative for integrating ITS in general.
- Because the data are already being collected for ITS control, other uses provide a value-added component to ITS.
- ITS is a rich data source for multiple uses, but not a panacea; traditional sources of data will continue to be important.
- As the focus of transportation policy shifts away from large-scale, long-range capital improvements and toward better management of existing facilities, ITS-generated data can support the creation and use of the system performance measures that are required to meet this new paradigm [Cambridge Systematics and Meyer, 2000].

Table 4.10 shows the data that could be generated from various intelligent transportation system (ITS) technologies and the possible uses of such data in transportation planning. The changing technology of the transportation system itself will provide new opportunities for data collection and analysis in future years. At the same time, these technologies could create new challenges.[1]

---

[1] In some cases, advances in data collection have created significant hurdles for database management and analysis. Engine computers, for example, can monitor between 74 and 140 data items every second of vehicle operation, depending on vehicle make. This data has been used by researchers to determine better ways of estimating vehicle emissions. The capability of linking into engine operation can result in millions of data points that have to be structured in a way to permit interpretation.

Table 4.10   Intelligent transportation system technologies and data collection

ITS Data Source	Primary Data	Collection Equipment	Real-Time Uses	Planning Uses
Freeway traffic flow surveillance	Volume Speed Occupancy	Loop detectors Video imaging Acoustic Radar/microwave	Ramp-metering timing Incident detection Congestion location	Congestion monitoring Link speeds for models AADT, capacity factors Flow rates Traffic center plans
Ramp meter and traffic signal preemptions	Time of preemption Location	Field controllers	Priority to transit HOV and emergency vehicles	Network details for traffic simulation
Vehicle counts from toll collection	Time Location Vehicle counts	Electronic toll collection equipment	Automatic toll collection	Traffic counts by time of day
Traffic management center traffic flows	Link congestion Stops/delay	Traffic management center software	Incident detection Traveler information Preemptive control strategies	Congestion monitoring Effectiveness of prediction methods
Parking management	Time Lot location Available spaces	Field controllers	Real-time information to travelers	Parking utilization and needs studies
Transit usage	Vehicle boardings Origin/destinations	Electronic fare systems	Electronic payment of fares	Route planning Ridership reporting
Transit route deviations	Route number Time of advisory Routes taken	Traffic management center software	Transit route revisions	Transit route and schedule planning
Ride-share requests	Time of day Origin/destinations	Computer software	Dynamic ride-share matching	Travel demand estimation Transit route planning

4.3 DEVELOPING A COMMUNITY VISION AND GOALS SET

The transportation planning database discussed up to this point has focused primarily on the characteristics of the transportation system and on individual trip making. As noted in Chap. 1, however, the role of the planner is one of planning with interested officials and stakeholders rather than planning for a perceived unitary general public. If planners are to adopt this more "open" process of planning, they must have information on the desires and attitudes of the community with respect to both general directions for the planning process and specific reactions to plan and project proposals.

Table 4.10 Intelligent transportation system technologies and data collection (continued)

ITS Data Source	Primary Data	Collection Equipment	Real-time Uses	Planning Uses
Incident logs	Location and type of incident Police accident record	Computer software	Incident response and clearance	Incident response evaluations Congestion monitoring Safety reviews
Emergency vehicle dispatch records	Time Origin/destinations Route	Computer software	Coordination of emergency response	Emergency response studies on routes
Construction and work zone identification	Location Date and time Lanes blocked	Traffic management center software	Traveler information	Congestion monitoring
Hazardous materials identifiers	Type Carrier Route and time	Commercial vehicle operations systems	Identifying hazardous materials in spills Identifying routes	Hazardous transport studies
Emissions management systems	Time and location Pollutants Wind conditions	Specialized sensors	Identification of hot spots	Trends in emissions Special air-quality studies
Weather data	Location and time Precipitation Wind conditions	Environmental sensors	Traveler information	Congestion monitoring Freeze/thaw cycles for pavement models
Vehicle probes	Vehicle ID Segment location Travel time	Probe vehicles GPS on vehicles	Coordinate traffic control strategies Congestion locations Transit schedules Electronic tolls	Congestion monitoring Link speeds for models Transit schedules Origin/destination

SOURCE: As modified from Cambridge Systematics and Meyer, 2000

As was indicated in Fig. 2.9, the urban transportation planning process begins with the development of a community vision. The vision describes the desired states and/or directions that the community wants to achieve in the future. This vision is then further refined with more specific guidance in the form of goals and objectives, referred to as a goals set, which ranges from general statements of desired policy and decision-making directions to standards that define in very specific terms minimum or maximum levels of acceptable outcomes. Transportation planners have important roles to play in providing the support to the very public process of developing both a vision and a goals set.

Although closely tied together from a process perspective, the articulation of a community vision and the development of a goals set will be treated separately in this chapter. This is done primarily because the type and level of information produced is slightly different in each case, and thus the planning efforts to support them can differ as well. This section also describes market research, an information-gathering method that has become very important to transportation planning.

4.3.1   Articulating a Vision

Chapter 1 defined the transportation system as an enabler of activities. It strongly influences urban form, a region's economy, environmental health, and community quality of life. One way of establishing these linkages in the transportation planning process is to begin with a vision of what the community wants to be, then define how the transportation system, and more importantly how changes to this system, relate to this vision [Ames, 1993; Ewing, 1997]. Because such a vision should reflect the collective opinions of system users, stakeholders, and the general public, this initial step in the planning process provides for intensive public involvement [Gayle, 1999]. The process of articulating a vision should also occur within a clearly understood organizational and decision-making structure so that a vision statement surfaces from what will inevitably be a pluralistic process. Usually, the decision on a final vision statement rests with the policy board or executive committee of the MPO or regional planning commission.

Four metropolitan areas provide examples of how a vision can be defined—Atlanta, Georgia; Albany, New York; Baltimore, Maryland; and Seattle, Washington.

Atlanta   The Atlanta Regional Commission (ARC), the MPO for the Atlanta region, undertook an extensive effort in 1996 to develop a vision for the metropolitan area. This visioning process was based on extensive public involvement, including the use of focus groups, newspaper supplements, television coverage on public television, surveys, public hearings, and newsletters. This effort was the most extensive public outreach program ever implemented in the Atlanta region. The final document adopted by the ARC policy board included desired directions in several policy areas of regional significance, including water supply, public services, education, and transportation. This vision statement became the basis for the goals that were used in the development of the year 2025 regional transportation plan. Table 4.11 shows the goals and objectives adopted for this plan update and their relationship both to the vision statement and to the planning factors identified in new federal transportation legislation.

Albany   The Capital District Transportation Committee (CDTC), the MPO for the Albany, New York, metropolitan area, created nine task forces to direct a community dialogue on desired directions for the region. These task forces focused on demographics/land use/growth futures, infrastructure renewal, transit futures, special transportation needs, expressway management, arterial corridor management, goods movement, bicycle and pedestrian issues, and urban issues [CDTC, 1995a]. Each group was asked to accomplish five tasks:

1. Identify issues relevant to both the near-term (year 2000) and the long-term (year 2015) regional vision, policy, and/or investment strategy. Modify the definition of these issues in response to public feedback.

Table 4.11   Atlanta's planning goals/objectives and relationship to vision

Goal	Objective	Vision Statement	TEA-21 Planning Factor
Accessibility and mobility for people and goods	• Develop intermodal passenger connections and equalize accessibility	X	
	• Implement transit/land-use changes to support transit/pedestrian development		X
	• Increase the accessibility and mobility options available to people and for freight		X
	• Enhance the integration and connectivity of the transportation system across and between modes for people and freight		X
Attain air-quality goals	• Meet air-quality attainment target for NO_x	X	
	• Meet air-quality attainment target for VOC	X	
	• Protect and enhance the environment, promote energy conservation, and improve quality of life		X
Improve and maintain system performance and system preservation	• Improve connections between truck, rail, and air freight facilities	X	
	• Promote efficient system management and operation		X
	• Emphasize the preservation of the existing transportation system		X
	• Promote energy conservation		X
	• Preserve historic resources	X	
	• Minimize community and environmental impacts	X	
	• Create incentives and regional policies to promote livable cities	X	
	• Protect and enhance the environment, promote energy conservation, and improve the quality of life		X
	• Support the economic vitality of the metropolitan area, especially by enabling global competitiveness, productivity, and efficiency		X
	• Increase the safety and security of the transportation system for motorized and nonmotorized users		X
	• Improve connectivity between low-income and minority populations and major employment and activity centers	X	
	• Improve social and environmental equity for all the region's citizens	X	

| SOURCE: Atlanta Regional Commission, 2000

2. Propose planning and investment principles that could provide implementation guidance on creating a regional vision.

3. Objectively analyze the issues and options, using a set of core performance measures as a yardstick. Supplemental performance measures could also be developed if they helped to capture the costs and benefits of alternative approaches.

4. Outline feasible and potentially desirable actions. Identify strategies where an apparent consensus existed. Where the Capital District faced a major policy choice, no recommendations were to be made; instead, performance-based information was provided to guide public dialogue.

5. Recommend what should happen to the task force beyond the development of the vision statement.

Note in this list the emphasis on system performance and consensus seeking. The MPO policy board did adopt a vision for the region that was built upon the common themes that emerged from the task forces. Such themes included the preservation of the unique nature of the area (small, connected, and walkable communities); the importance of vital urban centers to the health of the region; and a reliance on transportation system function, rather than on facility ownership, as a basis for regional investment decisions.

Baltimore   A long-range transportation plan developed for the Baltimore region in 1992 was strongly criticized by a citizens advisory committee because of its premise that land-use patterns would remain constant even with significant differences in transportation investments. In response, the Baltimore MPO established a process for the subsequent plan update in which the land-use and transportation linkage would be specifically analyzed [BMPO, 1995]. Three initial metropolitan development scenarios were developed based on the assumption that 10 percent of the expected growth in the region occurring in outlying areas could be redistributed to other locations. As was noted, "this 10 percent jurisdictional growth change limit which was established for all scenarios represents a figure on which the Subcommittee could reach consensus because it was considered feasible. This figure is not based on targets for performance standards" [BMPO, 1995].

The technical analysis in support of this effort shifted population and employment forecasts from outlying traffic analysis zones (or "provider" zones) to other traffic analysis zones (or "receiver" zones). Such a shift was deemed feasible because many of the jurisdictions in the region had policies in place that could be used to encourage such a change (Table 4.12). The regional travel forecasting model was then run for each scenario to determine resulting travel patterns. The three initial development scenarios were as follows:

Inside Beltway Scenario: An estimated 20,100 households, 50,300 people, 29,900 employees, and 23,000 jobs were reallocated from zones outside a perimeter highway to zones inside.

Fixed Transit Scenario: An estimated 19,700 households, 49,400 people, 28,300 employees, and 21,700 jobs were reallocated to zones located near rail stations.

Community Development Scenario: An estimated 16,000 households, 40,600 people, 23,000 employees, and 18,000 jobs were reallocated to zones designated for conservation and/or concentration of community growth.

As is usual in approaches such as this, a compromise or composite scenario resulted, which included the most acceptable elements of the three individual scenarios. The composite scenario assumed the inside-the-beltway scenario for regional population shifts but then assumed that those communities outside the beltway would shift up to 90 percent of the projected growth in the rural areas to more densely developed community centers.

The type of information produced by the transportation planning process in support of this effort is found in Table 4.13. The analysis included the year 2020 baseline transportation network without planned improvements; the 2020 transportation network with planned improvements but without transportation control measures (TCMs), such as parking cost increases and supply decreases, transit, and

Table 4.12   Overview of growth management techniques and their application in Baltimore

	Growth Boundary	Density Zoning	Overlay and Floating Zones	Transfer of Development Rights	Mandatory Clustering	Agricultural Zoning	Adequate Public Facilities Ordinance	Planned Unit Development	Designation of Sensitive Areas
Anne Arundel County	•		•	•		•	•	•	•
Baltimore City	•	•						•	•
Baltimore County	•	•	•		In RC-4	•		•	•
Carroll County	•	•	•	•		•	Schools, traffic, sewer	•	•
Harford County	•	•		•	In AG zones	•	•	•	•
Howard County	Water and sewer	•	•	•	•		•	•	

SOURCE: BMPO, 1995

Table 4.13   Land-use scenario evaluation in Baltimore

Scenario	2020 Base	2020 Plan	2020 Plan with TCMs	2020 Composite
Transportation Supply				
Lane miles	8,100	8,410	8,410	8,100
Difference from base		310	310	0
Miles of rail	120	147	147	120
Difference from base		27	27	0
Costs (millions)	$942	$4,676	$4,703	$942
Difference from base		$3,734	$3,761	$0
Transportation Demand				
Vehicle miles traveled	65,198,500	66,399,000	64,872,100	64,350,500
Difference from base		1,200,500	−326,400	−848,000
Vehicle trips	7,068,000	7,067,200	6,880,100	7,017,600
Difference from base		−800	−187,900	−50,400
Transit ridership	201,300	201,000	298,700	220,400
Difference from base		−300	97,400	19,100
Performance				
Average speed (mph)	42.0	42.6	NA	42.1
Difference from base		0.6	NA	0.1
Percent roads congested	14.9	13.4	NA	14.2
Difference from base		−1.5	NA	−0.7
Vehicle hours traveled	728,200	697,800	NA	712,200
Difference from base		−30,400	NA	−16,000
Average trip length (miles)	9.22	9.40	NA	9.17
Difference from base		0.18	NA	−0.05
User cost per day (1,000s)	$19,560	$19,920	$19,462	$19,305
Difference from base		$360	−$98	−$255
Air Pollutant Emissions				
Hydrocarbon (tons/day)	42.6	NA	NA	42.0
Difference from base				−0.6
Oxides of nitrogen	95.2	NA	94.0	94.3
Difference from base			−1.2	−0.9
Carbon monoxide	462.4	NA	NA	454.6
Difference from base				−7.8
Land Use				
Residential land consumed (acres)	609,500	609,500	609,500	590,600
Difference from base	0	0	0	−18,900

| SOURCE: BMPO, 1995

employer commute programs; the 2020 transportation network with these TCMs; and the composite development scenario applied to the 2020 base transportation network. The composite scenario showed positive results in most categories.

Seattle   Based on the most extensive regional public involvement program ever conducted in the region, the Puget Sound Regional Council (PSRC), the MPO for the Seattle region, adopted Vision 2020 to serve as the policy framework for the development of local comprehensive plans, state actions, and regional programs. This vision was based on the analysis of five alternative growth and transportation strategies: no action, implementing existing plans, focusing development in major urban centers, focusing development in multiple centers, and allowing growth to disperse throughout the region. The public outreach effort resulted in over 90 percent of those responding choosing either the major centers or multiple centers alternative (Fig. 4.6). In each of these scenarios, the regional transportation forecasting model was used to assess system performance with differing land-use patterns. As in the case of Baltimore, a hybrid scenario consisting of elements from the major and multiple centers scenarios became the "Preferred Alternative." As noted in the vision statement,

> Vision 2020 supports a new order of more compact, people-oriented living and working places, thereby reversing trends that have created increased numbers of low-density, auto-dependent communities. It limits the expansion of the urban area and focuses a significant amount of new employment and housing into approximately 15 mixed-use centers that are served by a more efficient, transit-oriented, multimodal transportation system [PSRC, 1990].

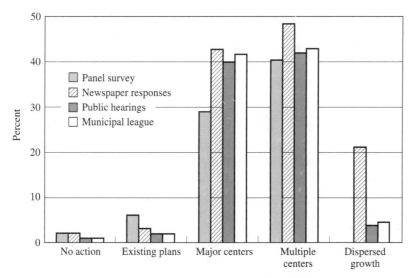

Figure 4.6   Results of polling on five visions for the future of Seattle
| SOURCE: PSRC, 1990

A five-part strategy was to serve as the basis for the future of the region.

1. Create a regional system of central places framed by open space.

2. Strategically invest in a variety of mobility options and demand management to support the regional system of central places.

3. Maintain economic opportunity while managing growth.

4. Conserve environmental resources.

5. Mitigate potential adverse effects of concentrating development by early action.

Transportation investments were considered an important means of implementing Vision 2020. A regional rapid and high-capacity transit system was proposed to connect urban centers. An interconnected system of more than 300 miles of high-occupancy vehicle lanes, additional passenger ferry service, local transit improvements, and comprehensive transportation demand management programs were also recommended as part of the transportation implementation strategy.

In summary, developing a community vision is a necessary first step to the transportation planning process. Critical to this process is the provision of data and information to the participants such that the resulting vision statement is relevant to the circumstances facing the region. The vision statement itself becomes important information that guides subsequent planning, including the next step, formulation of goals and objectives.

4.3.2   Planning Goals and Objectives

Goals and objectives provide specific guidance to the transportation planning process. The importance of goals and objectives lies in the fact that community attitudes with respect to transportation and the type of community desired could be different from the image of the community held by planners. Also, attitudes and desires can change over time, potentially invalidating many of the explicit and implicit assumptions inherent in previous planning studies. Thus, goals and objectives become an important means of guiding the planning process in the directions desired by the community.

The development of a goals set, however, can be hindered by difficulties in defining what exactly is meant by values, goals, objectives, measures of effectiveness (or criteria), and standards and in understanding the interrelationship between them. For the purpose of this discussion, these terms will be defined as follows [Wachs and Schofer, 1969; Thomas and Schofer, 1970]:

Values: Basic social drives that govern human behavior. They include the desire to survive, the need to belong, the need for order, and the need for security. Because values are assumed to be shared by most groups in a culture, one can speak of societal values.

Example: Need for order.

Goals: Generalized statements that broadly relate the physical environment to values, but for which no test for fulfillment can be readily applied.

Example: Maintain and/or improve the quality of transportation.

Objectives: Specific and measurable statements that relate to the attainment of goals.

> *Example:* Improve the reliability of the movement of persons and goods on the existing transportation system.

Measures of Effectiveness: Measures or tests that reflect the degree of attainment of particular objectives in the context of plan or project evaluation.

> *Example:* Degree of schedule adherence of bus trips.

Standards: Minimum acceptable level for the criterion measure (i.e., a fixed level of attainment of an objective).

> *Example:* The number of buses arriving more than 5 minutes late should not exceed 10 percent of the total bus trips on that route during an 8-hour period.

As can be seen from these examples, the degree of specificity increases as one proceeds from values to standards. One value can also lead to more than one goal, each goal can lead to one or more objectives, and the attainment of each objective can be judged with one or more measures of effectiveness. One objective could also satisfy different goals. For example, an objective of reducing travel costs for elderly and low-income persons could satisfy the achievement of two goals: improve the quality of transportation services and provide for equitable distribution of public services.

In general, the stated goals for transportation planning are similar from one metropolitan area to the next. These goals usually relate to accessibility, mobility, economic development, quality of life, environmental and resource conservation, safety, operational efficiency, and system condition and performance [Gayle, 1999; Cambridge Systematics and Meyer, 2000]. In order for a goals set to be useful for transportation planning and decision making, several criteria have to be met [JHK and Associates and Peat, Marwick, Mitchell and Co., 1977]:

- Goals and objectives must be clear, concise, unambiguous, and understandable to all actor groups.
- Objectives must logically follow from applicable goals.
- Goals and objectives must reflect the views, perceptions, and aspirations of the community.
- Each objective must be measurable by at least one measure of effectiveness (MOE).
- The MOEs must be measurable with reasonable effort.
- Goals and objectives must be developed independent of specific transportation plans and not be mode specific.

Table 4.14 illustrates the concept of a goals set as applied to the planning of a suburban transit service. Weights were assigned to each of the goals and objectives by a community task force that oversaw plan development. This goals set became the basis for evaluating alternative service configurations [Parsons et al., 1994].

Table 4.14   Goals set for a suburban transit plan

GOAL 1:   TRANSIT SERVICE IN COUNTY MUST BE SUPPORTIVE OF, AND BE FULLY INTEGRATED WITH, THE ECONOMIC GROWTH OF THE COUNTY.

Objective 1.1:   *Provide major employment centers with transit services that reflect the local and regional demand patterns for each center.*

    MOE 1.1.1:   Number of employment concentrations of 7,500 employees or more served by the transit alternative

    MOE 1.1.2:   Number of individual employment sites of 3,000 employees or more served by the transit alternative

    MOE 1.1.3:   Number of employment sites linked by transit service to other county activity centers

Objective 1.2:   *Provide transit service to special activity centers, other than major employment centers, which support the county's economy (e.g., major retail areas and convention center).*

    MOE 1.2.1:   Number of special activity centers served by transit service

    MOE 1.2.2:   Number of major transit markets connected by the transit alternative with special activity centers

Objective 1.3:   *Provide good connections to the regional transportation network that strengthens the county's position as a regional center of economic activity and that provide strong linkages to international marketplaces.*

    MOE 1.3.1:   Percent of trips in a transit service market area destined to outside the county that can be served by the transit alternative

    MOE 1.3.2:   Percent of trips destined to county in a transit service area that can be served by transit alternative

Objective 1.4:   *Relate transit service improvements to desired land-use patterns and characteristics.*

    MOE 1.4.1:   Degree to which transit service improvement reinforces adopted land-use plan

    MOE 1.4.2:   Percent higher density county population found within transit service area

    MOE 1.4.3:   Percent higher density county employment found within transit service area

GOAL 2:   TRANSIT SERVICE MUST PROVIDE MOBILITY OPTIONS FOR MAJOR TRANSIT MARKETS IN THE COUNTY.

Objective 2.1:   *Provide transit services that promote efficient trip making within the county and that interconnect with the regional transportation system and ultimately link to state, national, and international markets.*

    MOE 2.1.1:   Degree to which transit service serves major origin–destination patterns

    MOE 2.1.2:   Person-miles traveled on transit service

    MOE 2.1.3:   Person-miles per revenue vehicle mile for transit service

Objective 2.2:   *Provide transit services that are accessible to those with limited mobility.*

    MOE 2.2.1:   Degree to which limited mobility population live within transit service area

Objective 2.3:   *Provide transit services that are accessible to those with disabilities.*

    MOE 2.3.1:   Degree to which disabled population live within transit service area

    MOE 2.3.2:   Degree to which disabled have access to service

Objective 2.4:   *Provide transit service to special activity centers, other than major employment or retail centers (e.g., educational institutions).*

    MOE 2.4.1:   Number of special activity centers served by transit service

    MOE 2.4.2:   Number of major transit markets connected to the transit alternative with special activity centers

Table 4.14   Goals set for a suburban transit plan (continued)

Objective 2.5: *Integrate transit services with other aspects of transportation system, especially those that provide good access to transit services (e.g., sidewalks).*

   MOE 2.5.1: Degree to which transit service is integrated into total transportation system with associated transit-friendly infrastructure

GOAL 3: TRANSIT SERVICE MUST IMPROVE THE QUALITY OF THE ENVIRONMENT FOR ALL COUNTY CITIZENS AND CONTRIBUTE TO SOLVING REGIONAL ENVIRONMENTAL PROBLEMS.

Objective 3.1: *Target transit service investments in areas that will reduce roadway congestion and vehicle emissions.*

   MOE 3.1.1: Reduction of vehicle miles traveled due to service

   MOE 3.1.2: Reduction in vehicle emissions

   MOE 3.1.3: Reduction in congestion levels at key locations in service area

Objective 3.2: *Provide transit services that are safe, convenient, and affordable in order to attract automobile users.*

   MOE 3.2.1: Percent population served in market area

   MOE 3.2.2: Ratio of transit travel time to auto travel time for representative trip

Objective 3.3: *Provide good connections to the regional transit network to make long-distance transit travel attractive for these types of trips.*

   MOE 3.3.1: Percent of trips in a transit service market area destined to outside county

   MOE 3.3.2: Percent of trips destined to county in a transit service area that can be served

Objective 3.4: *Encourage policies in the public and private sectors that will promote high-occupancy vehicle use.*

   MOE 3.4.1: Degree to which success of alternative depends on proactive stance on single vehicle use reduction

GOAL 4: TRANSIT SERVICE MUST BE APPROPRIATE AND COST-EFFECTIVE FOR THE MARKETS SERVED.

Objective 4.1: *Provide transit service that meet goals 1, 2, and 3 in the most cost-effective manner possible.*

   MOE 4.1.1: Dollars per rider, initial capital and operating

   MOE 4.1.2: Dollars per emission-ton reduced

   MOE 4.1.3: Percent of project financing required by local sources

Objective 4.2: *Provide transit service that provides the greatest benefit to the citizens of the county.*

   MOE 4.2.1: Degree to which transit service serves major origin–destination patterns

   MOE 4.2.2: Number of employment concentrations of 7,500 employees or more served

   MOE 4.2.3: Number of employment sites of 3,000 employees or more served

   MOE 4.2.4: Number of employment concentrations connected to county activity centers

   MOE 4.2.5: Percent population served in market area

Objective 4.3: *Design transit services that meet the specific needs of the different markets in the county.*

   MOE 4.3.1: Degree to which transit serves major origin–destination patterns

   MOE 4.3.2: Degree to which transit service characteristics relate to perceived willingness of potential customers to use service

Objective 4.4: *Design transit services in a flexible way that preserves future service options.*

   MOE 4.4.1: Degree to which transit alternative permits future expansion or fits into other transit service options

SOURCE: Parsons et al., 1994

The effort transportation planners put into developing goals and objectives can vary according to the political, social, and demographic changes occurring in the urban area. In cases where little or no change is occurring, verifying that the previous goals set for the community represents its true desires could be a simple task. In cases where rapid growth and immigration are changing the basic characteristics of a community, a major effort might be needed to develop a new vision and corresponding goals and objectives. In either case, the effort of thinking about and defining such statements provides useful information to the planning process.

Several techniques have been used to collect this kind of information, including citizen advisory committees, newspaper mail-back coupons, public hearings, charrettes, surveys, transportation fairs, newsletters, focus groups, Internet surveys, and referenda [Arnstein, 1969; Hoover, 1994; Portland METRO, 1995; U.S. DOT, 1996; Lorenz and Ingram, 1999; Keever et al., 1999]. Through the use of these techniques, a consensus will usually develop around a set of goals and on their relative importance. A more extensive discussion of how goals, objectives, and measures of effectiveness can be used in the evaluation of alternative projects or systems is found in Chap. 8.

4.3.3   Market Research Information

Understanding consumer preferences and the characteristics of consumer demand is important information for all organizations providing products or services. This is certainly true for transit agencies and is increasingly being recognized as an important prerequisite for a customer focus in highway agencies as well. Market research is one way of obtaining this type of information. This approach identifies customer market characteristics or market segments and defines the attributes of transportation services that reflect the desires of these different groups. Market research can relate to many different kinds of decision making (Fig. 4.7). Figs. 4.8 and 4.9 show the results of a study that examined how transit agencies were using market research techniques in agency activities [Elmore-Yalch, 1998]. As shown, the most-used technique was onboard surveys, and the most frequent use of the data was for service operations planning.

Market research to determine what motivates a traveler's decision has occurred for many decades. For example, attitudinal surveys conducted since the 1960s have shown fairly consistent findings on what travel characteristics are most important to travelers [Charles River Associates, 1997]. Travel time, reliability, personal comfort, and convenience have consistently ranked high. Out-of-pocket costs usually ranked low in importance (note, however, that those interviewed were usually those traveling and thus presumably were able to afford transportation) [Paine et al., 1967; Abt Assocs. 1968; McMillan and Assael, 1969].

For the purpose of this discussion, a market segment will be defined as a subset of the population having specific characteristics distinguishing it from other population groups. Criteria for selecting market segments can include demographic and socioeconomic characteristics (e.g., age, income, sex, race, occupation, and life-cycle stage), attitudes (e.g., lifestyle and personality), use of existing services (e.g.,

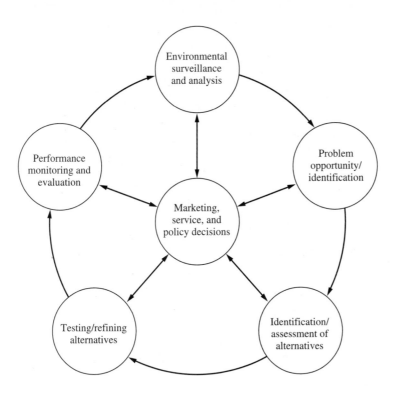

Figure 4.7　Market and customer research for decision making
| SOURCE: Elmore-Yalch, 1998

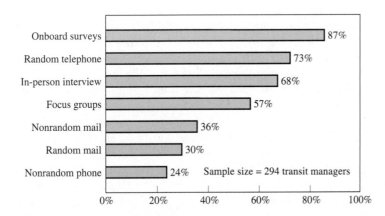

Figure 4.8　Market research techniques used in transit agencies
| SOURCE: Elmore-Yalch, 1998

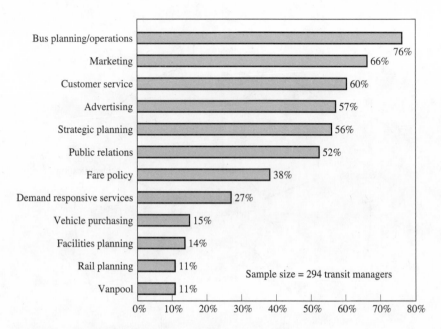

Bus planning/operations — 76%
Marketing — 66%
Customer service — 60%
Advertising — 57%
Strategic planning — 56%
Public relations — 52%
Fare policy — 38%
Demand responsive services — 27%
Vehicle purchasing — 15%
Facilities planning — 14%
Rail planning — 11%
Vanpool — 11%

Sample size = 294 transit managers

0% 10% 20% 30% 40% 50% 60% 70% 80%

Figure 4.9   Decisions informed by market research methods
I   SOURCE: Elmore-Yalch, 1998

auto versus transit), and perceptions of different services and preferences (e.g., sensitivity to price and importance of service attributes). In defining market segments, it is important to determine those characteristics that affect both the ability and willingness of individuals within selected population groups to use transportation services. For transit work commute trips, for example, likely markets include workers with low incomes and/or no car, workers with college education, minority groups, workers aged 17 to 29, women, and immigrants [Rosenbloom, 1998].

Once market segments are defined, the analyst can determine what service characteristics are most attractive to potential users. Market research data on perceptions and preferences for different services are often gathered by using questionnaires based on psychological scales (Fig. 4.10). The most common form is the Likert scale in which the respondent reacts to a strongly worded statement about a service attribute or policy goal by indicating level of agreement or disagreement on a 5- or 7-point scale. Although easily administered and readily understood by respondents, the Likert scale measures attributes only on an ordinal, rather than a cardinal scale. That is, the planner cannot infer from the results the magnitude of importance of service attributes as identified by the respondents. For example, planners noticing a movement from "extremely poor" to "very poor" on item (e) in Fig. 4.10 could conclude that driver friendliness is increasing, but one could not conclude that a movement from "extremely poor" to "poor" meant twice the improvement in friendliness as the movement from "extremely poor" to "very poor" [Urban and Hauser, 1980]. However, such scales do provide useful information to decision

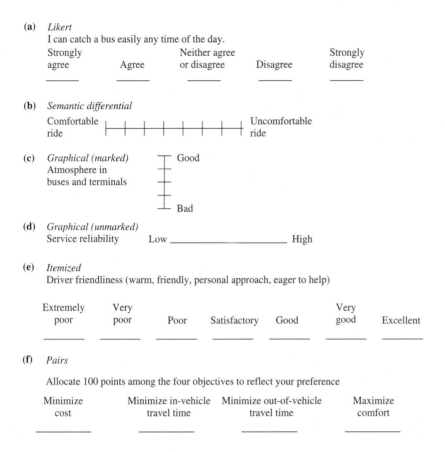

(a) *Likert*
I can catch a bus easily any time of the day.

Strongly agree	Agree	Neither agree or disagree	Disagree	Strongly disagree
_____	_____	_____	_____	_____

(b) *Semantic differential*
Comfortable ride ├─┼─┼─┼─┼─┼─┤ Uncomfortable ride

(c) *Graphical (marked)*
Atmosphere in buses and terminals
Good — Bad

(d) *Graphical (unmarked)*
Service reliability   Low _____ High

(e) *Itemized*
Driver friendliness (warm, friendly, personal approach, eager to help)

Extremely poor	Very poor	Poor	Satisfactory	Good	Very good	Excellent
_____	_____	_____	_____	_____	_____	_____

(f) *Pairs*

Allocate 100 points among the four objectives to reflect your preference

Minimize cost	Minimize in-vehicle travel time	Minimize out-of-vehicle travel time	Maximize comfort
_____	_____	_____	_____

Figure 4.10   Example of interview rating scales
SOURCE: Urban and Hauser, 1980

makers on how well existing and new services perform with respect to specific attributes or how specific locations compare to one another based on perception.

The process of turning scale measures into numerical values is called *scaling*. Approaches to scaling attitude measures range from relatively complicated methods, such as factor analysis or linear regression, to the development of a scoring function. For the scoring function method to be used, the respondents must specify not only their perceptions of service attributes, but also the weight or importance attached to each attribute. Such information can be obtained for a sample of individuals and then used to derive an initial indication of the importance of various attributes and the feasibility of different alternatives.

Market research differs from traditional transportation data gathering in that it seeks opinion, attitudes, and preferences from consumers, along with the more customary behavioral and socioeconomic data. Most importantly, market research provides useful diagnostic information on consumer perceptions. This type of information is important when transportation improvements are designed to achieve objectives that are not easily measured (e.g., the use of transportation investment to enhance the retail "climate" of downtown areas).

In summary, market research techniques are useful tools for transportation planners, especially given the service orientation of transportation. Service-oriented actions, the focus of much of transportation planning, can be effective only when efforts have been made to understand consumer behavior and preferences. Market research can provide transportation planners and managers with (1) a model of how consumers *process information* to form perceptions of transportation alternatives; (2) explicit measures of *consumer perceptions* of each transportation alternative; (3) identification and measures of *consumer feelings,* such as biases toward specific modes, personal expectations, and perception of societal norm; (4) measures of the *relative importance* of perceptions and feelings as they influence consumer preferences toward transportation alternatives; and (5) an understanding and measurement of how situational constraints, such as availability, combine with preference to *influence behavior,* such as choice of transportation mode [Hauser et al., 1981]. Because the validity of the information obtained from market research methods is dependent on the effectiveness of the tools used, transportation planners should be extremely careful in the design and use of techniques such as mail/telephone surveys and focus groups [Comsis Corp. et al., 1992; Hagler Bailly and Morpace International, 1999].

4.4 MONITORING TRANSPORTATION PERFORMANCE: THE FEEDBACK LOOP

Comprehensive system monitoring is an important part of the transportation planning process. A monitoring program can be designed to identify where *problems* occur (or are likely to occur) in the transportation system, where *opportunities* exist for improving the effectiveness and efficiency of current services even though they might not be related to identifiable problems, and how well the transportation *program goals* are being achieved.

4.4.1 Measures to Diagnose Problems and Identify Opportunities

One of the most effective ways of identifying problems in system operations is to use a set of measures or indicators that show areas of deficiency or possibilities for improvement. Such measures should relate to the goals and objectives established previously for the planning process or service operation. For example, a transit agency having an objective of providing the most cost-efficient service might adopt a measure of net cost per revenue passenger for monitoring the economic performance of individual routes. Those routes that do not meet a certain *performance standard* could be considered "problem" routes and thus become candidates for more detailed analysis (Fig. 4.11).

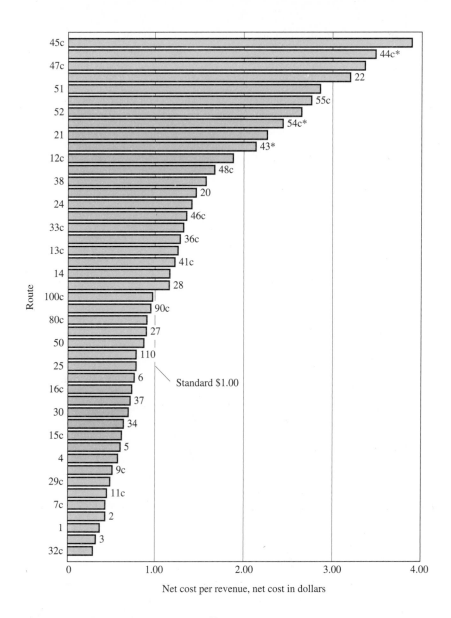

Figure 4.11   Diagnostic measurement of bus routes

A problem with this approach is that such standards are often developed in relationship to one objective. This unidimensional problem identification thus does not reflect the many objectives served by transportation. For example, one explanation for a low revenue/cost ratio of a transit route might be a high percentage of elderly riders who pay lower fares. Performance standards should therefore be used only as a screening tool to identify candidate routes meriting further analysis. Another approach to overcoming this problem is to develop a multidimensional ranking scheme that identifies problem routes along several different dimensions.

Similarly, highway agencies concerned with traffic congestion or accidents might use a series of diagnostic measures to identify problem sites. Transportation officials have been interested in such measures for many years. As urban road networks expanded and became more congested during the 1960s, measures of congestion were needed to determine if system performance was improving or deteriorating over time. Suggested measures of congestion related to

Operational characteristics of traffic flow, which included speeds, delays, and overall travel times.

Volume to capacity characteristics, which compared actual volumes to theoretical capacity.

Freedom of movement characteristics, which required a determination of the percentage of vehicles restricted from free movement and the durations of such restrictions.

Performance measures based on these types of factors have dominated professional practice for decades. Coded network maps were often used to indicate the locations of problem areas (e.g., the average speed or volume-to-capacity ratio of each segment, high accident locations, or areas of high-pollutant emissions). This systematic approach to problem identification, although undertaken by planners for many years, was required formally for the first time by ISTEA and TEA-21 for those metropolitan areas over 200,000 population not in attainment of air-quality standards. Each metropolitan area was required to develop a congestion management system, a process for identifying congestion and mobility problems in a region's transportation system and for selecting strategies to address them. Performance or diagnostic measures used to identify such problem areas were the foundation of successful congestion management systems.

Regardless of which diagnostic measures are chosen for problem identification, several issues should be kept in mind:

1. The required data must be collected on a periodic basis to allow updating of the problem identification process.
2. Many measures are related to one another, meaning surrogates can be used to identify closely related problems.
3. Standards used to identify the level of system or facility performance above (or below) which the performance is considered problematic must be carefully defined to relate to the problems being faced by the organization or community.
4. Diagnostic measures should be related to the planning and agency objectives.
5. Diagnostic measures only identify where problem areas exist; they do not indicate what types of corrective actions might be required.

Although not used as often as the approach for problem identification, the process of identifying *opportunities* for system improvement is likely to receive increased attention in future years as budget limitations make increasing the efficiency of existing public services an attractive alternative to large capital expenditures. Increasing the efficiency of such services could include implementing innovative service characteristics, restructuring management control systems, or using different funding arrangements. The reason for taking such actions is not to solve a problem, but rather a perception that by taking such actions, some improvements can be made in system operation. In some sense, identifying opportunities for improvement reflects the adoption of a proactive perspective on planning rather than a reliance on the reactive process of responding to problems.

Transit service planning is a good example of a planning process that can systematically examine the opportunities for improving service characteristics [Wilson and Gonzalez, 1982]. In this process, general actions are identified that can be applied to any part of the transit network during any time period. These generic actions, along with the conditions of bus routes that might suggest their applicability, are shown in Table 4.15. In such instances, data on system performance and trends in travel behavior are collected as part of a regional monitoring program [Meyer, 1980]. The task of the planner is to identify the operating conditions of transit routes and pinpoint those routes where there is potential for implementing a generic action.

Table 4.15   Generic actions and appropriate route conditions

Generic Action	Route Conditions
A. Holding strategy	Schedule adherence problem Long route Point on route with low through ridership
B. Increase running layover time	Schedule adherence problem Low loads
C. Increase frequency	Unacceptable crowding Moderate, rather than high ridership Even load profile
D. Decrease frequency	Low productivity and loads Time between arrivals below policy levels
E. Split route	Low productivity Uneven load profile Long route
F. Short turns	Tapering load profile Long route High ridership
G. Express or zonal service	High ridership Tapering load profile Long route Large time differentials local and express zone

Table 4.15   Generic actions and appropriate route conditions (continued)

Generic Action	Route Conditions
H. Partial returning of bus empty	Large imbalance in flows Large time differential in service and empty running time High frequencies
I. Eliminate route segment	Low ridership generation on segment Vehicle savings possible from elimination Higher frequency possible from elimination
J. Eliminate trips	Low ridership on trips High cost savings from elimination

SOURCE: Wilson and Gonzalez, 1982

There are two principal advantages of this process over the problem-centered approach. First, a wider set of actions than those directed toward resolving specific problems can be considered. Second, some routes not classified as problems will still be the subject of planners' attention. Any improvements made to these routes will potentially increase system efficiency and effectiveness.

4.4.2   Measures to Evaluate System Performance

The diagnostic measures discussed previously can be used in service or site-specific planning. However, transportation officials are often concerned with the overall performance of the transportation system, especially with its linkage to such things as economic development and environmental quality. The foundation of such planning is a set of performance measures that relate characteristics of system operation to program goals and objectives. Similar to the measures of effectiveness used in the analysis and evaluation of selected transportation actions, performance measures can range from those directed toward system operations (such as average speed) to the much broader consequences of system performance (such as ecosystem health and equitable distribution of benefits and costs). This type of planning is referred to as performance-based planning [Meyer, 1993, 1995; Cambridge Systematics and Meyer, 2000].

Performance-based planning exhibits several key characteristics [Meyer, 1995]:

System performance linked to fundamental roles of transportation: The measure of whether the transportation system is performing as expected should relate to a broad perspective on what role transportation plays in a metropolitan area. Congestion on individual links in the network does not say much about how the system performance is affecting quality of life, economic development, or environmental quality. Performance-based planning must thus consider a broader range of issues than just operational efficiency of the modal networks.

Outcomes as well as outputs: Initial experience with performance-based planning suggests that agencies measure success by the level of output produced. For example, number of lane-miles maintained or constructed or number of revenue bus-hours provided shows how productive an agency can be. These *are* important indicators of the amount of service provided in a region. However, in keeping with the characteristic described previously, *outcome* measures are also important indicators of system performance. Outcome measures relate to the ultimate effect of the transportation system on a community, such as quality of life, environmental health, equitable distribution of benefits and costs, economic development, and safety/security. Outcome measures should be part of the performance-based planning process.

Mobility and accessibility: Providing individual mobility and accessibility to urban activities is an important goal for transportation planning. Many MPOs have defined measures that indicate the degree to which the transportation system is providing acceptable levels of performance. However, in both cases, measures of mobility and accessibility beg the question, mobility and accessibility for whom? The distributional effects of transportation investment on different socioeconomic groups and on different geographic areas of a metropolitan region strongly suggest that performance-based planning should be based on a market segmentation approach that identifies existing and future travel markets, as well as who will benefit and who will pay for changes to mobility and accessibility.

Multimodal performance measures: Performance-based planning focuses on the ability of people and goods to achieve desired travel objectives and does so without modal bias (in fact, in a society substituting actual trip making with telecommunications, a mode of transportation in a traditional sense might not be needed to satisfy the objectives). Performance measures should thus not be modally biased. One of the ways of doing this is to focus on generic characteristics of trip making, such as travel time, and on the total trip experience of the traveler or goods mover. Bottlenecks in the system, and thus delay to the user, can often occur at access, egress, or transfer points that most likely will not be under the control of the agency responsible for the line-haul portion of the trip. Defining performance measures from a total trip perspective will provide opportunities for identifying these congestion points.

Performance measures tied to project evaluation criteria: Given that performance measures reflect what decision makers consider to be important indications of system success, they should be closely tied to the evaluation criteria used to select among plan alternatives and projects. This relationship becomes an important system performance linkage to the stated purpose of transportation investment. If job creation has been identified by decision makers as an important performance measure for system impact, then the evaluation of plan and project alternatives should have such a criterion as well.

Strategic data collection and management plan: The success of performance measurement relies heavily on the availability of data. For example, many of the MPOs responding to the federal requirement to establish a congestion management system developed performance measures that could be defined only with existing data. This was especially true for smaller MPOs, which did not have the resources to pursue a new and expensive data collection effort. A critical element of performance-based planning is thus the development of a strategic data collection and management plan. The term *strategic* implies this plan should encompass the entire spectrum of data that needs to be collected, which agencies will be the source of such data, and the frequency of data collection. Section 4.5, which follows, will discuss such an effort in more detail.

New data management and analysis techniques: The technology of data collection and management is evolving, with techniques used today that were unavailable several years ago. Video and machine vision recognition of vehicular movement, aerial and satellite photography, automatic vehicle identification, instrumented vehicles, and advance passenger information systems could be very useful in providing the data necessary to conduct performance-based planning. New analysis tools such as geographic information systems (GIS) have made some performance measures much easier to estimate and thus feasible in the context of providing information to decision makers. The best example of this is the use of GIS to estimate accessibility measures (e.g., how many people can access x square feet of retail space within a certain travel time). As the evolution in analysis tools allows transportation planners to become more sophisticated in their analysis efforts, this sophistication can relate to performance measurement as well.

The usefulness of performance measures depends on several characteristics of the measures themselves [Cambridge Systematics, 1980]:

1. *Measurability* requires that the data be available and that the tools exist to perform any required calculations.

2. *Pertinence* relates to the degree to which performance measures reflect the policies or objectives for which they were developed.

3. *Clarity* implies that the measure should be easily understood by planners and decision makers.

4. *Sensitivity and responsiveness* indicate the level of change that can occur in the transportation or activity systems and still be detected by the performance measure.

5. *Appropriate level of detail* addresses the issue of whether the measure is specified at a level of detail applicable to its intended use.

6. *Insensitivity to exogenous factors* requires that the performance measure not be influenced by nontransportation events that could distort a true indication of performance.

7. *Comprehensiveness* means the degree to which the performance measure can indeed measure across all the market segments and locations for which it is intended to be used.

8. *Discrimination between influences* assesses the degree to which one can differentiate among individual components affecting the performance of a system.

In some cases, performance measures cannot totally satisfy these characteristics. For example, it is often difficult to separate the effect of changes in the transportation system from the general impacts of the state of the economy. However, these characteristics do provide a useful checklist for the development of an initial set of performance measures that can be modified as experience dictates. An illustrative list of performance measures is shown in Table 4.16. As shown, these measures are linked to the types of goals that are often part of a transportation planning process.

Table 4.16   Example performance measures for different goals

---

Accessibility

Average travel time from origin to destination	Number of bridges with vertical clearance less than x feet
Average trip length	Percent of population within x minutes of y percent of employment sites
Accessibility index	
Mode split by region, facility, or route	Percent of region's mobility impaired who can reach specific activities by public transportation
Percent of employment sites within x miles of major highway	

Mobility

Origin–destination travel times	Frequency of transit service
Average speed or travel time	Mode split
Vehicle-miles traveled (VMT) by congestion level	Transfer time between modes
Lost time or delay due to congestion	Customer perceptions on travel times
Level of service or volume/capacity ratios	Delay per ton-mile
Vehicle-hours traveled per capita or VMT per capita	Person-miles traveled per capita or per worker
Person-miles traveled (PMT) per vehicle mile traveled	Person-hours traveled
Percent transit on-time performance	Passenger trips per household
	Percent walking or using bike by trip type

Economic Development

Economic costs of crashes	Jobs created or supported (directly and indirectly)
Economic cost of lost time	Percent of region's unemployed or low income that cite transportation access as a principal barrier to seeking employment
Percent wholesale/retail/commercial centers served with unrestricted (vehicle) weight roads	

Quality of Life

Lost time due to congestion	Customer perception of safety and urban quality
Accidents per VMT or per PMT	Average number of hours spent traveling
Tons of pollution generated	Percent of population exposed to noise above threshold

Table 4.16   Example performance measures for different goals (continued)

Environmental and Resource Consumption

Overall mode split or by facility or route
Tons of pollution
Number of days in air-quality noncompliance
Fuel consumption per VMT or per PMT

Sprawl—difference between change in urban household density and suburban household density
Number of accidents involving hazardous waste

Safety

Number of accidents per VMT, per year, per trip, per ton-mile, and per capita
Number of high accident locations
Response time to incidents
Accident risk index
Customer perception of safety

Percent of roadway pavement rated good or better
Construction-related fatalities
Accidents at major intermodal (e.g., railroad crossings)
Pedestrian/bicycle accidents

Operating Efficiency (system and organizational)

Cost for transportation system services
Cost/benefit measures
Average cost per lane-mile constructed
Origin–destination travel times
Average speed

Percent projects rated good to excellent in quality
Volume to capacity ratios
Cost per ton-mile
Mode split
Customer satisfaction

System Preservation

Percent of VMT on roads with deficient ride quality
Percent roads/bridges below standard condition
Remaining service life
Maintenance costs
Roughness index for pavement
Service miles between road calls for transit vehicles
Vehicle age distribution

Several efforts have been made to develop system-level indexes that can be used to monitor performance over time. Measures of mobility are of particular interest to the planning profession. Table 4.17 shows proposed mobility measures that could be applied at the metropolitan level [TTI, 2000]. Note that travel time plays a leading role in almost all of these measures. It is also of interest that one of the measures, the reliability factor, attempts to represent that characteristic of system performance, reliability, of most interest to system users.

With ITS technologies, the assessment of system performance can be undertaken in a real-time basis and this assessment given to individual travelers for their consideration in travel decisions. Figure 4.12, for example, shows an assessment of the expected travel time that could be expected in the transportation system. The travel times are based on the collection of travel time data from 240 surveillance cameras on the Atlanta freeway system. The different bands represent on average how long it will take to get from the location indicated, the dot in the middle, to

Table 4.17 Example mobility measures

Individual Measures

Travel rate (minutes per mile) $= \dfrac{\text{Travel time (minutes)}}{\text{Segment length (miles)}} = \dfrac{60}{\text{Average speed (mph)}}$

Delay rate (minutes per mile) $= \begin{array}{c}\text{Actual travel rate} \\ \text{(minutes per mile)}\end{array} - \begin{array}{c}\text{Acceptable travel rate} \\ \text{(minutes per mile)}\end{array}$

Relative delay rate $= \dfrac{\text{Delay rate}}{\text{Acceptable travel rate}}$

Delay ratio $= \dfrac{\text{Delay rate}}{\text{Actual travel rate}}$

Corridor mobility index $= \dfrac{\text{Passenger volume (persons)} \times \text{Average travel speed (mph)}}{\text{Optimum facility value* (person-mph)}}$

*125,000–freeways
25,000–streets

Travel rate index

$$= \dfrac{\left(\dfrac{\text{Travel rate}}{\text{Freeflow rate}} \times \text{Peak period VMT}\right) \times \left(\dfrac{\text{Travel rate}}{\text{Freeflow rate}} \times \text{Peak period VMT}\right)}{\left(\text{Freeway peak period VMT} + \text{Principal arterial street peak period VMT}\right)}$$

(Freeway) (Principal arterial street)

Reliability factor (RF_{10}) $=$ Percentage of the time that a person's travel time is no more than 10% higher than average

---

Total Mobility Measures

Accessibility (to opportunities) $=$ The sum of the number of jobs, shops, or other travel objectives that are within the acceptable travel time for each origin

---

Total delay (vehicle minutes) $=$ [Actual travel time (minutes) – Acceptable travel time (minutes)] $\times$ Vehicle volume (vehicles)

Congested travel (person miles) $=$ Sum of all [Congested segment length (miles) $\times$ Person volume]

Congested roadway (miles) $=$ Sum of all congested segment lengths (miles)

SOURCE: TTI, 2000

different locations in the metropolitan area. Such information could be widely available in the coming years.

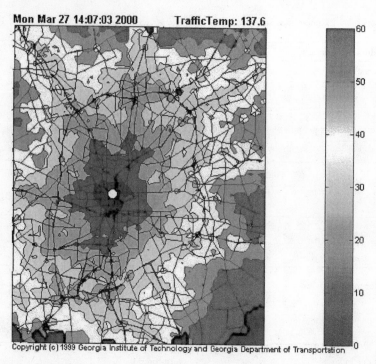

Mon Mar 27 14:07:03 2000   **TrafficTemp: 137.6**

Copyright (c) 1999 Georgia Institute of Technology and Georgia Department of Transportation

Figure 4.12   Real-time travel time estimates for Atlanta

The most challenging data collection effort for monitoring system performance is that needed to assess the performance of newly implemented services or facilities. Not only is it necessary in such cases to identify the outcome or impact of a new program or project, but it is also important to ascertain the reasons behind the outcome. Thus, the data collection efforts of many evaluation studies include activities to gather data on the physical changes caused by the new project, as well as on individuals' attitudes and behavior that might help to explain the results of new service implementation or facility construction. In such evaluation studies, it is extremely important to have an experimental design that identifies the type of information to be collected, describes the techniques to be used, and outlines where and when such data are to be collected. Most importantly, this design should carefully determine what data should be collected before a project is implemented so that useful comparisons can be made before and after project implementation.

4.5   DEVELOPING A DATA MANAGEMENT PLAN

Because a good database is essential for effective planning, designing a data management plan for a metropolitan area is an important role for transportation planning. Table 4.18, for example, shows the type of data collected for the development of the transportation plan in Milwaukee, Wisconsin. Figure 4.13 shows the different data sources for the use of travel forecasting models in Portland, Oregon. In both cases, the data to accomplish the required tasks had to be anticipated, data collection planned for, and budget resources allocated.

A data management plan should not only outline the method for, and frequency of, data collection, it should also identify which agency is responsible for what data [Meyer, 1980]. Given advances in database management (e.g., distributed systems and geographic information systems), it is important for a region to have a coordinated approach toward data management in order to gain the greatest efficiencies from these technologies. Figure 4.14 shows a proposed classification structure for data collection and storage. As noted in the report proposing this structure,

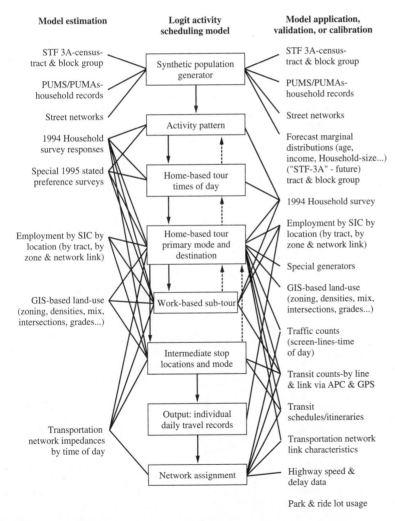

Figure 4.13   Data inputs for model system development in Portland, Oregon
SOURCE: Correspondence with Keith Lawton, Portland METRO

Table 4.18   Data collection strategy for plan development in Milwaukee

Inventory of Transportation Facilities and Network Utilization	Inventory of Transportation Movement and Behavioral Factors Affecting Travel Habits and Patterns
Highway facilities and service levels	Screen-line survey
Transit facilities and service levels	Home interview survey
Transportation terminal facilities	Truck and taxi survey
Automobile availability	External survey
Truck availability	Mass transit survey

SOURCE: SEWRPC, 1994

The manner in which data are grouped affects the efficiency and stability of the planning process. Efficiency is affected because practitioners rely on timely access to information in the development of plans and projects. Stability is affected because the time and cost of collecting data, as well as the need for systematic and reliable monitoring over time, work against constant modification of databases [Faucett, J. and Assocs., 1997].

Faced with limited data collection budgets, transportation planners will need to look very carefully at new and innovative ways to use existing data sources and to minimize the number of expensive data collection efforts. A data management plan

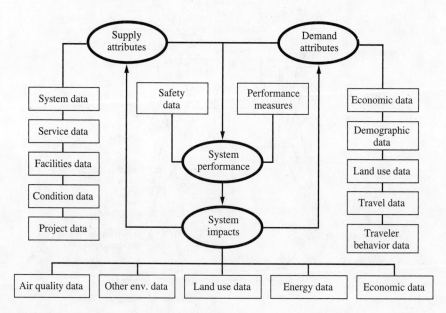

Figure 4.14   Hierarchy in data classification
SOURCE: Faucett, J. and Assocs., 1997

should provide a schedule of data collection activities over a specified period, identify likely unmet data needs, establish priorities among these needs, determine the level of resources to be devoted to each of these needs, and estimate the cost of the data collection efforts on an annual basis. With respect to annual cost estimates, the level of resources needed will depend on the type of data collected, the techniques used, and the sample size required. The cost of different types of data collection techniques will vary tremendously; however, the usefulness of these techniques also varies, with the most costly data collection techniques tending to provide much higher response rates. Planners should thus examine ways of combining the data collection activities that can be carried out under differing levels of funding. In addition, as noted earlier, ITS technologies provide important opportunities to "piggyback" data collection activities on top of system operations and control functions.

Table 4.19 shows a data collection and management plan adopted by the MPO in Albany, New York, to support its performance-based planning process. The program includes the data collection activities of other agencies as well as the desired frequency of collection.

Table 4.19   Data management program in Albany, New York

Data Item	Current Frequency	Desired Frequency
Traffic Volumes and Classification		
Freeway segments	3 years (some continuous)	Continuous
Cordon and screen lines	3 years	3 years
Other arterials	3 years	3 years
Other collectors	As needed	3 years
Local roads	As needed	As needed
Major signalized intersections	As needed	3 to 4 years
Other intersections	As needed	As needed
Roadway Characteristics		
Number of lanes, width	No schedule	When changes occur
Traffic control	With intersection counts	With intersection counts
Changes in bicycle/pedestrian accommodation	Not collected	When changes occur
Pedestrian Counts	With intersection counts	With intersection counts
Travel Speed		
Speed and delay; major arterials	As needed	As needed plus every 5 years
Frequency and extent of incident delays	Not collected	Daily
Vehicle Occupancy		
Cordon and screen lines	No schedule	2 years
Department of Motor Vehicles accident data	1991–1993 available	Annually

Table 4.19   Data management program in Albany, New York (continued)

Data Item	Current Frequency	Desired Frequency
Transit Ridership		
By route, bus trip	Daily	Daily, summarize annually
By demographic group	No schedule	3 years
Upstate transit by route, trip	Daily	Daily, summarize annually
Other transit by route	Unknown	Daily, summarize annually
Park and Ride Lot Usage		
Vehicles at designated lots	Not collected	Annual field survey
Persons per arriving vehicle	Not collected	Annual field survey
Commuter Register Usage (Carpool Matching)		
New commuter register entries	Monthly	Monthly
Carpools formed	2 months	2 months
Average carpool trip length	2 months	2 months
Carpool longevity	2 months	2 months
Arterial Management		
Traffic signal spacing, major arterials	Partial inventory	When changes occur
Driveway spacing (frequency), major arterials	1994–1995 inventory	When changes occur
Land Use Changes		
Building permit activity	Monthly	Monthly
New development	Not collected	Annually
New pedestrian/transit-oriented development	Not collected	Annually
New development with access management	Not collected	Annually
Closing or elimination of activity	Not collected	Annually
Goods Movement		
Truck volumes as percent of travel	With intersection counts	With intersection counts
Journey-to-Work Information		
Mode of trip	10 years (census)	5 years
Time of day of trip	10 years (census)	5 years
Origin-destination information	10 years (census)	5 years
Demographics of commuter	10 years (census)	5 years
Travel Behavior, All Purposes		
Mode of trip	20+ years	10–15 years
Trip purpose	20+ years	10–15 years
Time of day	20+ years	10–15 years
Occupancy	20+ years	10–15 years
Trip chaining	20+ years	10–15 years
Demographics of traveler	20+ years	10–15 years

SOURCE: CDTC, 1995b

The feasibility of any data management plan will depend on the available budget. With limited funds, an assessment will have to be made of essential data needs as well as those needs considered nonessential but potentially useful. A budget would most likely reflect the assessment of data needed to update planning models, to monitor system operation, to determine achievement of performance objectives, and to take advantage of complementary data from other sources (e.g., a census).

Although it is certainly preferable to have a comprehensive data management plan that includes all types of transportation data, it is often difficult within the institutional framework of most cities to coordinate such activities. Agencies collect data for their own use with little concern for what other agencies are doing. In such a context, it becomes important to at least develop a data management strategy for similar types of data collected by different agencies. Such a strategy should match the different objectives for collecting traffic counts with agency responsibilities and approaches to collecting data.

In sum, the technical database is a critically important component of an effective transportation planning process. Data are collected by a large number of agencies in a metropolitan area and reflect a variety of decisions made by agency and political officials. Such officials are not only concerned about the impact of new services and facilities, they are also interested in the transportation problems likely to occur in the future, the performance of the existing system, and the degree to which policy objectives are attained. A careful examination of how the existing database is providing the needed information for these issues is an important task of transportation planning. The types of techniques used, the level of accuracy desired, and the scheduling of data collection activities over time thus become important concerns for planners. As noted by Wachs,

> Transportation databases, information systems, and analytical models interact with one another and change over time as our understanding of transportation systems and their social and economic contexts evolve. New understandings both shape and are shaped by the data and models we use [Wachs, 2000].

With the rapid changes in society brought about by new technology and with a broadening of transportation-related concerns, this observation is likely to be even more relevant in the future.

4.6   CHAPTER SUMMARY

1. The origin–destination surveys and traffic volume counts used for years by highway planners were supplemented in the 1960s by more extensive data collection efforts, notably home interview surveys and transportation–land-use inventories. As a result, the database in most metropolitan areas today includes many different data items collected by numerous agencies and serving diverse purposes. This database is useful in the application of travel behavior models and in monitoring the performance of the transportation system so that problems can be identified and the degree of attainment of objectives assessed.

2. The collection and processing of data in urban areas are often done on the basis of spatial units of analysis. Traffic zones are defined according to one or more criteria and can in turn be aggregated into corridors of travel or districts of the metropolitan area. Also of importance is the functional classification of urban roads, transit services, and other modes of transportation.

3. The use of *sampling methods* in data collection enables planners to select a small percentage of the entire population or available database for closer examination. Procedures designed to help the planner select a sample representative of the characteristics of the targeted population include *simple random* sampling, *sequential* sampling, *stratified random* sampling, and *cluster* sampling. No matter which technique is employed, the cost of survey sampling must be weighed against the degree of accuracy required in the data collected.

    Closely related is the concept of *sample size,* which also influences the accuracy of the survey. Choosing an appropriate sample size involves making assumptions about how the elements to be surveyed are actually distributed, as well as determining the desired limits of error for the sample. Based on the specifics of a given situation, one of the equations described in Appendix B can then be used to calculate the necessary sample size.

4. Of the many data collection techniques in use, *household travel behavior surveys* are the most detailed and most costly. Personal in-home interviews, telephone interviews, and mail-back questionnaires can all be used to collect data on the demographic characteristics and trip-making behavior of individual households. Each instrument has both advantages and disadvantages with respect to cost, complexity, survey bias, and response rate; thus, planners should use a strategy that maximizes the usefulness of the data collected. *Panel* and *stated preference surveys* are used for special data-gathering activities; the first collects data from the same individuals over a long period of time, and the second collects attitudinal data on transportation characteristics that do not yet exist.

    Origin–destination (O–D) surveys are used to identify the movement of persons and vehicles in an urban area, helping planners to estimate travel demand on existing and planned transportation facilities. Home interview surveys, roadside interviews, postcard surveys, and license-plate surveys can be used to collect O–D data.

5. The physical characteristics of land use and transportation systems can be cataloged by undertaking *inventories* of highway assets or transit system capability. Asset management systems are being developed that use much of this inventory data and estimate future needs based on underlying deterioration relationships. Such systems will become more important in future years as metropolitan areas begin to manage their assets in a more cost-effective manner.

6. *Highway and transit counts* can supplement the data collected in regional inventories and can be done through either manual or automatic methods. *Cordon counts* estimate the total number of vehicle movements within a specific geographical area, while *screen-line counts* validate the accuracy of the cordon

counts by accounting for every vehicle crossing a specified boundary over a given time period. Traffic counts may also be used for more specific purposes, such as determining the level of congestion at a particular intersection or the number of passengers making use of a particular transit route.

7. Freight transportation data are becoming more important as inputs into the transportation planning process. However, such data are often difficult to collect because of their proprietary nature. In such cases, planners use secondary sources.

8. Transportation technology has changed dramatically over the past 20 years, such that many of the techniques used today for data collection and analysis have only recently been applied in practice. Intelligent transportation system (ITS) technologies offer great promise as a means of collecting continuous, real-time data. This data can be used not only for planning, but also for system monitoring and control.

9. Because the planner's role is to work *with* interested officials and representatives of the public, information is required concerning the desires and attitudes of the community in general. The creation of a community *vision* and of *goals* and *objectives* for transportation planning can help guide planners in analyzing and evaluating proposals. Clear statements of the goals and objectives for transportation planning are especially important in situations where community attitudes toward transportation tend to change rapidly. The effort of defining such statements can ensure that the type of community envisioned by the public coincides with the objectives pursued by transportation planners.

10. The attitudes and preferences of consumers can be measured through the use of *market research* techniques that identify distinct *market segments* and the transportation service attributes that reflect the preferences of each segment. Questionnaires containing attitudinal questions are a common method of gathering market research data. These questions can involve the use of a *psychological scale,* which requires the respondent to specify the strength of a particular attitude or perception. *Scaling* is the process of transforming these scale measures into numerical values. It can involve complex statistical techniques, such as factor analysis or linear regression, or more simple approaches, such as the development of a weighted scoring function to relate responses to different service attributes. These methods can determine the relative importance of different attributes from the responses of a sample of individuals. Market research techniques are therefore important in understanding consumer preferences and behavior, particularly with a focus of transportation planning on service-oriented transportation actions.

11. The data collected by transportation planners can be used to identify *problems* in the transportation system and *opportunities* for making improvements. *Diagnostic performance measures* can be used to locate areas of deficiency and to provide an indication of the overall performance of the transportation system. Diagnostic measures help the planner identify the specific services or facilities that do not meet some minimum *performance standard.*

The identification of *opportunities* for system improvement is likely to grow in importance as planners focus on increasing the efficiency of existing services in the face of budgetary constraints. The development of a framework that matches observed conditions in the transportation system with possible improvement strategies is one approach to making opportunity identification more systematic.

In situations where transportation officials are more concerned about the overall performance of the transportation system and the degree of attainment of goals and objectives, performance measures can relate system characteristics to specified planning goals. For such measures to be useful to planners, they should possess several characteristics, including measurability, clarity, sensitivity and responsiveness, and comprehensiveness.

12. Because transportation problems are closely related to public perceptions of transportation system performance, public involvement should be an integral part of a planning program's data collection effort. Public participation can incorporate the views of diverse interests into the formal planning process. Public meetings and surveys of community attitudes can give planners an indication of problem priorities that could be used for allocating planning resources and as input into the definition of community goals and objectives. Public involvement, then, is an important source of information that can be used to complement the data collected by planners.

13. The importance of a good database for effective urban transportation planning indicates the need for the development of a data collection and management plan that details the data to be collected, the collection methods, as well as agency responsibilities. Such a plan should schedule data collection activities over a 5- to 10-year period, identify any unmet data needs and establish priorities among them, determine the level of resources to be devoted to each of the needs, and estimate the annual costs to be incurred.

QUESTIONS

1. For an urban transportation plan with which you are familiar, identify the major data items likely used in developing the plan. What data collection techniques were likely used in obtaining this information?

2. Summarize how the types of data collected in urban transportation planning changed with the transition in such planning discussed in Chap. 2. What general social and political trends have directly affected the type of data collected?

3. Obtain a copy of the latest census information for your urban area. What data found in the census results are directly useful for urban transportation planning? How does the scale of analysis affect the usefulness of this census information?

4. Given three origins and destinations *a, b,* and *c,* and the estimated frequency of trips shown in the table on the next page, determine the required sample size

with a precision for all matrix cells of ±10 percent and a 95 percent confidence interval. Use Appendix B to make these estimates.

		From	
To	a	b	c
a		0.05	0.20
b	0.15		0.10
c	0.15	0.05	

5. For the types of trips shown in Fig. P4.5, identify the techniques that could be used to collect relevant trip data. For those trips internal to, or entering or leaving, the "tight cordon," identify the techniques that could be used to collect data on transit trips.

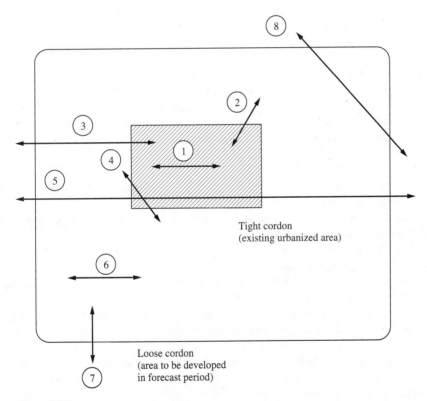

Tight cordon
(existing urbanized area)

Loose cordon
(area to be developed
in forecast period)

Figure P4.5

6. Assume that a new shared-ride taxicab service is to be implemented in your urban area. You have been asked by the city planning director to outline a detailed data collection effort to identify what impact such service will have on urban travel, especially with respect to trips by elderly persons. Further, your

budget for data collection has been limited to $40,000. Describe a before-and-after data collection plan that meets both the needs of the study and the specified resource constraints. How would your plan change if the budget was doubled?

7. Develop a goals and objectives statement for a particular transportation planning process in your urban area. What measures of effectiveness can be used to measure attainment of the goals and objectives you identified?

8. Develop a questionnaire that can be applied for a specific transportation issue in your community. What types of information do you want to collect? Why? How many questionnaires do you want to send out?

9. Identify diagnostic measures that can be used in your community to pinpoint transportation problem spots. What effort would be needed to collect the required data to use these measures? How important do you think such information is in the everyday activities of transportation agencies?

REFERENCES

Abt Associates, Inc. 1968. *Qualitative aspects of urban personal travel demand: A report of the U.S. Department of Housing and Urban Development's study in new systems of urban transportation.* Cambridge, MA: Abt Associates, Inc.

American Association of State Highway and Transportation Officials. 1992. *AASHTO guidelines for traffic data programs.* Washington, D.C.

_____.1994. *A policy on geometric design of highways and streets.* Washington, D.C.

Ames, S. (ed.). 1993. *Guide to community visioning.* Portland, OR: American Planning Association, Oregon Chapter.

Arnstein, S. R. 1969. A ladder of citizen participation. *Journal of the American Institute of Planners.* July.

Atlanta Regional Commission. 2000. *Transportation solutions for a new century. Vol 1. 2025 regional transportation plan.* Draft. Atlanta, GA. Jan.

Attanucci, J., I. Burns, and N. H. M. Wilson. 1981. *Bus transit monitoring manual.* Vol. 1. Report UMTA-IT-09-9008-81-1. Washington, D.C. Aug.

Baass, K. G. 1981. Design of zonal systems for aggregate transportation models. *Transportation Research Record 807.* Washington, D.C.: National Academy Press.

Baltimore Metropolitan Planning Organization. 1995. *Growth management and transportation: Alternative futures for the Baltimore region.* Baltimore, MD. Nov. 28.

Boyle, D. 1998. Passenger counting technologies and procedures. *TCRP Synthesis of Transit Practice 29.* Washington, D.C.: National Academy Press.

Bureau of Transportation Statistics. 1999. *Transportation statistics annual report.* Washington, D.C.: U.S. Department of Transportation.

Cambridge Systematics, Inc. 1980. *Performance measures and travel behavior inventory study. Final report. Phase 1—Development of performance measurements.* Prepared for Metropolitan Council, St. Paul, MN. Dec. 31.

_____. 1996a. *Data collection in the Portland, Oregon metropolitan area case study.* Report DOT-T-997-09. Washington, D.C.: U.S. DOT. June.

_____. 1996b. *Scan of recent travel surveys.* Report DOT-T-97-08. Washington, D.C.: U.S. DOT. June.

_____. 1996c. *Travel survey manual.* Washington, D.C.: U.S. Department of Transportation and U.S. Environmental Protection Agency. July.

_____. Apogee, Inc., and Jack Faucett, Inc. 1995. *Intermodal freight transportation, Volume 1.* Report DOT-T-96-04. Washington, D.C.: Federal Highway Administration. Dec.

_____ and Barton Aschman. 1996. *Scan of recent data research.* Report DOT-T-97-07. Washington, D.C.: U.S. DOT. Sept.

_____ and M. Meyer. 2000. *Performance-based planning manual.* NCHRP Report 8-32(2). Washington, D.C.: National Academy Press.

Capelle, R. 1999. Commodity flows and freight transportation. In J. Edwards (ed.) *Transportation planning handbook.* Washington, D.C.: Institute of Transportation Engineers.

Capital District Transportation Committee. 1995a. *New visions for Capital District transportation.* Albany, NY. Dec.

_____. 1995b. *Congestion management systems development.* Task 2 report. Albany, NY. Dec.

Center for Urban Transportation Research. 1996. *Demonstration of video-based technology for automation of traffic data collection: Travel time, origin–destination, average vehicle occupancy.* Tampa, FL: University of South Florida. Jan.

Charles River Assocs. 1997. Building transit ridership, An exploration of transit's market share and the public policies that influence it. *TCRP Report 27.* Washington, D.C.: National Academy Press.

Cochran, W. G. 1977. *Sampling techniques.* 3d ed. New York: John Wiley.

Comsis Corp., M. Meyer, K. Bhatt, and R. Pratt. 1992. *Effective travel demand management measures: Guidance manual, Marketing research for transportation demand management.* Washington, D.C.: Federal Highway Administration. Nov.

Crabtree, L. and G. Krause. 1982. Vehicle origin survey. *Transportation Research Record 886.* Washington, D.C.: National Academy Press.

Dilman, D. A. 1978. *Mail and telephone surveys—The total design method.* New York: John Wiley.

Doherty, S. et al. 1999. Moving beyond observed outcomes: Integrating global positioning systems and interactive computer-based travel behavior surveys. *Proceedings of the TRB Conference on Personal Travel: The Long and Short of It.* Washington, D.C.: Transportation Research Board. July.

Elmore-Yalch, R. 1998. A handbook: Integrating market research into transit management. *TCRP Report 37.* Washington, D.C.: National Academy Press.

Ewing, R. 1997. *Transportation and land use innovations.* Chicago, IL: Planners Press.

Faucett, J. and Assocs. 1997. Guidance manual for managing transportation planning data. *NCHRP Report 401.* Washington, D.C.: National Academy Press.

Federal Highway Administration. 1970. *Guide for traffic volume counting manual.* Transmittal 96. Washington, D.C. March.

_____ and American Association of State Highway and Transportation Officials. 1997. *Asset management: Advancing the state of the art into the 21st century through public–private dialogue.* Report FHWA-RD-97-046. Washington, D.C.

_____. 1998. *Gender issues in transportation.* Washington, D.C.

_____. 1999. *Asset management primer.* Report FHWA-JF-00-010. Washington, D.C. Dec.

Ferlis, R. 1980. *Guide for estimating urban vehicle classification and occupancy.* Report DOT-FH-11-9249. Washington, D.C.: U.S. DOT. March.

Ferlis, R., L. Bowman, and B. Cima. 1981. *Guide to urban traffic volume counting.* Report FWHA-PL-81-091. Washington, D.C.: Federal Highway Administration. Sept.

Furth, P. 2000. Data analysis for bus planning and monitoring. *TCRP Synthesis of Transit Practice 34.* Washington, D.C.: National Academy Press.

Gayle, S. 1999. Urban transportation studies. In J. Edwards (ed). *Transportation planning handbook.* Washington, D.C.: Institute of Transportation Engineers.

Grant, A. and A. Lemer (eds.). 1993. *In our own backyard.* Washington, D.C.: National Academy Press.

Hagler Bailly, Inc. and Morpace International, Inc. 1999. Guidance for communicating the economic impacts of transportation investments. *NCHRP Report 436.* Washington, D.C.: National Academy Press.

Hauser, J. R., A. M. Tybout, and F. S. Koppelman. 1981. Consumer-oriented transportation service planning: Consumer analysis and strategies. *Applications of Management Science* 1: 91–138.

Highway Research Board. 1944. *Proceedings of the twenty-fourth annual meeting of the Highway Research Board.* Washington, D.C.

Hoover, J. 1994. Post-Intermodal Surface Transportation Efficiency Act public involvement. *Transportation Research Record 1463.* Washington, D.C.: National Academy Press.

Hudson, R. and S. Hudson. 1994. Pavement management systems lead the way for infrastructure management systems. *Proceedings Third International Conference on Managing Pavements.* Washington, D.C.: Transportation Research Board. May.

_____., R. Haas, and W. Uddin. 1997. *Infrastructure management.* New York: McGraw-Hill.

JHK and Associates and Peat, Marwick, Mitchell and Co. 1977. *Basic TSM goals and objectives: Working paper no. 3.* Prepared for Federal Highway Administration. Washington, D.C. June.

Kalfs, N. et al. 1997. *New data collection methods in travel surveys. Activity-based approaches to travel analysis.* Oxford: Pergamon Press.

Karimi, H., J. Hummer, and A. Khattak. 2000. Collection and presentation of roadway inventory data. *NCHRP Report 437.* Washington, D.C.: National Academy Press.

Keever, D., G. Frankoski, and J. Lynott. 1999. In the possibilities are the solutions: Assessment and implications of the public involvement process during the environmental impact study of the Woodrow Wilson Bridge. *Transportation Research Record 1685.* Washington, D.C.: National Academy Press.

Kish, S. and M. Meyer. 1996. Developing management systems for statewide transportation planning: Early lessons. *Transportation Research Record 1552.* Washington, D.C.: National Academy Press.

Kyte, M., A. Khan, and K. Kagolanu. 1993. Using machine vision (video imaging) technology to collect transportation data. *Transportation Research Record 1412.* Washington, D.C.: National Academy Press.

Lau, S. 1995. *Truck travel surveys: A review of the literature and state-of-the-art.* Oakland, CA: Metropolitan Transportation Commission.

Levy, D. and L. Lawrence. 1991. *The use of automatic vehicle location for planning and management information.* STRP Report 4. Toronto: Canadian Urban Transit Association.

Lorenz, J. and R. Ingram. 1999. It's not just for projects anymore: Kansas Department of Transportation's innovative, agency-wide public involvement program. *Transportation Research Record 1685.* Washington, D.C.: National Academy Press.

Mahmassani, H., T. Joseph, and R-C Jou. 1993. Survey approach for study of urban commuter choice dynamics. *Transportation Research Record 1412.* Washington, D.C.: National Academy Press.

McLeod, D. 1996. Integrating transportation and environmental planning: Extending applicability of corridor and subarea studies and decisions on design concept and scope. *Transportation Research Record 1552.* Washington, D.C.: National Academy Press.

McMillan, R. and H. Assael. 1969. National survey of transportation attitudes and behavior. *NCHRP Report 82.* Washington, D.C.: National Academy Press.

McNeil, S. 1996. *Asset management: State-of-the-art, issues and practice.* Prepared for the executive seminar on asset management. Washington, D.C.: Federal Highway Administration. Aug. 20.

Metropolitan Transportation Commission. 1998. *1998 regional transportation plan: Draft environmental impact report.* Oakland, CA. Aug.

Meyer, M. 1980. Monitoring system performance: A foundation for TSM planning. *Special Report 190.* Washington, D.C.: National Academy Press.

_____. 1993. Jumpstarting the start toward multimodal planning. *Special Report 237.* Washington, D.C.: National Academy Press.

_____. 1995. *Alternative performance measures for transportation planning: Evolution toward multimodal planning.* Report FTA-GA-26-7000. Washington, D.C.: Federal Transit Administration.

Mid-Ohio Regional Planning Commission. 1994. *Transportation infrastructure improvement study for central Ohio inland port program.* Columbus, OH.

Miller, K., T. Harvey, P. Shuldiner, and C. Ho. 1993. Using video technology to conduct 1991 Boston region external cordon survey. *Transportation Research Record 1412.* Washington, D.C.: National Academy Press.

Multisystems, Inc. 1985. *Transit data collection design manual.* Report DOT-I-85-38. Washington, D.C.: Urban Mass Transportation Administration. June.

North Central Council of Governments. 1996. *2010 mobility plan update: The regional transportation plan for North Central Texas.* Arlington, TX.

Okunieff, P. 1997. AVL systems for bus transit. *TCRP Synthesis of Transit Practice Report 24.* Washington, D.C.: National Academy Press.

Paine, F., A. Nash, S. Hille, and G. Brunner. 1967. *User determined attributes of ideal transportation systems.* College Park, MD: University of Maryland Department of Business Administration.

Parsons, Brinckerhoff, Quade, and Douglass, 1994. *Multimodal strategic plan for public transportation in Cobb County.* Final report. Atlanta, GA. Sept.

_____. 1995. Guidelines for development of public transportation facilities and equipment management systems. *TCRP Report 5.* Washington, D.C.: National Academy Press.

Pedersen, N. and D. Samdahl. Highway Traffic Data for Urbanized Area Project Planning and Design. *National Cooperative Highway Research Program Report 255.* Washington, D.C.: Transportation Research Board. Dec.

Polak, J. and P. Jones. 1995. Using stated preference methods to examine traveler preferences and responses. In *Understanding travel behavior in an era of change.* Oxford: Pergamon-Elsevier.

Portland Metro. 1995. *Local public involvement policy.* Portland, OR. July.

Pratt, R. 1991. Collection and application of ridership data on rapid transit systems. *NCTRP Synthesis of Transit Practice 16.* Washington, D.C.: National Academy Press.

Puget Sound Regional Council. 1990. *Vision 2020.* Seattle, WA. Oct.

Quiroga, C. and D. Bullock. 1999. Travel time information using GPS and dynamic segmentation techniques. *Transportation Research Record 1660.* Washington, D.C.: National Academy Press.

Rosenbloom, S. 1998. Transit markets of the future. *TCRP Report 28.* Washington, D.C.: National Academy Press.

Rutherford, S. 1994. Multimodal evaluation of passenger transportation. *NCHRP Synthesis of Highway Practice 201.* Washington, D.C.: National Academy Press.

Sarasua, W. and M. Meyer. 1996. New technologies for household travel surveys. *Conference Proceedings 10.* Washington, D.C.: National Academy Press.

Sen, X., C. Wilmot, and T. Kasturi. 1998. Household travel, household characteristics, and land use: An empirical study from the 1994 Portland activity-based travel survey. *Transportation Research Record 1617.* Washington, D.C.: National Academy Press.

Smith, S. 1999. Guidebook for transportation corridor studies: A process for effective decision-making. *NCHRP Report 435.* Washington, D.C.: National Academy Press.

Southeast Wisconsin Regional Planning Commission. 1994. *A regional transportation system plan for southeastern Wisconsin, 2010.* Waukesha, WI. Dec.

Stopher, P. and C. Stecher. 1993. Blow up: Expanding a complex random sample travel survey. *Transportation Research Record 1412.* Washington, D.C.: National Academy Press.

Texas Transporation Institute. 2000. *Urban mobility study. Keys to estimating mobility in urban areas.* College Station: TX.

Thomas, E. N. and J. L. Schofer. 1970. Strategies for the evaluation of alternative transportation plans. *National Cooperative Highway Research Program Report 96.* Washington, D.C.: National Academy Press.

Transportation Research Board. 1994. Characteristics of bridge management systems. *Transportation Research Circular 423.* Washington, D.C.: National Academy Press. April.

_____. 1997a. Multipurpose fare media: Developments and issues. *TCRP Research Results Digest. No. 16.* Washington, D.C.: National Academy Press.

_____. 1997b. Decennial census data for transportation planning: Case studies and strategies for 2000. *Conference Proceedings 13.* Washington, D.C.: National Academy Press.

_____. 1997c. Information needs to support state and local transportation decision making into the 21st century. *Conference Proceedings 14.* Washington, D.C.: National Academy Press.

U.S. Department of Transportation. 1973. *Urban origin–destination surveys.* Transmittal 143. Vol. 20. Appendix 34. Washington, D.C.

_____. 1995. *Traffic monitoring guide.* 3d ed. Washington, D.C.: Federal Highway Administration. Feb.

_____. 1996. *Community impact assessment, A quick reference for transportation.* Report FHWA-PD-96-036. Washington, D.C.: Federal Highway Administration. Sept.

U.S. Government Accounting Office. 2000. *Highway infrastructure: FHWA's model for estimating highway needs is generally reasonable, despite limitations.* Report GAO/RCED-00-133. Washington, D.C. June.

Urban, G. L. and J. R. Hauser. 1980. *Design and marketing of new products.* Englewood Cliffs, NJ: Prentice-Hall.

Wachs, M. 2000. New expectations for transportation data. *TR News.* No. 206. Washington, D.C.: Transportation Research Board. Jan/Feb.

_____ and J. L. Schofer. 1969. Abstract values and concrete highways. *Traffic Quarterly.* Jan.

Wagner, D. 1997. *Lexington area travel data collection test, GPS for personal travel surveys.* Final Report for OHIM, OTA, and FHWA. Washington, D.C. Sept.

Wilson, N. and S. Gonzalez. 1982. *Methods for service design.* Paper presented for a workshop on short-range transit operations, planning, and management. Atlanta, GA. March 7–10.

Wolf, J. 2000. *Using GPS data loggers to replace travel diaries in the collection of travel data.* Unpublished Ph.D. dissertation. School of Civil and Environmental Engineering. Georgia Institute of Technology. Atlanta, GA.

Yen, J-R, H. Mahmassani, and R. Herman. 1994. Employer attitudes and state preferences toward telecommuting: An exploratory analysis. *Transportation Research Record 1463.* Washington, D.C.: National Academy Press.

Zimkowski, M., R. Tourangeau, and R. Ghadialy. 1997. *Introduction to panel surveys in transportation studies.* Report DOT-T-98-03. Washington, D.C.: U.S. DOT. Oct.

_____ and S. Pedlow. 1997. *Nonresponse in household travel surveys.* Report DOT-T-98-4. Washington, D.C.: U.S. DOT. Oct.

chapter

5

Demand Analysis

5.0   INTRODUCTION

Estimating the demand for transportation facilities and services is one of the most important analysis tasks in urban transportation planning. This demand includes not only passenger travel, but also the movement of goods. Section 5.1 discusses the role of demand analysis in the transportation planning process. Section 5.2 defines some basic terms and concepts that underlie demand analysis. Section 5.3 introduces a major approach to estimate demand, the use of simple techniques, including trend analysis, elasticity-based and pivot-point techniques, and various types of manual techniques (e.g., nomographs, work sheets). In general, such techniques employ a number of limiting assumptions that simplify the problem under consideration to the point where it can be analyzed by using simple calculations typically performed manually or with a spreadsheet program.

A second major approach to the analysis of transportation demand involves the use of a relatively standardized set of models that have evolved over the last 40-plus years of transportation planning—the so-called urban transportation modeling system (UTMS). Although originally developed, and probably best suited, for long-range, comprehensive planning, this system of models has been used in one form or another in a wide range of planning applications. The terminology and structuring of the travel demand question employed by UTMS (i.e., the "paradigm" of demand analysis it represents) permeate the entire field of urban transportation demand analysis, regardless of specific problem contexts or actual analysis methods employed. For both of these reasons, no discussion of travel demand analysis would be complete without a description of UTMS. Such a description is presented in Sec. 5.4.

While UTMS has been extensively employed within the transportation planning field for over 40 years, it has also been seriously criticized from many points of view for almost the same length of time. Most fundamentally, UTMS is not behavioral in nature; that is, it is not based in any real sense on a coherent theory of travel behavior. This characteristic, in turn, results in certain inconsistencies within the modeling system. For example, zone-to-zone travel times assumed by one component of the system may not be the same as those assumed by a subsequent system component.

Since the 1970s, two broad streams of model research and development have been undertaken to improve upon UTMS. One is the development of *individual choice* or *random utility* models. Random utility models have evolved to a point where they represent an extremely powerful and technically sophisticated tool for modeling a wide variety of decision processes (travel related and otherwise). The basic features of the random utility modeling approach are introduced in Sec. 5.5. The second, somewhat related but ultimately quite distinct, stream of model development has been in the area of *activity-based* models. These are discussed in Sec. 5.6. Sections 5.7 and 5.8 describe approaches that can be used to estimate the demand for two transportation movements that have been largely ignored over the past decades, nonmotorized travel and goods movement, respectively.

5.1   THE ROLE OF DEMAND ANALYSIS IN TRANSPORTATION PLANNING

Whether conducting a regional transportation planning study or examining the likely transportation impacts of a new development site, estimating expected travel demand at some future date is a critical point of departure for transportation planning. Considerable research has focused on how to best describe and predict travel demand and in conjunction with system performance, to develop demand forecasting methods that provide realistic estimates of future travel flows on transportation networks. Without demand analysis, transportation planning could not occur. However, developing realistic predictions of future travel is problematic. Figure 5.1, for example, shows the origins of trips in the Atlanta metropolitan area that used two links in the transportation network during a 12-hour period. As can be seen, the spatial dispersal of trip origins, which in a demand model would have to be replicated, creates significant methodological challenges.

Demand analysis consists of several tasks that are largely independent of the actual techniques used to predict travel flows. Figure 5.2 shows six basic tasks that are part of any demand analysis. Although depicted in a relatively linear fashion, the tasks involved are, in fact, highly interrelated. In particular, problem definition, choice of analysis technique, data collection, and model calibration are very much interconnected in a majority of cases. Data availability, for instance, often determines the analysis technique employed rather than the converse. Similarly, model calibration issues often help determine model functional forms and specifications. Figure 5.2 is thus a simplification of the iterative and circular nature of most demand analyses and is intended to illustrate some of the key aspects of the demand analysis process. Each of the major tasks shown in this figure is discussed in turn in the following subsections.

5.1.1   Problem Definition

Problem definition is largely determined outside of the demand analysis per se in that problems are usually identified in the diagnosis phase of the planning process. Such problem definitions, however, tend to be fairly general and abstract in nature, e.g., "examine short-term TDM strategies for relieving road congestion," or "assess long-term regional transit needs." In order to perform a demand analysis,

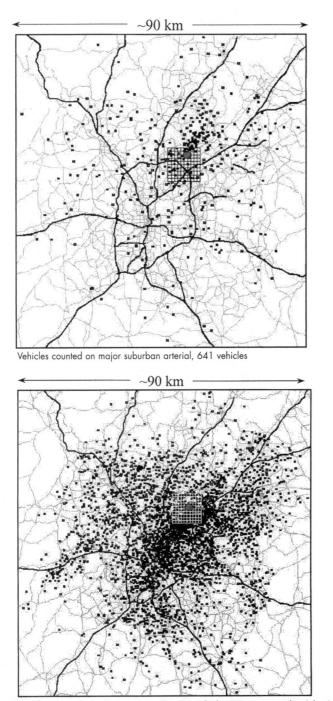

~90 km

Vehicles counted on major suburban arterial, 641 vehicles

~90 km

Vehicles counted on interstate ramp, 13,481 vehicles (10 percent within 3 kms)

Figure 5.1 Home location of vehicles observed during A.M. peak period, Atlanta

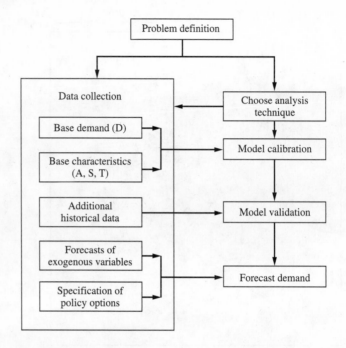

Figure 5.2 The demand analysis process

the problem definition typically needs to be made more specific. In particular, analysts must define, or have defined for them the following:

1.   *The specific planning period for the analysis.* The planning period is typically defined in terms of a base year and a horizon year. Many analyses provide only a horizon year forecast. If interim forecasts are also required (e.g., yearly ridership forecasts for the next 5 years), then it must be made explicit at the outset, given that this will have implications for the data requirements and, possibly, for the analysis techniques used.

2.   *The study region and zonal structure for the analysis.* The boundary of the study region, the zonal system within the study region, and whether trips crossing the study boundary to and from "external" zones are of interest, must all be defined. These decisions obviously depend on the particular problem context and, typically, on the existence of data for previously defined zone systems (census tracts, traffic analysis zones, planning district, etc.).

3.   *The temporal unit of demand.* Are demand estimates desired for the peak period, the typical day, a weekend day, and/or the year? The answer again depends on the problem context and data availability.

4.   *The policy variables.* What are the policy options available to the decision maker and how do these translate into observable and measurable variables? Clearly, data must be available to characterize the alternatives of interest in terms of

the relevant policy variables, and the demand technique used must be able to incorporate these variables if the analysis is to be useful in judging the relative merits of the various alternatives. For example, a demand model that does not include travel cost will not be of use in analyzing potential impacts of increasing energy costs or congestion pricing on urban trip patterns.

5. *The measures of effectiveness required for evaluation.* Just as the policy variables characterizing the policy options or alternatives can be viewed as "inputs" into the demand analysis, the output of the analysis is information on travel demand that relates to the impacts these alternatives will have on the urban system (e.g., average travel time within the system, volume–capacity ratios on major roadways, or annual transit ridership by route). Such effectiveness measures are then used within the evaluation phase to assist with the choice among the proposed alternatives (see Chap. 8). A particularly important consideration is whether this information is required for different market segments. That is, are systemwide totals or averages sufficient, or should these be broken down by socioeconomic characteristics, such as sex, income level, ethnicity, and age; or by geographic location, such as downtown or suburb, and so on, so that the *distribution* of impacts within the system can be assessed?

5.1.2   Choice of Analysis Technique

The choice of an analysis technique will be heavily influenced by the problem definition issues previously discussed—the analysis time frame, spatial scale, policy variables, and measures of effectiveness. A typology of "appropriate techniques" for various spatial scales and time frames is difficult to construct because a given technique can typically be applied at various spatial scales. Models that predict *individual* travel behavior can be employed at virtually any level provided that a suitable aggregation procedure is available. UTMS-type aggregate models are used in a wide range of planning activities, from assessing the immediate impacts of the opening of a new shopping center to estimating peak-period freeway flows over a 20- or 30-year horizon. As a generalization, two trends seem to exist. The first is that as the spatial scale and/or analysis time frame increases, so does the complexity and comprehensiveness of the analysis technique required to adequately address the planning problem. That is, as the "boundaries" of the problem expand, additional factors and relationships can no longer be considered to hold constant, and the analysis techniques and assumptions used must reflect this expansion of the problem context.

The other trend reflects the fact that, as spatial scale and time frame increase, so do the data, cost, and time requirements of the analysis. More fundamentally, the ability to observe detailed interactions, to hypothesize behavioral relationships, and to predict future system states decreases as spatial scale and time frame increase. Hence, a sort of social Heisenberg principle seems to exist in which as the need to be more comprehensive increases, the ability to do so decreases because of the volume, quality, and, ultimately, the availability of the data required.

5.1.3   Parameter Estimation

Virtually all transportation demand models consist of a dependent variable—the demand for transportation—expressed as a function of one or more independent or explanatory variables. If D denotes demand and X denotes the vector of explanatory variables, then the demand model can generally be represented by the equation

$$D = f(X,\theta) \qquad\qquad [5.1]$$

where θ is a vector of parameters, or coefficients (the two terms are used interchangeably within this book), that determine the "shape" of the demand curve and the relative "weights" of the various terms within the demand function. As will be discussed in subsequent sections, the elasticity value in a pivot-point analysis, the travel time exponent in a gravity model, and the utility function coefficients in a logit model are all examples of model parameters.

In the derivation of a demand model, the model parameters are assumed to be known. In actual fact, these parameters are not known; are likely to vary from city to city, as well as over time; and must be estimated for a given problem application from historical data. This estimation process, known as model calibration, involves a comparison of observed demand levels with the levels predicted by the model, given an assumed set of parameter values. These parameter values are then adjusted until the predicted demand levels match the observed demand as best as possible.

Statistical estimation procedures should be used whenever possible to determine model parameter values because of their rigor, efficiency, and the capability that they provide for making statistical statements about model goodness of fit and parameter significance. Occasionally, a model is developed that cannot be statistically estimated. One very common example of this is where graphical factors are used in place of a mathematical function to represent some system characteristic, for example, travel impedance.

A final note concerning parameter estimation is that given the dependence on historical data to calibrate a model, one is inevitably making the major assumption that the past weights, preferences, attitudes, and all other social, psychological, and spatial factors that somehow combine to determine the calibrated values of the model's parameters will remain constant over time. One can hope that the more behavioral or causal the model is, including, possibly, dynamic submodels of how attitudes, and so on, might change over time, the more fundamental or intrinsic these parameter values are likely to be and hence the less likely they are to change significantly over time. But this is ultimately only a hope that may or may not be realized.

5.1.4   Validation

Parameter estimation produces a model that best reproduces the historical data used in the estimation process. Before using this model to predict future demand, the analyst should be satisfied as best as possible that this model is, indeed, capable of predicting reasonably well. Testing the model's predictive capabilities is known as *validation*. At least three procedures exist for model validation, and each is discussed briefly in the following.

At a minimum, a demand model should be tested in terms of its reasonableness. Much of this sort of testing is performed during the calibration phase, when coefficients are examined for statistical significance and a priori expected signs (+ or −). For example, travel time and cost always have negative impacts on travel demand; that is, as travel time or cost go up, travel demand goes down. A demand model with a positive time or cost coefficient would not be considered a reasonable or valid model. Other tests are possible as well. Model outputs can be computed to see if they appear to lie within reasonable ranges; sensitivity tests can be performed on hypothetical data to explore the model's performance over expected ranges of input data, and so on. The objective within all such tests is simply to ensure (1) that the model does not violate theoretical expectations and, if it does, to try and determine which is "wrong"—the model or the expectations; (2) that it does not exhibit any pathological tendencies; and (3) that it is internally consistent (i.e., its predictions do not violate any assumptions used to generate them).

A far more rigorous test of the model's predictive capabilities involves using the model to predict demand for some time period other than that used for the calibration and then comparing this prediction with the demand actually observed during this second time period (using, for example, the same goodness-of-fit measures used during calibration). Whenever data for two or more time periods are available, this procedure provides the best test for the model.

Table 5.1 summarizes the results of one such validation exercise, in which the forecasts from 44 transportation planning studies in Great Britain for the time period 1962 to 1971 were compared with the actual observed system characteristics [Mackinder and Evans, 1981]. For each variable shown, the table indicates the number of observations (studies) used to compile the data, the average observed percentage increase in the variable 10 years after the study's base year, the average forecasted percentage increase for the same 10-year time horizon, and the average percentage error in these forecasts (all percentages are calculated relative to base-year values). It is clear from this table that these studies significantly overpredicted future system values, although this result is largely due to the overprediction of the land-use variables—population, employment, and so on—rather than of the transportation variables per se. This is perhaps not surprising given the relatively optimistic outlook among planners and others during the 1960s about future conditions—an outlook that did not include the staggering increases in the price of energy and the economic stagnation that were actually experienced in the 1970s.

The third, more restricted validation test that can be performed when multiple time period data are not available involves randomly splitting the one-period data set into two groups, using one group to calibrate the model and then using the calibrated model to predict the second group's demand. Thus, the model's predictive capability is validated against an independent set of observations, although the validation is clearly limited by the lack of temporal variation between the calibration and validation groups. This approach is particularly useful in the case of individual choice models in that one almost always has a large enough sample to split into two subsamples.

A major criticism of transportation demand modeling is the general lack of concern for, and effort put into, the validation phase of the analysis. To a certain extent,

Table 5.1   Comparison of predicted and observed changes for a sample of British transport studies

Forecast Item	Number of Observations	Average Observed Increase, %	Average Forecast Increase, %	Average Forecast Error, %	Root-Mean-Square Forecast Error, %
Population	39	5	15	10	14
Households	21	12	17	5	11
Employed residents	16	2	11	9	13
Employment	25	4	14	11	14
Cars per head	38	40	67	21	27
Household income	9	13	34	20	24
Highway trips	18	37	72	30	44
Public transport trips	11	−29	−4	35	41
Total person trips (private and public modes combined)	9	5	32	27	30
Screen-line trips	34	41	55	13	28
External trips	14	49	76	24	38
Through trips	8	17	59	49	76

SOURCE: Mackinder and Evans, 1981

this lack of concern is justifiable because of the time, budget, and data constraints that analysts typically experience in the course of any given demand analysis. Ultimately, however, the improvement in the predictive capabilities of transportation demand models and in the credibility of these models with decision makers, rests to a large extent on the analyst's ability to validate the procedures used. It can only be hoped that greater emphasis in the future will be placed on this phase of the analysis.

5.1.5   Forecasting

The final stage in demand analysis is the use of the calibrated and validated analysis technique to generate demand forecasts for each policy alternative and forecast year under consideration. This, of course, is the ultimate purpose of the whole demand analysis process. As such, the time and effort dedicated to this task should be commensurate with its importance within the overall analysis process. Unfortunately, this often does not prove to be the case: So much time and money are often spent in data collection and model development that very little is left for the forecasting phase. This problem was particularly typical of large-scale comprehensive modeling efforts, some of which never made it to the forecasting stage at all.

Ideally, one would like to generate a wide range of forecasts corresponding to a full set of possible alternatives and future scenarios. The sensitivity of the forecasts to key modeling assumptions should also ideally be examined. The extent to which this is actually done depends upon the problem under study, the budget and staff available to perform the work, and how successful the analysis team has been in saving budget and staff for the forecasting phase. The need to investigate a wide range of alternatives and scenario assumptions represents a major force underlying the movement toward the development of sketch-planning techniques, quick response methods, and so on, that are intended to eliminate or at least significantly reduce the time and effort spent on model development and model execution. Such techniques maximize the time and effort spent actually analyzing alternatives and investigating alternative scenarios.

5.1.6   Data collection

Two major types of data are needed for most demand analyses:

1. *Historical data* on travel behavior and the associated socioeconomic, activity system, and transportation system variables of interest.
2. *Forecasts* or specification of future values of socioeconomic, activity system, and transportation system variables required to predict future demand for each alternative under consideration.

Historical data are required for the calibration and validation phases of demand analysis. Chapter 4 dealt in some detail with data collection procedures and issues, and little more needs to be said at this point about the topic. Of greater immediate interest is the generation of data characterizing expected future system states required by the forecasting phase of the analysis. These data can be divided into two types: policy variables characterizing the alternatives under consideration and non-policy variables.

Policy variables, typically the transportation variables in the model and possibly some of the activity system variables, are not forecasted; they are simply specified as part of the process of defining the alternatives (e.g., links 400 to 425 inclusive will be a four-lane, controlled-access highway with a capacity of 2000 vehicles per lane per hour and a free-flow speed of 65 miles per hour). As such, they pose no special difficulty to the demand analyst. Far more problematic are the nonpolicy variables, such as the socioeconomic and activity system variables required by the demand model as exogenous inputs (e.g., population and employment by income level for each zone in the study region). Such variables typically must be themselves forecasted by some process. As with any forecasting exercise, the difficulty and complexity of this task depend on the forecast horizon year. In the very short run, the spatial distribution of people, buildings, and activities will change very little, and hence forecasting consists, at most, of minor updating of current conditions. In the very long run, all of these factors may well change dramatically in terms of their magnitude and spatial distribution, in which case forecasting may require an extensive and complex analysis of activity system trends and interactions. Chapter 6 presents an overview of land-use forecasting techniques. A major rationale for the continuing evolution and application of these models is to be able to use them to generate the detailed demographic and economic inputs required by current and emerging travel demand models.

5.2 ANALYSIS OF TRANSPORTATION DEMAND

Planners often use *models* of the transportation system and its relationship to socio-economic activities to analyze the consequences of changes to the system. A model is an abstraction and a simplification of a "real world" system. As such, it can be used to "experiment" with the system, to make conditional forecasts about what might occur within the system if certain changes to that system are introduced. In general, there are three basic assumptions in transportation system models that have significant impact on model validity. These are

1.   The key characteristics of the system may be specified or described in terms of a set of observable variables.

2.   There is an assumed explicit functional relationship between these observed variables and the observed behavior of the individuals. In other words, it is assumed that direct "cause-and-effect" relationships exist and that they may be at least approximately represented within the theory or model. It is this assumption that often prompts modelers to speak of "behavioral" models in that they believe the model truly "explains" behavior, that it captures the essence of the cause-and-effect interaction. It is more correct to say, however, that most models "explain" only in a statistical, correlative sense.

3.   It is assumed that the functional structure or nature of this relationship is the same for all individuals and is constant over time (or, if it does vary with time or across individuals, it does so in some specifiable way).

Given these assumptions, it is necessary to describe some of the theories or behavioral propositions that underlie many of the models of transportation demand that are found in practice today.

5.2.1 Economic Theory and Consumer Behavior

Some of the most important theoretical concepts that form the basis of demand analysis come from economic theory. In this section, four economic concepts that are especially important for transportation analysis—the theory of consumer behavior, derived demand, the supply curve, and equilibrium—are discussed. More detailed discussion of the derivation and further use of these concepts can be found elsewhere [Wohl, 1972; Mohring, 1976; Nicholson, 1978; Layard and Walters, 1978].

Consumer Travel Behavior   The basic premise of the theory of consumer behavior is that an individual will select a bundle of goods over all other affordable bundles if it yields the greatest *utility*, that is, satisfaction. Formally, the individual's decision-making process consists of maximizing a utility function U subject to a budget constraint, namely,

$$\max U = U(X_1, \ldots, X_n) \qquad \text{[5.2]}$$

$$\text{subject to } Y = P_1X_1 + \ldots + P_nX_n$$

where

$X_1, X_2, \ldots, X_n$ = goods that are consumed
$P_1, P_2, \ldots, P_n$ = prices of goods
Y = income

The solution to this problem can be illustrated graphically by using an indifference curve and a budget line. An indifference curve represents the various combinations of particular goods that will enable the individual to maintain a given level of utility. Thus, for the case of two goods, Fig. 5.3 characterizes the consumer's utility maximizing consumption of X_1 and X_2, given income level Y.

The utility maximizing values of X can, in general, be expressed in a functional relationship as

$$X_1^* = f(P_1, \ldots, P_n, Y)$$
$$X_2^* = f(P_1, \ldots, P_n, Y) \qquad \textbf{[5.3]}$$
$$X_n^* = f(P_1, \ldots, P_n, Y)$$

These functions are known as demand functions, and they denote the utility maximizing quantity of a good that an individual will purchase, given the prices of all goods that can be consumed and the individual's income.

The equilibrium level of consumption that was obtained for the goods in Fig. 5.3, denoted by E, has an important economic interpretation. A consumer has reached equilibrium in consumption when the individual's valuation of the goods is the same as the market's valuation. The consumer's valuation of the goods is represented by a trade-off among the commodities that he or she is willing to make along his or her indifference curve,[1] while the market's valuation of the commodities is represented by their price.[2]

Thus far it has been assumed that utility is simply a function of the quantity of the good(s) consumed. However, many, if not most, goods, including transportation services, are not consumed for their sheer quantity, but rather for their attributes [Lancaster, 1966]. That is, food is consumed for its nourishment and its taste; clothes for their comfort, aesthetic value, and durability; and so on. In other words, it is the attributes of goods that generate their utility, not the quantity of the goods per se. This implies that, in general, the demand for a good depends on its price, its characteristics relative to other goods, and the characteristics of the consumer who purchases the good.

In the case of transportation, the "good" being demanded is transportation services by various modes between points in space for particular purposes. The "price" of this good is generally taken to be not simply the monetary cost of the trip, but other costs perceived by the user as well, including the time spent traveling. Generally this cost is defined as the short-run or out-of-pocket or perceived cost of the trip rather than the long-run or "true" cost, on the assumption that tripmakers rarely, if

---

[1] This trade-off is technically referred to as the marginal rate of substitution.

[2] In technical terms, consumer equilibrium occurs in the two-good case when the marginal rate of substitution on the goods is equal to their relative prices.

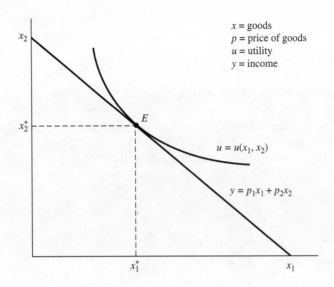

Figure 5.3 Consumer utility maximizing behavior

ever, use long-run costs in evaluating their travel options. If the monetary value of time is known, then time and price can be combined to yield a generalized cost of travel (other factors may be included also). This, however, is not a prerequisite for most transportation demand analyses, and it is very common for price to be represented by a vector of service measures, such as times and costs. The utility of a trip, and hence the demand for such a trip, depends upon a wide range of trip characteristics, characteristics of the available modes, and characteristics of the individuals making the trips.

Derived Demand   The concept of *derived* demand is the basis of most transportation demand analyses. That is, people do not travel for the sake of the traveling experience itself, the occasional "Sunday drive," bicycle ride, or walk in the park aside. Rather, they travel so as to participate in various activities located at their destinations. Homes, workplaces, stores, and schools all occupy finite amounts of space and are located at separate, discrete points in the urban landscape. The transportation system "mediates" between these locations of activity by physically connecting them and enabling people to move from one activity to another.

Two major implications can be drawn from the derived nature of transportation demand. First, no transportation demand analysis can be performed without explicitly considering the socioeconomic *activity system*—people and activities of various types and quantities distributed over space—that is served by the transportation system and that "generates" travel demand. This interaction between the activity and transportation systems, which is discussed at some length in Chaps. 3 and 6 (see, for example, Figs. 3.7 and 6.5), occurs both over the short and long term. Over longer periods of time, the accessibility provided by the transportation system can influence where people live and where economic activities occur. Thus, for example, forecasting travel demands over a 20- to 30-year time frame necessarily requires one to predict land-use patterns in the forecast year. In the short term, however, these land-use patterns are in place and serve as the basic representation of the locations of activities that will attract or generate trips. This chapter will focus on the analysis of transportation demand, given a particular activity system.

Second, because travel occurs for the sake of "getting there," it can be characterized in terms of the time, monetary cost, inconvenience, discomfort, and so on, associated with the trip. These characteristics represent the *disutility* or *generalized cost* of travel. That is, all else being equal, one would always presumably prefer to spend less time traveling, incur less expense, and be more comfortable. Further, and related to the concept of utility introduced in Sec. 5.2.1, one can speak of the utility associated with participation in the activity that generates the trip. Thus, it is reasonable to assume that travel decisions are based on the potential trip maker's assessment of the "pros and cons" or net utilities associated with the various travel modes that are part of the mobility options available. This concept of the utility of travel is an extremely powerful one and provides the starting point for a number of transportation demand modeling techniques.

Another characteristic of transportation demand relates to changes in transportation supply; for example, additional capacity or different pricing. When such changes occur, demand will shift in response. The term used in transportation planning to describe these shifts is *induced demand* when occurring over the long run and *induced traffic* when occurring over the short run.

The Supply Curve   The supply curve expresses the quantity of a given good (Q_s) that will be supplied or produced as a function of the price of the good, expressed mathematically as $Q_s = Q_s(P)$. The supply function is usually upward sloping (or at least nondecreasing) to the right, indicating that, all else being equal, greater quantities of goods will be produced only in the short run if the price of the good rises. This is because increased production is achieved in the short run through a more intensive use of existing capital (e.g., existing plant and equipment) and labor (e.g., production bonuses and overtime), meaning higher marginal operating costs must be incurred. In the long run, capital, labor, and production techniques can all change in order to reduce these marginal costs. Such a change in production will result in a change in the short-run supply function Q_s to a new curve Q_s', which will generally lie below and to the right of Q_s, representing the producer's ability to supply a greater quantity of the good at any given price, rather than a movement along the existing supply curve Q_s. On the other hand, a long-run supply curve, representing optimal adjustment of capital and labor at each level of production, is often found to be decreasing over a substantial range of output. Thus, as is the case with the demand function, the supply function ultimately depends on a range of factors other than just the good's market price, including the prices of the "input factors" (labor, materials, plant, and equipment) and the technology used to produce the good.

In the analysis of transportation services, *supply* is defined along at least three dimensions. The first of these is the concept of system *performance*, that is, the travel times, headways, and capacities that the transportation system provides for a given capital investment (representing the infrastructure and vehicles comprising the system), operating strategy, and demand level. As such, a performance function is an inverse supply function of the form $P_s = P_s(Q)$ where "price," P_s, is interpreted in a very generalized way (e.g., travel times and costs). The classic example of a performance function is the volume–delay curve for a section of roadway (discussed in Chaps. 3 and 7), in which the performance of the roadway—as measured by the travel time required to traverse it—is a function of the volume of flow.

Equilibrium   The point of intersection between the demand and supply curves is known as the *equilibrium point* (Fig. 5.4). At this point, the price of the good is such that $Q_D = Q_s = Q_E$; that is, the quantity demanded is equal to the quantity supplied. At equilibrium there is no "pressure" within the system for demand or supply to move to another price–quantity operating point.

Under most conditions, markets can be expected to move toward the equilibrium point, provided that shifts in the demand and supply curves do not occur. Note that if excess demand were to exist (i.e., $Q_D > Q_S$), then prices would likely rise as consumers "bid up" the price of the available goods. This, in turn, would stimulate both an increase in Q_S, because more producers would enter the market or those in the market would produce more intensively, and a decrease in Q_D, as fewer people would be willing to purchase the good at the higher prices, thus driving the market toward the equilibrium point. A similar argument can be developed for the excess supply case $Q_S > Q_D$.

Equilibrium is a fundamental concept in transportation systems analysis in that it is what links supply and demand. The equilibrium level of traffic, Q_E, must give rise to a level of performance, P_S, on the supply side that in turn gives rise to a level

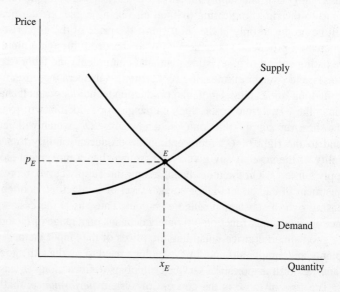

Figure 5.4 Equilibrium of supply and demand

of demand, Q_D, on the demand side (which itself equals Q_E). That is, while it is often convenient to take demand as fixed and exogenously given within an analysis of transportation supply (and vice versa within a demand analysis), ultimately this assumption is relaxed, and an equilibrium is established between demand and supply in order to achieve a description of the expected system state under a given set of operating conditions. This, of course, presupposes that transportation systems tend to be in a state of equilibrium or, at least, would arrive at such a state if left undisturbed for a sufficient period of time. Transient disequilibria clearly exist: a freeway traffic jam caused by an accident or a stalled vehicle is but one example of such a phenomenon. Nevertheless, the assumption that travel markets tend toward equilibrium positions has proved to be a very workable and useful hypothesis and forms the basis upon which all of the demand and supply analysis techniques discussed in Chaps. 5 and 7 are built.

5.2.2 Trip-Making Characteristics

Given a particular socioeconomic activity system, the demand for travel manifests itself in terms of *trips* made at given times by individuals from point to point within the urban area. These trips can be characterized in terms of a number of attributes or dimensions, including (see also Sec. 3.3)

1. The *purpose* of the trip (for individuals, this includes work, shop, social, etc.).
2. The *time of day* of the trip.
3. The *origin* of the trip.
4. The *destination* of the trip.
5. The *mode(s)* of travel used to make the trip (e.g., auto, bus, or bicycle for individuals; truck, rail, or intermodal for goods).
6. The *route* from origin to destination through the chosen mode's network.
7. The *frequency* (i.e., number of trips per unit time) with which such trips are made.

A trip is implicitly defined as consisting of movement between a single origin and a single destination for a single purpose. A *trip end* is, therefore, simply defined as one end (i.e., an origin or a destination) of a trip. In addition to their purposes, modes, and so on, person trips are usually also defined as being *home based* (if they either begin or end at home) or *nonhome based* (if they neither begin nor end at home). This latter distinction is maintained because home-based trips typically constitute a majority of person trips and are, in addition, conceptually easier to analyze and model.

A *trip chain* or *tour* is defined as a connected sequence of trips where the chain's origin and destination are the same point in space (see Fig. 3.11). Thus, a morning trip from home to work, combined with an afternoon's return trip from work to home, constitutes a home–work–home trip chain. Similarly, if the worker stopped at a shopping plaza on the way home, the trip chain would be home–work–shop–home and would consist of three trips (home to work, work to shop, and shop to home). In most current models, travel demand is typically analyzed at the level of the trip rather than the tour or trip chain, either because in practice little difference exists between the two concepts (e.g., whether one models a home–work–home tour or two home-based work trips is essentially an arbitrary decision) or because trip modeling is typically "easier" to do than trip-chain modeling, although theoretically less elegant.

It has long been recognized, however, that the trip-based approach to travel demand modeling is not well suited to representing complex travel patterns involving multipurpose, multistop travel or to representing traveler responses to the complex range of policies typically of interest to today's planners (pricing, HOV and car-pooling options, telecommuting, other TDM measures, etc.). *Activity-based* modeling methods, which explicitly derive travel demand as resulting from the need to participate in activities and which are typically operationalized in terms of tour-based models, have been emerging for some time as viable alternatives to the more traditional trip-based models. Activity-based modeling is discussed further in Sec. 5.6 (see p. 303).

In addition to identifying the attributes of travel demand itself, one can characterize individual trip makers in terms of various social, physical, and economic descriptors (e.g., the person's age, sex, income, occupation, and education—see Table 3.20). Each characteristic can be thought of as a dimension along which each individual can be measured or identified, such as a traveler who is 30 to 39 years old, male, and having more than one year postsecondary education. The number of these dimensions determines the extent to which individuals can be differentiated from one another; that is, they define the extent of our knowledge about people acting within the system under study. However, the trip-maker characteristics of potential relevance to the description of a traveler's behavior are not necessarily clearly defined and are potentially large in number. The criteria for choosing these characteristics are first, assumed relevance in meaningfully categorizing the individual in question, and second, observability (and, implicitly, measurability).

Similarly, one can observe various physical and economic descriptors of both the transportation system and the socioeconomic activity system, for example, the physical characteristics and level of service of the transportation system and the number, type, and size of stores at a given location. As is the case with the descriptors of trip makers, system characteristics are chosen based on an assessment of their impact on travel decision making and on their measurability.

As described in Chap. 4, data can be collected on the number and types of trips being made within a metropolitan area, the characteristics of the people or goods making these trips, and the characteristics of the activity and transportation systems that generate and serve these trips. One cannot, however, observe the decision-making process by which people decide to make a given trip or set of trips. This process is a mental one, involving the perception of needs and opportunities for travel and evaluation of these needs and opportunities given one's preferences or attitudes, leading ultimately to a choice of an action or pattern of behavior based upon this evaluation. This choice of travel behavior may be constrained to a greater or lesser extent by physical, temporal, social, or economic factors. It may also vary in terms of temporal stability. For example, the journey to work is not a trip choice that is reevaluated on a day-to-day or even necessarily a month-to-month basis, whereas an individual's recreational trip making over time can show considerable variation.

Nevertheless, the concept of *choice* as a mechanism for analyzing travel behavior has proved to be extremely useful and will be discussed in greater detail in Section 5.5.

5.2.3 Aggregation

The previous discussion of the characteristics of travel demand and of the people and systems that determine this demand illustrates the enormous and potentially overwhelming complexity and level of detail inherent in travel demand analysis. The demand for travel in a metropolitan area is the result of the decisions of thousands or even millions of individuals—based upon a variety of motives, perceptions, and preferences—made within a complex physical, social, and economic environment. In order to achieve a conceptually and analytically tractable formulation of the travel demand problem, it is typically necessary to work at a more *aggregate* level of system representation than that of the individual trip maker. Individuals are, in principle and in fact, exactly that: individual, unique, and, for all practical purposes, unpredictable with respect to the intricacies of their behavior. Aggregates of people, however, will tend to exhibit common tendencies and behave in similar ways. In other words, in the aggregate, statistical regularities emerge that are sufficiently strong, stable, and theoretically reasonable to be useful in the analysis and prediction of travel demand. Further, typically aggregate values are required in planning (e.g., peak-hour link volumes, total vehicle miles traveled, average fleet fuel consumption, average peak-period link speed) as opposed to predictions of individual activities or experiences.

Aggregation is performed in at least three dimensions: the spatial, the temporal, and the socioeconomic. As discussed in Sec. 4.1.1, spatial aggregation is performed by dividing the metropolitan study area into a set of *zones* and then treating these zones as the basic units of analysis. Thus, rather than dealing with *trips* made by individuals from point to point, the analysis is concerned with total *flows* of people from zone to zone. Zonal characteristics used to "explain" these flows typically take the form of zonal totals (e.g., total number of workers living in the zone) or zonal averages (e.g., average zonal household income). Even in those cases where microsimulation allows analysis at the individual vehicle level, some means of aggregating the resulting sum of individual vehicle's performance on the network is necessary. The transportation system is also spatially aggregated into a network of *links* and *nodes* that may or may not (but often not) constitute a one-to-one correspondence to the actual transportation network.

Temporal aggregation is performed by grouping travel flows that will tend to occur at varying levels over time into discrete time periods. Typical time periods include the weekday peak period (e.g., 6 to 9 A.M. and/or 3:30 to 6:30 P.M.), the weekday off-peak period, the weekday, and the year. Thus, one might analyze total weekday peak-period flows between zones in order to identify deficiencies in network capacity. Alternatively, total yearly transit ridership might be analyzed in order to estimate transit revenues. The temporal distribution of flows within the time period is not of interest. Temporal aggregation reduces the complexity of the analysis by converting a continuous variable (time of trip) to a discrete, nominal variable; that is, the trip occurs during a given time period or it does not.

Finally, socioeconomic aggregation occurs whenever individuals are categorized into "homogeneous" groups. It is common, for instance, to group households according to their income level, auto-ownership level, and family size for the purpose of trip generation analysis (see Sec. 5.5.1). In such cases, the explicit assumption is that all members of a given group (e.g., two-car, four-person, middle-income households) behave in the same way (e.g., generate the same number of daily trips), or at least the variance in their behavior is small relative to the differences in behavior observed between their group and other groups.

In principle, the choice of aggregation level for a given analysis will depend on the identification of an appropriate "behavioral unit" of analysis—the individual, the household, the firm, the urban area, and so on. In addition, this choice will be based on the analyst's understanding of the functional relationships that exist within the system under study and on the scale and complexity of the particular problem application. In practice, the level of aggregation employed typically depends as much or more on the quality and level of detail of the available data, on the time and monetary constraints imposed on the analysis, and on the analytical techniques available for use as it does upon strict behavioral or theoretical arguments.

5.3 SIMPLIFIED DEMAND ESTIMATION TECHNIQUES

Simplified demand analysis techniques are used in situations where, because of budget, time or data collection requirements, more complex methods are inappropriate. Examples of such situations include

1. Analysis of simple, typically small-area, short-range, planning issues, such as transit route extensions and site-specific land-use development impacts on transportation (e.g., impacts of a new shopping center on the local street system).

2. Severe data deficiencies that make the application of more complex techniques impossible and that cannot be remedied by collecting new data, typically because of budget or time constraints. This is a very common problem facing urban transportation planners in developing countries but can also be a problem elsewhere where a metropolitan area's travel database has not been maintained.

3. "Screening" of a wide range of alternatives in order to select a short list of alternatives for more detailed analysis. This type of analysis, known as *sketch planning,* allows the planner to examine a large number of alternatives quickly and cheaply, thus permitting

> . . . a two-step planning sequence in which the first step uses simplified, macro-transportation planning techniques (sketch planning) to carry out broad strategic-level planning on the land-use and transportation system options, and the second step uses conventional, urban transportation planning models to perform the detailed tactical-level planning leading to final alternative choice and functional design [McCoomb, 1982].

In this section, two simple demand analysis techniques are discussed in turn: trend analysis and elasticity-based models.

5.3.1   Trend Analysis

The simplest modeling approach estimates demand by plotting historical trip-making levels versus time and then extrapolating the plotted trend into the future. Trend analysis is extremely common both within and outside of transportation planning. Whenever growth rates are used to project future growth or whenever past and current experiences are extrapolated into the future, one is either explicitly or implicitly engaging in a trend analysis.

A range of functional forms can be assumed in trend analysis, from a simple linear trend curve (implying a constant increment of growth per unit time) to the so-called S curve, or logistic curve, which projects accelerated growth over a certain time period, followed by a period of decreasing growth rates, ultimately leading to a steady state of little or no growth. The choice among these and other functional forms depends upon which one appears to best fit the historical data. This choice is inevitably arbitrary, given a lack of additional information about the system being analyzed. In particular, since trend analyses only attempt to establish a relationship between system demand and time, they implicitly assume that all other factors and/or relationships affecting demand are constant. Hence, if key factors or relationships do vary significantly over time, then actual demand will "leave the trend curve," and the trend projection can, as a result, be seriously in error.

These points are illustrated in the example shown in Fig. 5.5. In this figure, it is assumed that the analyst has 7 years worth of traffic flow data for a given urban freeway for the years 1990 to 1996, inclusive. Two different trend curves were fitted to the data, both of which were plausible given the observed data. Each curve was extrapolated to project traffic flow on the freeway for the years 1997 to 2000. Actual annual flows for the facility over the forecast period were also plotted. As shown in the figure, different assumptions about the functional form can lead to very different future projections, particularly as one moves further into the future. Second, note that in this case, both projections seriously overestimated freeway flow, presumably due to some change in the system occurring around 1996 that significantly altered the growth in freeway travel (e.g., perhaps a parallel freeway opened, or perhaps the urban area entered an economic recession).

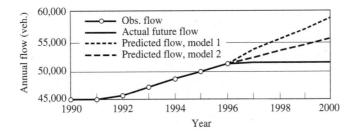

Figure 5.5   Predicted flow versus actual volumes

Despite limitations, trend analysis can play at least two important roles in transportation demand analysis. First, in the absence of better information, an examination of past trends and a careful judgmental assessment of how these trends might continue or change over time are probably the best that a planner can do. The key in such instances, however, is not simply to extrapolate past trends, but rather to think carefully about the likely causal factors underlying these trends and how they might change over time.

Second, trend analysis, while an unreliable *predictive* tool, is a very useful *diagnostic* tool for characterizing what has historically occurred within the system and what is likely to happen if changes do not occur. It can be used to help identify likely future problem areas (e.g., increasingly congested roadways) and/or future opportunities for system change (e.g., rising gasoline prices that may make public transit increasingly attractive to choice riders). It should not, however, be further used to examine the impacts of alternative strategies for addressing these problems or exploiting these opportunities. Such a predictive task requires more elaborate models and techniques.

5.3.2   Elasticity-based Models

In order to progress beyond simple trend extrapolations of demand, the analyst must possess or assume some knowledge of the demand function. As a minimum, the analyst should know the key variables within the demand function, whether they have a positive or negative impact on demand, and the sensitivity of demand to changes in these variables. These three issues are all captured within the concept of *demand elasticity.*

The elasticity of demand with respect to a certain variable, such as travel cost or transit frequency, is defined as the rate of change of demand with respect to that variable, normalized by the current levels of demand and the variable in question. Mathematically, if D_0 is the current demand level and x_0 the current value of the system variable of interest, then the elasticity of demand with respect to x, denoted as e_{Dx} is defined as

$$e_{Dx} = \frac{\delta D/\delta x}{D_0/x_0} = \frac{\delta D/D_0}{\delta x/x_0}$$

[5.4]

where $\delta D/\delta x$ is the partial derivative of D with respect to x.

A demand elasticity is thus a measure of the sensitivity of demand to changes in system conditions. Equation [5.4] expresses elasticity as a point estimate in that it is defined for a given "operating point" (D_0, x_0) and uses the derivative at this point to measure rate of change. It should be noted that elasticity will generally vary from one operating point to another. In other words, it is not a constant but a variable that characterizes how the sensitivity of demand to system conditions varies with these conditions.

The concept of elasticity, as defined previously, is of little use to planners unless the demand function D is a known and differentiable function of x so that the partial derivative $\partial D/\partial x$ can be evaluated. This requirement is met when models assuming explicit demand functions are employed (such as those discussed in Secs. 5.4 and 5.5). However, planners often encounter situations in which such models are either inappropriate or unavailable. In such cases, if some information about how demand has changed in response to a specific change in the system is available, then the *arc elasticity,* defined as the percentage change in demand given a percentage change in an explanatory variable, can be of use. The arc elasticity of demand with respect to the variable x, denoted as $\bar{e}_{Dx}$ is defined as

$$\bar{e}_{Dx} = (\text{percent change in } D)/(\text{percent change in } x)$$

$$= \frac{\Delta D/D_0}{\Delta x/x_0} = \frac{\Delta D/\Delta x}{D_0/x_0} \qquad \text{[5.5]}$$

where ΔD is the change in demand level from the original value of D_0 or $(D_1 - D_0)$, which occurs when x_0 varies by the amount Δx or $(x_1 - x_0)$. The point elasticity e_{Dx} and the arc elasticity $\bar{e}_{Dx}$ are not equal, except in the very special case when D is represented as a linear function of x.

D is said to be *elastic* with respect to x if the absolute value of the elasticity is greater than 1, that is, in the case where a 1 percent change in x results in a greater than 1 percent change in D. Similarly, D is *inelastic* with respect to x if a 1 percent change in x results in less than a 1 percent change in D. In the event that the absolute value of e_{Dx} is exactly 1, then D possesses unit elasticity with respect to x.

Direct demand elasticities are those that involve variables relating "directly" to the demand in question. The elasticity of transit demand with respect to transit fare, transit travel time, transit service headway, and so on, would all be direct elasticities. Similarly, an indirect or cross elasticity typically relates to variables characterizing other modes of travel. For example, if travel times by auto between two points increase, all else being equal, one might expect transit ridership to increase somewhat. The relationship between these two changes could be measured in terms of the cross elasticity of transit demand with respect to auto travel time.

Demand elasticity models can be computed in three ways [U.S. Department of Transportation, 1980]:

1. Quasi-experimental approaches, which include demonstration projects or practical experiments where fares or services are altered under relatively controlled conditions, resulting in changed ridership levels.

2. Time-series analyses of demand levels that are not related to a specific fare or service change, usually involving some form of regression analysis of the time-series data (e.g., Gaudry, [1975]).

3. Derivation of elasticities from cross-sectional demand models.

Of these three methods, the quasi-experimental approach is probably the one that is most readily usable by planners under most circumstances. Given data from such a study of demand changes, demand elasticities are generally constructed in one of two ways. The first assumes that elasticity is constant over the range of the service variable that is of interest. This is equivalent to specifying a demand function of the form

$$D = ax^b \qquad [5.6]$$

where it can be shown that if the system can be observed at two "operating points," (D_0, x_0) and (D_1, x_1), then

$$e_{Dx} = b = \frac{\log D_1 - \log D_0}{\log x_1 - \log x_0} \qquad [5.7]$$

and the expected demand level D_2, given a change in the service variable to a new value of x_2, is given by

$$D_2 = D_0 \left(x_2 / x_0 \right)^b \qquad [5.8]$$

where D_0, x_0, and x_2 are all known quantities, and b is given by Eq. [5.7].

Alternatively, one can assume that the demand function is approximately linear over the range of interest. In this case, the arc elasticity can be computed directly from Eq. [5.5], given the observation of the two operating points (D_0, x_0) and (D_1, x_1). Note that since elasticities are not constant along a linear demand curve, the value of the arc elasticity will vary depending on whether it is computed with the point (D_0, x_0) or the point (D_1, x_1) as its base. Either point can be used in the calculation as the base point, *providing that all subsequent calculations retain this point as the base*. That is, the new expected demand level D_2, given a new service variable value of x_2, can be obtained by replacing D_1 and x_1 in Eq. [5.5] with D_2 and x_2 and rearranging to yield

$$D_2 = D_0 \left[1 + e_{Dx} \left(x_2 - x_0 \right) / x_0 \right] \qquad [5.9]$$

If D_1, and x_1, had been used as the base point in Eq. [5.5], then they would replace D_0 and x_0, in Eq. [5.9]. This is illustrated in Fig. 5.6, in which a new demand level D_2 is computed using both point 0 (1000 riders/day, \$1.50 fare) and point 1 (950 riders/day, \$1.75 fare) as the base point. Analysis based on Eq. [5.9] is often referred to within the literature as pivot-point analysis and represents the most common approach to the use of elasticities in transportation demand analysis.

Note that if a constant elasticity assumption had been made for the system shown in Fig. 5.6, then from Eq. [5.7], the estimated elasticity would have been -0.333, and the estimated new demand level for the \$2.00 fare, as calculated by Eq. [5.8], would have been 909. As illustrated by this example, the constant elasticity assumption generally will not generate the same result as the linear demand assumption. Further, it will generate a more conservative (i.e., smaller) estimate of the expected system change (in Fig. 5.6, a 41-passenger-per-day decrease rather than a 50-passenger-per-day decrease) than will the linear demand assumption.

Elasticity-based models are extremely useful in analyzing incremental system changes, particularly when limited data and time are available for the analysis. Changes in a mode's ridership due to price increases or frequency changes, for instance, are often predicted using some form of elasticity calculation. If the data needed to calculate a local elasticity are not available, elasticities are often "borrowed" from other locations and applied to the local problem context.

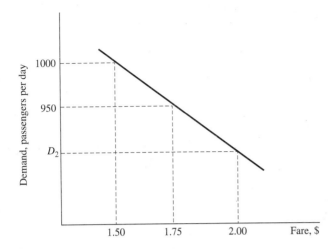

Elasticity at point 0: $E_0 = \{(950 - 1000)/(1.75 - 1.50)\} \times (1.50/1000) = -0.30$

Estimate of D_2 using point 0 as base: $D_{2,0} = E_0(1000/1.50) \times (2.00 - 1.50) + 1000 = 900$

Elasticity at point 1: $E_1 = \{(1000 - 950)/(1.50 - 1.75)\} \times (1.75/950) = -0.368$

Estimate of D_2 using point 1 as base: $D_{2,1} = E_1 (950/1.75) \times (2.00 - 1.75) + 950 = 900$

Figure 5.6 Example of an elasticity analysis

A number of issues exist in the use of elasticity-based models, which the analyst should keep in mind. These include the following:

1. An individual's travel elasticity will, in general, depend upon a host of factors, including income, trip purpose, time of day of the trip, availability and level of service of competing modes, and service characteristics other than the one being measured in the elasticity calculation. Elasticities are often computed from fairly aggregate statistics with little or no market segmentation. Thus, considerable potential for "aggregation bias" exists in most elasticity calculations. Application of an elasticity computed for another system or city obviously compounds this potential problem.

2. Because elasticities assume that all factors other than the one in question are being held constant, they are useful only for short-run predictions, because in the long run, many of these factors are apt to change.

3. Most elasticity analyses assume either the constant elasticity case or the linear demand case as the basis for their calculations. Either assumption is most tenable for small changes in the system. The larger the projected system change, the more likely the demand response will be a nonlinear, nonconstant-elasticity response, which is poorly predicted by the elasticity analysis.

4. Considerable confusion exists over the correct use of pivot-point techniques. As discussed previously, pivot-point analysis assumes a linear demand function that

permits a new demand to be estimated based on an elasticity that has been computed at a particular base or pivot point. Correct use of this technique requires consistency in the definition of the pivot point. That is, the same point must be used to estimate the elasticity and to predict new demand levels. Pivot-point analysis does not assume a constant elasticity. If an assumption of constant elasticity is preferred, then Eq. [5.8] should be used rather than Eq. [5.9]. In particular, if the elasticity has not been computed locally but has instead been "borrowed" from elsewhere and, hence, the appropriate pivot point is not known, it is preferable to assume constant elasticity and thus use Eq. [5.8], which is typically the assumption implicit within this approach.

5.4 THE URBAN TRANSPORTATION MODELING SYSTEM

The approach to urban travel demand modeling commonly employed by the transportation planning profession is embodied in a type of model generally known as the urban transportation modeling system (UTMS). UTMS is used to predict the number of trips made within an urban area by type (work, nonwork, etc.); time of day (peak period, daily, etc.); zonal origin–destination (O–D) pair; the mode of travel used to make these trips; and the routes taken through the transportation network by these trips. The final product of UTMS is a predicted set of modal flows on links in a network. As such, it represents an "equilibrium" procedure in which the demand for transportation, represented by zonal O–D flows by mode, is assigned to the modal networks constituting the transportation system, where this assignment is a function of the networks' performance characteristics. The major inputs to UTMS are a specification of the activity system generating these flows and the characteristics of the transportation system that will serve these flows.

UTMS consists of four major stages and hence is often referred to as the four-stage or four-step model, as shown by Fig. 5.7:

1. *Trip generation* is the prediction of the number of trips produced by and attracted to each zone, that is, the number of trip ends "generated" within the urban area. In other words, the trip generation phase of the analysis predicts total flows into and out of each zone in the study area, but it does not predict where these flows are coming from or going to.

2. *Trip distribution* is the prediction of origin–destination (O–D) flows, that is, the linking of the trip ends predicted by the trip generation model together to form trip interchanges or flows.

3. *Modal split* models predict the percentages of travel flow that will use each of the available modes (auto, transit, walk, etc.) between each origin–destination pair.

4. *Trip assignment* places the O–D flows for each mode on specific routes of travel through the respective modal networks.

The four stages of UTMS thus correspond to a sequential decision process (Fig. 5.8) in which people decide to make a trip (generation), decide where to go (distribution), decide what mode to take (modal split), and decide what route to use

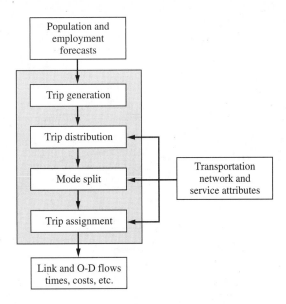

Figure 5.7 The urban transportation modeling system

(assignment). For most trips, this is undoubtedly a highly unrealistic representation of travelers' decision making. UTMS, however, makes no claim of representing individual trip-making behavior. Rather, it represents a pragmatic approach to reducing the extremely complex phenomenon of travel behavior into analytically manageable components that can be dealt with using relatively simple techniques and reasonable amounts of data.

The major role that UTMS has played in transportation demand analysis is well documented in the literature [see, for example, Hutchinson, 1974; Kanafani, 1983; Ortuzar and Willumsen, 1994]. In this section, only a limited description of each component is presented in order to provide a brief overview of UTMS.

5.4.1   Trip Generation

Trip generation models are used to predict the trip ends generated by a household or a zone, usually on a daily or a peak-period basis. Trip ends are classified as being either a *production* (defined as the home end of a home-based trip or the origin of a nonhome-based trip) or an *attraction* (the nonhome end of a home-based trip or the destination of a nonhome-based trip). Separate models are used to predict productions and attractions. Variables used as predictors of trip productions include household income, auto ownership and size, number of workers per household, residential density, and distance of the zone from the central business district (CBD). Trip attraction predictors include zonal employment levels (possibly disaggregated by occupation type), zonal floor space (disaggregated by business type), and accessibility to the work force (i.e., some weighted accessibility measure).

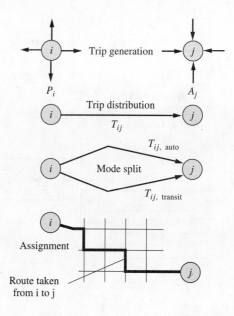

Figure 5.8 Steps in the sequential modeling of transportation demand

Trip generation, as applied in regional transportation planning, is usually based on mathematical relationships between trip ends and socioeconomic or activity characteristics of the land use generating or attracting the trips. However, impact analyses of new development sites often use vehicle or person trip rates obtained from historical studies to estimate the number of trips that will be generated. At the site-specific level, the trip rates are expressed as the number of trips per x, where x is a descriptor of the land-use activity. Thus, for residential uses, trip generation might be expressed as the number of trips per dwelling unit; for hospitals, the number of trips per patient beds; and for retail stores, the number of trips per 1000 square feet of gross leasable space. Table 5.2 shows typical vehicle trip rates for different land uses that are found in the Institute of Transportation Engineers' (ITE) *Trip Generation* manual, one of the most widely used references for determining trip rates. Such rates are often modified with factors that reflect transit use, TDM activities, or multiuse projects; thus, the reader should not use the trip rates shown in Table 5.2 without referring to [ITE, 1997]. For regional or subarea studies, two general classes of trip generation models have traditionally been used in practice: linear regression models and cross-classification models.

Regression Models Given the high correlations that typically exist between trip generation and the variables listed previously, ordinary least-squares regression is often used to estimate models that predict trip generation as a linear function of one or more of these variables. Three examples of typical trip generation regression models are

Table 5.2 ITE trip generation rates for sites

Land Use	Daily Vehicle Trip Rate	Per	Percent of Total Daily Vehicle Trips A.M. Peak	P.M. Peak
Residential				
Single-family	9.55	Dwelling Unit (DU)	8.0%	10.7%
Apartment	6.47	DU	8.6	10.7
Condo/townhouse	5.86	DU	7.5	9.2
Mobile-home park	4.81	Occupied DU	8.9	12.1
Planned unit development	7.44	DU	7.8	9.7
Retail[1]		Gross floor area (GFA)		
Shopping center				
<100,000 sq. ft.	70.7	1000 sq. ft. GFA	2.3	9.2
100,000 to 500,000	38.7	1000 sq. ft. GFA	2.1	9.5
500,000 to 1 million	32.1	1000 sq. ft. GFA	2.0	9.3
>1 million	28.6	1000 sq. ft. GFA	1.8	9.1
Office[2]				
General	11.85	1000 sq. ft. GFA	13.8	13.1
Medical	34.17	1000 sq. ft. GFA	10.0	13.0
Office park	11.42	1000 sq. ft. GFA	16.1	13.2
Research/development	7.70	1000 sq. ft. GFA	16.0	13.9
Business park	14.37	1000 sq. ft. GFA	11.3	10.3
Restaurant				
Quality restaurant	96.51	1000 sq. ft. GFA	6.6	10.1
High turnover	205.36	1000 sq. ft. GFA	8.7	15.5
Fast food w/out drive	786.22	1000 sq. ft. GFA	9.7	13.7
Fast food w/ drive	632.12	1000 sq. ft. GFA	9.5	7.3
Bank				
Walk-in	140.61	1000 sq. ft. GFA	13.7	0.4
Drive-through	265.21	1000 sq. ft. GFA	13.3	19.3
Hotel/Motel				
Hotel	8.7	Occupied room	7.5	8.7
Motel	10.9	Occupied room	6.7	7.0
Parks and recreation				
Marina	2.96	Berth	5.7	7.1
Golf course	37.59	Hole	8.6	8.9
City park	2.23	Acre	NA	NA
County park	2.99	Acre	NA	NA
State park	0.50	Acre	NA	NA
Hospital				
General	11.77	Bed	10.0	11.6
Nursing home	2.6	Occupied bed	7.7	10.0
Clinic	23.79	1000 sq. ft. GFA	NA	NA

(continues)

Table 5.2   ITE trip generation rates for sites (continued)

Land Use	Daily Vehicle Trip Rate	Per	Percent of Total Daily Vehicle Trips A.M. Peak	P.M. Peak
Educational				
Elementary school	10.72	1000 sq. ft. GFA	25.6	23.2
High school	10.90	1000 sq. ft. GFA	21.5	17.8
Community college	12.57	1000 sq. ft. GFA	17.2	8.2
University/college	2.37	Student	8.4	10.1
Airport				
Commercial	104.73	Ave. flights/day	7.8	6.6
General aviation	2.59	Ave. flights/day	10.4	12.7
Industrial				
General light industry	6.97	1000 sq. ft. GFA	14.5	15.5
General heavy industry	1.5	1000 sq. ft. GFA	34.0	45.3
Warehouse	4.88	1000 sq. ft. GFA	11.7	12.3
Manufacturing	3.85	1000 sq. ft. GFA	20.3	19.5
Industrial Park	6.97	1000 sq. ft. GFA	11.8	12.3

[1] Rates given are for high end of indicated range.
[2] Rate for a 200,000 sq. ft. general office building.
SOURCE: Martin and McGuckin, 1998

$$T = 1.229 + 1.379V \qquad [5.10]$$

$$T_w = 0.135P + 0.145DU - 0.253C \qquad [5.11]$$

$$A = 61.4 + 0.93E \qquad [5.12]$$

where
T = number of person trips per day per household
V = number of vehicles per household
T_w = work-trip productions per zone
P = zonal population
DU = number of dwelling units per zone
C = total number of automobiles owned in the zone
A = peak-hour work-trip attractions per zone
E = total employment in the zone

Eq. [5.10] predicts total daily trip productions per household, while Eq. [5.11] predicts daily work-trip productions for an entire zone. Finally, Eq. [5.12] is a trip attraction model, which predicts the total number of work trips attracted to a given zone as a function of the number of employees working in that zone.

Typically, trip generation equations are developed by trip purpose, given the logic that trip-making behavior would be linked to different explanatory variables for different trip types. For example, Table 5.3 shows equations that predict the number of person-trips by trip type and by central business district (CBD) or non-CBD location.

Table 5.3   Person-trip attraction equations

HBW attractions = 1.45 × total employment

HBO attractions CBD = (2.00 × CBD RE) + (1.7 × SE) + (0.5 × OE) + (0.9 × HH)

HBO attractions NCBD = (9.00 × NCBD RE) + (1.7 × SE) + (0.5 × OE) + (0.9 × HH)

NHB attractions CBD = (1.40 × CBD RE) + (1.2 × SE) + (0.5 × OE) + (0.5 × HH)

NHB attractions NCBD = (4.10 × NCBD RE) + (1.2 × SE) + (0.5 × PE) + (0.5 × HH)

Where	CBD RE	= Retail employment in central business district zones
	NCBD RE	= Retail employment in noncentral business district zones
	SE	= Service employment
	OE	= Other employment (basic and government)
	HH	= Households

SOURCE: Martin and McGuckin, 1998

Regression models are very easy and inexpensive to construct from data that are typically available in planning studies. Problems with the use of such models, however, include

1. Correlation among explanatory variables (particularly income and auto ownership) may create estimation problems.
2. The assumption that the explanatory variables have linear, and additive impacts on trip generation may be wrong.
3. "Best fit" equations may yield counterintuitive results [e.g., the negative auto ownership coefficient in Eq. [5.11] implies that as auto ownership levels increase, trip productions decrease—something that one would not normally expect].
4. By using zonal averages, important socioeconomic variations within the zone may be obscured or may yield spurious results.

Category Analysis   Rather than grouping households spatially (i.e., by zones) as in regression models, cross-classification analysis groups individual households according to common socioeconomic characteristics (auto-ownership level, income, household size, etc.) so as to create relatively homogeneous groups. Average trip production rates are then computed for each group from observed data. Figure 5.9 provides an example of a cross-classification trip production model in which total daily home-based trip rates are predicted as a function of household size and auto ownership. Figure 5.9a presents the data required to *construct* the cross-classification trip rates: total number of trips and total number of households for the study area for each household-size-auto-ownership category. Figure 5.9b shows the daily trip rates that result from dividing daily trips by the number of households for each category. Figure 5.9c presents an example of the data required to *forecast* trip productions for a given zone, that is, the number of households in each size-auto ownership category expected to be living in the zone in the forecast year. Finally, Fig. 5.9d shows the forecasted number of daily home-based trips generated by this zone,

(a) Number of Households and Total Trips Made, Categorized by Household Size and Auto-Ownership Level

	Automobile ownership					
	0		1		2 or more	
Family size	No. of households	No. of trips	No. of households	No. of trips	No. of households	No. of trips
1	925	1,098	1,872	4,821	121	206
2	1,471	2,105	1,934	6,129	692	1,501
3	1,268	1,850	3,071	13,989	4,178	19,782
4 or more	745	1,509	4,181	18,411	4,967	25,106

(b) Household Trip Rates

	Automobile ownership		
Family size	0	1	2 or more
1	1.19	2.57	1.70
2	1.43	3.16	2.17
3	1.45	4.55	4.74
4 or more	2.02	4.40	5.05

(c) Forecasted Number of Households in One Zone, Categorized by Household Size and Auto-Ownership Level

	Automobile ownership		
Family size	0	1	2 or more
1	24	42	8
2	10	51	107
3	11	31	158
4 or more	3	17	309

(d) Forecasted Number of Trips from This Zone

	Automobile ownership			
Family size	0	1	2 or more	Total
1	29	106	14	151
2	14	161	232	407
3	16	141	749	906
4 or more	6	75	1564	1645
Total	65	485	2559	3109

Figure 5.9 A cross-classification trip generation analysis

SOURCE: Morlok, 1978

which is obtained by multiplying the number of households in each category (Fig. 5.9c) by the corresponding trip rate for the category (Fig. 5.9b). The trips generated by each household category are then summed to yield the total number of trips generated by the zone (3,109 trips in this example).

Cross-classification analysis can be similarly performed for trip attraction calculations. In such cases, classification is generally done with respect to employment type (e.g., manufacturing, retail, office) and possibly employment density (i.e., number of employees per acre).

Category analysis avoids the regression model's assumption of a linear, additive relationship between trip generation and its explanatory variables, as well as the pitfalls inherent in spatially aggregated models. It does, on the other hand, require considerably more detailed data than typical regression models, both to initially construct and, more critically, to use in predicting future trip generation. As with regression models, the stability of the estimated rates over time may also be a concern.

As noted earlier, ITE's *Trip Generation* is a commonly used reference for determining trip generation rates. A typical page from this handbook is shown in Fig. 5.10. Note that the information provided includes both vehicle rates and a corresponding regression equation, in addition to such data as the number of data points in the sample, the range of values, directional distribution, and standard deviation.

Control Totals   Because trip productions and attractions are calculated separately, one must ensure that the areawide production and attraction totals are the same. In general they will not be. This can be corrected by multiplying each zone's trip attractions by the ratio of total productions to total attractions. This approach to the problem is based on the expectation that trip production models are better predictors of trip rates than the somewhat cruder trip attraction models. In addition, the balancing procedure must take into account the number of trips attracted to external zones. The trip generation approaches described previously produce estimates for trips internal to the study area (so-called I–I trips), trips generated internal to the study area but destined to outside or external zones (I–E trips), and vice versa (E–I trips). (See [Martin and McGuckin, 1998] for an approach to estimate through (E–E) trips). The control total for the study area can thus be represented as

$$CT_p = \sum P_z + \sum P_e - \sum A_e \qquad \text{[5.13]}$$

where

CT_p = control total of productions
P_z = trip productions for each zone
P_e = trip productions at each external station
A_e = trip attractions at each external station

The factor that is used to balance productions and attractions is thus

$$\text{Factor} = \frac{CT_p}{\sum A_z} \qquad \text{[5.14]}$$

where A_z is the number of trip attractions at each zone by purpose.

An example of this approach is shown in Fig. 5.11.

Single-family detached housing
(210)

Average vehicle trip ends v. dwelling units on a weekday

Number of studies: 348
Avg. number of dwelling units: 198
Directional distribution: 50% entering, 50% exiting

Trip generation per dwelling unit

Average rate	Range of rates	Standard deviation
9.57	4.31- 21.85	3.69

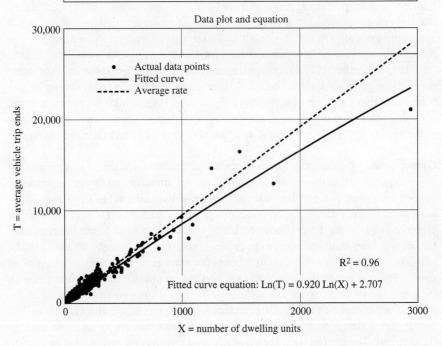

Data plot and equation

- Actual data points
— Fitted curve
----- Average rate

$R^2 = 0.96$

Fitted curve equation: $\mathrm{Ln}(T) = 0.920\ \mathrm{Ln}(X) + 2.707$

T = average vehicle trip ends

X = number of dwelling units

Figure 5.10 Typical trip generation data, Institute of Transportation Engineers
I SOURCE: ITE, 1997

5.4.2 Trip Distribution

The task of a trip distribution model is to "distribute" or "link up" the zonal trip ends, that is, the productions and attractions for each zone as predicted by the trip generation model in order to predict the *flow* of trips T_{ij} from each production zone i to each attraction zone j.

Many types of trip distribution models exist. These include *growth factor* techniques such as the Fratar method, which were used in early transportation studies, but which are now used mostly for short-term updating of trip tables and estimation of "through trips" for urban areas [Fratar, 1954; Hutchinson, 1974; Ortuzar and

	Unbalanced work trips		Balanced work trips	
Zone	Productions	Attractions	Productions	Attractions
1	100	4000	100	4540
2	300	3000	300	2305
3	500	5000	500	5675
4	1000	1000	1000	1135
5	1500	1200	1500	1362
6	1000	1500	1000	1703
7	5000	500	5000	568
8	7500	100	7500	114
9	3000	1500	3000	1703
10	1000	2000	1000	2270
Subtotal	20,900	19,800	20,900	22,475
External stations				
11	500	50	500	50
12	1000	100	1000	100
13	250	25	250	25
Subtotal	1750	175	1750	175
Total	22,650	19,975	22,650	22,650

$$\text{Balancing factor} = \frac{20,900 + 1750 - 175}{19,800} = 1.1351$$

Figure 5.11   Example of balancing productions and attractions for the work trip
SOURCE: Martin and McGuckin, 1998

Willumsen, 1994]; *intervening opportunities* models, which have seen limited use over the years, are cumbersome to calibrate, and have never enjoyed generalized acceptance [Stouffer, 1940; Schneider, 1960; Golding and Davidson, 1970]; *disaggregate destination choice* models (discussed in Sec. 5.5); and, finally, the most commonly used *gravity model*.

The gravity model, in one form or another, has been in existence for over 100 years. It received its name from its earliest derivation as an analogy drawn between the "spatial interaction" of trip making and the gravitational interaction of physical bodies distributed over space. The most typical version of the gravity model used in transportation planning applications is

$$T_{ij} = \frac{P_i \left[A_j f_{ij} k_{ij} \right]}{\displaystyle\sum_{n=1}^{m} A_j f_{ij} k_{ij}} \qquad \text{[5.15]}$$

where

P_i = total number of trips produced in zone i
A_j = total number of trips attracted to zone j
f_{ij} = friction factor
k_{ij} = adjustment factor for trip interchanges between zones i and j

The friction factor is an inverse function of the "cost" of travel between zones i and j, for example, travel time, distance, monetary out-of-pocket cost, "generalized cost," and so on. This friction factor, often known as a travel impedance factor and denoted as f_{ij}, can have different functional forms including

$$f_{if} = c_{ij}^{-b} \qquad\qquad\qquad \text{[5.16a]}$$

$$f_{ij} = e^{-bc_{ij}} \qquad\qquad\qquad \text{[5.16b]}$$

or simply f_{ij} represented as a graphical function of c_{ij}.

where c_{ij} is the cost of travel between zones i and j, and b is a model parameter.

In all cases, the function f_{ij} must be empirically calibrated for any given metropolitan area in order to derive the value of the parameter b if either Eqs. [5.16a] or [5.16b] are used, or the locus of the graphical function if this latter method is used. In such a case, an ad hoc, trial-and-error procedure must be employed. The graphical friction factors are often manually adjusted until the observed and predicted trip-length distributions for the system being modeled are "close" to one another, and the observed and predicted average trip lengths differ by no more than some prespecified percentage. Average trip lengths in minutes can be estimated from travel surveys. A similar distribution is produced by the gravity model, and the parameters that produce a best fit to the survey results are used for predictive purposes. Figure 5.12 shows an example of a "fitted" gravity model trip length frequency distribution. As in the trip generation step, such calibration is done for different trip purposes.

Equation [5.15] automatically satisfies the constraint that the total number of trips predicted to leave any zone i is equal to the observed productions P_i. It does not, in general, satisfy the converse constraint that the total number of trips predicted to enter zone j is equal to the observed attractions A_j. The latter requirement is accomplished through an iterative "balancing" procedure in which the trip attractions used in Eq. [5.15] are systematically adjusted until predicted and observed attractions are equal for all zones in the system [Ortuzar and Willumsen, 1994; see also Question 6 at the end of this chapter].

Equation [5.15], in combination with the balancing procedure mentioned previously, represents the standard gravity model formulation that has been used throughout North America for more than 2 decades. Despite its widespread use, the gravity model suffers from a number of shortcomings, perhaps most notably its lack of a credible theoretical basis. This shortcoming can be partially remedied by deriving essentially the same mathematical form as a so-called entropy model, either through an analogy with statistical mechanics [Wilson, 1967; Ortuzar and Willumsen, 1994] or through the use of information theory [Webber, 1977]. The latter approach, in particular, serves to make clear the theoretical underpinnings of the model, which in essence consist of two main points. The first is that the model is derivable from an explicit set of constraints, that is, those already discussed—that predicted and observed trip ends for each zone must match—plus any other "information" that one might have about the system being modeled. Second, the model is no more and no less than a logical, self-consistent procedure for *describing* an observed set of flows. Its *predictive* capabilities are unclear, especially in light of its

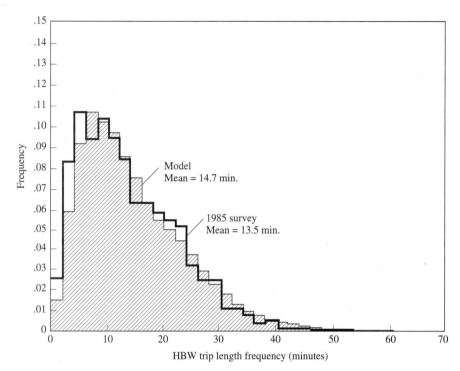

Figure 5.12 Fitted gravity model trip length distribution
| SOURCE: Portland METRO, 1996

explicit lack of behavioral assumptions (aside from the recognition that travel distance or time is an important determinant of spatial interaction). Indeed, the amount of error involved in gravity model predictions has been shown to be large, even in "good fitting" models [Smith and Hutchinson, 1979].

One final step is necessary in trip distribution before proceeding to mode split. Because productions are defined as including the home end of the trip no matter which "direction" the trip is occurring, and attractions indicate work, school, or shop trip ends, the production and attraction interchanges must be converted to an origin–destination (O–D) trip table. In most software packages, this is done automatically. A manual procedure for accomplishing this for home-based trips is to add one-half of the production–attraction interchange in one direction to one-half the production–attraction interchange for the opposite direction [Martin and McGuckin, 1998]. Figure 5.13 shows an example of how this is done. The origin–destination trip interchange between zones 1 and 2, for example, is 1/2 (100) + 1/2 (30) = 65 trips. Note in this case that the trip table is for 24 hours and thus it is assumed that the number of trips originating in a zone equals the number of trips destined for the zone, that is, everyone ends up at home during a 24-hour period. The process of estimating trip tables for different time periods is discussed in Sec. 5.4.5. Also note in this example that nonhome-based production–attraction interchange table would be the same as the O–D trip table without adjustment.

Example production–attraction interchange table

	Attraction zones			
Production zones	*1*	*2*	*3*	*Total*
1	50	30	20	100
2	100	70	30	200
3	250	200	50	500
Total	400	300	100	800

Example origin–destination trip table

	Destination zones			
Origin zones	*1*	*2*	*3*	*Total*
1	50	65	135	250
2	65	70	115	250
3	135	115	50	300
Total	250	250	300	800

Figure 5.13   Converting production–attraction table to origin–destination table
SOURCE: Martin and McGuckin, 1998

5.4.3   Modal Split

Modal split models predict the percentage of trips using each of the modes available to the given trip makers. Figures 5.7 and 5.8 show modal split as occurring after trip distribution in the UTMS structure, in which case the model is known as a *trip-interchange* modal split model. In some versions of UTMS, however, modal split is performed prior to distribution. In this latter case, the model is known as a *trip-end* model, since it splits trip ends (i.e., productions and attractions) rather than flows. Both types of models are discussed briefly as follows.

Trip-End Models   Trip-end modal split modeling is based on the assumption that transit ridership is primarily a function of socioeconomic variables; that is, virtually all transit riders are assumed to be "captive" riders—people who have no other choice but to ride transit. This assumption is most valid in areas that possess relatively low transit service levels, such as small urban areas or in cities in which the socioeconomic factors are such that virtually all transit users are "captive users" of the system. This latter case holds in many developing countries in which the vast majority of transit riders cannot afford an automobile and in which virtually anyone who can afford to buy a car does so and then ceases to use transit.

The major advantages of such models are that they are simple to apply and require relatively little data for calibration or prediction, in particular, since the trips have not yet been distributed, and hence the modal service characteristics associated with the trips are not yet known. The only variables that can be used in these models are those that were used in the trip generation stage: auto ownership, income, distance from the CBD, household size, zonal population density, and so on. The major disadvantage of these models is that they are generally insensitive to transportation policy changes.

Trip-Interchange Models   Because trip-interchange models are used after trip distribution, they can utilize the service characteristics (travel times, costs, etc.) of the modes available for the given trip, along with any relevant socioeconomic characteristics, such as income or auto ownership, to determine the modal splits. This is the preferred and overwhelmingly typical approach for urban areas in which significant transit service exists and in which the "competition" between auto, transit, and other modes of travel must be explicitly considered. Almost universally, this is accomplished through the use of some type of disaggregate, random utility modal choice model, in particular, the multinomial logit model. These models are discussed separately in Sec. 5.5.

5.4.4   Assignment

The last step in the UTMS sequence is the assignment of the predicted modal flows between each origin–destination pair to actual routes through the given mode's network. Although manual assignment techniques are possible for very small networks, the networks involved in practical-sized problems usually require the use of computer-based analysis. All such computerized approaches are based on the assumption that the underlying principle determining route selection is what has been labeled *user equilibrium*. In a user-equilibrated network, no user can improve the trip travel time (cost) by unilaterally changing routes [Wardrop, 1952]. Thus, trip assignment procedures are based on the assumption that each individual chooses the route perceived as being the best; that is, each individual minimizes or optimizes travel time or cost. This approach can be contrasted with the concept of *system optimization,* in which the system users would be assigned to routes so as to minimize the systemwide average cost of travel. User equilibrium does not, in general, yield the same route assignments as system optimization. This means a system in user equilibrium will typically not have the lowest possible average system cost. The need to generate user equilibrium rather than system optimal solutions has important ramifications in that the former generally involves far more cumbersome and costly techniques than do the latter.

Traffic assignment techniques include

1.   Minimum path (all-or-nothing) assignment.
2.   Equilibrium assignment.
3.   Stochastic assignment.
4.   Dynamic assignment.

Each of these techniques is discussed briefly in turn.

Minimum Path (all-or-nothing) Assignment   In this approach, ideal, that is, uncongested, minimum travel time paths or routes are computed for each O–D pair, and all flows between these pairs are loaded onto these routes (Chap. 7 discusses procedures for calculating minimum time paths). A given route receives "all or nothing" of a given O–D pair's flow. Advantages of this approach are that it is simple and inexpensive to use; it depicts the routes most travelers would be expected to use in the absence of capacity and/or congestion effects; and the results are easy to understand and interpret. The major disadvantage of the approach is that it clearly generates unrealistic flow patterns in situations where capacity constraints and congestion effects do exist.

Equilibrium Assignment   Equilibrium assignment techniques explicitly recognize that transportation network link costs generally depend on the volume using that link. Thus, for example, the more congested a highway link, the more time or "cost" will be associated with travel along it, which in turn will tend to discourage more travelers from using it, assuming that other options exist. Hence, these techniques search for a user-equilibrium solution in which link flows and costs are simultaneously solved for.

Figure 5.14 illustrates the user-equilibrium concept in the simple case of flow between a single O–D pair where two paths are available, each of which consists of a single link. Each link has an assumed *volume-delay function,* which estimates the average travel time (t_1) on link i as a function of the flow volume or level of congestion on the link (i.e., $t_1 = \mathrm{f}(V_1)$ and $t_2 = \mathrm{f}(V_2)$; see Chap. 7 for further discussion of volume-delay functions). If we plot the volume-delay functions for the two links on the same graph (Fig. 5.14b) so that each point along the graph's horizontal axis represents a *feasible* assignment of the total O–D flow (V_{ab}) to the two link flows (V_1, V_2), such that $V_1 + V_2 = V_{ab}$, then the equilibrium solution will be the set of flows (V_1^*, V_2^*) that results in the travel times on the two paths to be equal. This is the equilibrium solution, since at any other flow combination, the travel time on one link is higher than that on the other link, and thus some of the users of the first link would be able to improve their travel time by switching to the other link. This process will continue until the travel times are equal on both routes, at which point no one has any incentive to switch to another route, and so the equilibrium point is maintained.

Extension of this simple case to real-world urban networks involving tens of thousands of links providing hundreds of thousands of paths between possibly millions of O–D pairs involves the application of complex algorithms derived from the field of operations research. While the detailed description of these algorithms is beyond the scope of this text (for a detailed and accessible treatment of equilibrium assignment methods, see, Sheffi, [1985]), it is important to note that user-equilibrium assignment procedures capable of handling large, real-world networks are routinely available within most commercially available transportation modeling software packages.

Stochastic Assignment   The equilibrium assignment method briefly sketched in the preceding discussion is more properly referred to as a *deterministic user equilibrium*

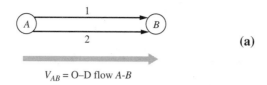

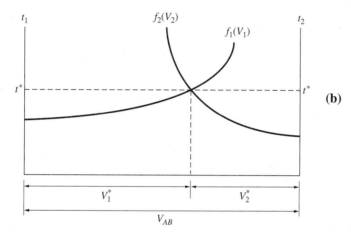

Figure 5.14   User-equilibrium assignment example
| (a) Flow for a Single O—D Pair, with Choice of Two Routes
| (b) Equilibrium Solution, Two-Route, Single O—D Pair Problem

(DUE) method, since it assumes that all users in the system have perfect information about the travel times on alternative paths within the network and that they make perfectly correct route choice decisions based on this information. Thus, no randomness or "error" exists in their route choices. In practice, of course, users generally do not have perfect information about the travel times that will actually occur on all feasible paths under all circumstances, although with the advent of ITS traveler information systems, travelers in many cities now have more information than they ever did on their options. Further, in practice, computerized representations of the actual road network always are abstractions and simplifications of the real-world network and so introduce error into the calculations relative to users' actual perceptions.

Given these observations, various *stochastic* approaches to traffic assignment have been proposed and sometimes used. These procedures recognize that several routes between an origin and a destination might be *perceived* to have equal travel times or otherwise be equally attractive to a traveler and, as a result, might be equally likely to be used by that traveler. Or, in other words, these procedures treat link costs as random variables that can vary among individuals given their individual preferences, experiences, and perceptions, rather than deterministically as is done within the DUE approach. Suggested procedures include use of an incremental assignment in a stochastic simulation procedure [Burrell, 1968], use of a multinomial logit model to predict route choice probabilities [Dial, 1971], and the use of discrete choice models (see Section 5.5) within a "stochastic user-equilibrium" (SUE) framework [Sheffi, 1985]. Stochastic assignment procedures have seen limited applications relative to DUE procedures, both due to a relative lack of commercially available software and due to their considerably greater theoretical and computational complexity. Two notable examples of the use of stochastic assignment procedures are (1) many commercial software packages use some form of stochastic *multipath* assignment procedure for transit assignment, in which transit O–D flows are probabilistically allocated across a set of feasible transit routes; and (2) the MEPLAN and TRANUS integrated transportation–land-use modeling packages (discussed further in Chap. 6) use stochastic assignment procedures for both road and transit assignments.

Dynamic Assignment   Both DUE and SUE methods generally are developed in a *static* way in which the flows of vehicles or people through the network are not explicitly represented within the time dimension. That is, while these procedures compute O–D and link travel times as model outputs, they do not attempt to estimate *when* a given vehicle will actually travel over a given link. Indeed, these procedures actually assume that each vehicle is simultaneously located on every link on its chosen path, and so, in effect, assign all flow "simultaneously" to all links on the chosen paths. This is obviously an unrealistic assumption because each vehicle can be on only one link at a time and each vehicle must travel through time as well as space as it moves from its origin to its destination.

For many regional transportation planning applications, the static assignment assumption is acceptable and, with a properly validated network, can yield very useful results. For many other applications (real-time system control under ITS; detailed emissions modeling; traffic operations modeling; etc.), however, the static representation of network performance is not sufficiently accurate. In such cases, a *dynamic* representation of route choice behavior and resulting network performance (congestion, speeds, etc.) is required in which the movements of vehicles along their chosen paths is explicitly simulated through time. At each point in simulated time t, a given vehicle i will have a computed location x_{it}, a speed v_{it}, etc. Dynamic assignment models may be either probabilistic in terms of the simulation of users' route choices and/or determination of vehicles' travel times along given links, or they can be deterministic. They may solve for an equilibrium traffic pattern or simply generate a "single outcome" from a distribution of possible flow patterns, and they can be developed at various levels of spatial and temporal aggregation. Spatially, dynamic assignment procedures range from modeling aggregate flows over networks constructed at a level of resolution similar to that used by DUE models, to modeling the movements of each individual vehicle through a network representation that includes every street in the real-world system. Similarly, temporally models can range from dealing with flows occurring within (typically) 15-minute "time slices," to simulating individual vehicle movements on a second-by-second or even a fraction of a second level.

Detailed discussion of dynamic assignment methods are beyond the scope of this text. For further discussion, see, for example, [Bernstein, et al., 1995; Boyce et al., 1998; Mahmassani, 1997; Miaou and Summers, 1999]. Many dynamic assignment software packages are emerging at varying levels of commercial availability. Notable examples (but certainly not an exhaustive list) include DYNASMART [Hu and Mahmassani, 1995], INTEGRATION [Van Aerde and Yager, 1988a,b], PARAMICS [Quadstone, 1999], and TRANSIMS [Barrett et al., 1995].

No matter what approach is used for assignment, the result of this modeling step is an estimate of trip volumes on network links. One of the ways of validating the accuracy of this assignment process is to compare estimated volumes based on a current trip table with volume counts taken at key locations in the network. These locations are determined by using screen lines that divide travel corridors into segments that must be crossed by those traveling. The planner can then compare actual and estimated volumes to determine the level to which the model replicates actual travel volumes. Figure 5.15 illustrates this process for a transit study in Seattle.

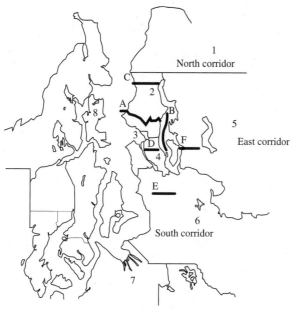

1985 and 1990 daily transit volumes at selected screen lines

Screen line	1985 Estimated	1990 Observed	1990 Estimated	1990 Estimated/observed
A Ship canal	68,300	68,200	73,500	1.08
B Lake Washington	18,100	18,400	21,800	1.18
C N. 185th Street	10,000	11,900	12,400	1.04
D S. Spokane Street	38,200	41,700	41,800	1.00
E S. 188th Street	12,500	15,100	16,200	1.07
F I-405 Newport	1,900	2,300	1,900	0.83

Figure 5.15   Screen line analysis
SOURCE: Seattle METRO, 1996

5.4.5   Time-of-Day Models

In many cases, trip rates and thus total trip generation are estimated on a 24-hour basis. The reason for this is that data collection often did not include "time-of-departure" information, although this data is increasingly being collected in today's surveys. However, trip rates do vary by time of day, and for purposes of air quality, capacity, and congestion analyses, travel estimates by time periods other than 24 hours are important. A typical approach for determining these estimates was to apply a time-of-day factor, usually a percentage, to a 24-hour trip table to produce a trip table for a different time period. Table 5.2, for example, showed percentages that represented the level of trip making that occurred in the A.M. and P.M. peaks. The 24-hour trip generation would be multiplied by these percentages to produce an A.M. and P.M. peak trip generation and, ultimately, an A.M. and P.M. peak trip assignment. Other time-of-day approaches have been to apply such factors after trip distribution, mode choice, or trip assignment [Cambridge Systematics, 1997]. The most common approach is for the adjustment to occur post-trip assignment (Fig. 5.16).

Several new approaches have been tried to take into account the adjustments that travelers make in time of travel given congested conditions (which would vary

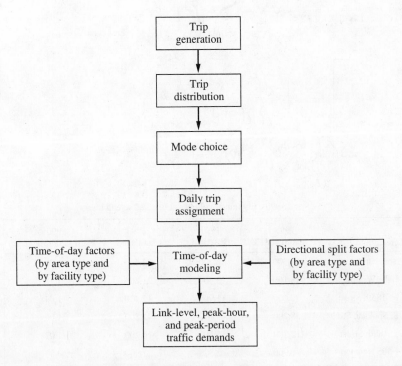

Figure 5.16   Time-of-Day modeling after trip assignment
SOURCE: Cambridge Systematics, Inc., 1997

by trip purpose). In most cases, these approaches have been applied within the UTMS framework and have attempted to "spread" peak travel into the time periods on either side of the peak period. The assumption of these approaches is that some subset of travelers will leave earlier or later to avoid congestion. Approaches to represent this have focused on individual link adjustments (i.e., spread trips to other times for those trips using congested links); trip-based adjustments (i.e., trip tables are adjusted to reduce trip origin–destination interchanges using congested links); and systemwide adjustments (i.e., distributing excess demand systemwide among different hours during the peak period). For more detail on these approaches, see [Cambridge Systematics, 1997].

New time-of-day modeling approaches will likely be developed in the near future. The level of detail necessary to examine policies such as congestion pricing and TDM requires a good assessment of time of travel. With new data collection techniques, such as panels and travel diaries, time-of-departure data will be increasingly available. It is likely that such data will be used to develop models that predict time of travel as it relates to system performance.

5.4.6 Summary

UTMS was originally developed during the 1950s and 1960s as the basic modeling framework for comprehensive, long-range transportation planning. The primary purpose of these studies was to plan the major transportation facilities, principally highways, required by metropolitan areas to cope with postwar growth. Since that time, planning issues and alternatives have evolved dramatically, to the point where concerns exist about the ability of UTMS to deal with the full range of issues and policies of current and emerging interest.

This does not mean, however, that UTMS as an *approach* to urban transportation demand analysis is not still of great importance. A considerable amount of the terminology and the "world view" adopted by most demand analysts is drawn directly from UTMS. In the next section, for instance, the discussion of individual choice models is presented within the four-stage framework of generation, distribution, mode split, and assignment, although spoken of in terms of trip frequency, destination choice, and route choice. Such consistency is useful in terms of ensuring a common language and understanding among demand analysts. It does, however, also pose potential problems in terms of limiting one's ability to think imaginatively about transportation demand, particularly in terms of achieving a better behavioral understanding of some of the processes involved.

Further, UTMS remains the standard modeling tool for the vast majority of metropolitan areas around the world. A wide variety of commercially available software packages is available to support UTMS-based modeling, some of the more common of which in North America include EMME/2, MINUTP, System II, TRANPLAN, and TRANSCAD. Thus, while new techniques and analysis tools are emerging, UTMS-based models will still by and large constitute the most common approach employed in practice in the foreseeable future.

5.5   DISCRETE CHOICE MODELS

5.5.1   Overview of Choice Theory

Transportation demand can be characterized as the aggregation of the decisions of all individual trip makers within a metropolitan area. A number of techniques for predicting demand directly at the *aggregate* level were discussed in previous sections. An alternative approach, which first emerged in the 1970s, is to model directly the decision process of *individual trip makers* and then sum over all trip makers in order to obtain the aggregate demand predictions typically required by the evaluation process.

The most common starting point for individual choice models is the notion of *utility maximization* that was introduced in Sec. 5.2.1. That is, decision makers are assumed to assign at least an ordinal ranking to the trip alternatives available in terms of their relative desirability or utility. Being a rational person, the decision maker will choose the alternative with the maximum utility—the one that maximizes benefits. Utility maximization is central to microeconomic theory but is not restricted to it in its applications. In particular, given the derived nature of transportation demand, it seems reasonable that travelers will want to minimize travel time and cost, maximize comfort and convenience, and so on, whenever possible. In this context, utility simply represents a convenient generalized function that accounts for the positives and the negatives involved in trip making and that forms the basis for a traveler's decision.

Conventional microeconomic theory assumes that the traveler is able to use perfectly all of the trip information available and relevant to the decision and to make a completely rational, consistent decision given this information. A major relaxation of these assumptions is possible by introducing the concept of *random utility*. Primarily originating in the field of psychology [Thurstone, 1927; Luce and Suppes, 1965], such models represent an attempt to retain the analytical tractability provided by the economic assumption of a human being as a rational utility maximizer within a more flexible or realistic world view. These models recognize that, in practice, people do not always choose the objectively best course of action, nor do they necessarily exhibit consistent choices over time (similar to the satisficing model of decision making discussed in Chap. 2). That is, random utility theory still assumes that an individual will choose the alternative that *appears* to maximize his or her utility when the choice is being made. This utility is assumed to consist of two components: (1) the systematic, observable utility that is identical to the conventional microeconomic utility function; and (2) a random term that is intended to capture such effects as variations in perceptions and tastes of individual trip makers, misspecification of the utility function by the analyst, and measurement errors on the part of the analyst [Manski, 1973].

If one can assume that this random term enters the utility function additively, then the utility of some course of action *i* for an individual *t* can be expressed as

$$U_{it} = V_{it} + \epsilon_{it} \qquad [5.17]$$

where

    U_{it} = random utility of alternative i for individual t

    V_{it} = systematic (observable) portion of utility

    ϵ_{it} = random portion of utility

Further, the systematic utility V_{it} is assumed to be a function of the attributes of the alternative X_i and the characteristics of the individual S_t. In particular, it is typically assumed for reasons of analytical tractability that V_{it} is given by

$$V_{it} = b_1 Z_{it1} + b_2 Z_{it2} + \ldots b_n Z_{itn} \qquad \text{[5.18]}$$

where

    b = row vector of parameters

    $Z_{it} = f(X_i, S_t)$ **[5.19]**

If the modeler could observe the value of the random terms for any given decision maker, these values would be incorporated within the systematic or observable portion of the utility function and would no longer be treated as random. However, with randomness incorporated into the decision-making formulation, the modeler cannot say with certainty which alternative will have the maximum utility for a specific decision maker and thus which alternative will be chosen. What can be assessed is the *probability* that a given alternative i from a set of alternatives available to individual t will be the maximum utility alternative for that individual and hence be chosen. That is, given Eq. [5.18] and given a set of alternatives C_t, the probability of individual t choosing alternative i from this set of alternatives (P_{it}) is

$$P_{it} = P(U_{it} \geq U_{jt}) \quad \forall j \in C_t \qquad \text{[5.20]}$$

Or, substituting Eq. [5.17] into [5.20],

$$P_{it} = P(V_{it} + \epsilon_{it} \geq V_{jt} + \epsilon_{jt}) \quad \forall j \in C_t$$
$$= P(\epsilon_{jt} - \epsilon_{it} \leq V_{it} - V_{jt}) \quad \forall j \in C_t \qquad \text{[5.21]}$$

Equation [5.21] is an expression for the joint cumulative distribution function of the random variable $\epsilon_{jt} - \epsilon_{it}$ evaluated at the points $V_{it} - V_{jt}$. Thus, if the distribution of the ϵ's is known or assumed, this equation can be used to compute the probability of an individual making a given choice. Perhaps the most obvious assumption to make is that the ϵ's are distributed multinomially normal. This assumption generates what is known as a *probit* model. Unfortunately, multinomial probit models cannot be expressed easily in an analytically closed form [Daganzo, 1979] and hence are computationally cumbersome and expensive to use.

An alternative assumption concerning the distribution of the ϵ's is that they are each independently and identically distributed (*iid*) with a Gumbel Type I distribution whose cumulative distribution function is given by

$$F(w) = e^{-e^{-w}} \qquad \text{[5.22]}$$

Choosing this particular distribution is motivated entirely by considerations of analytical convenience, since when Eq. [5.22] is integrated in order to evaluate

Eq. [5.21], it can be shown [Domencich and McFadden, 1975; Hensher and Johnson, 1981] that the final expression for P_{it} is the multinomial *logit* model given by

$$P_{it} = \frac{e^{V_{it}}}{\sum_j e^{V_{jt}}} \qquad j \in C_t \qquad\qquad [5.23]$$

As an example of a multinomial logit model, consider a three-mode choice situation in which a worker must choose between auto, bus, and walking for the journey to work. The systematic utility functions associated with these alternatives might take the form

$$V_{auto} = 1.0 - 0.1(TT_{auto}) - 0.05(TC_{auto}) \qquad\qquad [5.24a]$$

$$V_{bus} = -0.1(TT_{bus}) - 0.05(TC_{bus}) \qquad\qquad [5.24b]$$

$$V_{walk} = -0.5 - 0.1(TT_{walk}) \qquad\qquad [5.24c]$$

where
TT_i = travel time by mode i, minutes
TC_i = travel cost by mode i, dollars

Assume that a given individual is faced with travel times of 5, 15, and 20 minutes for the auto, bus, and walk modes, respectively. Similarly, out-of-pocket travel costs by auto and bus are \$1.60 and \$1.50, respectively. In this case, the values of the systematic utility functions given by Eq. [5.24] are

$$V_{auto} = 0.42 \quad V_{bus} = -1.575 \quad V_{walk} = -2.5 \qquad\qquad [5.24a]$$

Substituting these values into Eq. [5.23], the probability of this worker choosing the auto mode is

$$P_{auto} = e^{0.42} / \left(e^{0.42} + e^{-1.575} + e^{-2.5} \right) = 1.522/1.811 = 0.841$$

Similarly, $P_{bus} = 0.207/1.811 = 0.114$, and $P_{walk} = 0.082/1.811 = 0.045$

5.5.2   Characteristics of the Logit Model

The logit model has a tractable and convenient functional form. In particular, its parameters can be statistically estimated relatively easily and efficiently using fairly standardized maximum likelihood techniques. Major characteristics and issues associated with the use of this model include

1. The independence of irrelevant alternatives assumption.
2. Representation of the individual's decision-making structure.
3. Specification of the utility function.
4. Aggregation of predictions.
5. Data requirements.
6. Model transferability.

Each of these topics is discussed briefly in the following. For a more detailed discussion of these and other issues concerning individual choice models, the reader is referred to Ben-Akiva and Lerman [1985] or Ortuzar and Willumsen [1994].

Independence of Irrelevant Alternatives   The logit model belongs to a class of models that possesses the so-called independence of irrelevant alternatives (IIA) property. This property can be illustrated most easily by observing from Eq. [5.23] that the relative probability of an individual t choosing alternative i rather than j, another alternative in the choice set, is

$$\left(P_{it}/P_{jt}\right) = \frac{e^{V_{it}}}{e^{V_{jt}}}$$ [5.25]

The key point to note about Eq. [5.25] is that the relative probability of choosing i rather than j depends only on the characteristics (utility) of the alternatives i and j. That is, it is independent of any other alternative that might be available. Further, as long as the values of V_{it} and V_{jt} do not change, this relative probability will not change, regardless of whether other alternatives are added or deleted from the choice set.

The IIA property is both one of the strengths of the logit model and its major weakness. The property is advantageous in that it means the model can be estimated based on one choice set and then used to predict choices from a modified choice set. Thus, for example, a mode split model can be estimated based on currently available modes and then used to examine the impact of the introduction of a new mode into the system. The property can also be exploited in cases where the choice set is potentially very large (e.g., shopping destination choice, residential location choice, etc.) to eliminate the need for explicitly including the entire choice set in the calculations. That is, a subset randomly selected from the overall choice set can be used to generate statistically consistent estimation and prediction results [McFadden, 1978].

The problem with the IIA property is that care must be taken to ensure the alternatives included in the choice set are, indeed, independent of each other. Figure 5.17 provides a case in which the independence assumption is violated, with disastrous results [Daganzo and Sheffi, 1977]. Figure 5.17a presents a simple route choice problem in which two routes with equal travel times are available and in which the probability of either route being chosen is clearly 0.5. Figure 5.17b presents a modification of the first problem in which one route has been split into two subroutes that are identical except for an arbitrarily small link at one point. Travel times on all three routes remain equal. Obviously, this arbitrarily small change in the network should have no practical effect on the system state: there are still essentially two "real" routes available, and the traffic should split evenly between them. As shown by Fig. 5.17b, however, a simpleminded application of the logit model to the second case results in a prediction of 0.33, 0.33, 0.33 for the three routes, or a 1/3 to 2/3 split between the two "real" routes. This is a direct result of the IIA property (note that the ratio P_1/P_2 equals 1.0 in both cases; that is, it is independent of what other alternatives are available) or rather a direct result of applying the logit model to a choice set that clearly violates the IIA assumption. Alternatives 2 and 3 are *not*

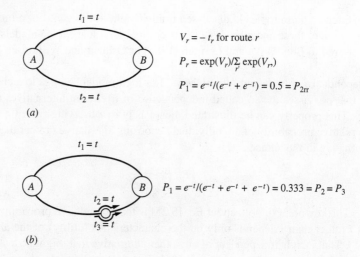

$$V_r = -t_r \text{ for route } r$$

$$P_r = \exp(V_r)/\sum_{r'} \exp(V_{r'})$$

$$P_1 = e^{-t}/(e^{-t} + e^{-t}) = 0.5 = P_{2rr}$$

$$P_1 = e^{-t}/(e^{-t} + e^{-t} + e^{-t}) = 0.333 = P_2 = P_3$$

Figure 5.17   Example violation of the independence of irrelevant alternatives property: (a) sequential decision process; (b) joint decision process
| SOURCE: Daganzo and Sheffi, 1977

independent of each other; rather they are highly dependent, and the probability of choosing one is highly correlated with the probability of choosing the other.

Tests are available for identifying violations of the IIA assumption [Ben-Akiva and Lerman, 1985]. If the choice set does appear to violate the IIA assumption to the extent that the logit model is untenable, then modeling options include using a probit model, which is capable of handling correlation among alternatives, or restructuring the decision structure assumed (discussed further in the following) so as to reduce or eliminate the dependence between alternatives.

Decision Structure   One approach for resolving the IIA violation in the route-choice problem previously discussed is to consider the problem as a two-stage decision process in which choice is first made between the two major routes, and then a second choice is made, if required, between the two subroutes. Figure 5.18 presents the *decision tree* representation of this two-stage, or *sequential*, process and contrasts it with the corresponding one-stage, or *joint*, decision process previously discussed.

In any complex choice situation, a number of decision structures are generally conceivable. In Sec. 5.4, it was observed that UTMS implicitly assumes a sequential process consisting of decisions concerning whether to make a trip, where to go given that a trip is made, what mode to use given the trip destination, and what route to use through the chosen mode's network to reach the chosen destination. An alternative decision structure is to assume that the decisions of whether to make a trip, where to go, and what mode to use to get there are all made simultaneously; that is, a joint decision process exists.

Bowman and Ben-Akiva [1997] suggest the *choice hierarchy* shown in Fig. 5.19 for representing travel decisions. Higher-level decisions in the choice hierarchy are

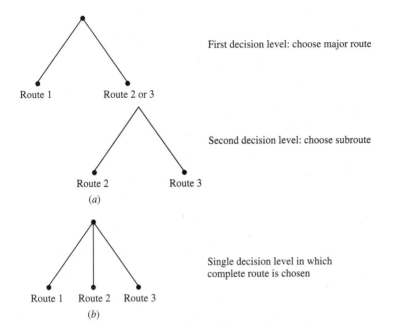

First decision level: choose major route

Second decision level: choose subroute

Single decision level in which
complete route is chosen

Figure 5.18 Alternative decision structures for a route choice problem: (a) sequential decision
process; (b) joint decision process

made prior to lower-level decisions, which in turn are *conditional* decisions based on
the higher-level choices. Thus, nonwork travel decisions are assumed to depend on
prior work-trip decisions that, among other things, determine the number of household autos that will be available for nonwork trips. Decisions within each level are
generally assumed to be made jointly, although subhierarchies are conceivable.

The determination of what choice structure to adopt is primarily based on theoretical grounds, although data availability, problem context, and calibration issues
can also play a role. The key point, however, is that an assumption of a decision
structure is exactly that—a behavioral assumption concerning the trip maker's decision-making process. As such, its validity should be tested to the extent that this is
possible.

One approach to empirically testing decision structure hypotheses, as well as to
providing an alternative decision structure "in between" the pure joint and pure
sequential structures discussed to this point, is to adopt the so-called *nested* decision
structure [McFadden, 1979]. In a nested structure, decisions are still made sequentially, but a higher-level decision (i.e., one made early in the decision process) may
include in its calculations expectations concerning subsequent lower-level decisions
(i.e., ones made later in the decision process). In particular, the *expected maximum
utility* associated with the next stage in the decision process is included in the current stage's utility function.

Mathematically, the nested logit model decomposes into two ordinary logit
models [Ben-Akiva and Lerman, 1985]. If, for example, one models the choice of

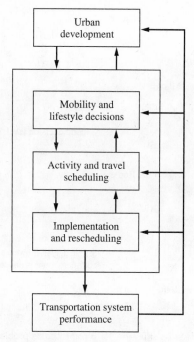

Figure 5.19   Choice hierarchy for travel decisions
SOURCE: Bowman and Ben Akiva, 1997

mode m and destination d for shopping trips, a typical nested logit model for this process would be:

$$P_{m\mid d} = \frac{e^{\dfrac{V_{m\mid d}}{\mu}}}{\sum e^{\dfrac{V_{m'\mid d}}{\mu}}} \qquad m' \in M_d \qquad\qquad \text{[5.26a]}$$

$$P_d = \frac{e^{(V_d + \mu I_d)}}{\sum e^{(V_{d'} + \mu I_{d'})}} \qquad d' \in D \qquad\qquad \text{[5.26b]}$$

where

$\quad P_d \quad$ = probability of choosing destination d from the choice set D

$\quad P_{m/d} \quad$ = probability of choosing mode m from the choice set M_d, given that d has been chosen as the trip's destination

$\quad I_d \quad$ = *inclusive value* for destination d = expected maximum utility associated with choosing a travel mode, given the choice of destination d

$\qquad\quad = \log_e \left\{ \sum e(V_{m\mid d}/\mu) \right\} \qquad m \in M_d \qquad\qquad \text{[5.26c]}$

$\quad \mu \quad$ = *scale parameter* $(0 \leq \mu \leq 1)$

Equations [5.26a] and [5.26b] can be estimated sequentially as two separate logit models, using ordinary logit model estimation software (with some loss of statistical efficiency), or simultaneously, using commercially available software such as ALOGIT and LIMDEP. Nested logit models are very commonly used for modeling mode choice, both for implementation within urban UTMS modeling systems and for use in intercity travel demand modeling applications [Miller and Fan, 1992].

Nested logit models are actually special cases of an even more general class of models known as *generalized extreme value* (GEV) models [Ben-Akiva and Lerman, 1985]. Considerable experimentation with various, more complex GEV models is underway, since they possess the potential to deal with a variety of more complex choice situations for which even nested logit models are inadequate. The mathematical complexity of these models, however, is quite high, and software to support their application in practical planning contexts does not yet exist.

Utility Function Specification In the theoretical development of the choice model, it is simply assumed that a systematic utility function V_{it} exists for each individual t and alternative i. In practice, the specification of this utility function constitutes a major task in the model-building process. While it is possible to develop utility functions for each individual [Lerman and Louviere, 1978], conventional practice involves either categorizing individuals into relatively homogeneous groups and then developing utility functions for each group or developing generalized utility functions within which an individual's socioeconomic characteristics enter directly.

Variables within a utility function can be either *generic* or *alternative-specific* in nature. A generic variable is one that is included in every alternative's utility function with exactly the same weight (i.e., the same parameter value). An alternative-specific variable, on the other hand, has different weights for different alternatives, including an a priori specified weight of zero (i.e., it does not enter into a particular alternative's utility function). A special type of alternative-specific variable is the alternative-specific constant or bias term that is often employed to capture systematic, "all else being equal" preferences exhibited within a sample.

Some of these concepts can be illustrated with a simple modal choice problem consisting of two modes, auto (a) and transit (t). Transportation variables chosen to characterize the system are in-vehicle travel time (*IVTT*), out-of-vehicle travel time (*OVTT*), and out-of-pocket travel costs (*OPTC*). Two socioeconomic variables are used to characterize each traveler: household income (*INC*) and household auto-ownership level (*AO*). A simple modal split model using these variables might be specified by the following utility functions:

$$V_a = \beta_1 + \beta_2 IVTT_a + \beta_4 OVTT_a + \beta_5\left(OPTC_a/INC\right) + \beta_6 AO \qquad \textbf{[5.27a]}$$

$$V_t = \beta_3 IVTT_t + \beta_4 OVTT_t + \beta_5\left(OPTC_t/INC\right) \qquad \textbf{[5.27b]}$$

Several points concerning Eqs. [5.27a] and [5.27b] should be noted:

1. *OVTT* and the composite variable *OPTC/INC* are generic variables because they enter both utility functions with the same parameter value (that is, β_4 and β_5, respectively).

2. *IVTT* is an alternative-specific variable because it enters the two equations with different weights (that is, β_2 and β_3). This reflects the hypothesis that a minute spent riding a bus is perceived (weighted) differently than a minute spent driving in a car.

3. While the utility functions are "linear in the parameters," nonlinear composite variables (such as the *OPTC/INC* term) can be included.

4. β_1 is an alternative-specific constant for the auto mode. No transit constant is specified (or, more correctly, the transit constant is arbitrarily set equal to zero) because in a choice set of n alternatives, at most $n-1$ alternatives are statistically identifiable.

5. Because socioeconomic characteristics for a given individual do not vary across alternatives, socioeconomic variables must enter the utility functions as alternative-specific variables (such as *AO*) or generically in functional combination with a system variable (e.g., the *OPTC/INC* term). That is, a generic socioeconomic variable will add exactly the same value to every alternative's utility function and will thus have absolutely no impact on the choice probabilities.

Table 5.4 summarizes home-based work mode choice coefficients that were estimated for different variables and in different cities. Generally, the coefficients are in a similar range for each variable.

It can also be shown mathematically [Ben-Akiva and Lerman, 1985; Koppelman and Wen, 1998] that the direct elasticity of the probability of an individual t choosing alternative i with respect to a change in some attribute that is an independent variable in the model (that is, *IVTT, OVTT*, etc.) having 1 to k parameters is

$$\text{Direct elasticity} = \left[1 - P_i\right]\left(x_{itk}\right)\left(\beta_k\right) \qquad \textbf{[5.28a]}$$

The cross elasticity of the probability alternative i that is selected with respect to an attribute of alternative j is

$$\text{Cross elasticity} = -P_i\left(x_{itk}\right)\left(\beta_k\right) \qquad \textbf{[5.28b]}$$

For nested logit models, the direct elasticity can be estimated with

i not in nest $\quad \text{Direct elasticity} = \left(1 - P_i\right)\beta_k x_{itk}$ **[5.28c]**

i in nest m $\quad \text{Direct elasticity} = \left[\left(1 - P_i\right) + \left(1/\mu_m - 1\right)\left(1 - P_{i/m}\right)\right]\beta_k x_{itk}$ **[5.28d]**

and the cross elasticity can be estimated with

i and i' not in the same nest $\quad \text{Cross elasticity} = -\left(P_i\right)\beta_k x_{jtk}$ **[5.28e]**

i and i' in nest m $\quad \text{Cross elasticity} = -\left[P_i + \left(1/\mu_m - 1\right)P_{i/m}\right]\beta_k x_{jtk}$ **[5.28f]**

where μ_m is an index of dissimilarity of alternatives included in nest m, which varies between 0 and 1.

Aggregation Individual choice models generate predictions of the probabilities associated with given individuals choosing a particular outcome from a set of

Table 5.4 Home-based work mode choice coefficients of selected cities

Coefficients on Service-Level Variables from a Sample of Home-Based Work Mode Choice Models

City	In-Vehicle Time	Transit Drive-Access Time	Out-of-Vehicle Time	Auto Terminal Time	Transit Walk Time	Initial Transit Wait Time	Transit Transfer Time	Total Cost	Auto Operating Cost	Transit Fare	Parking Cost
New Orleans	-.015	-.100			-.033	-.077	-.032	-.008			
Minn/St. Paul	-.031				-.044	-.030	-.044	-.014			
Chicago	-.028			-.030	-.114	-.023	-.114	-.0121			
Los Angeles	-.020		-.112					-.0144			
Seattle	-.040	-.286			-.044	-.030	-.044	-.014			
Cincinnati	-.019		-.028					-.0045			
Washington	-.017		-.058						-.004	-.004	-.009
San Francisco	-.025		-.058				-.059	-.0039			
Dallas	-.030			-.055	-.055	-.055			-.005	-.005	-.012
Shirley (low)	-.022	-.055	-.035					-.0037			
Shirley (high)	-.034		-.044					-.0046			
Salt Lake City	-.019		-.037					-.0059			
Portland	-.034		-.072					-.01384			

[1] SOURCE: Martin and McGuckin, 1998

alternatives. As such, these probabilities are of little direct use for planning purposes. That is, a planner is rarely interested in the probability that a specific individual will choose transit, but rather the total number of people in a zone or in a study area likely to choose transit. Thus, some procedure must be employed to aggregate the individual choice predictions of the model to yield the total demand predictions required for planning purposes.

In principle, the simplest aggregation procedure is to enumerate all individuals within the study area and sum their probabilities of choosing a given alternative. While such a *total enumeration,* as this procedure is called, is being experimented with within emerging microsimulation models (see Sec. 5.6), it does not currently represent a practical approach within most operational planning environments. Some other aggregation procedure is generally required. While a range of procedures exist [Koppelman, 1975], the three most commonly considered are

1. "Naive" aggregation.
2. Classification with naive aggregation.
3. Sample enumeration.

Naive aggregation involves treating the individual choice model as if it were an aggregate model by using zonal average values for the utility function variables in order to compute an "average" zonal probability. Because logit model probabilities are nonlinear functions of the utility function variables, however, such use of average values will not generate the correct average probability. The errors that can occur through the use of naive aggregation can be substantial, and one should avoid the use of this technique whenever possible.

Aggregation errors can generally be reduced if the population is classified into relatively homogeneous groups with respect to one or more key variables prior to the use of naive aggregation. Koppelman identifies auto ownership and transit availability as two key variables to be used in classification for the modal split case [Koppleman, 1975], although other researchers have found that the key variables may differ from one urban area to another, as does the accuracy of the procedure [Reid, 1978].

While total enumeration is generally impractical, if not infeasible, very often a representative sample of the population is available to the analyst. In such cases, this sample can be enumerated, and the resulting sample prediction can be used as an estimate of the population prediction (with appropriate "grossing up," if required). For short-run analyses, the calibration data set or some similar sample is generally available and can be used. For longer-run analyses, or in the absence of a current sample, a sample can often be synthesized from census data, zonal population forecasts, and so on, if reasonable assumptions about the distributions of sample characteristics can be made. Provided that a representative sample is available or can be reliably generated, sample enumeration is generally the preferred aggregation procedure.

Table 5.5 provides a numerical example illustrating the use of the enumeration, naive, and classification approaches to aggregation for the bimodal logit model specified by Eq. [5.27]. The table presents the coefficient values assumed for the example (Table 5.5a) and the values of the explanatory variables in the utility functions for 10 observations (Table 5.5b). The final column of Table 5.5b presents the probabilities of choosing the auto mode for each observation. To aggregate by enumeration, one simply adds up these individual probabilities and divides by the total number of observations (10). This yields the true expected aggregate probability of 0.88635.

Table 5.5c illustrates the naive aggregation procedure. In this table, the average values for the utility function variables are shown. Using these average values in Eq. [5.27] results in the "average" probability of choosing auto (0.91505), shown in the last column of the table—a value about 3 percent higher than the true expected value computed in Table 5.5b. This result is typical of all naive aggregation predictions in that it represents an overprediction of the "dominant" alternative, in this case, the auto mode.

Finally, classification with naive aggregation is illustrated in Table 5.5d. This table shows the number of observations falling into each income–auto-ownership category, the average values for the utility function variables *within* each category, and the naive "average" probability of choosing auto for each category (computed by substituting the within-category average values into Eq. [5.27]). The aggregate "average" value is then computed by taking an average of these probabilities, weighted by group size. That is, the aggregate "average" probability computed by the classification procedure is equal to

$$\frac{3(0.81038) + 2(0.93913) + 1(0.97851) + 2(0.85516) + 2(0.96390)}{3 + 2 + 1 + 2 + 2} = 0.89260$$

Thus, while the classification procedure still overpredicts auto usage, it performs much better than the pure naive approach in that it differs, for this test example, by less than 1 percent from the true average value.

The need to aggregate individual choice model predictions is sometimes viewed as a liability of the technique. In fact, it represents a major strength in that it means the model can be applied at any level of spatial aggregation. Conventional aggregate models can be used only in conjunction with the zonal system for which they were calibrated, since this calibration includes (in unknown ways) all the idiosyncrasies of the spatial aggregation implicit in the zonal system adopted. Thus, once calibrated, an aggregate model can perform analysis at only one level of spatial aggregation, regardless of whether this level is too coarse or too fine for the given problem application. Individual choice models, on the other hand, can be aggregated to any desired level, depending upon data availability and the problem context.

Data Requirements   Individual choice models are more efficient than corresponding aggregate models in their use of data. This is because aggregate models typically employ zonal averages, which require a fair number of observations to construct, whereas individual choice models (and disaggregate models in general) employ every observation directly in their calibration. Thus, individual choice models require fewer observations to construct, for a given level of accuracy. Given the high cost of data collection and the very large samples typically gathered in the past

Table 5.5   Numerical example of aggregation procedures

(a) Utility Function Coefficients		
$b_1 = 0.25$	$b_3 = -0.11$	$b_5 = -0.0007$
$b_2 = -0.10$	$b_4 = -0.20$	$b_6 = 0.25$

(b) Utility Function Variables and Choice Probabilities for Individual Observations

Obs. No.	$IVTT_a$	$IVTT_t$	$OVTT_a$	$OVTT_t$	$OPTC_a$	$OPTC_t$	INC	AO	P_{auto}
1	20	25	5	10	250	50	1	1	0.89187
2	25	35	5	15	300	50	3	2	0.98274
3	15	18	3	8	225	50	2	2	0.89741
4	30	40	5	15	400	50	3	1	0.97851
5	20	30	5	10	300	50	2	1	0.93776
6	10	12	3	5	150	50	1	1	0.75951
7	15	25	8	5	100	50	2	2	0.79939
8	35	40	5	10	600	75	3	2	0.92605
9	30	40	5	10	450	50	2	1	0.94048
10	10	15	5	5	125	50	1	1	0.74979
Average P_{auto}									0.88635

(c) Naive Aggregation

	$IVTT_a$	$IVTT_t$	$OVTT_a$	$OVTT_t$	$OPTC_a$	$OPTC_t$	INC	AO	P_{auto}
Average values:	21.0	28.0	4.9	9.3	290.0	52.5	2.0	1.4	0.91505

(d) Classification with Naive Aggregation

No. of Obs.	$IVTT_a$	$IVTT_t$	$OVTT_a$	$OVTT_t$	$OPTC_a$	$OPTC_t$	INC	AO	P_{auto}
3	13.3	17.3	4.3	6.7	175.0	50	1	1	0.81038
2	25.0	35.0	5.0	10.0	375.0	50	2	1	0.93913
1	30.0	40.0	5.0	15.0	400.0	50	3	1	0.97851
0							1	2	
2	15.0	21.5	5.5	6.5	162.5	50	2	2	0.85516
2	30.0	37.5	5.0	12.5	450.0	62.5	3	2	0.96390
Weighted average P_{auto}									0.89260

NOTE: All times are in minutes; all costs are in cents; income is expressed as a code (1, 2, 3)

to construct aggregate models, this is a very important strength of the individual choice model technique.

The nature and the detail of the data required by individual choice models, however, are often greater than that collected for aggregate models. A wider range of socioeconomic information and a more detailed representation of the level of service variables (including service characteristics for "unchosen" alternatives) experienced by the observed travelers are typically required. Further, this more detailed information must be available for the future situations for which predictions are required. As a rule, then, while individual choice models require less quantity of data (in terms of the number of observations in the sample), they often require higher quality data (in terms of the information obtained per observation).

Model Transferability   Because individual choice models are not tied to a specific zone structure within a specific city and because they possess the potential for a relatively rich representation of the factors affecting a traveler's decision making, it has been argued that such models should be capable of being transferred from one geographical location to another. The benefits of such transferability would be enormous in that it would significantly reduce in a number of cases the data requirements, calibration time and costs, and detailed analytical expertise required by planners to perform demand analyses. Considerable research has gone into investigating the transferability properties of discrete choice models within practical planning contexts. While no universally transferable model exists, logit mode split models can, with care, be transferred from one "context" (e.g., one city or one time period) to another, particularly if some updating of the model parameters is performed, using either available aggregate data that is used to adjust the modal constants within the model or a small local sample, in which case all or most of the model parameters can be adjusted [Badoe and Miller, 1995].

5.6 ACTIVITY-BASED METHODS

In order to develop more behaviorally based travel models, researchers in recent years have focused on the fact that travel arises out of the need to participate in out-of-home activities (work, shop, school, etc.). This directly leads to the conclusion that what one should study in the first instance is not travel per se, but rather the participation in the activities that ultimately generate travel. Potential advantages of this approach with respect to improving transportation policy analysis capabilities include

1. An improved capability for modeling nonwork, nonpeak-period travel, which involves complex interactions among household members (Who will do the shopping?), activities/trip purposes (If I do the shopping, I won't have time to pick my child up from day care), time periods (Do I do the shopping before or after work?), sharing of household vehicles, and so on.

2. An improved capability to deal with the trade-offs between in-home and out-of-home activities, telecommuting, "e-commerce" impacts, and so forth.

3. Greater potential for moving beyond traditional explanatory variables, such as travel time and cost in our explanations of travel behavior.

4. Greater potential for dealing with the effects of household structure, life-cycle stage, lifestyles, and so on, on travel behavior.

Activity-based methods do not have a single, unified theoretical base. Rather, they represent a collection of partial theories and various modeling approaches that share the common theme of focusing on activity participation as the motivating force behind travel. A full discussion of these theories and methods is well beyond the scope of this text. Excellent reviews of the field can be found in TTI [1997] and Ettema and Timmermans [1997], as well as in Jones et al. [1983, 1990] and Axhausen and Gärling [1992].

A key starting point for the development of activity-based models, however, is to simply observe that activities occur at specific points in space and occur over a certain duration in time. Locations of some activities are highly constrained in space for most people (work at the office, sleep at home, etc.), while others may occur at a wide variety of places (e.g., shopping). Similarly, some activities are highly constrained in terms of when they can occur (fixed work hours, school class times, etc.), while others can be much more flexibly scheduled. Figure 5.20 illustrates these concepts with a simple time–space diagram (this type of diagram was first applied in this context by Hagerstrand [1970]) in which the location, timing, and duration of activities are explicitly depicted, along with the travel required to move from one activity location to another. In this diagram, a relatively simple *activity pattern* is observed, in which a person leaves home in the morning for work, spends the day at work, and then returns home in the evening, stopping off along the way to do some shopping. In this diagram it is assumed that the work location is fixed, while the shopping location presumably is at least somewhat flexible, depending, of course, on what type(s) of good or service is being sought.

If it is assumed that there is a minimum feasible departure time from home in the morning (e.g., the person can't leave until the children leave for school at 8:30

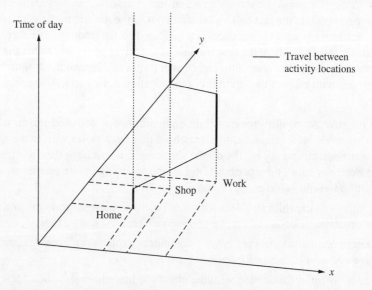

Figure 5.20 Example of a time–space diagram for an individual

A.M.), a maximum feasible arrival time at work (must be there for a 9:00 A.M. meeting), a minimum feasible work departure time (4:30 P.M.), and a maximum feasible arrival time back home again in the evening (must be home to drive a child to soccer practice at 6:15 P.M.), then a more detailed representation of this person's day can be drawn, as shown in Figure 5.21. Note that for simplicity of presentation, this figure assumes a one-dimensional space, wherein the person travels back and forth only along a line. In Fig. 5.21 *time–space prisms* are drawn that indicate the feasible *activity space,* which is defined by the locations and maximum durations for activities in which it would be feasible for the person to engage, given their known time–space constraints. That is, in the prism connecting home and work in the morning, the person could choose to leave the home as early as 8:30; travel to any location contained within the prism and spend up to the maximum time defined by the vertical height of the prism at this location, engaging in an activity at this point; and still be able to reach work by 9:00 A.M. (the latest feasible work arrival time). A similar prism defines feasible activity location–duration combinations in the afternoon postwork period, one of which was actually chosen by the person in order to do some shopping at the point indicated.

The potential power of an activity-based approach relative to conventional demand modeling methods can be illustrated with a very simple example. Suppose that the person represented in Figs. 5.20 and 5.21 is induced to switch to transit for the trip to and from work (say, due to an increase in workplace parking charges or perhaps a transit incentive program of some sort). Suppose also that because of this change in travel mode, the shopping stop on the way home from work is no longer feasible (perhaps the store is not within convenient walking distance of the transit route). In this case, the activity/travel pattern shown in Fig. 5.22 may result in which

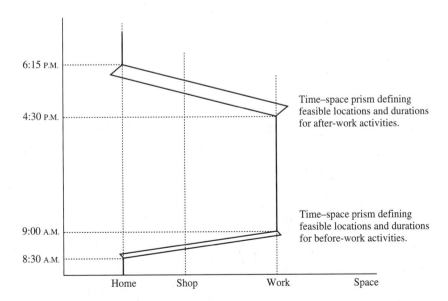

Figure 5.21   Time–space prisms

the shopping activity is now satisfied by an auto trip in the evening from home to the store and back again. Conventional UTMS models are not able to represent linkages between work and nonwork trip making and so would not be able to predict the ripple effect that the change in work trip mode has in this example on non-work travel behavior. The most likely result in this case would be an overestimate of the reduction in VMT, energy use and emissions derived from the work trip mode shift, since the model would have missed the prediction of the new nonwork auto trip in the evening period that was generated by the change in work trip mode.

Although very simple in concept, these figures provide the starting point for developing detailed models of activity-based travel. Bowman and Ben-Akiva [1997] identify two broad classes of activity-based models: *econometric* models and *hybrid simulation* models. Econometric models consist of systems of equations that compute probabilities of possible outcomes (i.e., activity/travel choices). These models are usually utility based and usually consist of multinomial or nested logit models. Bowman and Ben-Akiva further identify three subclasses of econometric models:

1.   *Trip-based* models, in which individual trips within a daily activity schedule are directly modeled. The first (and to this day, one of the few) example of an integrated, trip-based modeling system is the model developed for the San Francisco Bay Area's Metropolitan Transportation Commission (MTC) in the mid-1970s [Ruiter and Ben-Akiva, 1978]. While some would argue that a trip-based model cannot be truly considered an activity-based model, the MTC model certainly represented a pioneering effort to develop an operational model that was able to at least begin to address activity-based concerns and ideas.

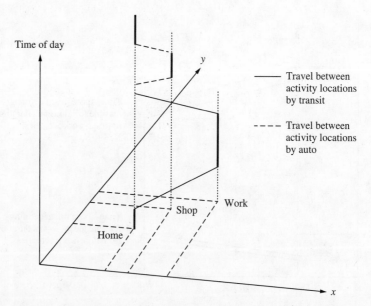

Figure 5.22 Impact of work trip mode change on a daily activity pattern

2. *Tour-based* models, in which home-based (and, sometimes, work-based) tours are explicitly modeled. Several tour-based models have been developed, principally in Europe, including the Netherlands [Daley et al., 1983; Hague Consulting Group, 1992], Sweden [Algers et al., 1995], and Italy [Cascetta et al., 1993].

3. *Daily schedule system* [Bowman and Ben-Akiva, 1997], which extends the tour-based approach by explicitly modeling the choice of a daily activity pattern that provides the framework from which individual tours are then selected. It adds choice of the time of day when tours/trips occur as an explicit component of the choice process. Figure 5.23 illustrates the daily schedule system approach, which has been implemented within an operational planning setting in Portland, Oregon [Bradley and Bowman, 1998].

Bowman and Ben-Akiva's "hybrid simulation" models are more formally known as *computation process models* (CPM's) [Pas, 1997]. CPM's derive from psychological models of decision making [Newell and Simon, 1972] and generally consist of rule-based algorithms that try to model choice processes in a way that (it is hoped) more closely mimics actual decision making [Gärling et al., 1994]. These models typically focus on the *activity scheduling process,* which includes the choice of activity type, duration, sequencing, location, and mode of travel within a daily activity schedule. Examples of CPM-type models include CARLA [Jones et al., 1983], STARCHILD [Recker et al., 1986a, 1986b], SCHEDULER [Gärling et al., 1989; Golledge et al., 1994], AMOS [RDC Inc, 1995], SMASH [Ettema et al., 1995], and PCATS [Kitamura, 1997]. CPM models developed to date tend to be experimental research prototypes rather than operational planning tools, although the methodology is rapidly maturing as more experience is gained with the approach. A partial exception to this generalization is AMOS, which was developed to analyze the likely impacts on travel demand of a range of TDM initiatives in the Washington, D.C. area.

The data required to develop activity-based models must include information on the activities in which people engage, the attributes of these activities such as duration, location, and timing, as well as on actual trips. This is not as difficult as it may first appear. To a large extent, this involves shifting the fundamental question

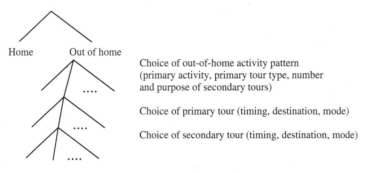

Figure 5.23 Daily schedule system used in Portland, Oregon
| SOURCE: Bradley and Bowman, 1998

within a survey from "Where did you go?" to "What did you do?" A number of activity-based surveys have been undertaken in recent years, including ones in Boston, San Francisco, Portland, and Raleigh-Durham (see Chap. 4 for a review of activity-based surveys). These typically involve mail-out activity diaries (for 1- or 2-day survey periods) that are then retrieved via telephone interview [Lawton, 1997]. Computer-based methods, however, are also emerging that enable the user to directly record their activities *as they are scheduled,* with subsequent updating to reflect actual execution of the planned schedule [Doherty and Miller, 2000]. A particular advantage of this approach is that it provides insight into the activity scheduling process itself, rather than just on the pattern of activities that are actually engaged in.

Many activity-based models are being implemented within a *microsimulation* framework, within which the behavior of each individual is dynamically simulated over time [Goulias and Kitamura, 1996; Miller, 1996; Miller and Salvini, 2000]. Figure 5.24 provides a simple, generic flowchart of the microsimulation process. Microsimulation can be used with either the CPM approach (in which case it is typically the only means by which the model can be operationalized) or econometric models (in which case the econometric model choice probabilities are used within a Monte Carlo process to randomly generate choice outcomes). Despite the computational intensity required by microsimulation, potential advantages of the approach include the following [Miller, 1996]:

1. It is a natural framework within which to implement disaggregate models of human decision making.

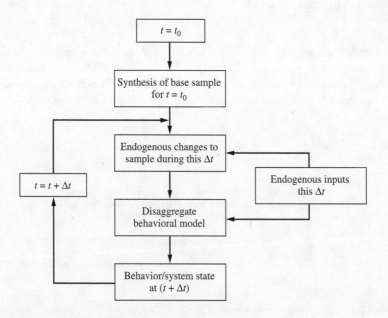

Figure 5.24 Microsimulation process for demand analysis

2. It provides outputs at the disaggregated level required by current and emerging microsimulation network models such as TRANSIMS.

3. It may be the computationally most efficient method for modeling behavior within a highly heterogeneous population.

4. It provides a mechanism for endogenously generating the detailed demographic and socioeconomic inputs required by complex disaggregate decision models.

5. Microsimulation models possess the potential to generate *emergent behavior,* that is, behavior that is not explicitly "hard wired" into the model based on its calibration to past, observed behavior.

6. Given their "high fidelity" with real-life processes, microsimulation models may well be more readily explainable to, and therefore have more credibility with, decision makers than current highly abstract "black boxes" such as UTMS.

5.7 ESTIMATING NONMOTORIZED TRAVEL DEMAND

One of the critical deficiencies in the way transportation demand modeling has occurred during the past several decades has been the relative paucity of attention given to nonmotorized transportation. Much of the travel demand forecasting methodology was oriented to the highway or transit networks. Very little attention was paid to pedestrian or bicycle travel. Several factors often not considered in the traditional four-step modeling process could influence nonmotorized travel demand (Fig. 5.25), but most models do not include such factors. For example, models have not included climate/weather as a factor in mode choice or the friendliness of the network as a consideration of whether a particular mode would be chosen. Yet studies have shown that such factors are important in the decision to walk or bicycle.

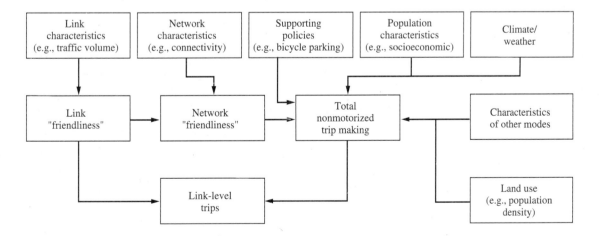

Figure 5.25 Relationship of factors influencing non-motorized travel
SOURCE: Schwartz et al., 1999

The approaches for estimating nonmotorized travel demand range from comparative studies to regional travel demand models. Table 5.6 describes each of these methods and important characteristics of each. Examples of such methods include the following:

Aggregate behavior studies. Nelson and Allen [1997] conducted cross-sectional analysis of 18 U.S. cities to predict bicycle mode split based on weather, terrain, number of college students, and per capita miles of bikeway facilities.

Sketch planning methods. Matlick [1996] developed a method to estimate potential walking trips within 0.8 kilometers of a selected corridor. The method identified traffic generators such as housing units, average persons per unit, and average number of trips per person. Trip attractors were similarly identified. All potential walking trips (i.e., all trips less than 0.8 kilometers in length) were multiplied by a mode split percentage obtained from census data.

Discrete or individual choice models. Wilbur Smith and Associates [1997] developed a set of transit access models to predict the impacts of transit access mode on rail ridership. Surveys and visual simulation exercises were used to determine whether people would shift to nonmotorized access modes if improvements such as bicycle parking, paths and lanes, and debris removal were made. Pedestrian improvements included sidewalks, traffic calming, and intersection enhancements.

Regional models. Several metropolitan areas have included nonmotorized modes as part of the travel demand forecasting methodology. Table 5.7 shows how these modes have been included. Of some interest is an environmental factor, which is a composite index of factors that have been shown to influence the attractiveness of an area for walking or bicycling. For example, Portland's pedestrian environment factor includes zonal descriptors relating to (1) sidewalk availability, (2) ease of street crossing, (3) connectivity of street/sidewalk system, and (4) terrain. Many of these model enhancements were funded by environmental groups (see, for example, [1,000 Friends of Oregon, 1991–1997; Chesapeake Bay Foundation, 1995]).

It seems likely that future model developments will include in a more serious way the ability to predict nonmotorized travel demand. Those interested in an overview of these methods are referred to [Schwartz et al., 1999a; 1999b].

5.8   ESTIMATING GOODS MOVEMENT DEMAND

Historically, planning agencies have devoted more attention to passenger movement than they have to urban goods movement. This is primarily so because goods movement was considered the responsibility of the private sector, while at the same time, the underlying phenomena influencing freight demand were more complex

Table 5.6 Available methods for estimating nonmotorized travel demand

Purpose/Method	Description
Demand Estimation	*Methods That Can be Used to Derive Quantitative Estimates of Demand*
Comparison studies	Methods that predict nonmotorized travel on a facility by comparing it to usage and to surrounding population and land-use characteristics of other similar facilities
Aggregate behavior studies	Methods that relate nonmotorized travel in an area to its local population, land use, and other characteristics, usually through regression analysis
Sketch plan methods	Methods that predict nonmotorized travel on a facility or in an area based on simple calculations and rules of thumb about trip lengths, mode shares, and other aspects of travel behavior
Discrete choice models	Models that predict an individual's travel decisions based on characteristics of the alternatives available to them
Regional travel models	Models that predict total trips by trip purpose, mode, and origin/destination, and distribute these trips across a network of transportation facilities, based on land-use characteristics such as population and employment and on characteristics of the transportation network
Relative Demand Potential	*Methods That Do Not Predict Actual Demand Levels but Which Can be Used to Assess Potential Demand for, or Relative Levels of, Nonmotorized Travel*
Market analysis	Methods that identify a likely or maximum number of bicycle or pedestrian trips that may be expected, given an ideal network of facilities
Facility demand potential	Methods that use local population and land-use characteristics to prioritize projects based on their relative potential for use
Supply Quality Analysis	*Methods That Describe the Quality of Nonmotorized Facilities (Supply) Rather Than the Demand for Such Facilities; these may be Useful for Estimating Demand if Demand Can be Related to the Quality of Available Facilities.*
Bicycle and pedestrian compatibility measures	Measures that relate characteristics of a specific facility such as safety to its overall attractiveness for bicycling or walking
Environment factors	Measures of facility and environment characteristics at the area level that describe how attractive the area is to bicycling or walking
Supporting Tools and Techniques	*Analytical Methods to Support Demand Forecasting*
Geographic information systems	Emerging information management tools, with graphic or pictorial display capabilities, that can be used in many ways to evaluate both potential demand and supply quality
Preference surveys	Survey techniques that can be used on their own to determine factors that influence demand and that also serve as the foundation for quantitative forecasting methods such as discrete choice modeling

SOURCE: Schwartz et al., 1999a

Table 5.7 Inclusion of nonmotorized modes in regional travel models

Region	Modes				Steps of Modeling Process			Network Characteristics	
	Bicycle	Pedestrian	Combined	Walk to Transit	Mode Choice	Route Choice	Other	Environment Factor	Bike/Ped Facilities
Edmonton, Canada	X	X			X	X	X[1]		X
Portland, OR			X		X			X	
Sacramento, CA	X	X			X			X	
Montgomery Co., MD				X	X			X	
San Francisco, CA	X	X			X				
Los Angeles, CA			X	X	X				
Albany, NY	X	X			u	X		u	X
Leicester, UK (TRIPS/START)	X					X			X
Ipswich, UK (Quovadis)	X					X			X

[1] Utilities based on mode and route characteristics feed back to affect trip generation and trip distribution.

NOTE: u = under development

SOURCE: Schwartz et al., 1999b

and interdependent than those influencing passenger demand. For example, a 1996 manual listed the following reasons for this complexity:

- Decisions by shippers, carriers, and receivers affect whether or not a particular shipment is made and, if so, by what mode and route.

- There are many different types of commodities that make up the freight traffic, and these commodities have a wide range of prices or values associated with them (also some are perishable while others are not).

- Freight movements are measured in various units such as dollar value, quantity, weight, volume, container, carload, truckload, and so on.

- The cost of moving freight is much harder to determine compared to the cost of moving passengers because more specialized services are required for freight (i.e., handling, loading, unloading, classifying, storing, packaging, warehousing, inventorying, etc.)[Cambridge Systematics, 1996].

With the increased emphasis on goods movement and intermodal freight that came from ISTEA, transportation planners have become more interested in methods to forecast goods movement [Coogan, 1996]. These methods have ranged from simple trend analysis to full-scale modeling. Examples of two approaches—growth factors and regional models—are presented as follows:

Growth factors. Using historical volume data, an annual growth factor can be calculated and then used to determine future volumes. A similar, although more complex, approach assumes freight traffic is proportional to some economic activity. Forecasting future freight traffic for specific commodities thus requires estimating the growth of future economic activity. The advantage of this latter approach, besides the behavioral linkage to economic activity (i.e., demand is derived from the needs of economic production and consumption), is that most states and metropolitan areas have detailed economic forecasts for their jurisdictions. Thus, forecasts of goods movement could be tied to the results of ongoing economic modeling.

Regional models. Incorporating goods movement into the UTMS modeling approach is another way of estimating freight demand. Figure 5.26 shows such an approach and the freight-specific characteristics of each step in the process. In most cases, this approach is identical to the four-step process for passenger travel only using freight trip data and characteristics of the freight network in the model. Table 5.8, for example, shows commercial vehicle trip generation rates by type of commercial activity [Cambridge Systematics et al., 1996]. The friction factors that could be used to distribute commercial vehicle trips by using the gravity model are

Four-tire commercial vehicles $\qquad F_{ij} = e^{-0.08t_{ij}}$ $\qquad$ **[5.29a]**

Single unit trucks $\qquad F_{ij} = e^{-0.1t_{ij}}$ $\qquad$ **[5.29b]**

Combination trucks $\qquad F_{ij} = e^{-0.03t_{ij}}$ $\qquad$ **[5.29c]**

where t_{ij} is the cost of travel between zones i and j.

Table 5.8   Trip generation rates for commercial vehicles

| | Commercial Vehicle Trip Destinations (or origins) Per Unit Per Day | | | |
Generator	Four-Tire Vehicles	Single-Unit Trucks	Combinations	Total
Per Employment				
• Agriculture, mining, and construction	1.110	0.289	0.174	1.573
• Manufacturing, transportation, communications, utilities, and wholesale trade	0.938	0.242	0.104	1.284
• Retail trade	0.888	0.253	0.065	1.206
• Office and services	0.437	0.068	0.009	0.514
Per Household	**0.251**	**0.099**	**0.038**	**0.388**

SOURCE: Cambridge Systematics et al., 1996

The trip assignment process allocates commercial vehicle trips to the network with the same underlying principle as passenger assignment—minimize travel time. Due to the larger size, commercial vehicles will likely have a larger impact on congestion than passenger cars. Thus, capacity analyses usually equate commercial vehicles to "passenger car equivalents (PCEs)." The PCEs for different commercial vehicle types are multiplied by the number of vehicles of each type in the traffic stream, and the resulting values are assigned to the highway network.

A goods movement study in Portland, OR demonstrates the several steps that would occur in such a study. The objectives of this study were to

1. Develop an interconnected intermodal and multimodal transportation network using existing arterials to serve the Columbia corridor employment centers and residential and recreation areas.

2. Determine if the transportation network will be able to accommodate the planned levels of development, based on comprehensive plan designations. Based on that analysis, determine whether land-use designations should be modified to reflect the capacity of the network.

3. Improve efficiency and access along and between NE Columbia Boulevard and NE Portland Highway to primarily serve intermodal goods movement using these arterials.

4. Determine environmental impacts and neighborhood mitigation/protection for residential areas close to NE Portland Highway/Lombard Avenue, which may result from increased truck traffic.

5. Develop a strategy to improve NE Marine Drive, which will enhance regional recreational opportunities in the Columbia corridor region [City of Portland, 1994].

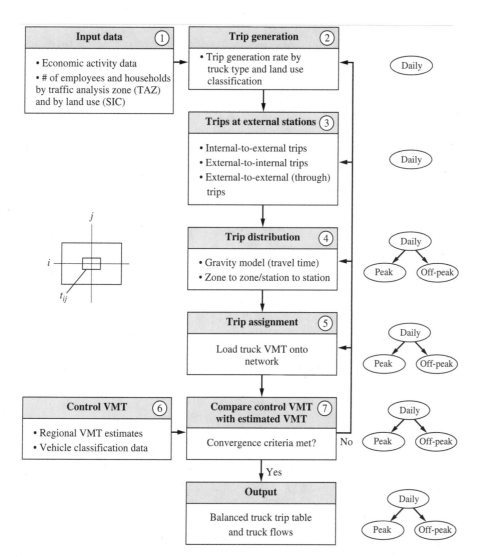

Figure 5.26 Simplified quick-response freight-forecasting procedure
SOURCE: Cambridge Systematics, Inc., 1996

Figure 5.27a shows the study area boundaries; Fig. 5.27b identifies the origin–destination zones of the major industrial areas; Fig. 5.27c presents the data showing the study area's truck peak-hour volumes; Fig. 5.27d is an example of the assigned truck trips to the road network; and Fig. 5.27e demonstrates how policy analysis can occur in such a study. The model was used to examine different capital improvements and truck-routing scenarios. Truck counts were taken at 27 locations in the study area, representing 320 directional counts and turning movements. Truck surveys, over a 2-day period, originating from or destined to, two port terminals were used to get a detailed picture of truck movements. The regional traffic analysis zone system was divided into industrial and nonindustrial areas. Based on the truck survey results, the known number of truck trips were distributed and assigned to the road network and adjusted until truck link volumes matched actual counts.

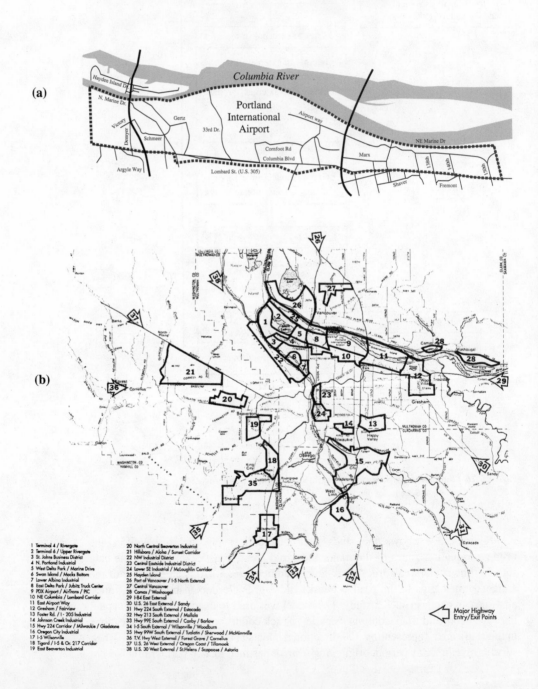

(a)

(b)

1 Terminal 4 / Rivergate	20 North Central Beaverton Industrial
2 Terminal 6 / Upper Rivergate	21 Hillsboro / Aloha / Sunset Corridor
3 St. Johns Business District	22 NW Industrial District
4 N. Portland Industrial	23 Central Eastside Industrial District
5 West Delta Park / Marine Drive	24 Lower SE Industrial / McLoughlin Corridor
6 Swan Island / Mocks Bottom	25 Hayden Island
7 Lower Albina Industrial	26 Port of Vancouver / I-5 North External
8 East Delta Park / Jubitz Truck Center	27 Central Vancouver
9 PDX Airport / AirTrans / PIC	28 Camas / Washougal
10 NE Columbia / Lombard Corridor	29 I-84 East External
11 East Airport Way	30 U.S. 26 East External / Sandy
12 Gresham / Fairview	31 Hwy 224 South External / Estacada
13 Foster Rd. / I- 205 Industrial	32 Hwy 213 South External / Molalla
14 Johnson Creek Industrial	33 Hwy 99E South External / Canby / Barlow
15 Hwy 224 Corridor / Milwaukie / Gladstone	34 I-5 South External / Wilsonville / Woodburn
16 Oregon City Industrial	35 Hwy 99W South External / Tualatin / Sherwood / McMinnville
17 I-5 Wilsonville	36 T.V. Hwy West External / Forest Grove / Cornelius
18 Tigard / I-5 & Or. 217 Corridor	37 U.S. 26 West External / Oregon Coast / Tillamook
19 East Beaverton Industrial	38 U.S. 30 West External / St.Helens / Scapoose / Astoria

Time Period	Truck Volumes	Total Vehicle Volumes
8:00-9:00	6,095	23,790
8:15-9:15	6,301	
8:30-9:30	6,387	
8:45-9:45	6,359	
9:00-10:00	6,185	22,197
9:15-10:15	6,220	
9:30-10:30	6,070	
9:45-10:45	6,062	
10:00-11:00	6,071	24,183
10:15-11:15	6,054	
10:30-11:30	6,017	
10:45-11:45	6,004	
11:00-12:00	6,008	27,083
11:15-12:15	5,829	
11:30-12:30	5,740	
11:45-12:45	5,717	
12:00-13:00	5,632	28,677
12:15-13:15	5,693	
12:30-13:30	5,861	
12:45-13:45	5,933	
13:00-14:00	6,039	28,279
13:15-14:15	6,207	
13:30-14:30	6,324	
13:45-14:45	6,284	
14:00-15:00	6,213	30,983
14:15-15:15	6,120	
14:30-15:30	5,960	
14:45-15:45	5,756	
15:00-16:00	5,568	38,886
15:15-16:15	5,324	
15:30-16:30	4,949	
15:45-16:45	4,719	
16:00-17:00	4,509	38,337
16:15-17:15	4,304	
16:30-17:30	4,151	
16:45-17:45	3,982	
17:00-18:00	3,753	34,547

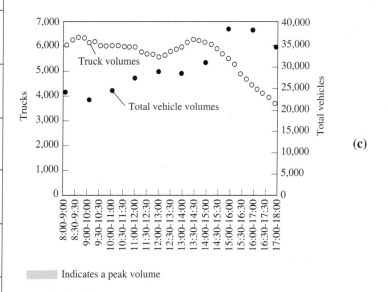

(c)

▓ Indicates a peak volume

Note: 1. Truck data was aggregated from 45 count locations
2. Total vehicle data was aggregated from 29 corresponding streets

(continued)

Figure 5.27 Example steps from truck study, Portland, Oregon
SOURCE: City of Portland, 1994
(a) Study Area Boundary
(b) Origin—Destination Zones
(c) Truck Peak-Hour Determination
(d) Truck Assignment (next page)
(e) Policy Analysis of Alternative Routes (next page)

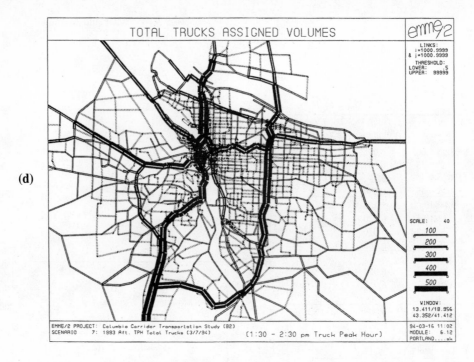

(d)

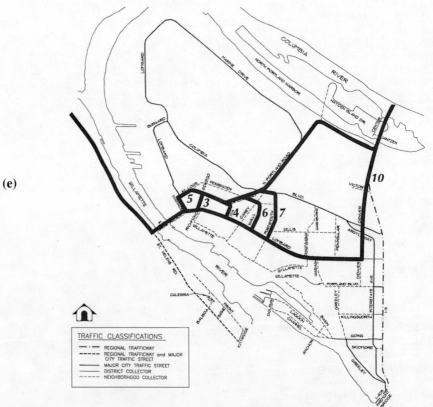

(e)

Figure 5.27 Example steps from truck study, Portland, Oregon (continued)

5.9 CHAPTER SUMMARY

1. Demand analysis, predicting how many people, what type of people, what purposes they will use the transportation system for, and the expected flow of goods in a metropolitan area, is an important task in transportation planning.

2. There are several important concepts related to the analysis of travel demand. First, demand for transportation services is a *derived demand* in that people do not travel for the sake of the traveling experience itself. Rather, people travel so that they can participate in activities located at various locations within the urban area. Second, travel has associated with it a *disutility* or *generalized cost,* measured in terms of time, monetary cost, inconvenience, and so on. Third, the demand for travel manifests itself as *trips* that can be characterized along several dimensions, such as trip purpose, origin, destination, and so forth. Fourth, demand analysis is usually conducted at the *aggregate* level of trip making, rather than at the individual trip-maker level. This aggregation can occur along spatial, temporal, and socioeconomic dimensions. Fifth, demand analysis methods can be classified into four broad categories: *simplified techniques,* the *urban transportation modeling system* (UTMS), *individual choice models,* and *activity-based methods.* Figure 5.28 qualitatively illustrates how these methods relate to the level of analysis complexity. Typically, more complex methods are used in response to increasing problem scale, but then this complexity ultimately decreases, reverting to simpler techniques as limitations in available data and theory begin to dominate. The techniques discussed in this chapter have been distributed along the curve so as to approximate their typical "scale-

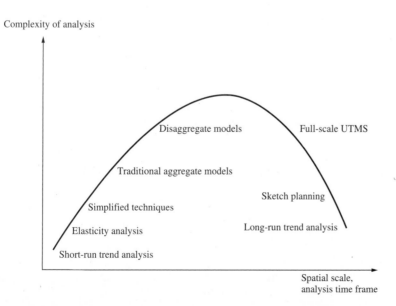

Figure 5.28   Relation between problem scale and analysis complexity

complexity" placements, although, as discussed previously, this can vary for a given technique from problem to problem.

3. Simple demand analysis techniques are often used in small-area, short-range planning issues, in preliminary "order-of-magnitude" analyses, in situations where more detailed analysis techniques are not available, or in the "screening" of a wide range of alternatives to select a smaller set of options for more detailed analysis (called sketch planning). Examples of simple demand analysis techniques include *trend analysis* and *elasticity-based models.*

4. The urban transportation modeling system (UTMS) has been, and continues to be, the approach most commonly used in regional transportation studies. The UTMS consists of four major stages, each having its own modeling approach. These stages are *trip generation, trip distribution, modal split,* and *trip assignment.*

5. Individual choice models model the decision process of individual trip makers. These models are based on the concept of utility maximization, that is, on the assumption that an individual traveler will choose an alternative that maximizes the benefits (e.g., provides minimum travel time and cost, maximum comfort and convenience, etc.). Multinomial logit and nested logit choice models are extensively used across a broad variety of travel demand modeling applications, including mode choice, trip destination choice, automobile holdings choice, and residential location choice.

6. Activity-based methods are emerging as viable alternatives to UTMS for a variety of planning applications. These methods model the scheduling and participation in activities directly from which the need to travel is derived. Two broad classes of activity-based travel models exist: *econometric models,* which typically use individual choice models to analytically compute the choice probabilities associated with different activity patterns, and *computational process models,* which are rule-based procedures for replicating decision-making behavior.

7. The demand analysis process consists of several steps, including problem identification, choice of analysis technique, data collection, model calibration, model validation, and forecasting.

8. Because the environment of planning changes over time, demand analysis must also be flexible in addressing new policy issues in a timely, responsive, and cost-effective manner. It is also important for the planner to understand the underlying assumptions and limitations of demand analysis in order to place it in proper perspective with respect to the overall transportation planning process.

QUESTIONS

1. Traditionally, many travel demand models have been predicated on the assumption of one-worker households. Comment on the implications of the existence of large numbers of multiworker households for

   (a) Characteristics of travel demand.

(b) Modeling assumptions and techniques.

(c) Analysis data requirements.

2. The following trip generation equation has been suggested for use in a regional model.

$$P_i = 9.1 - 1.65\,(HH_i) + 0.095\,(AO_i) - 24.3\,(ADULT_i) + 0.23\,(CHILD_i)$$

where

P_i	= number of trips produced per household i
HH_i	= number of people in household i
AO_i	= number of autos in household i
$ADULT_i$	= number of adults in household i
$CHILD_i$	= number of children in household i

Critique this model. Are the magnitude and signs of the coefficients reasonable?

3. The following data are available from a survey of urban residents in your city. Estimate the number of of trips generated for your study area given the forecasted number of households.

Survey Data

HH Income	Household Size = 1		Household Size = 2		Household Size = 3+	
	#HH	#Trips	#HH	#Trips	#HH	#Trips
< 20,000	500	1220	450	1300	500	1950
20,001–40,000	600	1860	700	2950	800	3700
> 40,000	500	2125	800	4500	750	3600

Forecasted # Households

	Household Size		
	1	*2*	*3+*
< 20,000	35	69	47
20,001–40,000	50	83	29
> 40,000	71	23	16

4. Given that the demand for travel is a derived demand, what variables do you think might be of importance in explaining the demand for grocery shopping trips in an urban area? Explain your reasoning.

5. A trip production regression equation was developed using the data given in Fig. 5.9. The resulting equation is

$$O_i = 0.091 + 0.735(SIZE_i) + 0.945(AO_i)$$

where

O_i	= total daily trips per household produced in zone i
$SIZE_i$	= average household size in zone i
AO_i	= average auto-ownership level in zone i

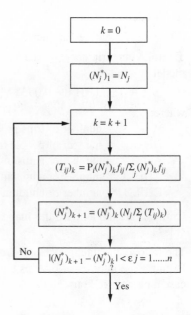

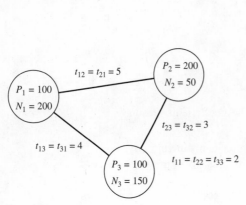

Figure P5.6a   Three-zone system

Figure P5.6b   Gravity model balancing procedure

Using this equation and the data from Fig. 5.9c, compute the total daily trip productions for the given zone. Compare your results with those of Fig. 5.9d. Can you explain the differences?

6. Figure P5.6a presents a simple three-zone system, the link travel times for this system, and the zonal productions and attractions.

Assume a gravity model of the form

$$T_{ij} = \frac{P_i A_j^* (t_{ij})^{-b}}{\sum_{j'} A_{j'}^* (t_{ij})^{-b}}$$

where A_j^* is a "modified attraction term" defined by the algorithm shown in Fig. P5.6b which ensures that the predicted trips to a given zone $\sum_i T_{ij}$ equal the true zonal attractions A_j. Compute O–D flows for this system for values of b of 1.0, 1.5, and 2.0. Table P5.6 presents the observed O–D flows for this system. Which values of b provide the "best fit"? Define explicitly how you are measuring goodness of fit.

Table P5.6   Observed O–D flows for three-zone system

From	To		
	1	2	3
1	80	5	15
2	80	40	80
3	40	5	55

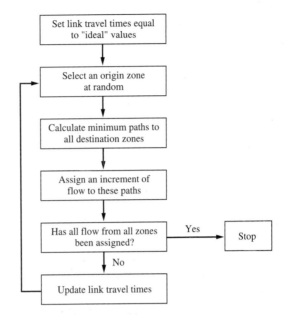

Figure P5.7a   Five-link network

Figure P5.7b   Capacity restraint trip assignment

7. For the network shown in Fig. P5.7a, the O–D flow matrix shown in Table P5.7a, and the incremental assignment process shown in Fig. P5.7b, compute link flows using

(a) All-or-nothing assignment.

(b) Incremental assignment, using the procedure shown in Fig. P5.7b, modified so that instead of choosing an origin zone at random and assigning flow to all destinations, you choose individual O–D pairs at random and assign flows for each pair in turn. Assume a volume-delay curve of the form

$$t_1 = t_{01}/[1 - (V_1/C_1)]$$

where

t_1 = travel time on link 1
t_{01} = travel time on link 1 under zero flow conditions
V_1 = volume flow on link 1
C_1 = capacity of link 1

Table P5.7a   Observed O–D flow matrix

From	To			
	1	**2**	**3**	**4**
1	0	100	100	100
2	0	0	50	50
3	0	0	0	100
4	0	0	0	0

Table P5.7b provides the values of t_{0l} and C_l for the links in the network. Assign 50 percent of an O–D pair's flow per iteration. Table P5.7c provides the "random" order in which you should choose O–D pairs for assignment.

Table P5.7b   Link characteristics

Link	1	2	3	4	5
t_{0l}	10	15	3	5	4
C_l	300	500	150	200	200

Table P5.7c   "Random" order for O–D flow assignment

O–D pair	Assignment order for iteration no. 1	2
(1,2)	1	3
(1,3)	6	6
(1,4)	2	4
(2,3)	3	2
(2,4)	5	5
(3,4)	4	1

8. The buses in the numerical example of a logit mode split model presented in Sec. 5.5.1 (i.e., Eq. [5.23]) are operated by the Red Bus Transit Co. Assume that a new operator, Blue Coach Lines, introduces a service that is identical to that of the Red Bus Transit Co. (i.e., 15-minute travel time, $1.50 fares), except that the service is provided by blue buses rather than red. We now have four "modes" in the choice set: auto, red bus, blue bus, and walk. Compute the probabilities of each of these four modes being chosen using Eq. [5.23] (and assuming that the blue bus utility function is the same as the red bus function). Compare these probabilities with those for the three-mode example in Sec. 5.5.1. Are these results reasonable, especially given that the model ignores frequency of service? Why or why not?

9. Ignoring data limitations (at least within "reasonable" limitations!), how and why would you modify the mode split model defined by Eq. [5.24] in Sec. 5.5.1 so as to improve its predictive capability and/or its policy sensitivity?

10. What options exist for structuring the decision process for an individual's choice of mode for work and access mode if transit is used (i.e., walk, park-and-ride, kiss-and-ride, etc.)? If you were modeling this process, what approach would you adopt?

11. The logit model was applied to the choice of mode for work trips in the San Diego area. Two models were formulated. The first model was estimated for CBD-oriented trips where relatively good transit service is available, while the second model was estimated for non-CBD-oriented trips. For both models, the utility functions were defined as

For auto passenger: $U_p = cX3_p - dL3_p - eCH_p$

For auto driver: $U_D = cX3_D - dL3_D - eCH_D + aTI + b$

For transit passenger: $U_T = cX3_T - dL3_T - eCH_T + f$

where
$X3_i$ = excess time for mode i
$L3_i$ = line-haul time for mode i
Ch_i = cost for mode i
TI = transformed household income = $100[1 - \exp(-0.035 \times INC)]$
a, b, c, d, e, f = model parameters

Table P5.11 presents the calibrated parameter values for both models.

Table P5.11 San Diego logit model parameters

Parameter	CBD Model	Non-CBD Model
a	0.0295	0.0268
b	−1.4809	−0.5441
c	0.0916	0.1314
d	0.0563	0.0192
e	0.0106	0.0184
f	1.1635	1.6600

(a) Discuss the specification of this model. How would you modify it?

(b) Evaluate the estimated models:

  i. Does each coefficient have the proper sign?

  ii. What statistical tests would you perform on the model? What do you expect to learn from these tests?

  iii. Outline the procedure you would follow in computing direct and cross elasticities in this model.

(c) Compare the estimated coefficients of the two models and discuss the differences in terms of your a priori expectations. Can you suggest a more general model specification that can be applied to both CBD- and non-CBD-oriented trips? Explain.

12. Record all your in-home and out-of-home activities for a 24-hour period for one weekday. For each activity, record the type of activity (eating, attending class, etc.), the duration of the activity (i.e., how long you spend on the activity, exclusive of any travel time involved in going to the activity), the timing of the activity (i.e., when the activity started), whether travel was required to participate in the activity (and, if so, record the mode of travel used and the travel time involved), any temporal, spatial or other constraints associated with the activity (e.g., the class is held at a fixed address at a fixed time), and any other constraints which existed with respect to your ability to travel to/participate in the activity (e.g., was a car available for the trip?). For in-home activities, only record details of the activity other than duration if the in-home activity could have been substituted with an out-of-home activity (e.g., eat dinner at home instead of go out to a restaurant).

13. Given your activity diary from Question 12, discuss the extent to which the UTMS structure would be able to model your travel behavior accurately. What might it "miss" or not be able to adequately represent? Describe what you would like to see in an activity-based model in order to achieve a more behavioral and more policy-sensitive model of your daily behavior.

REFERENCES

Algers, S., A. Daley, P. Kjellman, and S. Widlert. 1995. Stockholm model system (SIMS): Application. Paper presented at the 7th World Conference of Transportation Research. Sydney, Australia.

Axhausen, K.W. and T. Gärling. 1992. Activity-based approaches to travel analysis: Conceptual frameworks, models, and research problems. *Transport Reviews*. Vol. 12. No. 4.

Badoe, D.A. and E.J. Miller. 1995. Comparison of alternative methods for updating disaggregate logit mode choice models. *Transportation Research Record 1493*. Washington, D.C.: National Academy Press.

Barrett, C., K. Berkbigler, L. Smith, V. Loose, R. Beckman, J. Davis, D. Roberts, and M. Williams. 1995. *An operational description of TRANSIMS*. Report LA-UR-95-2393. Los Alamos, New Mexico: Los Alamos National Laboratory.

Ben-Akiva, M. and S.R. Lerman. 1985. *Discrete choice analysis: Theory and application to predict travel demand*. Cambridge, MA: MIT Press.

Bernstein, D., T.L. Friesz, R.L.Tobin, and B.W. Wie. 1995. A comparison of system optimum and user equilibrium dynamic traffic assignment with schedule delays. *Transportation Research. Part C, Emerging Technologies*. Vol. 3. No. 6.

Bowman, J.L. and M.E. Ben-Akiva. 1997. Activity-based forecasting. In Texas Transportation Institute (eds.) *Activity-based travel forecasting conference, June 2–5, 1996. Summary, recommendations, and compendium of papers*. Washington, D.C.: Travel Model Improvement Program. U.S. Department of Transportation and U.S. Environmental Protection Agency.

Boyce, D.E., B.N. Janson and D.H. Lee. 1998. *Analytical dynamic traffic assignment models: Features and solution experience.* Report No. UCI-ITS-WP-98-12. Irvine, CA: University of California at Irvine, Institute of Transportation Studies.

Bradley, M. and J.L. Bowman. 1998. *A system of activity-based models for Portland, Oregon.* Travel Model Improvement Program. Washington, D.C.: U.S. Department of Transportation. May.

Burrell, J. E. 1968. Multipath route assignment and its application to capacity restraint. Presented at the Fourth International Symposium on the Theory of Traffic Flow. Karlsruhe, Germany.

Cambridge Systematics, Inc. 1996. Forecasting freight transportation demand: A guide for planners and policy analysts. *NCHRP Report 8-30.* Washington, D.C.: National Academy Press.

_____. 1997. *Time-of-day modeling procedures: State-of-the-art, state-of-the-practice.* Report DOT-T-9-01. Washington, D.C.: FHWA. October.

_____, Comsis, and University of Wisconsin-Milwaukee. 1996. *Quick response freight manual.* Report DOT-T-97-10. Washington, D.C.: U.S. DOT. Sept.

Cascetta, E., A. Nuzzolo, and V. Velardi. 1993. *A system of mathematical models for the evaluation of integrated traffic planning and control policies.* Salerno, Italy: Laboratorio Richerche Gestione e Controllo Traffico.

Chesapeake Bay Foundation et al. 1996. *A network of livable communities: Evaluating travel behavior effects of alternative transportation and community designs for the national capital region.* Washington, D.C.: May.

City of Portland. 1994. *Columbia corridor transportation study.* Technical Report 2: Truck Routing Model. Portland, OR: Office of Transportation. April.

Coogan, M. 1996. Freight transportation planning practices in the public sector. *NCHRP Synthesis of Highway Practice 230.* Washington, D.C.: National Academy Press.

Daganzo, C. F. and Y. Sheffi. 1977. On stochastic models of traffic assignment. *Transportation Science.* Vol. 11. No. 3.

Daganzo, C. F. 1979. *Multinomial probit.* New York: Academic Press.

Daley, A.J., H.H.P. van Zwam, and J. van der Valk. 1983. Application of disaggregate models for a regional transport study in the Netherlands. Paper presented at the 3rd World Conference on Transport Research. Hamburg, Germany.

Dial, R. B. 1971. A probabilistic multipath traffic assignment model which obviates path enumeration. *Transportation Research.* Vol. 5.

Doherty, S.T. and E.J. Miller. 2000. Tracing the household activity scheduling process using one-week computer-based survey. *Transportation.* Vol. 27.

Domencich, T. and D. McFadden. 1975. *Urban travel demand: A behavioral analysis.* Amsterdam: North Holland.

Ettema, D., A. Borgers, and H. Timmermans. 1995. SMASH (Simulation Model of Activity Scheduling Heuristics): Empirical test and simulation issues. *Activity based approaches: Activity scheduling and the analysis of activity patterns.* Eindhoven, The Netherlands.

Ettema, D. and H. Timmermans. 1997. Theories and models of activity patterns. In D. Ettema and H. Timmermans (eds.) *Activity-based approaches to travel analysis.* Oxford: Pergamon Press.

Fratar, T. J. 1954. Forecasting distribution of inter-zonal vehicular trips by successive approximations. *Proceedings, Highway Research Board.* Vol. 33. Washington, D.C.

Gärling, T., K. Brannas, J. Garvill, R.G. Golledge, S. Opal, E. Holm, and E. Lindberg. 1989. Household activity scheduling. In *Transport Policy, Management and Technology Towards 2001: Selected Proceedings of the 5th World Conference on Transport Research.* Vol. IV. Ventura, CA: Western Periodicals.

Gärling, T., M.P. Kwan, and R.G. Golledge. 1994. Computational-process modeling of household travel decisions: Conceptual analysis and review. *Transportation Research B.* Vol 28. No. 5.

Gaudry, M. J.I. 1975. *An aggregate time-series analysis of urban transit demand: The Montreal case.* Publication No. 6. Montreal, Canada: Center for Transportation Research. University of Montreal. April.

Golding, S. and K. B. Davidson. 1970. A residential land use prediction model for transportation planning. *Proceedings, Australian Road Research Board.* Melbourne, Australia.

Golledge, R.G., M.P. Kwan, and T. Gärling. 1994. Computational process modeling in household travel decisions using a geographical information system. *Papers in Regional Science.* Vol. 73. No. 2.

Goulias, K.G. and R. Kitamura. 1996. A dynamic model system for regional travel demand forecasting. Chapter 13 in T. Golob, R. Kitamura, and L. Long (eds.) *Panels for transportation planning: Methods and applications.* New York: Kluwer Academic Publishers.

Hagerstrand, T. 1970. What about people in regional science? *Regional Science Association Papers.* Vol. 24.

Hague Consulting Group. 1992. *The Netherlands national model 1990: The national model system for traffic and transport.* Ministry of Transport and Public Works. The Hague, The Netherlands.

Hensher, D. A. and L. W. Johnson. 1981. *Applied discrete-choice modeling.* London: Croom Helm.

Hu, T.Y. and H.S. Mahmassani. 1995. Evolution of network flows under real-time information: Day-to-day dynamic simulation assignment framework. *Transportation Research Record 1493.* Washington, D.C.: National Academy Press.

Hutchinson, B. G. 1974. *Principles of urban transportation systems planning.* New York: Scripta.

Institute of Transportation Engineers. 1997. *Trip generation.* Vols. 1–3. 6th ed. Washington, D.C.: ITE.

Jones, P., M. Dix, M. Clarke, and I. Heggie. 1983. *Understanding travel behaviour.* Aldershot, UK: Gower.

Jones, P., F. Koppelman, and J.P. Orfeuil. 1990. Activity analysis: State-of-the-art and future directions. In P. Jones (ed.) *Developments in dynamic and activity-based approaches to Travel.* Aldershot, UK: Avebury.

Kanafani, A. 1983. *Transportation demand analysis.* New York: McGraw-Hill.

Kitamura, R. 1997. Applications of models of activity behavior for activity-based demand forecasting. In Texas Transportation Institute (eds.) *Activity-based travel forecasting conference, June 2–5, 1996. Summary, recommendations, and compendium of papers.* Washington, D.C: Travel Model Improvement Program. U.S. Department of Transportation and U.S. Environmental Protection Agency.

Koppelman, F. S. 1975, *Travel prediction with models of individual choice behavior.* Center for Transportation Studies Report No. 75-7. Cambridge, MA: MIT. June.

_____ and C. H. Wen. 1998. Alternative nested logit models: Structure, properties, and estimation. *Transportation Research B.* Vol. 32. No. 5.

Lancaster, K. J. 1966. A new approach to consumer theory. *Journal of Political Economy.* Vol 34. pp. 132–157.

Lawton, T. K. 1997. Activity and time use data for activity-based forecasting. In Texas Transportation Institute (eds.) *Activity-based travel forecasting conference. June 2–5, 1996. Summary, recommendations, and compendium of papers.* Washington, D.C.: Travel Model Improvement Program, U.S. Department of Transportation and U.S. Environmental Protection Agency.

Layard, P. R. G. and A. A. Walters. 1978. *Microeconomic theory.* New York: McGraw Hill.

Lerman, S. R. and J. J. Louviere. 1978. On the use of functional measurement to identify the functional form of the utility expression in travel demand models. *Transportation Research Record No. 673.* Washington, D.C.: National Academy Press.

Luce, R. D. and P. Suppes. 1965. Preference, utility and subjective probability. Chapter 19 in R. D. Luce, R. R. Bush, and E. Galanter (eds.). *Handbook of mathematical psychology.* Vol . 3. New York: Wiley.

Mackinder, I. H. and S. E. Evans. 1981. *Predictive accuracy of British transport studies in urban areas.* Transport and Road Research Laboratory SR 699. Crowthorne, U.K.

Mahmassani, H. S. 1997. Dynamic traffic simulation and assignment: Models, algorithms and application to ATIS/ATMS evaluation and operation. *Proceedings of the NATO Advanced Study Institute on Operations Research and Decision Aid Methodologies in Traffic and Transportation Management.* Balatonfured, Hungary.

Manski, C. F. 1973. The stochastic utility model of choice. Unpublished Ph.D. thesis. MIT, Department of Economics, Cambridge, MA.

Martin, W. and N. McGuckin. 1998. Travel estimation techniques for urban planning. *NCHRP Report 365.* Washington, D.C.: National Academy Press.

Matlick. J. 1996. *If we build it, will they come?* # 69 Forecasting Pedestrian Use and Flows, Forecasting the Future. Washington, D.C.: Bicycle Federation of America—Pedestrian Federation of America. Pro Bike/Pro Walk '96.

McCoomb, L. A. 1982. Simplified urban transportation planning procedures using census data. Unpublished Ph.D. thesis. University of Toronto, Dept. of Civil Engineering, Toronto.

McFadden, D. 1978. Modeling the choice of residential location. *Transportation Research Record 673.* Washington, D.C.: National Academy Press.

_____. 1979. Qualitative methods for analyzing travel behavior of individuals: Some recent developments. In D. A. Hensher and P. R. Stopher (eds.). *Behavioral travel modeling.* London: Croom Helm.

Miaou, S.P. and M.S. Summers. 1999. Laboratory evaluation of real-time dynamic traffic assignment systems. *Journal of Computer-Aided Civil and Infrastructure Engineering.* Vol. 14. No. 4.

Miller. E.J. 1996. Microsimulation and activity-based forecasting. In Texas Transportation Institute (eds.) *Activity-based travel forecasting conference, June 2–5, 1996. Summary, recommendations, and compendium of papers.* Washington, D.C.: Travel Model Improvement Program, U.S. Department of Transportation and U.S. Environmental Protection Agency.

_____ and K.S. Fan. 1992. Travel demand behaviour: Survey of intercity mode-split models in Canada and elsewhere. Chapter 19. Vol. 4. *Directions, The final report of the Royal Commission on National Passenger Transportation.* Ottawa: Ministry of Supply and Services.

_____ and P.A. Salvini. 2000. Activity-based travel behavior modeling in a microsimulation framework. Invited resource paper, forthcoming in the proceedings of the Eighth Meeting of the International Association of Travel Behaviour Research, Austin, TX.

Mohring, H. 1976. *Transportation economics.* Cambridge, MA: Ballinger.

Nelson, A. and D. Allen. 1997. *If you build them, commuters will use them: Cross sectional analysis of commuters and bicycle facilities.* City Planning Program, Georgia Institute of Technology. Presented at the 1997 Annual meeting of the Transportation Research Board, Washington, D.C.

Newell, A. and H.A. Simon. 1972. *Human problem solving.* Englewood Cliffs, NJ: Prentice-Hall.

Nicholson, W. 1978. *Microeconomic theory: Basic principles and extensions.* Hinsdale, IL: Dryden.

1,000 Friends of Oregon. 1993. Making the land use/transportation/air quality connection. Vols. 1–9. Portland, OR.

Ortuzar, J. de D. and L.G. Willumsen. 1994. *Modelling transport.* New York: John Wiley and Sons.

Pas, E.I. 1997. Recent advances in activity-based travel demand modeling. In Texas Transportation Institute (eds.) *Activity-based travel forecasting conference, June 2–5, 1996. Summary, recommendations, and compendium of papers.* Washington, D.C.: Travel Model Improvement Program. U.S. Department of Transportation and U.S. Environmental Protection Agency.

Portland METRO. 1996. *Travel demand forecasting methods report.* Portland. May 20.

Quadstone. 1999. *Paramics modeller V3.0 reference manual.* Edinburgh, UK: Quadstone Ltd.

RDC, Inc. 1995. *Activity-based modeling system for travel demand forecasting.* Report DOT-T-96-02. Washington, D.C.: U.S. Department of Transportation.

Recker, W.W., M.G. McNally, and G.S. Root. 1986a. A model of complex travel behavior: Part I: Theoretical development. *Transportation Research A.* Vol. 20. No. 4.

_____. 1986b. A model of complex travel behavior: Part II: An operational model. *Transportation Research A.* Vol. 20. No. 4.

Reid, F. A. 1978. Systematic and efficient methods for minimizing error in aggregate predictions from disaggregate models. *Transportation Research Record 673.* Washington, D.C.: National Academy Press.

Ruiter, E. R. and M. E. Ben-Akiva. 1978. Disaggregate travel demand models for the San Francisco Bay Area. *Transportation Research Record 673.* Washington, D.C.: National Academy Press.

Schneider, M. 1960. *Panel discussion on inter-area travel formulas.* Highway Research Board Bulletin No. 253. Washington, D.C.

Schwartz, W. et al. 1999a. *Guidebook on methods to estimate non-motorized travel: Overview of methods.* Report FHWA-RD-98-165. Washington, D.C.: FHWA. July.

_____. 1999b. *Guidebook on methods to estimate non-motorized travel: Supporting documentation.* Report FHWA-RD-98-166. Washington, D.C.: FHWA. July.

Seattle METRO. 1996. *Metro modeling methodology.* Seattle: METRO.

Sheffi, Y. 1985. *Urban transportation networks: Equilibrium analysis with mathematical programming models.* Englewood Cliffs, NJ: Prentice-Hall.

Smith, D. P. and B. G. Hutchinson. 1979. Goodness of fit statistics for trip distribution models. Waterloo, Ontario: Dept. of Civil Engineering, University of Waterloo.

Stouffer, S. A. 1940. Intervening opportunities: A theory relating mobility and distance. *American Sociological Review.* Vol. 5. No. 6.

Texas Transportation Institute (eds.). 1997. *Activity-based travel forecasting conference, June 2–5, 1996. Summary, recommendations, and compendium of papers.* Washington, D.C: Travel Model Improvement Program. U.S. Department of Transportation and U.S. Environmental Protection Agency.

Thurstone, L. 1927. A law of comparative judgement. *Psychological Review.* Vol. 34.

U.S. Department of Transportation. 1980. *Patronage impacts of changes in transit fares and services.* UMTA Report RR 135-1. Washington, D.C. September.

Van Aerde, M. and S. Yager. 1988a. Dynamic integrated freeway/traffic signal networks: Problems and proposed solutions. *Transportation Research A.* Vol. 22A, No. 6.

_____. 1988b. Dynamic integrated freeway/traffic signal networks: A routing-based modeling approach. *Transportation Research A.* Vol. 22A. No. 6.

Wardrop, J. G. 1952. Some theoretical aspects of road traffic research. *Proceedings of the Institute of Civil Engineers.* Vol. 1.

Webber, M. 1977. Pedagogy again: What is entropy? *Annals of the Association of American Geographers.* Vol. 67, No. 2.

Wilbur Smith and Assocs. 1996. *Non-motorized access to transit.* Final report. Prepared for Regional Transportation Authority, Chicago, IL. July.

Wilson, A. G. 1967. A statistical theory of spatial distribution models. *Transportation Research.* Vol. 1.

Wohl, M. 1972. *Transportation investment planning.* Lexington, MA: D.C. Heath.

6

Urban Activity System Analysis

6.0 INTRODUCTION

As discussed in Chap. 3, the demand for personal transportation is derived from the need to participate in activities of various types (e.g., work, shop, visit friends) that occur in places dispersed throughout the metropolitan area. Similarly, freight transportation providers move goods from point to point within, into, and out of the metropolitan area in support of the regional economy. In order to estimate the demand for transportation services, one must first understand the urban activity system that generates this demand. A definition of land-use activities and patterns in a metropolitan area and their relationship to the generation of trips is thus an important prerequisite of demand analysis, which was discussed in detail in Chap. 5.

It takes considerable time to develop land and construct buildings. Employment and other activities grow, decline, or change radically in nature, and large numbers of people change where they live or work over decades. For short-term analyses, therefore, transportation planners can usually take the current activity system as being "fixed and given." In such cases, an inventory of the current activity system for the study area is sufficient. Predicting how this activity system will change over time is not needed. In the long run, however, where the long run is probably anything greater than 5 years in the future, the urban activity system clearly does change. Neighborhoods gain or lose population or employment of various types. New areas on the metropolitan fringe may be developed, while older, developed areas may decline in quality, be renovated, or undergo redevelopment. As a result, travel demand patterns and transportation system requirements will also change. Hence, transportation planning efforts focused on the long term must explicitly consider expected changes in the urban activity system in order to predict future travel demand.

Further, proposed transportation system improvements can potentially influence future activity patterns by altering the accessibility levels of various locations and activities. As discussed in Chap. 3, the impact of the transportation system on activity patterns is complex and contingent on a host of other factors. Nevertheless, concern over the nontransportation impacts of transportation policies—including the explicit use of transportation to achieve nontransportation objectives (e.g., enhance the commercial viability of a downtown)—necessitates a capability to analyze and predict transportation impacts on urban activity, in both the short and long term.

Transportation planners can thus potentially face two major analytical tasks with respect to the urban activity system: (1) forecasting future states of this activity system as a first step in forecasting the future demand for transportation; and (2) predicting the impacts that proposed transportation system changes are likely to have in either the short or the long run on urban activities. Section 6.1 presents some basic concepts and issues associated with the analysis of the urban activity system. This section also discusses issues concerning the role of urban activity analysis within the transportation planning process. Sections 6.2 and 6.3 discuss a range of analytical techniques and issues associated with each of these two tasks. For a more detailed discussion of urban activity system models, see, among others, Miller et al. [1998]; Southworth [1995]; and Wegener [1994, 1995, 1998].

6.1 BASIC CONCEPTS AND THE ROLE OF URBAN ACTIVITY ANALYSIS IN TRANSPORTATION PLANNING

The terms "urban activity system" or "urban land use" (the two terms will be used interchangeably within this chapter) refer to the spatial distribution of people and activities within a metropolitan area. Given the modeling state of the art, the spatial distributions of population and employment constitute primary inputs into transportation demand models. Implicit in this concept of spatial distribution are several other concepts that are important in characterizing an urban activity system. One of these is the *building stock*, which physically occupies the land. People occupy and use buildings of specific sizes, types, and qualities, located on specific sites within a given topography. These buildings are costly to construct, maintain, or demolish. Thus, where people live and where firms locate not only depend on their preferences or *demand* for certain locations, but also upon the availability or *supply* of suitable building stock at these locations. Hence, a land-use *market* exists in which people and activities compete for the available locations and buildings within a metropolitan area and in which the supply of these locations and buildings presumably adjusts over time to meet these demands.

A major factor linking the demand and supply sides of this market together is the *price* or *rent* associated with the buildings and the land being bought and sold. Developers, builders, and landlords will presumably wish to charge the most they can to maximize their profits. They are constrained (in principle at least) by the existence of other suppliers who will underprice them if they charge too much. Similarly, home owners and renters will wish to minimize the prices they pay for accommodation. Because land and buildings are scarce commodities, however, other home owners and renters will "bid up" the price in an effort to obtain a particular location, with the location most likely going to the person or activity to whom it is most valuable (and, hence, to the person who is willing to pay the most for it). The end result of this bidding process is an "equilibrium bid rent surface," which defines the equilibrium price or rent of land or building stock at each point in the metropolitan area and which represents both the profit-maximizing result for the suppliers of the land and the utility-maximizing result for the purchasers [Alonso, 1964].

In general, it is expected that these rents will decline as one moves farther from the city center, which has generally been assumed to be the most attractive or desirable location for most activities. The concept of declining rents with distance is also applicable in urban subcenters, in which the densest economic activity is not necessarily at the center of the city. Figure 6.1 depicts a typical rent surface, in this case for 1996 housing prices in the Greater Toronto Area, in which the central peak occurs in "midtown," within highly attractive residential neighborhoods with good access to the central core business area. A ridge of high prices runs north, up Yonge Street (the traditional "main street" of Toronto), and other local peaks represent other locations of high accessibility and attractiveness for residential housing. In the United States, such bid rent surfaces have more pronounced spikes in suburban centers that have become very valuable residential and employment locations.

It is not clear, however, whether most urban systems ever reach a state of equilibrium. Exogenous "shocks" to the urban activity system happen continuously: in- and out-migration of people and jobs, changes in interest rates and capital availability, changes in the transportation system, and changes in zoning and other controls and guidelines for development. These and other factors will alter the status quo and cause the urban activity system to adapt and evolve over time. Further, considerable inertia and durability exist in the system, which may prevent it from adjusting quickly to changing conditions. While not invalidating the notion of an urban market and its interaction between demand and supply mediated by price, these observations do highlight the need to recognize that this interaction is, in fact, a dynamic one and hence one that cannot be adequately modeled or analyzed from a static perspective.

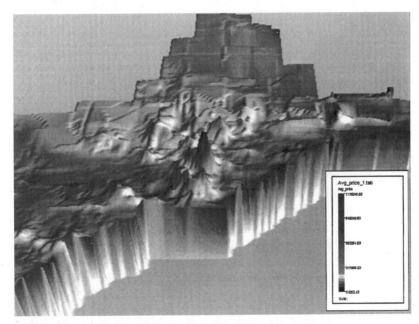

Figure 6.1   Rent surface for Toronto; 1996 housing prices

As is explicit in its name, the urban activity system consists of people participating in a range of activities at different locations and times. People do not occupy space or buildings for the sake of mere occupation; they do so for some purpose—to live in a home or to produce goods or services (and in so doing, to provide employment opportunities to the resident population). In general, people and activities will choose among locations based on a location's *attractiveness* for a particular type and scale of activity and on the location's *accessibility* to other activities.

The attractiveness of a location for a given activity depends upon a wide variety of factors. For example, the attractiveness of a neighborhood as a residential location depends on such characteristics as the price, size, type, age, and quality of the available housing; the quality and proximity of schools (if the household contains school-age children); the availability of parks and other recreational facilities; the extent to which the neighborhood is "hazard free" (where hazards might include busy streets, noxious factories, etc.); and the social-ethnic-racial composition of the neighborhood (and perceived trends in this composition). Similarly, the attractiveness of a location for a retail store depends upon such factors as the availability of a suitable building for the store; the location of the building relative to the street; pedestrian flows and parking; the rent to be paid for the building; the expected market at the location for the goods being sold; and the mix of retail stores currently located in the neighborhood.

In principle, therefore, attractiveness and, hence, location choice depends upon a relatively complex set of attributes, which are perceived by the household decision makers and which enter into their preference or choice considerations. In modeling practice, however, the analyst is limited in terms of the number of attraction attributes that can be specified and observed. The result is that gross surrogates, generally "size" variables (e.g., total retail floor space in a zone as a measure of retail attractiveness or total number of single-family housing units in a zone as a measure of residential attractiveness), are used in place of more specific "behavioral" variables.

The concept of accessibility typically provides the basis for the transportation component of urban activity modeling. Accessibility is generally defined as some aggregate measure of the size and closeness of activity opportunities of a given type to a particular location. For example, if one were interested in characterizing the accessibility of a residential zone i to retail shopping opportunities, a common measure used is

$$A_i = \sum_{j=1}^{n} F_j^{\alpha} e^{-\beta t_{ij}}$$ [6.1]

where

A_i = accessibility of zone i to shopping opportunities
F_j = amount of retail floor space in zone j
t_{ij} = travel time from zone i to zone j
n = number of zones with retail stores
α = parameter indicating the relative sensitivity of accessibility to store size ($\alpha \geq 1$)
β = parameter indicating the sensitivity of trip making to travel time (i.e., the larger β is in magnitude, the less likely people are to travel long distances to shop)

In general, it is assumed that location choice is positively correlated with accessibility. That is, people would presumably like to have more accessibility than less to shopping and employment opportunities, retail stores would like to be highly accessible to high-income households, and so on. For a more detailed discussion of the concept of accessibility, and particularly of its definition through the use of individual choice models, see Ben-Akiva and Lerman [1985].

The previous discussion of the basic concepts of urban activity system analysis suggests that there is an important linkage between the urban activity system and the transportation system. One would expect, therefore, that characteristics of both would be fully integrated in urban transportation planning. However, the role of urban activity system analysis within the urban transportation planning process as a whole is a rather ambiguous and, at times, even a contradictory one. In principle, that is, as presented in transportation planning texts—including this one—the transportation–land-use interaction is of fundamental importance to planning and should represent the starting point for the analysis of many transportation policies and issues. In practice, however, the planning and analysis of the urban transportation and activity systems proceed essentially independent of one another. Further, even when land-use considerations are incorporated within transportation planning analyses, the results have been very mixed.

The large amount of independence that has existed between urban transportation and activity system planning stems primarily from the fact that, institutionally and professionally, different agencies and people are responsible for the respective activities. A city planning department may well be concerned with both transportation and land-use issues, but typically, even in such cases, a relatively autonomous transportation group will exist within the planning department to deal with transportation issues. This institutional dichotomy encourages planners and analysts to think of transportation problems or land-use problems as being independent from one another. Thus, for example, improving transportation energy efficiency typically involves encouraging the use of more energy-efficient energy modes (transit or car pool) and improving the technical efficiency of the vehicles and systems in operation (e.g., improved automobile fuel economy). They rarely include policies for encouraging a land-use pattern that might facilitate improved transportation energy efficiency (through shorter trip lengths, use of high-efficiency modes, reducing the need to travel, etc.).

This separate focus is further manifested in analytical techniques, which typically model the transportation system given the land-use system (or vice versa) rather than the dynamic interaction between the two. This leads not only to a potentially myopic or partial analysis of the problem of interest, but can also result in the development of models that contain serious internal inconsistencies. The classic example of this is the urban transportation modeling system (UTMS), discussed in Chapt. 5, which predicts the distribution or flow of work trips from residential zones to employment zones, given the zonal population and employment levels as generated separately by a land-use forecasting model. The land-use forecasting model typically used to generate these numbers distributes residential population among zones based on place of employment. In other words, a zone-to-zone work-trip flow pattern is implicit in the land-use forecasts but is ignored by UTMS, which forecasts its own flow pattern—one which might bear little resemblance to the original pattern implicit in the land-use forecasts. At the same time, these land-use forecasts are generated based on an assumed future transportation system that might bear little relationship to any of the actual transportation system alternatives under consideration. It is thus often unclear whether the land-use or the transportation forecasts are in any way compatible with one another or with the system alternatives under consideration.

The disconnect between land-use and transportation analysis is further exacerbated by the simple fact that relatively few urban areas currently possess a land-use modeling capability. As is discussed further in the next section, formal land-use modeling first began during the 1960s as part of the large-scale, comprehensive, long-range planning studies that typified the period. The shift in the 1970s from a long-range to a short-range planning focus was associated with a near-total abandonment (particularly in North America) of land-use modeling efforts. This movement away from formal modeling methods largely reflected the failure of the models available at the time to meet the needs of the planning process, as has been documented in Lee's often quoted paper "Requiem for Large-Scale Models" [Lee, 1973].

Since the 1980s, long-range planning has reemerged as a valid component of urban transportation activities, with an attendant resurgence of interest in longer-run forecasting models. This has primarily been manifested in efforts to improve travel demand modeling capabilities, but over the past decade, increased interest in and application of land-use models has also occurred. This resurgence of interest in land-use modeling can be traced to several sources. First, in the United States, both the 1990 Clean Air Act Amendments (CAAA) and the 1991 ISTEA placed explicit emphasis on formally considering land-use–transportation impacts within the transportation planning process. ISTEA, in particular, required planning agencies to consider:

> The likely effect of transportation policy decisions on land-use and development and the consistency of transportation plans and programs with provisions of all applicable short- and long-term land-use and development plans [cited in Shunk et al., 1995].

TEA-21 subsequently weakened the language with respect to legislative requirements concerning consideration of land-use–transportation interactions (which has always been held to be a responsibility of local government), but the overall need to consider such interactions remained clear.

A second impetus to renewed interest in land-use modeling is the significant improvements in the technical capabilities to support such models, including

- Computer hardware (both processing speed and data storage capabilities),
- Computer software (GIS, relational database management systems [RDMS], statistical analysis packages, object-oriented programming capabilities, etc.),
- Databases (GIS-based, census, special-purpose surveys, etc.),
- Modeling methods (random utility-based models, activity-based models, advanced econometric methods for model estimation, etc.).

These have all improved significantly since land-use models were first attempted in the 1960s. The pioneering efforts of the 1960s were at a severe disadvantage due to the lack of data, computing power, and modeling techniques available at the time. Today, the technical capabilities to undertake large-scale urban modeling with reasonable expectations concerning both the efficiency and the effectiveness of the exercise are available [Miller and Salvini, 1998].

Finally, increased operational application of land-use models reflects the growing recognition that the land-use–transportation interaction *does* exist, and it must be understood, modeled, and accounted for in planning analysis. Metropolitan areas have physical definition only through the transportation system: space has little meaning otherwise. The interconnections between points (activities) in space are perceived through the medium of the transportation system. Build the transportation system differently, and people will use it differently, and they will organize themselves over space differently. Build the city differently, and the transportation needs will be different. Achieving a better understanding and, eventually representation, of the land-use–transportation interaction is therefore essential to the urban transportation policy debate worldwide, whether dealing with roads, transit, or nonmotorized modes of travel.

One of the most important examples of this is the question of whether constructing a new urban expressway has a net beneficial environmental impact (due to congestion relief and associated reductions in stop-and-go traffic) or a net negative impact (due to induced sprawl of land-use and travel patterns, increased auto dependency, etc.). A blue ribbon panel in the United States came to no definitive conclusion on this issue [TRB, 1995], while a similar study in the United Kingdom clearly endorsed the negative impact case [SACTRA, 1994]. The fact that significant scope appears to exist for considered professional opinion to differ so dramatically and the fact that so much of the debate over such issues is based on subjective arguments (often with strong ideological overtones) points to the failure of analysis methods available in the mid-1990s to provide more definitive insights into such problems.

Transit planning is particularly linked to land-use questions for at least two reasons. First, conventional transit is viable only within certain types of land-use/urban forms. In particular, certain minimum levels of trip-end densities are required before fixed-route transit services can be cost-effectively operated [Pusharev and Zupan, 1977, 1980]. Second, transit is often viewed as part of the solution with respect to urban sustainability. Yet, this is a very difficult proposition without a direct tie, behaviorally and with respect to policy, between transit and land-use.

If improved analytical capabilities are to be developed to address such fundamental questions, then they must clearly be much more holistic in nature in that they must address the entire urban system consisting of land-use activities, transportation, and environmental impacts in an integrated and comprehensive fashion [Wegener, 1995]. In particular, they must be able to

1. Explore the full range of system responses (short and long run) to a given policy.

2. Address a full suite of policies (transportation, land-use, etc.).

3. Incorporate the demographic/socioeconomic inputs needed by disaggregate behavioral models, which are generally required to properly address the behaviors/policies of interest within urban systems.

In addition to their role as forecasting/policy analysis tools, integrated transportation–land-use models also have the potential of providing an experimental platform/laboratory for exploring transportation–land-use interactions so that the dynamics/relationships can be better understood and hence better policies developed. At least one reason why more is not known about this interaction is that insufficient data have been available to analyze the problem in rigorous ways. In particular, the "without" case in any particular city is never observed (e.g., What would have happened in New York City without the subway system? Or in Toronto without the Yonge Subway?). Likewise, other "alternative histories/futures" are never observed (e.g., What would have happened if tight development controls had been implemented in city X twenty years ago, combined with a proactive transit investment program?). A credible urban simulation model (if one can be constructed) would be of great benefit in exploring such questions and thereby learning more about how cities work and about the likely efficacy of various policies.

Finally, land-use models do more than predict land-use patterns. In particular, they are also ideally demographic simulators, generating not only the population, but the attributes of the population required for behavioral location choice and travel demand modeling. Goulias and Kitamura [1992], among others, argue that this demographic simulation capability is critical to support the implementation of activity-based models that are generally considered the next evolutionary stage in demand modeling. A strong case can also be made that the relatively limited impact that even disaggregate logit mode choice models have had on travel demand modeling practice can be traced in large measure to the difficulty of projecting the population socioeconomic attributes required for these models to operate most effectively [Miller, 1996].

Even with "ideal" land-use models, uncertainty would still exist with respect to the exact nature of future activity systems. The existence of such uncertainty reinforces the need for a cyclical planning process of the sort outlined in Fig. 2.9, in which continuous monitoring of the system permits new trends and issues to be identified and diagnosed as they emerge and new policies to be formulated as required in response to these emerging trends. Even more fundamental, it argues for the need for flexibility within the transportation system itself. That is, one would ideally like to have a transportation system that is suitable for a range of likely future activity system states or that can adapt appropriately over time to the urban activity system as it evolves.

At the most general level, the challenge to planners is to understand the dynamics of the urban activity system and its relationship to the transportation system. To return to a theme that was introduced in the first chapter, it is most important for planners in a decision-oriented planning process to understand the decision-making process and the types of decisions that need to be made. Understanding the dynamics of the urban activity system requires an awareness of the key decision makers and the factors that influence their decisions. A number of actors can play critical roles: developers (who develop land, construct buildings, and sell or lease these buildings); firms (who build, buy, or lease buildings and provide employment); lenders (who provide the capital to build residential and employment units); residents (who buy, sell, and maintain houses or rent apartments); and governments (who regulate construction and land-uses, set taxes, provide services, provide the transportation infrastructure, etc.). In short, the previous observations reinforce the opening statement of this chapter—the transportation–land-use interaction is of fundamental importance to planning. If it is not well understood, then the analyses in subsequent stages of the planning process will suffer or be limited accordingly.

6.2   LAND-USE MODELS

The history of land-use modeling over the past 40 years can be roughly divided into at least three generations of modeling activity. The first occurred during the 1960s and very early 1970s. During this period, many different approaches to land-use modeling were tried as part of the general thrust of the time of developing mainframe computer-based forecasting models to support comprehensive, long-range planning activities. Approaches tested include

- Simultaneous equations systems to predict changes in time of zonal population and employment levels (as typified by the EMPIRIC model, [Hill, 1965; Hill et al., 1965; Brand et al., 1967]).

- Linear programming models allocating activities over space given an assumed objective function and set of constraints (e.g., Herbert and Stevens [1960]).

- Early simulation models (e.g., Chapin [1965]).

- The Lowry model.

Of these, the Lowry model has had the most lasting impact on the profession. Lowry-type models have been applied worldwide, with most current operational models having evolved out of the "Lowry model heritage"[Goldner, 1971], some still containing elements of the Lowry modeling approach. For these reasons, Sec. 6.2.1 provides a brief overview of the fundamental concepts involved in Lowry-type models.

Second generation land-use models emerged in the 1970s. These primarily consisted of large-scale, aggregate (i.e., zonal), mainframe-based simulation models that simulated the changes in the urban activity system over time through a set of discrete time steps (usually 5- or 10-year steps). Important advances in these models relative to the first generation models of the 1960s include a more explicit behavioral (typically microeconomic) theoretical foundation; explicit representation of building supply processes, building markets, and endogenous price formation; more disaggregate representations of households and other decision makers; and more complex database and modeling structures. All of these features are important in current models, and so some appreciation of these early large-scale simulation efforts is of use in understanding current models. The NBER and CAM models are perhaps the best known of this second generation of models, and these are briefly described in Sec. 6.2.2.

The third, and current, generation of operational models began to emerge in the 1980s, building on the lessons learned in the previous 2 decades, but also exploiting advances in computer hardware and software that have been continuously occurring since the emergence of the microprocessor and the personal computer in the early eighties. A representative sample of current models is discussed in some detail in Sec. 6.2.3.

As noted in the previous section, most metropolitan areas do not currently employ formal land-use models. To generate forecasts of zonal population and employment (and perhaps other attributes of the urban activity system), some form of scenario-based or judgmental approach is usually employed. Section 6.2.4 briefly discusses scenario-based methods.

Both formal models and more judgmental approaches generally assume that regional population and employment totals are exogenously determined; that is, they are forecasted separately, outside of the land-use forecast process itself, which concerns itself with distributing these totals spatially over the urban region. Procedures for forecasting population over time include [Steuart, 1977]

1. The "ratio-trend" method, which relates the population of a study area to the rising or falling ratio of that area's population to the population of a larger area, for which an accepted population forecast exists.

2. The "cohort-survival" method, which adds the effects of net natural increase and net-migration to the existing population.

3. The "economic-base" method, which gears population growth to a forecast of employment growth.

4. The application of a constant or gradually declining compounded annual rate or percentage increase in population.

5. A constant absolute rate of population increase per annum or per 5- or 10-year period.

Employment forecasts typically tend to be more difficult to perform. Techniques used include trend extrapolation, input-output analysis, and judgmental estimations. For more detailed discussions of population and employment forecasting, see, for example, Wilson [1974].

6.2.1   Heuristic Models: Lowry-Type Models

The Lowry model is named after Ira Lowry of the Rand Corporation, who first proposed the modeling approach in 1964 for application in Pittsburgh [Lowry, 1964]. The approach was later modified and further developed for applications throughout North America and Great Britain [Goldner, 1971; Batty, 1972]. The key concept underlying the model involves defining two classes of employment: retail and basic. Retail employment arises from all activities that

> are implicitly related to population and purchasing power. All activities for which a local market or service area can be identified for final products or services are in the category of "retail." The criterion is locational, flowing from the existence of a local market or service area [Goldner, 1971].

Basic employment is composed of everything else, that is, of all those activities that are site oriented in that their locations are dependent upon factors other than the size and location of residence-oriented local market areas. The model assumes that the basic employment in each zone of the urban area, E_b, is exogenously determined. Given E_b, the model allocates these workers to residential areas in the urban area using a work-to-residence distribution function. Given this residential distribution, the distribution of retail employment serving this population E_r is similarly allocated using a resident-to-shop distribution function. These workers, in turn, must be allocated to residences, which then generate additional retail activity (employment), and so on. Thus, the model incorporates a multiplier effect in which each new employee (basic or retail) generates further retail employment, until the entire process converges to an equilibrium state. In particular, if each worker requires m additional retail workers ($m < 1$), then for each initial basic worker there will exist at equilibrium $(1 + m + m^2 + m^n + \ldots) = 1/(1 - m)$ total workers. Figure 6.2 presents a flowchart for this basic Lowry model, while Table 6.1 summarizes the equations involved.

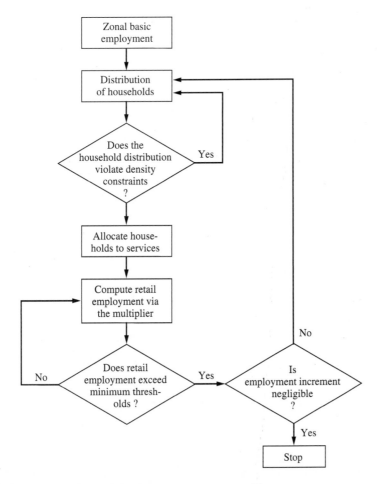

Figure 6.2   Lowry model flowchart

Table 6.1   The Lowry model equations

Land use

$$A_j = A_j^U + A_j^B + A_j^R + A_j^H \qquad\qquad\text{[6.2a]}$$

Retail sector

$$E^k = a^k N \qquad\qquad\text{[6.2b]}$$

$$E_j^k = b^k \left(\sum_{i=1}^{n} \frac{c^k N_i}{T_{ij}^k} + d^k E_j \right) \qquad\qquad\text{[6.2c]}$$

$$E^k = \sum_{j=1}^{n} E_j^k \qquad\qquad\text{[6.2d]}$$

$$E_j = E_j^B + \sum_{k=1}^{m} E_j^k \qquad\qquad\text{[6.2e]}$$

$$A_j^R = \sum_{k=1}^{m} e^k E_j^k \qquad\qquad\text{[6.2f]}$$

Household sector

$$N = f \sum_{j=1}^{n} E_j \qquad\qquad\text{[6.2g]}$$

$$N_j = g \sum_{i=1}^{n} \frac{E_i}{T_{ij}} \qquad\qquad\text{[6.2h]}$$

$$N = \sum_{j=1}^{n} N_j \qquad\qquad\text{[6.2i]}$$

Constraints

$$E_j^k \geq Z^k \quad \text{or} \quad E_j^k = 0 \quad \text{for all } j, k \qquad\qquad\text{[6.2j]}$$

$$N_j \leq Z_j^H A_j^H \qquad\qquad\text{[6.2k]}$$

$$A_j^R \leq A_j - A_j^U - A_j^B \qquad\qquad\text{[6.2l]}$$

Variables

A = area of land (thousands of square feet)
E = employment (number of persons)
N = population (number of households)
T = index of trip distribution
Z = constraints

Superscripts and subscripts

U = unusable land
B = basic sector
R = retail sector
H = household sector
k = class of establishments within retail sector ($k = 1, \ldots, m$)
i, j = tracts or zones within region ($i, j = 1, \ldots, n$)

SOURCE: Lowry, 1964

Garin [1966] reformulated Lowry's model using matrix notation and was able to demonstrate that the Lowry model converged to a unique equilibrium employment distribution. This is a comforting piece of information whenever one engages in an iterative solution procedure such as the one shown in Fig. 6.2. Garin also developed an equation to directly solve for the final total employment distribution (and hence the final residential population distribution as well), thus obviating the need for an iterative procedure. Implicit in these findings is a demonstration that the Lowry model is a static model whose predictions are entirely determined by the assumed basic employment distribution, the allocation functions, and a few key system parameters. That is, the way in which the iterative procedure shown in Fig. 6.2 appears to dynamically "grow" the city is entirely illusory and in no way truly dynamic or casual. Rather, it simply represents a mechanical, iterative procedure for computing the static equilibrium solution.

The Lowry model can be made quasi-dynamic by adding increments of basic employment over time to the existing distributions, thereby "building" the city over the forecast period rather than simply using the horizon year totals to generate a one-shot forecast. It should be noted, however, that unless the model contains other factors such as density constraints, which might be sensitive to the sequencing of these employment increments (and thus yield different horizon year forecasts depending on the order and magnitude of the increments), the incremental approach will yield the same horizon year forecast as the one-shot approach. Also note that the incremental approach implicitly assumes that all previously allocated employment and residential population remains fixed in place—clearly an important assumption given that a significant percentage of a metropolitan area's urban population changes residences and/or their place of employment each year.

Aside from the very important consideration of the static nature of the model, at least three other major issues exist with respect to the Lowry model.

1. The division of employment into basic and retail is conceptually interesting and important in that it permits prediction of at least some of the employment distribution to be "endogenized" within the model.

2. The Lowry model is clearly only as good as the allocation submodels used to distribute workers to residences and residences to shopping areas. The original Lowry model used an extremely crude allocation formula. Considerable work has occurred over the years in developing improved allocation submodels, based both on entropy maximization [Wilson, 1974; Batty, 1976; Putman, 1983,1991] and random utility maximization [Ben-Akiva and Lerman, 1985] methods.

3. The Lowry model ignores the entire issue of the market for land and buildings, that is, the demand, supply, and price of land and buildings, which are central to the allocation of people and firms over space. Aside from some simple and typically nonbinding constraints, people and firms are allowed by the model to locate at will over a flat, featureless plain. Thus, the model implicitly assumes that the supply of buildings perfectly adjusts to demand and that price is irrelevant to the process.

6.2.2   Simulation Models

Simulation models are characterized by their explicitly dynamic nature and their attempt to replicate, albeit in a simplified and abstract way, key events that occur over time. Simulation models represent the evolution of an urban area over several time periods as the outcome of a series of interrelated actions by the people and firms that constitute the metropolitan area. Such models tend to be complicated, large, and data-hungry. They typically consist of a relatively large number of submodels, each of which deals with one aspect or actor within the system, interconnected by the information about the urban system that they share.

The typical structure of a simulation model can be characterized as consisting of a set of information (current population, employment, and building stock characteristics, etc.) at some basic time period stored in a central data bank. Submodels characterizing the actors and their interrelationships draw upon this information as a basis for their actions (residential location shifts, employment changes by zone, building stock adjustments, etc.). Given these actions, the database is updated so that, at the end of the time period, there exists a new system state that serves as the basis for the decisions to be made in the next simulation time period. Finally, from time to time exogenous changes may be imposed on the system such as changes in migration rates into and out of the region, changes in total regional employment levels, or changes in interest rates.

In the 1970s, large-scale urban simulation models emerged for the first time. The best known of these is probably the National Bureau of Economic Research (NBER) model. A lesser-known but relatively successful model is the Community Analysis Model (CAM) developed by the MIT–Harvard University Center for Urban Studies in association with the U.S. Department of Housing and Urban Development. Although no longer in use, these models do illustrate the basic concepts of land-use simulation models and are thus discussed briefly as follows.

The NBER Model[1]   A major criticism of Lowry-type models is their lack of economic content. Noting this, NBER set out to develop a simulation model that was explicitly based on microeconomic concepts of the utility-maximizing household and the profit-maximizing firm. The model was originally developed by using Detroit as a test site but was then further developed using data from Pittsburgh, San Francisco, Minneapolis, and Washington, D.C. The NBER model identifies three main actors: *employers* who locate their industries in given zones within the region and who employ given numbers of employees; *households* who supply labor to the employers and who buy (or rent), maintain, and sell (or vacate) housing; and *suppliers* of housing who build (renovate, maintain, etc.) and sell (or rent out) housing. Households can locate in a zone only if vacant housing of a suitable type is available. The price of housing depends upon the competition for the available housing. In turn, the demand for and supply of housing over time depends, among other factors, on price.

Figure 6.3 displays the major submodels in the NBER system. Briefly, these are

---

| [1]   For a detailed description of this model, see Ingram, et al. [1972].

1. *Employment location submodel.* Employment changes for each of nine industrial groups for each zone in the region are specified for each time period. In the Detroit prototype, all employment data were exogenously supplied. In later versions of the model, the "retail" components were endogenously determined in a Lowry-like procedure. This employment by occupation types is then transformed into household types.

2. *Movers submodel.* Movers consist of in- and out-migrants for the region, newly formed households (resulting from children moving out, divorce, etc.), dissolved households (because of death, reconsolidation of families, etc.), and intraregional movement of households from one location to another. In general, simple moving rates are used to generate the various classes of movers.

3. *Demand allocation submodel.* This submodel predicts the probability that a household of a certain type with a worker employed in a given zone will wish to live in a housing unit of a given type.

4. *Vacancies submodel.* This is essentially a bookkeeping procedure, which simply keeps track of the vacant units by type and zone that were left over from previous time periods or were vacated by movers during the current time period being simulated.

5. *Filtering submodel.* Filtering consists of upgrading (through maintenance or renovation) or downgrading (through a lack thereof) the quality of the housing stock and hence shifting the existing housing stock from one category to another.

6. *Supply submodel.* This submodel generates the construction of new housing and the conversion of existing housing (conversion consists of changing the housing *type,* for example, from a single-family to a multiple-family dwelling, as opposed to filtering, which consists of changing the housing quality) during each time period.

7. *Market-clearing submodel.* This submodel links the demand and supply sides of the housing market by assigning households demanding housing of given types (as determined by the demand allocation submodel) to available units in specific zones (as determined by the vacancy, filtering, and supply submodels).

Despite its vintage, the NBER model has been discussed in some detail here in order to illustrate the complexity and the difficulties inherent in attempting to simulate a metropolitan area's housing market in a reasonably comprehensive fashion. Section 6.1 raised the issues of the dynamic nature of urban areas; the need to address the demand, supply, and market-clearing processes in order to understand the evolution of urban land-use patterns; and the need to explicitly consider the major actors at work within urban regions. The NBER model's major contribution to the field is that it represents an early, comprehensive attempt to address these issues in a pragmatic, yet theoretically rich, way.

The Community Analysis Model (CAM)[2] CAM is fairly similar to the NBER model in terms of its submodels and overall structure. The major differences between the

---

| [2] For a detailed discussion of this modeling approach, see Birch et al. [1974]; Birch [1974].

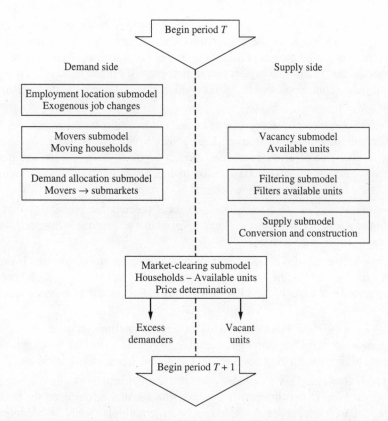

Fig. 6.3 The National Bureau of Economic Research (NBER) model
| SOURCE: Ingram, Kain, and Ginn, 1972

two models lie in what is "inside the boxes" and in the theoretical constructs used to generate the submodels. In contrast to the explicitly disciplinary approach adopted by the NBER in which a model was formulated based on economic theory and then calibrated with observed data, the builders of CAM adopted a far more inductive approach. They first started with comprehensive time-series data (in this case, for New Haven, Connecticut) and then experimented with a wide range of models in an attempt to understand and explain the trends and the interactions that were observed. Central to this understanding was the identification of a set of *actors* responsible for the major decisions affecting the urban area, their *roles* (i.e., what decisions or actions were available to them), what *information* they used in making these decisions, and how to best *characterize* these actors. Table 6.2 presents the 11 actors ultimately chosen, the decisions for which they are responsible, the variables used to characterize them, and the major variables or determinants influencing their decision making. Points to note from this table include

1.   Both *households* and *individuals* enter the model, reflecting the fact that some decisions (e.g., where and in what type of house to live) are made by the household as a whole. Others are more properly considered as being made by the individual

concerned (e.g., whether to go to school or get a job). This implies the need to be able to relate households and individuals to one another (or to "map" them onto one another) in order to keep track of how individual decisions affect the households in which the individuals live, and vice versa.

2. Any one individual in an urban area may well play several roles (i.e., represent several different actors) simultaneously. For example, in addition to being a member of a household, the individual may also be a home owner, a resident, an employer, or a landlord.

3. While several of its actors are the same as those in the NBER model (e.g., households and builders), CAM possesses a far richer cast of characters than other models to this point. These additional actors were considered necessary to represent adequately the key roles that such factors such as school quality, insurance rates, and availability of capital play in urban development.

Table 6.2   Actors within the community analysis model

Actor	Decision	Stratification	Major Determinants
Household	To move within region	Age By ethnic or racial background By education	Life cycle Race Educational level Racial change Forced moves
	Choice of neighborhood		Life cycle Race Educational level Available units Social class of neighborhood Job accessibility of neighborhood Location of neighborhood Racial transition of neighborhood
	Choice of unit		Housing preferences Financial capability Availability of mortgage credit
	Migrate in and out of region		Employment opportunities Unemployment Income levels in region Educational mix in region City size Proximity to other areas
Individual	Have children	Age By ethnic or racial background By education	Life cycle Race Level of education
	Obtain education		Ethnic/racial background

(continues)

Table 6.2   Actors within the community analysis model (continued)

Actor	Decision	Stratification	Major Determinants
Individual	Join workforce		Age Ethnic/racial background Educational level Growth rate of local economy
Home owner	Setting selling price of home	Age By education By price of home By ethnicity	Potential demand relative to available vacancies
	Investing in home maintenance		Characteristics of home owner (e.g., housing preferences) Characteristics of housing unit (e.g., age of unit) Characteristics of neighborhood (e.g., average housing condition)
Landlord	Setting rent levels on apartments	Rent level of apartments	Potential demand relative to available vacancies
	Investing or disinvesting in maintenance for apartments (including abandoning apartments)		Characteristics of tenants (e. g., age, education, ethnicity) Characteristics of apartment (e.g., age of unit) Characteristics of neighborhood (e.g., ethnic composition)
Builder	Constructing single-family homes under contract	Contract v. speculative	Vacancy rate in submarket and region
		Type of unit (tenure and price)	Availability of suitable vacant land Availability of credit
	Constructing apartments under contract		Absorption rate in submarket Excess demand in submarket Vacancy rate in submarket and region Zoning restrictions Availability of suitable vacant land Availability of credit
	Constructing single-family homes and apartments speculatively		Assessment of prospects for future demand in the neighborhood according to demand in nearby areas, rate of growth in nearby areas Availability of suitable land Availability of credit

(continues)

Table 6.2   Actors within the community analysis model (continued)

Actor	Decision	Stratification	Major Determinants
Lender	Lending for new construction and home mortgages	None	Security on loans: depends on characteristics of loan applicant (e.g., age, ethnicity, education), characteristics of housing unit (e.g., condition), and characteristics of neighborhood (e.g., ethnic composition, average housing condition) Availability of funds
	Refusing to grant loans in certain neighborhoods		Risk attached to loans: depends on characteristics of neighborhood (e.g., ethnic composition, average housing condition)
Insurer	Insuring homes and apartments	None	Expected profit (excess of premium revenue over insurance payments): depends on characteristics of insurance applicant (i.e., age, ethnicity, education), characteristics of housing unit (e.g., condition), and characteristics of neighborhood (e.g., ethnic composition, average housing condition)
	Refusing to write insurance in certain neighborhoods		Expected loss (excess of insurance payments over premium revenue): depends on characteristics of neighborhood (e.g., ethnic composition, average housing condition)
Zoning board member	Determining permissible land uses, granting variances	None	Density preferences of existing residents of neighborhood Extent of commercial activity in a neighborhood
Employer	Location of a new firm Expansion of employment Relocation Contraction of employment Closing a firm's operations	Major SIC code divisions	Vacant land in a neighborhood Characteristics of the population of a neighborhood Population density Proportion of jobs in a given industry located in a given area, called job concentration Proportion of jobs in a given area that are in a given industry, called job specialization The tenure of and value of occupied housing units

(continues)

Table 6.2   Actors within the community analysis model (continued)

Actor	Decision	Stratification	Major Determinants
School superintendent	Modify school characteristics	None	Residents' reactions to school characteristics Court decisions
Resident	Support (or not) existing school policies	Age By ethnic or racial background By education	Percent minority in school Teachers' qualifications in school Public image of school Curriculum in school Class size in school Resident's attitude toward education Social class of resident

SOURCE: Birch, 1974

Figure 6.4 shows how the various actors interact within the model. Boxes indicate the various submodels in which the actors' decisions are actually simulated, ovals represent key pieces of information used and/or generated by these submodels, and circles represent the key descriptors of the system state (distributions of jobs, households, housing, etc.) as they exist at any point in time. As is illustrated by Table 6.2 and Fig. 6.4, the key strength of CAM lies in its detailed representation of the actors and their interrelationships. The actual decision models tend, in general, to be relatively simple (simple interaction rates, simple probability models, etc.).

The hypothesis underlying this modeling approach is that behavior is complex because of the multitude of actors, motives, and interactions involved. Once one has sorted out the network of actors and their interactions and has done a reasonably good job of accurately identifying and characterizing these actors and interactions, the behavior of any given actor is relatively straightforward to represent and relatively constant over time. This modeling approach can be contrasted to that of the NBER model (and most other models discussed in this book) in which essentially the opposite approach is taken: a limited number of actors are considered, but a relatively complicated decision function is assumed in order to try to explain the complex set of observed actions. In many respects, CAM anticipated current and emerging object-oriented, agent-based modeling methods that are the basic approach for current model development.

After its initial development in New Haven, CAM was applied to Houston, Texas, and then subsequently to Dayton, Ohio; Worcester, Massachusetts; Rochester, New York; and Charlotte, North Carolina. Although not explicitly developed for use in transportation studies, CAM was used in all six cities in a wide range of policy applications.

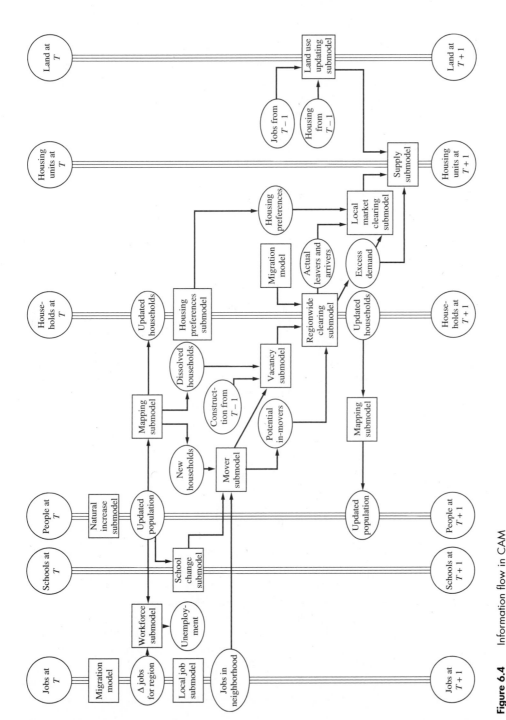

Figure 6.4 Information flow in CAM
SOURCE: Birch, 1976

6.2.3   Operational Models[3]

A significant number of integrated or semi-integrated urban models exist around the world, in varying degrees of completeness and usability. Wegener [1995] identified 20 active urban modeling centers (i.e., models) around the world, of which approximately 12 integrated urban models were sufficiently operational to have been used for actual research and/or policy analysis in particular urban areas. Southworth [1995] identified three more models.

This section briefly reviews six models that are representative of the current operational state-of-the-practice in integrated urban modeling. These are ITLUP (also often referred to as DRAM/EMPAL), MEPLAN, TRANUS, MUSSA, METROPOLIS, and UrbanSim. The six models generally have the following features in common:

- They are operational packages with an established history of use, most of which are available as commercially supported software.

- With one exception, each has been applied in North America in at least one practical setting (e.g., at the MPO or state level).

- Again with one exception, each contains a significant market representation (i.e., there is an explicit treatment of demand, supply, and prices in land development).

As indicated in the Wegener and Southworth reviews, many other integrated urban models exist. Noteworthy examples include examples of microsimulation models[4] [Mackett, 1990b; Wegener, 1993; Miller et al., 1987; Miller and Salvini, 1998, 2000; Oskamp, 1997]; optimization models [Brotchie et al., 1980; Caindec and Prastacos, 1995; Dickey and Leiner, 1983; Kim, 1989]; land accounting-type models [Landis, 1994; Yen and Fricker, 1997]; and other European, Japanese, and Australian models [Anderstig and Mattson, 1991; Eliasson and Mattsson, 1997; Gu et al., 1992; Mackett, 1983, 1985a, 1990a; Nakamura et al., 1983; Simmonds, 1998; Wegener, 1982a, 1982b, 1986; Wegener et al., 1991]. While each of these models is of potential interest in one or more ways, it was felt that the six models chosen for detailed review are sufficient for the purposes of this text in that they are representative of the state of operational practice.

ITLUP   ITLUP (Integrated Transportation and Land-Use Package) was developed by Dr. Stephen H. Putman at the University of Pennsylvania over the course of the last 25 years. ITLUP comprises a number of submodels, the best known of which are DRAM (Disaggregate Residential Allocation Model) and EMPAL (Employment Allocation Model). DRAM and EMPAL are Lowry-type models. The model allocates household categories (usually four income categories, though further categorizations are possible, e.g., by household structure), employment categories (usually by four types, though more detailed SIC groupings are possible), and travel patterns (public and private modes), using exogenous study-area forecasts of employment, population and trips, activity rates and household type.

---

[3] This section draws heavily upon a more detailed review of these models contained in Miller et al. [1998].
[4] Microsimulation modeling is discussed further in Chap. 7.

A simple example of the residential allocation process will provide the reader with a basic understanding of how these models work.[5] Assume that the number of employees living in zone i can be expressed as

$$N_i = \sum_j E_j P_{ij} \qquad\qquad \text{[6.3]}$$

where

N_i = number of employees living in zone i
E_j = number of employees working in zone j
P_{ij} = probability that an employee who works in zone j might choose to live in zone i (function of trip time)

The total number of employed residents in the area must equal the total employment. Thus,

$$\sum_i N_i = \sum_j E_j \qquad\qquad \text{[6.4]}$$

The probability of a work trip between two zones i and j is given by

$$P_{ij} = \frac{F(c_{ij})}{\sum_i F(c_{ij})} \qquad\qquad \text{[6.5]}$$

where $F(c_{ij})$ is a function of the travel time or cost between zones i and j. Assuming $F(c_{ij}) = (c_{ij})^{-2}$, for example, says that as the cost of travel between two zones increases, the employee is less likely to make the trip. Substituting Eq. 6.5 into Eq. 6.3

$$N_i = \sum_j \left[\frac{E_j (c_{ij})^{-2}}{\sum_i (c_{ij})^{-2}} \right] \qquad\qquad \text{[6.6]}$$

Suppose we have a simple five-zone system in which the employment for each zone is shown in Table 6.3a. Table 6.3b shows the estimated travel time between all of the zones. The first step is to calculate $(c_{ij})^{-2}$, which is shown in Table 6.3c. The column sums in Table 6.3c represent the denominator sums in Eq. 6.6. The total number of employees living in zone 1 can be estimated as

$$N_1 = \frac{E_1 (c_{11})^{-2}}{\sum_i (c_{i1})^{-2}} + \frac{E_2 (c_{12})^{-2}}{\sum_i (c_{i2})^{-2}} + \ldots + \frac{E_5 (c_{15})^{-2}}{\sum_i (c_{i5})^{-2}}$$

Substituting the values from Table 6.3c, we obtain

$$N_1 = \frac{(100)(1.00)}{2.29} + \frac{(200)(0.44)}{2.38} + \frac{(3.00)(0.25)}{2.10} + 0 + 0 = 116$$

| 5 This example is based on Putman, 1993.

Table 6.3

(a) Example allocation of residence units

Zone	Employment
1	100
2	200
3	300
4	0
5	0

(b) Travel "costs" between zones

Zone	1	2	3	4	5
1	1.0	1.5	2.0	1.5	2.5
2	1.5	1.0	1.5	2.0	2.0
3	2.0	1.5	1.0	2.0	2.5
4	1.5	2.0	2.0	1.0	1.5
5	2.5	2.0	2.5	1.5	1.0

(c) c_{ij}^{-2} values

Zone	1	2	3	4	5
1	1.0	0.44	0.25	0.44	0.16
2	0.44	1.00	0.44	0.25	0.25
3	0.25	0.44	1.00	0.25	0.16
4	0.44	0.25	0.25	1.00	0.44
5	0.16	0.25	0.16	0.44	1.00
Σ	2.29	2.38	2.10	2.38	2.01

SOURCE: Putman, 1993

This simple model estimates that 116 employed residents will reside in zone 1. Similar calculations result in $N_2 = 166$, $N_3 = 191$, $N_4 = 76$, and $N_5 = 51$. Note that the sum of the employed residents equals the total employment control number we started with. Also note that zones 4 and 5 were allocated employed residents even though there were no jobs allocated to these zones to begin with.

Detailed documentation of the model is provided in Putman [1983, 1991, 1994, 1995]. Useful summary descriptions are provided in Wegener [1995], Southworth [1995], Webster et al. [1988], and Putman [1996], from which this discussion is drawn. Notable features of ITLUP are

1.   ITLUP (more specifically, DRAM and EMPAL) is the most widely used spatial allocation model in the United States today. A recent count indicates over a dozen active U.S. applications [Putman, 1997], although over 40 calibrations have been performed across the United States and elsewhere.

2.   ITLUP contains a multinomial logit modal split submodel, as well as a trip assignment submodel that can support various (auto) assignment algorithms. Trip generation and distribution are developed within DRAM, simultaneously with household location. However, DRAM and EMPAL often are used separately and have been linked in actual applications with other commercial travel demand forecasting models (including EMME/2, TRANPLAN, and UTPS). Thus, considerable detailing of travel demand and travel costs can be provided through exogenous links.[6]

3.   The model has been applied to a wide range of situations, ranging from long-term transportation network planning, to the evaluation of different urban forms and, notably, environmental impacts and regionwide vehicle emissions. However, although in theory there appear to be no constraints to including transit travel "costs" in the accessibility definitions, most applications appear to have emphasized road network improvements.

4.   Compared with other models, DRAM/EMPAL is considered to have relatively parsimonious data requirements. Southworth [1995] notes that an important advantage of DRAM/EMPAL is their basis in generally available data (i.e., related to population, households, and employment). However, he also notes that this reflects a weakness of the approach, namely, that the model does not account for land market clearing processes (or, it follows, other market clearing processes).

MEPLAN   MEPLAN is a proprietary software package developed by Marcial Echenique and Partners Ltd. in the United Kingdom. The MEPLAN package draws on 25 years of practical integrated urban modeling experience, with work on the package itself beginning in 1985. It has been applied to over 25 regions throughout the world, including Sacramento, California. Detailed documentation can be found in Echenique [1985], Echenique et al. [1969, 1990], Echenique and Williams [1980], Hunt [1994], Hunt and Simmonds [1993], and Hunt and Echenique [1993].

MEPLAN is based on a general, highly flexible framework. This framework has an aggregate perspective: space is divided into zones; quantities of households and economic activities (called factors or sectors) are allocated to these zones, and flows of interactions among these factors in different zones give rise to flows of transport demand. Figure 6.5 illustrates the key components of MEPLAN, in which the demand for activity locations interacts with the supply of land and floor space within a land market model and the demand for and supply of transportation services interact within a transport market. Endogenously determined prices mediate between demanders and suppliers of floor space in the land market, with demanders responding to current market prices, while supply decisions in time period t are a function of the prices that occurred in the previous time period (t-1). Similarly, travel demand decisions are based on the levels of (dis)utility provided by the transportation system. The land and transport markets interact in that the spatial interaction between activities gives rise to travel demand (workers traveling to work, shoppers traveling to stores, etc.), and in that accessibilities defined by the transportation

---

| [6]   See Chap. 5 for discussion of these and other travel demand modeling terms and methods.

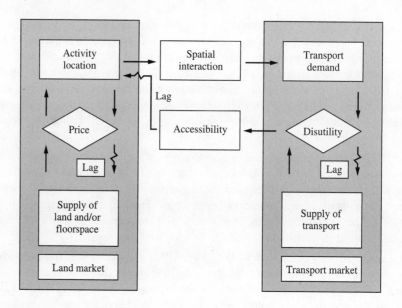

Fig. 6.5   MEPLAN model structure
SOURCE: Hunt, 1994

system influence activity location choice (i.e., people, stores, etc., are more likely to locate at points of higher, rather than lower, accessibility).[7]

The heart of the framework is a spatially disaggregated input–output matrix, or social accounting matrix, extended to include variable technical coefficients, labor sectors, and space sectors. All economic activities, including households, are treated as producing and consuming activities, with consumption patterns expressed using technical coefficients. Spatial disaggregation is accomplished by having the production arising to satisfy consumption allocated among the spatial zones according to discrete choice models reacting to the prices for such production. The resulting interactions among zones give rise to the demand for travel. In an example model, six industries (agriculture, heavy industry, etc.) and five household categories (defined in terms of income) interact in terms of economic exchanges (workers sell labor to firms; firms sell goods and services to other firms and households; etc.), consumption of land, and transport flows by mode. An important feature of MEPLAN is that goods movements by origin, destination, and mode are generated as well as the person movements, which are the usual focus of such models.

Temporal change is simulated by having the model consider sequential points in time. Space (both land and floor space) is "nontransportable" and must be consumed in the zone where it is produced. The supply of space in each zone is fixed at a given point in time. The technical coefficients for the consumption of space are elastic with respect to price, and prices for space that ensure demand equates with supply in each

[7]   At this level of abstraction Fig. 6.5 is reasonably representative of the macro structure of most of the models discussed in this section.

zone are established endogenously as part of an equilibrium solution established for each point in time considered. Prices for the outputs of other sectors are established endogenously running back along the "chains" of production–consumption. Travel demands arising for a given point in time are allocated to a multimodal network using logit functions representing mode and route choice, taking account of congestion. Transport disutilities feed back into the next time period, representing lags in response to transport conditions. Exogenous demand (analogous to the Lowry basic sector) provides the initial impetus for economic activity. Changes in studywide exogenous demand and in the quantity of space in each zone from one time period to the next fuel economic change, with these changes allocated among zones.

MUSSA MUSSA ("Modelo de Uso de Suelo de Santiago") is an operational model of urban land and floor-space markets developed by Dr. Francisco Martinez for Santiago, Chile. It is fully connected with a four-stage travel demand model (known as ESTRAUS). Together, the combined models are referred to as 5-LUT and provide equilibrated forecasts of land-use and travel demand for Santiago. The model has been used to examine various transportation and/or land-use policies, usually involving transit as a central component of the policy. Documentation of the model is provided in Martinez [1992a, 1992b, 1997a, 1997b] and Martinez and Donoso [1995]. Notable features of MUSSA include

1. It is consistently based throughout on an extremely rigorous and compelling application of microeconomic theory.

2. It is an equilibrium model of building stock supply and demand. Demand for building stock (whether by households or firms) is based on their willingness to pay (WP). Willingness to pay is defined as the maximum amount a household would be willing to pay for a given dwelling unit in order to achieve a given utility level and can be expressed using the random utility modeling concepts discussed in Chap. 5 [Martinez, 1997a]. Buyers attempt to maximize their surplus (WP less price actually paid), while sellers attempt to maximize price paid. Building stock is supplied by developers so as to maximize profits, given the apparent demand. Building stock prices are endogenously determined within the equilibration process.

3. The model solves for a static equilibrium in the forecast year by adjusting the amount of building stock supplied, a supply response, and consumers' expectation levels (expected utility to be obtained from their housing), a demand response, until demand and supply balance. The model end-state is path independent and does not require solution for intermediate year results, although such intermediate results can also be generated.

4. The model uses traffic analysis zones as its spatial unit of analysis (264 zones in the Santiago application), thereby providing a very fine level of spatial disaggregation relative to many other current models. In addition, extensions to more microlevels of spatial analysis are being investigated [Martinez, 1997b].

5. The model is highly disaggregated relative to most other currently operational models. The Santiago implementation has 65 household types and could be run using a large weighted sample of observed households (and their associated detailed attributes) in essentially a "static microsimulation" format.

6.   As with the METROPOLIS family of models (discussed in the following section), MUSSA provides operational evidence of the usefulness of integrated urban models in the analysis of both the impacts of major transit projects on land use and the impacts of land development on transit (among other policies).

METROPOLIS (NYMTC-LUM)   METROPOLIS (NYMTC-LUM) is the most recent of a series of land-use and housing market models developed by Dr. Alex Anas over the last 2 decades [Anas, 1982, 1992, 1994, 1995, 1998; Anas and Arnott, 1993, 1994; Anas and Brown, 1985]. NYMTC-LUM is a currently operational version of METROPOLIS that has been developed for the New York Metropolitan Transit Commission (MTC). Notable features of the NYMTC-LUM implementation of METROPOLIS include:

1.   It is consistently based throughout on microeconomic theory.

2.   It simultaneously models the interactions between residential housing, commercial floor space, labor, and nonwork travel markets, with explicit representations of demand and supply processes in each case.

3.   Housing prices, floor-space rents and workers' wages are all endogenously determined within the model and are used to mediate between demand and supply processes within their relevant markets.

4.   The model solves for a static equilibrium in the forecast year by finding the prices and wages that cause demand and supply in the markets being modeled to balance. The model end-state is path independent and does not require solution for intermediate year results.

5.   The model uses traffic analysis zones as its spatial unit of analysis (up to 3500 zones in the New York application), thereby providing a very fine level of spatial disaggregation relative to many other current models.

6.   In its current state of implementation, the model does not contain much disaggregation of its main "behavioral units" (households, employment, buildings).

7.   In its current implementation, the model is not integrated with a travel demand model. Rather, it is "connected" to the existing MTC travel demand model in terms of receiving as inputs model utilities from the MTC mode choice model. This is similar to the case for ITLUP and UrbanSim.

A particularly noteworthy point about the NYMTC-LUM implementation of the model is that it has been developed for a transit property, explicitly for transit-related planning. Features of the model that facilitate this type of application include use of small traffic zones as the spatial unit of analysis; access to detailed transit network representations and mode choice models in the MTC travel demand model; and the microeconomic structure of the model, which permits a range of economic evaluation measures to be computed (property values, consumers' surplus, producers' surplus, etc.). Earlier models have similarly been applied to the evaluation of the impacts of the proposed South Corridor rapid transit line in Chicago (using the CAT-LAS model; see Anas and Duann [1986]) and assessing the impacts of a range of road and transit service changes in New York (using NYSIM; see Anas [1995]).

TRANUS TRANUS is a proprietary software package developed by Dr. Tomas de la Barra and the firm Modelistica in Venezuela. It draws on much the same modeling experience as MEPLAN, with the elements of the package first coming together in the early 1980s. The focus of TRANUS is on a somewhat more restricted set of functional forms and modeling options within the framework, allowing a more "set" approach to model development relative to MEPLAN. It has been applied to a number of regions in Central and South America and in Europe. TRANUS models of the Sacramento and Baltimore regions and the state of Oregon have been completed or are under development. For detailed documentation, see de la Barra [1982, 1989], de la Barra et al. [1984], and Modelistica [1998].

UrbanSim UrbanSim is an operational model of urban land and floor-space markets developed by Dr. Paul Waddell (University of Washington) for the states of Hawaii, Oregon, and Utah. A prototype has been completed in Eugene–Springfield, Oregon. It is "fully connected" with a traditional four-stage model in Eugene–Springfield, Oregon, and is being integrated with a new activity-based travel model in Honolulu, Hawaii. The model and software have been placed in the public domain by the Oregon DOT, and the University of Washington will support its release and dissemination through the Internet, as part of an NCHRP project 8-32(3) "Integration of Land-use Planning and Multimodal Transportation Planning." Documentation of the model can be found in Waddell [1998a, 1998b, 1998c] and Waddell et al. [1998], as well as at <*www.urbansim.org*>.

Notable features of UrbanSim include:

1. It is consistently based throughout on an extremely rigorous and compelling application of microeconomic theory, following a similar framework to that used in MUSSA but differing in significant aspects such as the assumption of equilibrium.

2. It is a disequilibrium model of building stock supply and demand with annual time increments. Demand for building stock (whether by households or firms) is based on their willingness to pay (WP) or bid (prices actually paid are used rather than the unobservable WP). Buyers attempt to maximize their surplus (WP less price actually paid), while sellers attempt to maximize price paid. Building stock is supplied by developers so as to maximize profits, given the apparent demand. Building stock prices are determined within the market clearing process, which occurs at the submarket level of the traffic analysis zone and property type.

3. The model operates as a dynamic disequilibrium in each year, with the supply component developing and redeveloping individual land parcels on the basis of expected profits (expected revenue less costs). Expected revenue is based on prices lagged by 1 year, and new construction choices are not assumed to be available for occupancy until the subsequent year. Demand is based on lagged prices and current supply, and prices are adjusted based on the balance of demand and supply in each submarket in each year. The model end-state is path dependent and requires solution for each intermediate year.

4. The demand side of the model uses traffic analysis zones as its spatial unit of analysis (271 zones in the Eugene–Springfield application, 761 in Honolulu, and over 1000 in Salt Lake City), thereby providing a very fine level of spatial

disaggregation relative to many other current models. In the supply side, the model uses the individual land parcel as the unit of land development and redevelopment, making this the only model to date to use the parcel level as the fundamental unit of analysis.

5. The model is highly disaggregated relative to most other currently operational models. The Eugene–Springfield implementation has 111 household types and could be run using a large weighted sample of observed households and their associated detailed attributes in essentially a "static microsimulation" format.

6. The model can analyze policy scenarios that include comprehensive land-use plans, growth management regulations such as urban growth boundaries, minimum and maximum densities, mixed-use development, redevelopment, environmental restrictions on development, and development pricing policies; as well as the range of transportation infrastructure and pricing policies handled by the linked travel demand models;

Tables 6.4a, 6.4b, and 6.4c summarize the capabilities of the six models in responding to the typical range of policies that might be of interest within an urban area. The table is divided into land-use, transportation, and other policies. These policies are further categorized into pricing, infrastructure/services, regulatory, and education/marketing policies. Finally, specific examples of each type of policy are listed. For each policy, a model is given a check (✔) if the policy can (at least in principle) be explicitly represented and tested within the model, while an X indicates that the model structure/specification precludes testing the given policy. An asterisk (*) indicates that the policy can be "implicitly" addressed (either through a "proxy variable" or through straightforward extensions to the current modeling system). Finally, a plus sign (+) indicates that the only way the model can be made to respond to a given policy is through user-defined, exogenous changes to model parameters.

In general, the models can respond to a fair range of pricing and infrastructure/service-related policies. This is not surprising, given that these represent the traditional types of policies that have motivated model development over the years. Three important caveats to this general observation exist. First, because ITLUP does not model land-use demand–supply interactions explicitly, relative to the other models it is limited in its ability to address land-use pricing and infrastructure policies. Second, while ITLUP, MEPLAN, and TRANUS are all shown as being able to address transit infrastructure and service policies, the large zone system typically used in these models limits the sensitivity that these models can have with respect to transit. Third, the extent to which a given model is sensitive to transportation policies depends critically on the specific travel demand model being used. If a "best-practice" travel demand model is used, then the full set of check marks shown in Table 6.4b is appropriate. If, however, in a given application, a more limited travel demand model is used, then Table 6.4b may well overestimate the integrated model's capabilities.

All models reviewed in the preceding discussion are much less able to address regulatory policies, especially with respect to transportation. TDM and ITS impacts are generally not well addressed. These limitations primarily reflect well-known

Table 6.4 Policy capabilities of current models

	(a) Land Use					
Policy Category	**Specific Policy**	**ITLUP**	**MEPLAN/ TRANUS**	**MUSSA**	**NYMTC-LUM**	**UrbanSim**
Pricing	• Taxation	X	✔	✔	✔	✔
	• Subsidies	X	✔	✔	✔	✔
	• Development charges	+	✔	✔	✔	✔
Infrastructure and services	• Public housing	X	✔	✔	✔	✔
	• Servicing land (excl. transportation)(e.g., sewers, water, cabling)	+	✔	✔	✔	✔
	• Government buildings/other not-for-profit institutions	X	✔	✔	✔	✔
Regulatory	• Zoning (uses, densities)	+	✔	✔	✔	✔
	• Microdesign building/ neighborhood issues	X	X	✔	X	✔
Education/ marketing	• Changing/how to change attitudes and sensitivities (e.g., traveler "value of time" as opposed to deeply held values)	+	+	+	+	+

LEGEND:
✔ Explicit and normally/could be found
X No
* Implicit
+ Can respond, but only through exogenous change in parameters

	(b) Transportation					
Policy Category	**Specific Policy**	**ITLUP**	**MEPLAN/ TRANUS**	**MUSSA**	**NYMTC-LUM**	**UrbanSim**
Pricing	• Road tolls	*	✔	*	*	*
	• Gas taxes	*	✔	*	*	*
	• Subsidies (capital, operating)	X	X	X	X	X
	• Transit fares	+	✔	*	*	*
	• Parking pricing	+	✔	*	*	*
Infrastructure and services	• Build roads, HOV	✔	✔	✔	✔	✔
	• Build rail/dedicated transit ways	+	✔	✔	✔	✔
	• Operate transit services	+	✔	✔	✔	✔
	• ITS	X	X	X	X	X
	• Parking	X	X	X	X	X

(continues)

(b) Transportation (continued)

Policy Category	Specific Policy	ITLUP	MEPLAN/ TRANUS	MUSSA	NYMTC- LUM	UrbanSim
Regulatory	• Parking provision regulations (off-street)	X	X	X	X	X
	• Rules of the road (speed limits, on-street parking, HOV lanes etc.)	+	✔	✔	✔	✔
	• Nonpricing TDM	X	X	X	X	X
	• Vehicle/driver licensing	+	X	X	X	X
	• Inspection/maintenance	X	X	X	X	X
Education/ marketing	• Changing/how to change attitudes and sensitivities (e.g., traveler "value of time" as opposed to deeply held values)	+	+	+	+	+

LEGEND:
✔ Explicit and normally/could be found
X No
* Implicit
+ Can respond, but only through exogenous change in parameters

(c) Other

Policy Category	Specific Policy	ITLUP	MEPLAN/ TRANUS	MUSSA	NYMTC- LUM	UrbanSim
Pricing	• Auto purchase tax	X	X	X	X	X
	• License charges	X	X	X	X	X
	• Income redistribution (e.g., progressive taxation, welfare, etc.)	+	✔	✔	✔	✔
Infrastructure and services	N/A	N/A	N/A	N/A	N/A	N/A
Regulatory	• Air-quality standards	X	X	X	X	X
	• Emissions standards	*	✔	✔	*	*
	• Noise	*	*	*	*	*
	• Safety (accidents)	*	*	*	*	*
	• Vehicle technology standards	X	X	X	X	X
Education/ marketing	• Changing/how to change attitudes and sensitivities (e.g., traveler "value of time" as opposed to deeply held values)	+	+	+	+	+

LEGEND:
✔ Explicit and normally/could be found
X No
* Implicit
+ Can respond, but only through exogenous change in parameters

weaknesses in the four-stage travel demand modeling system, upon which these integrated models heavily rely, rather than any fundamental problem with land-use models *per se*. Thus, as operational travel demand models continue to improve, integrated models can be expected to directly benefit as well.

Not surprisingly, all the models can respond to education/marketing policies aimed at changing people's values/attitudes only through exogenous changes in the model parameter values, where these parameters are intended to capture decision makers' tastes and preferences. All models are calibrated against observed, historical data. Given this, at best, they capture the behavioral preferences manifested in the historical data. To use these models in a forecasting mode, we must assume that these behavioral preferences will hold in the future as well. While the increasing availability of time-series data (both panel and repeated cross sections) is beginning to provide us with the opportunity to study the evolution of tastes and preferences over time,[8] such research is a long way from providing us with operational models of parameter formation and evolution.

Some key points to note about currently operational integrated models include the following:

1.   Historically, integrated models have been spatially quite aggregate. The current trend among newer models is toward a finer spatial scale, typically the traffic-zone level. UrbanSim, the most disaggregate of the models reviewed, deals with land development at the parcel level and travel demand at the traffic-zone level.

2.   Most models are based on strong equilibrium assumptions. UrbanSim is a noteworthy exception to this rule.

3.   Most models are still temporally very aggregate, using at most 5-year time steps. Again, UrbanSim is the exception to this rule, in which a 1-year time step is used.

4.   Data requirements and implementation effort obviously vary, depending on model complexity. Any integrated urban model, however, requires significant investment in time and money to implement.

5.   Ongoing technical support is required to operate these models.

6.2.4   Scenarios

An alternative to the use of mathematical forecasting techniques such as the ones discussed in the previous subsections is the use of future scenarios constructed on the basis of professional judgment, current or expected trends, and so on. If this approach is adopted, it is typical to construct several scenarios, representing the expected range in likely future states.

Although not heavily based on analytical technique, scenario analysis can still be a complex process. The first task is to identify the key dimensions that are likely to affect future land-use. This requires the planner to develop a conceptual model of

---

[8]   For a good overview of the use of panel data in travel demand analysis, see Golob, Kitamura, and Long [1997].

the land development process and of its interaction with physical, economic, socio-logical, and political forces. For example, it might be hypothesized that future land development could be significantly impacted by the state of the economy (e.g., availability of capital), the types of transportation technologies likely to exist, the socioeconomic characteristics of urban residents, the availability and price of energy, and so on. Once the key variables have been identified, the next task is to determine how to predict the range in their future values. One of the most common approaches is to convene a conference of experts or futurists having some knowl-edge of past trends and future prospects. The purpose of this conference is to describe likely value ranges for the dimensions identified earlier. These ranges can then be used to identify land-use patterns, given the different values of the dimen-sions, with the resulting land-use scenarios serving as alternative sources of land-use information that can be input into subsequent analysis and evaluation. Because one does not know which scenario will occur, the components of the analyzed changes to the transportation system that work well under all or most of the scenar-ios become the best candidates for adoption, all else being equal.

Advantages of the use of scenarios include the following:

1. They are inexpensive and quick to construct relative to model forecasts.

2. Because they are less costly, a wider range of future states can typically be examined than in the case of model forecasts.

3. They often represent the only technique available when lack of data and/or the-ory renders model forecasting impossible.

4. They may encourage the analyst to think deeply and imaginatively about future system states and interactions in a way that the more mechanistic process of model forecasting may not.

On the other hand, scenario-based analyses are often viewed as being more hypothetical or conjectural in nature (and hence somewhat more suspect) than model-based forecasts. Furthermore, they may not be able to achieve the internal consistency and level of quantitative detail typically achievable by model forecasts.

A good example of the use of scenarios in urban activity system analysis is pro-vided by a growth management study in Montgomery County, Maryland [Mont-gomery County Planning Department, 1989]. The county had experienced signifi-cant growth over the previous decade and wanted to implement policies that would guide future growth. Four economic scenarios (JOBS, HOUSING, FAST, SLOW) were identified that represented different rates of growth for housing and employ-ment. These scenarios are shown in Fig. 6.6 as they relate to the 1990 base data and a simple trend analysis. In addition to economic scenarios, it was necessary to allo-cate jobs and housing to subareas in the county that reflected a degree of land-use concentration and the provision of transportation. These geographic scenarios were called AUTO, VAN, and RAIL and had the land-use effects as shown in Fig. 6.7. The AUTO scenario assumed the build-out of the county's master plan for high-ways. The VAN scenario added preferential lanes for car pools, vanpools, and buses. The RAIL scenario added to the highway plan new light rail rights-of-way.

CGPS County Control Totals

	J/H Ratio	Jobs	Housing Units
1990 BASE	1.0	455,000	200,000
TREND	1.7	750,000	450,000
JOBS OVER HOUSING	2.0	300,000	450,000
HOUSING OVER JOBS	1.3	750,000	800,000
FAST AND BALANCED	1.5	900,000	600,000
SLOW AND BALANCED	1.5	600,000	400,000

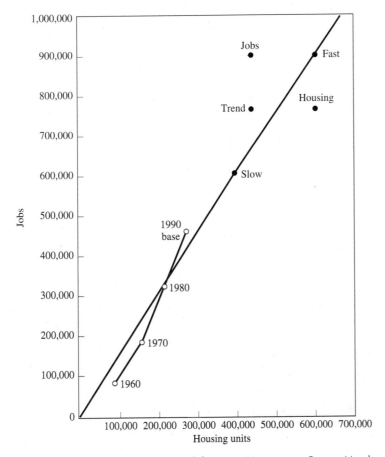

Figure 6.6   Employment housing scenario definition in Montgomery County, Maryland
⏐ SOURCE: Montgomery County Planning Department, 1989

The scenario analysis also assumed the existence of different incentives and enhancements to encourage or discourage the use of certain modes. These assumed influences on travel behavior are shown in Table 6.5.

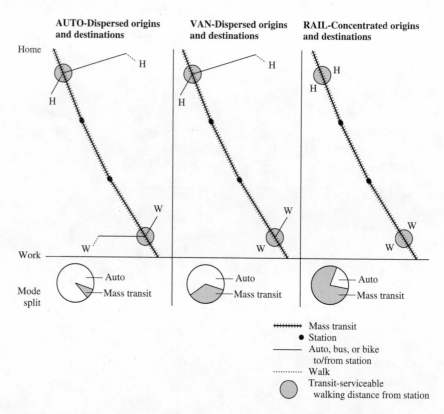

Figure 6.7   Transportation-related descriptors for alternatives analysis, Montgomery County
I SOURCE: Montgomery County Planning Department, 1989

The results of this scenario analysis are shown in Fig. 6.8. The vertical axis shows the average level of traffic congestion in the county as a whole, while the horizontal axis shows the average auto driver mode split for work trips. As noted in the report (and that reflects nicely the themes of this book),

> The virtue of the countywide average approach is that of simplicity. It provides a relatively easily understood gaming board, with which to conduct a public discussion that can wander off down byways of detail, but still retain a framework to keep the discussion from disintegrating into chaos, due to the complexity of the factors and measurements [Montgomery County Planning Department, 1989].

The report concluded that all of the AUTO scenarios showed unacceptably high levels of traffic congestion, regardless of the economic scenario. The report also observed that from a long term perspective, the pattern of urban growth (represented by AUTO, VAN, and RAIL) was much more important than either the pace of growth (represented by SLOW and FAST) or the proportion of growth between jobs and population (represented by JOBS and HOUSING).

The county also conducted a fiscal analysis of the different scenarios to determine whether it could afford the infrastructure and government services that would

Table 6.5 Transit incentives and enhancements package for scenario analysis, Montgomery County

Package/Policy Option	Parking Cost in Employment Centers	Automobile Operating Costs	User-Perceived Transit Fares	Quality of Pedestrian/Bike Transit Access	Household Auto Ownership Levels	Park and Walk Time at Destination
WEAK—current policies	1988 parking fees: Silver Spring, $4/day Shady Grove, $0/day Life Center, $0/day White Oak, $0/day	1988 cost ($0.15/mile)	1988 fares	Poor conditions in most of county except in central business districts	Somewhat higher than 1988 to reflect trends (2.2 cars per household)	1988 conditions (2–3 minutes from parking to door)
MODERATE— used with VAN scenario	Free everywhere for HOVs; higher for low-occupancy vehicles: Silver Spring, $12/day Shady Grove, $10/day Life Center, $8/day White Oak, $4/day	1988 cost ($0.15/mile)	1988 fares	Modest improvements in sidewalks, bike paths, and transit serviceable site planning	Same as 1988 (1.9 cars per household)	Same as 1988
STRONG—used with RAIL scenario	Much higher for all autos: Silver Spring, $12/day Shady Grove, $10/day Life Center, $8/day White Oak, $4/day	($0.30/mile) Higher gas tax and registration fees	1/2 of 1988 fares	Major enhancements in sidewalks, bike paths, and transit serviceable site planning in and near all growth nodes	Slightly lower than 1988 for areas within walking distance of transit stations (1.8 cars per household)	Longer walk times in all growth nodes to reflect lower parking supply

1 SOURCE: Montgomery County Planning Department, 1989

369

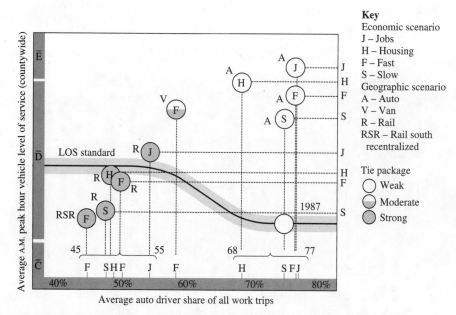

Figure 6.8   Traffic effects of alternative scenarios, Montgomery County
| SOURCE: Montgomery County Planning Department, 1989

be needed to support the expected growth. Once again, a scenario approach was used. Three fiscal scenarios were defined. A "sunny prospect" scenario assumed that real income and property would increase in value, independent of inflation, by about 1 percent per year. Also, federal and state government aid was assumed available to construct needed schools and transportation infrastructure. A "stormy prospect" scenario assumed the opposite. Both scenarios showed a net positive balance in revenues over costs (the likely consequence of a fairly affluent community). In order to focus on the fiscal comparisons among growth scenarios, a "hazy" scenario was developed that assumed the positive increases in income and property appreciation but very little federal or state financial assistance. This fiscal scenario analysis is shown in Fig. 6.9, where the vertical axis represents the same level of service value as before, but the horizontal axis now represents the net revenue balance to the county after expenditures for infrastructure. Interestingly, the AUTO scenarios generally show a more positive revenue to cost balance, presumably because of the higher costs for providing rail transit.

6.2.5 Summary

Of all the steps in transportation systems analysis, the representation of the urban activity system is perhaps the most important and yet the most complex. The various types of urban activity models previously discussed illustrate the major

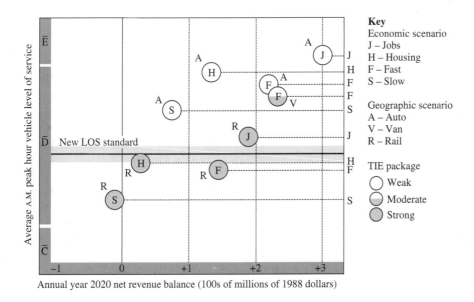

Figure 6.9   Fiscal effects of alternative scenarios, Montgomery County
| SOURCE: [Montgomery County Planning Department, 1989]

approaches that have been used by transportation and land-use planners to represent the land-use activities in an urban area, a prerequisite to the analysis of the demand for transportation. It is important to note that not only are the results of urban activity modeling used in transportation analysis, but that transportation is an important input variable in most urban activity models. The transportation component of the models relates to the influence of accessibility on the location decisions of households and other actors in an urban area, with the level of complexity in the representation of accessibility differing from one model to the next.

Although an important component of transportation analysis, urban activity modeling has often been viewed with skepticism by many transportation planners. Such modeling has usually required large amounts of data, extensive use of computers, and the expenditure of significant planning resources [Lee, 1973]. The result of this disenchantment with large-scale models, in combination with generalized trends within the planning process toward short-range planning (and hence, shorter-run, more limited analyses) was an abandonment, by and large, of large-scale land-use modeling efforts during the 1970s. Ever since the 1980s, land-use model applications have been gradually becoming more common (particularly in European applications). It is expected that this trend will continue in response to a growing recognition of the need for such planning analysis tools, as well as in response to our growing capabilities with respect to hardware, software, and data to support the application of such tools.

6.3   ASSESSMENT OF TRANSPORTATION IMPACTS ON THE URBAN ACTIVITY SYSTEM

Two major approaches exist for the assessment of transportation impacts on the urban activity system. One is the ex post evaluation of implemented transportation policies and actions. Section 3.2.4 presented several examples of such ex post assessments in its discussion of transportation–land-use interactions, while Sec. 8.8 will discuss issues associated with ex post project evaluation in general. Perhaps the only further observation required here is that in all ex post evaluations, one is primarily interested in a *with* and *without* comparison (i.e., what has occurred which would not have occurred if the transportation system change had not been implemented, and vice versa), whereas what one almost inevitably observes is a *before* and *after* comparison (i.e., what existed before and after the system change), which need not be the same thing. While this is potentially troublesome in any ex post evaluation, it is perhaps particularly so in the assessment of urban activity system impacts, given the wide range of other factors (typically uncontrollable, in either a policy or an experimental design sense, and possibly even unobservable) that may be simultaneously changing and affecting the urban activity system. This problem would appear to represent the major reason why the record is still so unclear and why debate still goes on as to just what the impacts of even major transportation system improvements (e.g., subways and freeways) really are on urban activity systems.

The second form of impact assessment involves an attempt to predict impacts prior to project implementation as one part of the analysis phase of the planning process. Such a priori assessments generally involve the development of a model of the urban activity system (or the portion of that system which is of immediate interest within the analysis) that is sensitive to the proposed transportation system changes. The assessments also involve the use of this model to predict and compare expected future activity system states *with* and *without* implementation of each of the proposed transportation alternatives. Thus, a priori assessment is focused directly at the *with* and *without* level of analysis, although it can estimate impacts only by using forecasted (rather than "real") data.

As large-scale, comprehensive land-use models, such as those discussed in the previous section, become more widely used, it is becoming more common to use them in the prior assessment of transportation system impacts on land development. Among others, recent examples of model applications to impact analyses include Anas et al. [2000], in which the NYSIM-LUM implementation of METROSIM was used to evaluate induced trip generation by a proposed expressway project in the New York area, and Abraham and Hunt [1999a, 1999b], in which MEPLAN was used to evaluate a range of transportation alternatives and their land-use impacts in the Sacramento region.

Figure 6.10 presents one example of the results of the Sacramento study, in which population and employment increases during the time period 1990 to 2015 for the region are compared between the "trend" scenario and a "transit-oriented development" scenario. The latter involves the construction of a light rail transit (LRT) line, rent subsidies for zones served by the LRT line combined with tax surcharges for zones not served by the line, use of an advanced transit passenger infor-

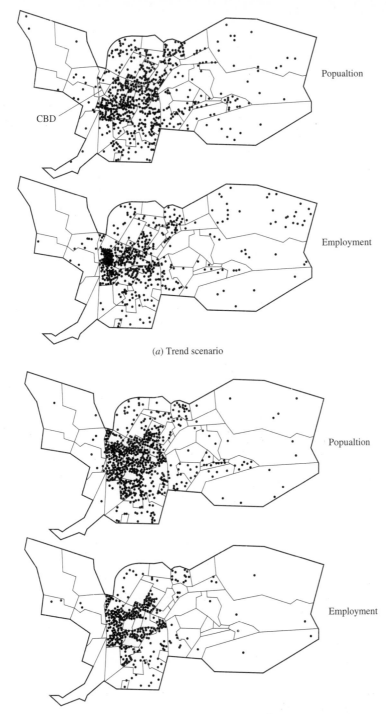

Popualtion

CBD

Employment

(*a*) Trend scenario

Popualtion

Employment

(*b*) Transit-oriented development scenario

Fig. 6.10 Comparison of predicted population and employment growth, 1990–2015, Sacramento, for "trend" and "transit-oriented development" scenarios using MEPLAN

SOURCE: Abraham and Hunt, 1999

mation system, and other upgrades to transit system performance. Comparison of the results of the two simulations indicates that the subsidized zones attract significant development (especially employment based) relative to the base trend case and that, in general, the combined effect of the coordinated transportation–land-use policies is to focus development within the transit corridors.

6.4   CHAPTER SUMMARY

1. An understanding of the urban activity system and its interaction with the urban transportation system is important to transportation planning because it is required to predict the effects that changes in the urban activity system will have on future transportation demand, as well as to assess the impacts that proposed transportation system changes are likely to have on the activity system.

2. Activities occupy both land and buildings. Thus, the location of activities depends on both people's preferences or demand for certain locations and on the supply of suitable building stock at these locations. The interaction between demand and supply within the land-use market results in a bid rent surface over the urban area that reflects the competition between activities for given locations.

3. In addition to price, factors affecting location choice include the attractiveness of the location for a given activity (as measured by a wide range of variables or characteristics) and the location's accessibility to other activities or groups of people.

4. At least three "generations" of land-use modeling efforts can be identified: experimentation in the 1960s with a variety of modeling methods, the emergence of large-scale simulation models in the 1970s, and currently operational models that have evolved over the past 20 years that have built upon and extended the earlier models.

5. Most planning agencies do not currently use formal land-use models, but rather use some form of scenario-based approach or other judgmental techniques.

6. Of the various first generation models, the Lowry model has had by far the most enduring impact on the field. It is a heuristic model that iteratively allocates households to residential locations and "retail" or "population-serving" workers to employment locations, based on an exogenously supplied distribution of "basic" employment. Lowry models in various forms and of varying levels of complexity have been widely applied, although they are subject to a number of criticisms, including lack of a dynamic structure and lack of a representation of urban land markets.

7. Early simulation models include the National Bureau of Economic Research (NBER) model and the Community Analysis Model (CAM). Simulation models attempt to replicate the evolution of an urban area over several time periods as the outcome of a series of interrelated actions by the people and firms that comprise the urban area. By their very definition, such models are extremely complicated, large, and data-hungry. They typically consist of a relatively large num-

ber of submodels, each of which deals with one aspect or actor within the system, interconnected by the information about the urban system that they share.

8. Currently operational models include ITLUP (also known as DRAM/EMPAL), MEPLAN, TRANUS, METROSIM, MUSSA, and UrbanSim. With the exception of ITLUP, these models all explicitly simulate demand–supply–pricing interactions within urban housing markets. All are capable of analyzing a variety of transportation and land-use policies of interest to planning agencies.

9. An increasing number of urban areas worldwide are implementing formal land-use models in response to the improved modeling methods available; the improved computer hardware, software, and databases available to support these models; and the growing recognition of the importance of explicitly analyzing transportation–land-use interactions within the transportation planning process.

10. Despite their interconnection, transportation planning and land-use planning are typically performed separately, usually by separate agencies and professionals. It can thus be institutionally cumbersome to perform integrated analyses and planning studies. Separation of the two functions, however, can result in "myopic" definitions of problems and their possible solutions and in potentially serious misspecifications in both transportation and land-use models.

11. Regardless of whether one is doing long-range or short-range forecasting, urban activity system analysis is complicated by three major factors: the dynamic nature of the urban area; the complexity of urban behavior; and the need for good-quality, detailed data.

QUESTIONS

1. Figure P6.1 presents a simple three-zone system. Using the basic Lowry model as given in Table 6.1 and the data provided in Table P6.1, compute the total employment for each of these zones after three iterations of the model. What were the changes in total employment for each zone between iterations 2 and 3? Are further iterations required to achieve convergence? Do not compute land area totals (you have insufficient data to do so), and do not worry about the residential density constraint. The retail employment threshold constraint, however, must be maintained.

2. In most land-use models, more attention is paid to the "demand side" of the housing market than to the "supply side." Describe the modeling procedure used by one of the models reviewed in this chapter to represent the supply side of the housing market. In particular discuss

   (a) Model methodologies used.

   (b) "Information flow" within the models.

   (c) Treatment of time.

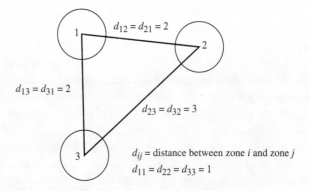

Figure P6.1   Three-zone system

Table P6.1   Data describing three-zone system

Basic Employment Data	Impedance Function	Model parameters
$E^B_1 = 100$	$T^k_{ij} = T_{ij} = d^k_{ij}$	$f = 1.0$ households/worker
$E^B_2 = 200$		
$E^B_3 = 100$		

Two retail sectors: $k = 1$	2
$a^k = 0.1$	0.05
$b^k = 1$	1
$c^k = 0.1$	0.5
$Z^k = 1010$	

(d) "Market clearing" (i.e., how the supply and demand sides are "brought together").

(e) Any other aspects of the procedure that you feel are relevant to its description or critique

3. Evaluate a large-scale land-use forecasting model of your choice in terms of

(a) Data requirements.

(b) Likely development costs.

(c) Theoretical assumptions.

(d) Usefulness in practical planning applications.

(e) Any other strengths or weaknesses of the model that are noteworthy.

4. For an urban area with which you are familiar, trace its history of land-use modeling. Was a large-scale modeling system developed and used? Is it still in use?

If not, what techniques, if any, are being used? What is the current state of the database? What are the implications of these findings for transportation planning in this urban area?

5. Figure P6.5 presents a three-zone region. Table P6.5 provides a travel time matrix for this system, the current distribution of retail floor space in the region, as well as retail floor space estimates for three future distributions, and the retail expenditures per residential zone for the current and future scenarios. A simple model for estimating the impact on retail sales of proposed shopping center development that is still often used is given by Lakshmanan and Hansen [1965]:

$$S_j = \sum_{i=1}^{n} \frac{C_i (F_i)^a (t_{ij})^{-b}}{\sum_{k=1}^{n} (F_k)^a (t_{ik})^{-b}}$$

where

S_j = total expenditure at retail stores in zone j
C_i = total retail expenditures generated by residential zone i
F_j = retail floor space in zone j
t_{ij} = travel time between zones i and j
n = number of zones
a, b = model parameters

Using the Lakshmanan and Hansen model and the data provided, do the following:

(a) Compute total retail expenditure at each retail center (i.e., in each zone) and the retail expenditure to floor-space ratio at each center for the current and three future scenarios. Assume a travel time exponent of 2.0 and a floor-space exponent of 1.0.

(b) Assuming that the number of shopping trips made is directly proportional to retail expenditure, compute average shopping trip lengths for the four retail systems.

(c) Determine which of the three alternative future systems you would recommend. Make explicit your criteria for this evaluation and the reasoning underlying your choice.

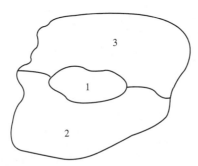

Figure P6.5   Three-zone system

Table P6.5   Data for three-zone region

Travel Time, Min				Retail Floor Space, Ft^2 (1000s)			
To					Scenarios		
From	1	2	3	Zone	0	1	2	3
1	10	20	25	1	3	3	3	3.25
2	20	15	30	2	0.5	1	1.75	1.75
3	25	30	20	3	0.5	1	1.25	1

	Retail Expenditure, $/Year, (1000s)	
	Current	Future
Zone	0	Zones 1-3
1	100	100
2	50	100
3	100	150
Total	250	350

6. For the example given in Table 6.3, what would the impact be on residential unit distribution if

    (a) a new transportation improvement is constructed between zones 3 and 5, reducing travel cost between them to 1.5?

    (b) 150 employees are relocated from zone 2 to zone 5?

7. Transportation access is usually assumed to play a significant role in determining the viability of a given location for retail activity. If you were going to analyze the impacts of transportation policies (e.g., parking control, energy prices, transit incentives, downtown auto-restricted zones) on the commercial viability of the downtown area, how would you go about doing so? Write your answer in the form of a research proposal. This proposal should include

    (a) Problem definition.

    (b) Theoretical framework and hypotheses.

    (c) Method of analysis.

    (d) Data requirements.

    In your answer, do not worry about time and budget constraints. Likely data problems (acquisition of data, quality of data, etc.), however, should be discussed if and when they are relevant.

REFERENCES

Abraham, J.E. and J.D. Hunt. 1999a. Firm location in the MEPLAN model of Sacramento. *Transportation Research Record 1685*. Washington, D.C.: National Academy Press.

Abraham, J.E. and J.D. Hunt. 1999b. Policy analysis using the Sacramento MEPLAN land-use–transportation interaction model. *Transportation Research Record 1685*. Washington, D.C.: National Academy Press.

Alonso, W. 1964. *Location and land-use*. Cambridge, MA: Harvard University Press.

Anas, A. 1982. *Residential location models and urban transportation: Economic theory, econometrics, and policy analysis with discrete choice models*. New York: Academic Press.

_____. 1992. *NYSIM (The New York Simulation Model): A model of cost-benefit analysis of transportation projects*. New York: Regional Planning Association.

_____. 1994. *METROSIM: A unified economic model of transportation and land-use*. Williamsville, NY: Alex Anas and Associates.

_____. 1995. Capitalization of urban travel improvements into residential and commercial real estate: Simulations with a unified model of housing, travel mode and shopping choices. *Journal of Regional Science*. Vol. 35. No. 3.

_____. 1998. *NYMTC transportation models and data initiative: The NYMTC land-use model*. Williamsville, NY: Alex Anas and Associates.

_____ and R.J. Arnott. 1993. Development and testing of the Chicago prototype housing market model. *Journal of Housing Research*. Vol. 4. No. 1.

_____ and R.J. Arnott. 1994. The Chicago prototype housing market model with tenure choice and its policy applications. *Journal of Housing Research*. Vol. 5. No. 1.

_____ and L.S. Brown. 1985. Dynamic forecasting of travel demand, residential location and land development. *Papers and Proceedings of the Regional Science Association*. Vol. 56.

_____ and L.S. Duann. 1986. Dynamic forecasting of travel demand, residence location, and land development: Policy simulations with the Chicago area transportation/land-use analysis system. In B.G. Hutchinson and M. Batty (eds.) *Advances in Urban Systems Modelling*. Amsterdam: North-Holland.

_____, K. Oryani, and J. Krantz. 2000. Application of the NYSIM model to evaluate induced trip generation by Staten Island expressway MIS alternatives. Paper presented at the 79th Annual Meeting of the Transportation Research Board. Washington, D.C. Jan.

Anderstig, C. and L-G Mattsson. 1991. An integrated model of residential and employment location in a metropolitan region. *Papers in Regional Science*. Vol. 70. No. 2.

Batty, M. 1972. Recent developments in land-use modelling: A review of British research. *Urban Studies*. Vol. 9. No. 2. June.

_____. 1976. *Urban modelling*. London: Cambridge University Press.

Ben-Akiva, M. and S. Lerman. 1985. *Discrete choice analysis: Theory and application to travel demand*. Cambridge, MA: MIT Press.

Birch, D. 1976. *Overview of the model*. Cambridge, MA: Dept. of Urban Studies, MIT.

_____, R. Atkinson, S. Sandstrom, and L. Stack. 1974. *The New Haven laboratory: A test-bed for planning*. Lexington, MA: D.C. Heath.

Brand, D., B. Barber, and M. Jacobs. 1967. Technique for relating transportation improvements and urban development patterns. *Highway Research Record 207*. Washington, D.C.: National Academy Press.

Brotchie, J.F., J.W. Dickey, and R. Sharpe. 1980. TOPAZ planning techniques and applications. *Lecture Notes in Economics and Mathematical Systems Series.* Vol. 180. Berlin: Springer-Verlag.

Caindec, E.K. and P. Prastacos. 1995. A description of POLIS. The projective optimization land use information system. *Working Paper 95-1.* Oakland: Association of Bay Area Governments.

Cambridge Systematics Inc. 1992. *Making the land-use air quality connection: Modelling practices.* Vol. 4. Portland, OR: 1000 Friends of Oregon.

Chapin, F.S., Jr. 1965. A model for simulating residential development. *Journal of the American Institute of Planners.* Vol. 31. May.

de la Barra, T. 1982. Modelling regional energy use: A land-use, transport and energy evaluation model. *Environment and Planning.* Vol. 9B.

_____. 1989. *Integrated land-use and transport modelling.* Cambridge, UK: Cambridge University Press.

_____, B. Perez, and N. Vera. 1984. RANUS-J: Putting large models into small computers. *Environment and Planning.* Vol. 11B.

Dickey, J.W. and C. Leiner. 1983. Use of TOPAZ for transportation–land-use planning in a suburban county. *Transportation Research Record 931.* Washington, D.C.: National Academy Press.

Echenique, M.H. 1985. The use of integrated land-use transportation planning models: The cases of Sao Paulo, Brazil and Bilbao, Spain. In M. Florian (ed.) *The Practice of Transportation Planning.* The Hague: Elsevier.

_____, D. Crowther, and W. Lindsay. 1969. A spatial model for urban stock and activity. *Regional Studies.* Vol. 3.

_____, A.D. Flowerdew, J.D. Hunt, T.R. Mayo, I.J. Skidmore, and D.C. Simmonds. 1990. The MEPLAN models of Bilbao, Leeds, and Dortmund. *Transportation Reviews 10.*

_____ and I.N. Williams. 1980. Developing theoretically based urban models for practical planning studies. *Sistemi Urbani* 1:13–23.

Eliasson, J. and L-G Mattsson. 1997. TILT—A model for integrated analysis of household location and travel choices. Paper presented at the 8th Conference of the International Association of Travel Behavior Research. Austin, TX. Sept.

Garin, R. A. 1966. A matrix formulation of the Lowry model for intrametropolitan activity allocation. *Journal of the American Institute of Planners.* Vol. 32.

Goldner, W. 1971. The Lowry model heritage. *Journal of the American Institute of Planners.* Vol. 37. No. 2.

Golob, T.R., R. Kitamura, and L. Long (eds.). 1997. *Panels for transportation planning: Methods and applications.* Dordrecht, the Netherlands: Kluwer Academic Publishers.

Goulias, K.G. and R. Kitamura. 1992. Travel demand forecasting with dynamic microsimulation. *Transportation Research Record 1357.* Washington, D.C.: National Academy Press.

Gu, Q., A. Haines, and W. Young. 1992. *The development of a land-use/transport interaction model.* Report 2. Melbourne: Monash University.

Herbert, J.D. and B.H. Stevens. 1960. A model for the distribution of residential activities in urban areas. *Journal of Regional Science.* Vol. 2. No. 2. Fall.

Hill, D. M. 1965. A growth allocation model for the Boston region. *Journal of the American Institute of Planners.* Vol. 31. May.

_____, D. Brand, and W. B. Hansen. 1965. Prototype development of statistical land-use prediction model for Greater Boston region. *Highway Research Record 114.* Washington, D.C.: National Academy Press.

Hunt, J.D. 1994. Calibrating the Naples land-use and transport model. *Environment and Planning.* Vol. 21B.

_____ and M.H. Echenique. 1993. Experiences in the application of the MEPLAN framework for land-use and transport interaction modeling. *Proceedings of the 4th National Conference on the Application of Transportation Planning Methods.* Daytona Beach, FL. May.

_____ and D.C. Simmonds. 1993. Theory and application of an integrated land-use and transport modelling framework. *Environment and Planning.* Vol. 20B.

Ingram, G. K., J. F. Kain, and J. R. Ginn. 1972. *The Detroit prototype of the NBER urban simulation model.* New York: National Bureau of Economic Research.

Kim, T.J. 1989. *Integrated urban systems modeling: Theory and practice.* Norwell, MA: Martinus Nijhoff.

Lakshmanan, T. R. and W. G. Hansen. 1965. A retail market potential model. *Journal of the American Institute of Planners.* Vol. 31. May.

Landis, J.D. 1994. The California urban futures model: A new generation of metropolitan simulation models. *Environment and Planning.* Vol. 21B.

Lee, D. A. 1973. Requiem for large-scale models. *Journal of the American Institute of Planners.* Vol. 39. No. 3. May.

Lowry, I. S. 1964. *A model of metropolis.* RM-4035-RC. Santa Monica, CA: Rand Corp.

Mackett, R.L. 1983. *The Leeds integrated land-use transport (LILT) model.* Supplementary Report 805. Crowthorne, UK: Transport and Road Research Laboratory.

_____. 1985a. Integrated land-use-transport models. *Transportation Reviews 5.*

_____. 1985b. Micro-analytical simulation of locational and travel behaviour. *Proceedings PTRC Summer Annual Meeting.* Seminar L: Transportation Planning Methods. London: PTRC.

_____. 1990a. The systematic application of the LILT model to Dortmund, Leeds, and Tokyo. *Transportation Reviews 10.*

_____. 1990b. *MASTER model (Micro-Analytical Simulation of Transport, Employment, and Residence).* Report SR 237. Crowthorne, UK: Transport and Road Research Laboratory.

Martinez, F.J. 1992a. Towards the 5-stage land-use–transport model: Land-use development and globalization. *Selected Papers of the 6th World Conference on Transportation Research.* Lyon, France.

_____. 1992b. The bid-choice land-use model: An integrated economic framework. *Environment and Planning.* Vol. 24A.

_____. 1997a. Towards a microeconomic framework for travel behavior and land-use interactions. Resource paper, 8th Meeting of the International Association of Travel Behavior Research. Austin, TX.

_____. 1997b. *MUSSA: A land-use model for Santiago City.* Santiago: Department of Civil Engineering, University of Chile.

_____ and P.P. Donoso. 1995. MUSSA Model: The theoretical framework. *Proceedings of the 7th World Conference on Transportation Research.* Sydney.

Miller, E.J. 1996. Microsimulation and activity-based forecasting. *Travel Model Improvement Program Activity-Based Travel Forecasting Conference, June 2-5, 1996, Summary, Recommendations and Compendium of Papers.* DOT-T-97-17.Washington, D.C.: U.S. Department of Transportation.

_____, D.S. Kriger, and J.D. Hunt. 1998. *Integrated urban models for simulation of transit and land-use policies.* Final project report to TCRP Project H-12.Toronto: University of Toronto Joint Program in Transportation [published online by the Transportation Research Board, Washington, D.C., as Web Document 9 at *<www4.nas.edu/trb/crp.nsf>*].

_____, P.J. Noehammer, and D.R. Ross. 1987. A micro-simulation model of residential mobility. *Proceedings of the International Symposium on Transport, Communications and Urban Form. Volume 2: Analytical Techniques and Case Studies.*

_____ and P.A. Salvini. 1998. The integrated land-use, transportation, environment (ILUTE) modeling system: A framework. Presentation at the 77th Annual Meeting of the Transportation Research Board. Washington, D.C. Jan.

_____. and P.A. Salvini. 2000. The integrated land-use, transportation, environment (ILUTE) microsimulation modelling system: Description and current status. Presented at the 9th International Association for Travel Behaviour Research Conference. Gold Coast, Queensland, Australia. July.

Modelistica. 1998. *TRANUS: Integrated land-use and transport modeling system; Version 5.0.* Available on the Oregon Department of Transportation website.

Montgomery County Planning Department. 1989. *Comprehensive growth policy study.* Silver Spring, MD. Aug.

Nakamura, H., Y. Hayashi, and K. Miyamoto. 1983. A land-use–transport model in metropolitan areas. *Papers of the Regional Science Association.* Vol. 51.

Oskamp, A. 1997. *Local housing market simulation: A micro approach.* Amsterdam: Thesis Publishers.

Pushkarev, B. and J. Zupan. 1977. *Public transportation and land-use policy.* Bloomington, IN: Indiana University Press.

_____. 1980. *Urban rail in America.* Bloomington, IN: Indiana University Press.

Putman, S.H. 1983. *Integrated urban models.* London: Pion Limited.

_____. 1978. The integrated forecasting of transportation and land-use. In W. F. Brown, R. B. Dial, D. S. Gendall, E. Weiner (eds.) *Emerging Transportation Planning Methods.* U.S. Dept. of Transportation, Office of University Research.

_____. 1978. Calibrating urban residential location models no. 3 empirical results for non-U.S. cities. *Environment and Planning.* Vol. 12A.

_____. 1981. Theory and practice in urban modelling: The art of application. Presented at the 13th annual conference of the Regional Science Association—British Section. University of Durham. Sept.

_____. 1991. *Integrated urban models 2.* London: Pion Limited.

_____. 1993. Integrating transportation and land-use analysis, planning and policy evaluation. In Lincoln Institute of Land Policy *Linking Transportation and Land-use Planning: A Key to Suburban Growth Management.* Cambridge, MA.

_____. 1994. Integrated land-use and transportation models: An overview of progress with DRAM and EMPAL with suggestions for further research. Presentation at the 73rd Annual Meeting of the Transportation Research Board, Washington, D.C. Jan.

_____. 1995. EMPAL and DRAM location and land-use models: An overview. Paper distributed at the TMIP land-use modeling conference. Dallas, TX. Feb.

_____. 1996. Extending DRAM model: Theory-practice nexus. *Transportation Research Record 1552*. Washington, D.C.: National Academy Press.

_____. 1997. *LINKAGES: Newsletter of integrated transportation and land-use analyses*. Townsend, DE: S.H. Putman Associates.

Quigley, J. M. 1976. Housing demand in the short run: An analysis of polytomous choice. *Explorations in Economic Research*. Vol. 3. No. 1.

SACTRA. 1994. *Trunk roads and the generation of traffic*. Standing Advisory Committee on Trunk Road Assessment. London: Department of Transport, UK.

Shunk, G.A., P.L. Bass, C.A. Weatherby, and L.J. Engelke (eds.). 1995. *Travel model improvement program land-use modeling conference proceedings*. Washington, D.C.: Travel Model Improvement Program.

Simmonds, D.C. 1998. *The design of the DELTA land-use modelling package*. Cambridge: David Simmonds Consultancy.

Southeastern Wisconsin Regional Planning Commission. 1981. *Alternative futures for southeastern Wisconsin*. Technical Report No. 25. Milwaukee, WI.

Southworth, F. 1995. *A technical review of urban land-use–transportation models as tools for evaluating vehicle travel reduction strategies*. Report ORNL-6881. Oak Ridge, TN: Oak Ridge National Laboratory.

Steuart, G. N. (ed.). 1977. *Urban transportation planning guide*. Toronto: University of Toronto Press.

Transportation Research Board. 1995. *Expanding metropolitan highways: Implications for air quality and energy use*. Special Report 245. Committee for Study of Impacts of Highway Capacity Improvements on Air Quality and Energy Consumption. Washington, D.C.: National Academy Press.

Waddell, P. 1998a. The Oregon prototype metropolitan land-use model. *Proceedings of the ASCE Conference on Transportation, Land-use and Air Quality: Making the Connection*. Portland, OR. May.

_____. 1998b. *Final report on the Oregon prototype metropolitan land-use model, Phase II*. Report to the Oregon Department of Transportation.

_____. 1998c. An urban simulation model for integrated policy analysis and planning: Residential location and housing market components of UrbanSim. Paper presented at the 8th World Conference on Transport Research. Antwerp, Belgium. July.

_____, T. Moore, and S. Edwards. 1998. Exploiting parcel-level GIS for land-use modeling. *Proceedings of the ASCE Conference on Transportation, Land-use and Air Quality: Making the Connection*. Portland, OR. May.

Webster, F.V., P.H. Bly, and N.J. Paulley (eds.). 1988. *Urban land-use and transport interaction: Policies and models*. Aldershot, England: Avebury (Gower Publishing).

Wegener, M. 1982a. A multilevel economic-demographic model for the Dortmund region. *Sistemi Urbani.* Vol. 3.

_____. 1982b. Modeling urban decline: A multilevel economic-demographic model of the Dortmund region. *International Regional Science Review.* Vol. 7.

_____. 1986. Transport network equilibrium and regional deconcentration. *Environment and Planning.* Vol. 18A.

_____. 1993. Microsimulation and GIS: Prospects and first experience. Presented at the Third International Conference on Computers in Urban Planning and Management. Atlanta, GA. July 23–25.

_____. 1994. Operational urban models: State of the art. *Journal of the American Planning Association.* Vol. 60.

_____. 1995. Current and future land-use models. In G.A. Shunk et al. (eds.) *Travel Model Improvement Program Land Use Modeling Conference Proceedings.* Washington, D.C.: Travel Model Improvement Program.

_____. 1998. Applied models of urban land-use: Transport and environment state of the art and future developments. In D. Batten, T.J. Kim, L. Lundqvist, and L.G. Mattson (eds.) *Network Infrastructure and the Urban Environment: Recent Advances in Land Use/Transportation Modelling.* Berlin: Springer-Verlag.

_____, R.L. Mackett, and D.C. Simmonds. 1991. One city, three models: Comparison of land-use/transport policy simulation models for Dortmund. *Transportation Reviews 11.*

Wilson, A. G. 1974. *Urban and regional models in geography and planning.* New York: Wiley.

Yen, Y.M. and J.D. Fricker. 1997. An integrated transportation land-use modeling system. Paper presented at the 76th Annual Meeting of the Transportation Research Board. Washington D. C.

chapter
7

Supply Analysis

7.0 INTRODUCTION

A transportation system can be characterized by its *performance* (e.g., travel times, frequency of service, safety, and reliability), the corresponding *impacts* on the environment, and the *costs* incurred in building, maintaining, and using the system. The networks, facilities, and services that are part of this system and their corresponding characteristics are referred to as *transportation supply.* This chapter begins with a discussion of the role of supply analysis within the planning process. The three sections that follow describe the methods and analysis tools that are used to estimate system performance, impacts, and costs. It is beyond the scope of this text to delve deeply into the mathematical foundations for these analysis approaches. For more rigorous, mathematical discussions of these techniques, the reader is referred to the following texts: [Manheim, 1979; Newell, 1980; Larson and Odoni, 1981; Vuchic, 1981; Sheffi, 1985; May, 1990; Ortuzar and Willumsen, 1994; Oppenheim, 1995; Taylor et al., 1996; Daganzo, 1997; Bell and Iida, 1997].

7.1 THE ROLE OF SUPPLY ANALYSIS IN TRANSPORTATION PLANNING

Developing and managing the supply of transportation is a primary focus of an effective transportation planning process. Such a focus can support different types of decisions, including the following:

> *Metropolitan-level network analysis for strategic investment:* This level of analysis generally involves examining alternative modal networks, typically at a regional or areawide scale. Decisions at this level relate to such things as adding new roads or transit facilities to the network, adopting regional land-use or taxation policies aimed at influencing travel behavior, or adding new ITS technologies that will affect individual trip-making choices. The planning time frame for these decisions is usually measured in years, if not decades. Supply analysis in such cases determines equilibrium flow patterns on a modal network using one of the assignment techniques discussed in Sec. 5.4.4. Given modal origin–destination flows as determined by demand analysis and a set of performance functions for the links comprising the network, models can estimate overall performance of the transportation system. The complexity of

such analyses stems from the size of the networks being assessed, which can involve thousands of nodes and tens of thousands of links.

Operational or tactical planning: The analysis of individual routes, links, or terminals is a very common focus of supply analysis, particularly for operating agencies. In these cases, only a small "network" is considered, usually consisting of an individual facility or service such as a transit route or a freeway and its linkages to the rest of the system. The planning time frame for such analyses is often short term, for example, 1 to 3 years. The types of decisions in this environment focus on operational changes, reflected in such questions as, "How should a transit route be modified to improve performance?" or "How can traffic signals be better coordinated to reduce vehicle delay?" If longer-term assessments are required, horizon-year demand estimates that forecast peak-period flows 20 years into the future are needed. The range of analysis techniques for operations planning is far broader than for strategic network analysis and includes everything from simple "back of the envelope" calculations to computer simulation programs. The choice of technique for any given application depends on the specific nature of the problem, the data available, the capabilities of the planner, the time and budget available for the analysis, and the accuracy and level of detail of the information required for the decision-making process.

Scheduling of transportation services: The scheduling of transit vehicles to routes and assigning drivers to vehicles are major tasks for transit operators. The major objective in both cases is to minimize operating costs while in the long run minimizing vehicle fleet size, subject to service standard constraints and labor agreements. The time frame for such scheduling processes is very short, generally every 3 to 6 months. The scale of the analysis is the entire transit network or at least large subnetworks, such as all routes serviced by a given bus garage. Service characteristics are represented in considerable detail. The technique used for scheduling ("run cutting," as it is often called) typically involves a combination of manual and computerized procedures for heuristically finding "good" schedules. (Note: This text does not discuss further the analysis of service scheduling. See [Pine et al., 1998] for a discussion of these procedures.)

The transportation system as described in Chap. 3 consists of five major components—the system user; the mode or technology of travel; the infrastructure that allows such travel to occur; the intermodal connections for transfers; and the stakeholders that either influence or are affected by the performance, impacts, and costs of the transportation system. Each of these individually or in combination can be an important focus in an analysis of transportation supply. For example, it would be difficult to decide how many buses to operate on a given route at a desired frequency without knowing service demand (i.e., system users), the round-trip travel time for a bus as a function of its operating characteristics (i.e., modal technology), the number and location of stops along the route (i.e., intermodal transfers), and typical congestion levels on the street network (e.g., infrastructure). Similarly, the demand for a transportation service generally cannot be estimated until the travel times and costs associated with this service are known; nor can the merits of this service be evaluated relative to other alternatives until the performance, impacts, and costs of all reasonable alternatives are analyzed.

Of the five components of a transportation system, transportation infrastructure is the most important for supply analysis. This infrastructure includes the rights-of-way over which travel occurs (e.g., roads, tracks, sidewalks, and in the case of telecommunications, the fiber optic network); the signal systems that control flows; terminals; and all other facilities required to operate and maintain the transportation system, in particular vehicle storage, servicing, and maintenance facilities. It also includes the routes and schedules (where applicable) that govern the operation of services and the procedures for operating the system, which include everything from government regulations, such as speed limits, licensing requirements and service standards, to the labor agreements concerning driver working conditions.

Figure 7.1 shows the role of supply analysis in the transportation planning process. Supply analyses relate to other parts of the process in four major ways:

Linkage 1. The performance and cost of transportation services as perceived by potential users are important determinants of the demand for these services. Transportation system performance in turn generally depends upon the level and character of travel demand. Demand and supply analyses are thus

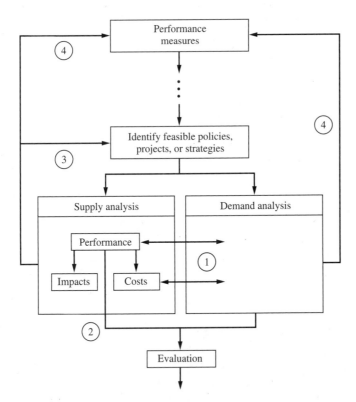

Figure 7.1   Supply analysis in transportation planning

inherently linked together. In travel demand modeling, this linkage is most explicit in the trip assignment stage (see Sec. 5.4.4) in which link flows (demand) and travel times (performance) reach equilibrium values after several iterations. System performance can be incorporated into other components of demand analysis as well, for example, trip generation, trip distribution, and mode split. In these cases, system performance and costs are used as variables that affect how many trips occur and where they are destined.

This influence of this linkage on actual travel behavior is dramatically shown in Table 7.1, where characteristics of U.S. metropolitan areas' trans-

Table 7.1   Effect of supply-side strategies on transit ridership

Metro Area	Mean Transit Rank	Transit Share	SOV Share	Car-Pool Share	% Within 1/4 Mile of Transit	Annual Congestion Costs Per Traveler
High-Transit Cities						
San Francisco	3.0	9.3	69.8	13.0	60.1	$760
New York	3.3	27.8	52.3	10.3	46.0	390
Chicago	4.7	13.7	67.5	12.0	47.1	300
Boston	5.0	10.6	70.2	10.3	46.4	495
Seattle	6.3	6.3	73.8	12.0	53.1	660
Philadelphia	6.3	10.2	69.2	12.2	39.7	270
Portland	7.3	5.4	74.1	12.3	50.0	330
Mean	5.1	11.9	68.1	11.7	48.9	458
Medium-Transit Cities						
Los Angeles	8.0	4.6	72.9	15.5	49.9	670
Buffalo	8.3	4.7	77.1	11.2	58.3	380
Pittsburgh	8.3	8.0	71.5	12.8	36.7	270
Denver	9.0	4.3	75.0	12.4	53.8	370
Miami	11.0	4.4	75.5	14.5	43.2	520
Milwaukee	13.7	4.9	77.3	10.9	26.2	160
Mean	9.7	5.2	74.9	12.9	44.7	395
Low-Transit Cities						
Providence	15.0	2.6	78.6	12.3	37.0	380
Cleveland	15.3	4.6	79.6	10.3	30.8	120
Houston	15.3	3.8	76.3	14.6	25.8	570
Cincinnati	16.0	3.7	79.3	11.4	32.1	160
Dallas	17.0	2.4	78.9	13.8	30.4	570
Hartford	17.3	1.6	78.5	13.3	28.3	220
Detroit	19.7	2.4	82.7	10.1	21.2	380
Mean	16.5	3.0	79.1	12.3	29.4	343
Overall mean		6.8	74.0	12.2	40.8	399

SOURCE: Dueker et al., 1998

portation supply are related to transit usage. Although such relationships are often much more complex than as shown in this table, the differences in how transportation supply is provided and managed, and the related transit usage, is telling [Dueker et al., 1998].

Linkage 2. Over and above their influence on the demand for transportation, measures of system performance, impacts, and costs constitute important evaluation criteria in their own right. As will be seen in Chap. 8, the evaluation of transportation projects and programs is based on measures of effectiveness that relate to these types of concerns.

% Change in Central City Population 1980–1990	Pay-to-Park Probability	Parking Maximums in CBD	Parking Tax > 10%	Maximum Meter Rate	Stand-Alone Parking Garages Allowed in CBD	Residential Parking Permit Programs
6.6	4.6	Yes	Yes	$1.50	No	Yes
3.5	5.5	Yes	Yes	1.50	No	No
−7.4	4.1	No	Yes	3.00	No	Yes
2.0	6.6	No	No	1.00	No	Yes
4.5	5.7	Yes	No	1.00	No	Yes
−6.1	4.4	No	Yes	1.00	No	Yes
18.8	7.7	Yes	No	0.90	No	Yes
3.1	5.5	57% yes	57% yes	1.41	100% no	86% yes
17.4	3.4	No	Yes	2.00	Yes	Yes
−8.3	4.5	No	No	1.00	No	No
−12.8	7.6	No	Yes	2.00	No	Yes
−5.1	5.8	No	No	1.00	No	Yes
3.4	1.4	No	No	1.00	Yes	No
−1.3	13.7	No	No	1.00	No	Yes
−1.1	6.1	0.0 % yes	33% yes	1.33	66% no	67% yes
2.5	4.8	No	No	0.75	Yes	No
−11.9	7.2	No	No	0.75	Yes	Yes
2.2	6.6	No	No	1.00	Yes	No
−5.5	5.2	No	No	0.50	Yes	No
11.3	6.1	No	No	1.00	Yes	Yes
2.5	2.5	No	No	0.75	Yes	No
−14.6	4.9	No	No	1.00	Yes	Yes
−1.9	5.3	0.0% yes	0.0% yes	0.82	0.0% no	43% yes
1.7	5.6	20.0% yes	30% yes	1.18	55% no	65% yes

Linkage 3. The estimation of current and forecasted system performance, impact, and cost characteristics leads to the identification of project alternatives. Systematically exploring supply relationships and predicting performance measures will provide planners with a better understanding of the operation of the transportation system and of feasible alternatives that could improve this operation.

Linkage 4. This linkage is relatively new to transportation planning and reflects the trend toward continual monitoring of transportation system performance. The feedback from both demand and supply analyses is used to determine where deficiencies exist in the transportation network that can then lead to the identification of strategies for solving potential problems. Most performance measures reflect supply-side characteristics of the transportation system.

Figure 7.1 indicates that in order to estimate system impacts and costs, one must first estimate system performance. For example, energy consumption or levels of air pollution on a section of highway depends on the speed of the vehicles using the highway, the number of speed-change cycles the vehicles undergo, and the number of vehicles using the highway per unit time. Similarly, the operating costs of providing transit service at a given frequency to meet a certain demand will depend on the route round-trip travel times, and in turn on the number of buses and drivers required to meet the schedule. It is therefore important to discuss first the analysis of transportation system performance, which is done in the next section, before discussing cost and impact estimation.

7.2   ANALYSIS OF TRANSPORTATION SYSTEM PERFORMANCE

The performance of the transportation system can be viewed from the perspective of the *user* of the system and of the *operator* of the system. Users of the system are interested in the characteristics of the service—travel cost, travel time, and the reliability of successfully completing a trip within a certain time constraint—that most directly affect their ability to accomplish their trip purpose. Travel time can be further disaggregated into *in-vehicle* time and *out-of-vehicle* time, which, in turn, consists of walk time, wait time, and transfer time. Service operators are concerned with route frequencies, vehicle cycle times, route capacities, and the operating cost associated with the provision of the service. Not surprisingly, all of these concepts are interrelated. Vehicle travel times over a route translate into stop-to-stop travel times for passengers. Average wait times at passenger stops are a function of route frequency, and passenger travel costs are related to the fares charged that are presumably based on the cost to provide the service.

This section discusses the methods used to predict system performance. Basic concepts are introduced in Sec. 7.2.1. Section 7.2.2 discusses a range of analysis techniques applicable to a wide variety of transportation performance problems. Finally, Sections 7.3 and 7.4 discuss some of the more common planning applications of these techniques.

7.2.1 Basic concepts

In general, facility or service performance and the choice of methods to analyze this performance relate to the following:

The type of technology used for achieving a trip purpose. Transportation modes exhibit different operating characteristics. Subways, for example, can travel faster than light rail trains, but light rail trains can negotiate tighter curves and are thus more appropriate for city streets. Other important technology characteristics include the type of guideway or suspension (e.g., steel wheel on steel rail for subways), the method of propulsion (e.g., internal combustion engine and direct current electric motor), and the means of controlling system operations (e.g., signalized intersections and automated control of rail operations).

The type of right-of-way. Different modes of transportation can share a right-of-way. For example, a typical urban street allows automobiles, buses, trucks, pedestrians, and bicycles to operate in close proximity to one another. The design of other facilities separates modal operations. For example, a subway line or an exclusive bus lane is designed only for one mode. Minimizing the conflicts between different modes results in higher capacity and safer operation.

The service configuration. Service can be provided on fixed routes or flexibly routed. Conventional transit, for example, is usually operated on specified routes, whereas the private auto, taxi, or demand-responsive public transit can respond to demands that occur anywhere in the metropolitan area.

The service operating rules. Safe and effective transportation requires rules of operation that provide guidance on how the transportation system is to operate. For example, speed limits represent the safe speed for driving on the road system, transit schedules identify how often a bus will arrive at a bus stop, and hours of operation dictate when goods are delivered or picked up.

Which method to use for supply analysis is also influenced by the underlying behavior of the service or operations being examined. A key distinction is whether this behavior is in a *steady-state* condition, and thus average system performance can be used, or whether the behavior is unpredictable, and thus expected values of system performance must be based on probabilities of certain system conditions occurring. Examples of the latter would be freeway conditions when one or more lanes are blocked by an accident or bus "bunching" due to operating conditions along a route that cannot be predicted with any degree of certainty. *Stochastic methods* based on probabilities are used to analyze these types of situations.

Network definition is perhaps the single most important step in the entire supply analysis process in that it determines the level of accuracy and detail of analysis achievable, the quantity and quality of data required to represent and analyze the system, and the type(s) of analysis techniques that are suitable. As noted in Sec. 3.3.3, a transportation network is represented with an "approximating network" [Newell, 1980] consisting of a connected set of *links* and *nodes*. As shown in Fig. 7.2, links can represent roadways, transit lines, walking paths, or other rights-of-way, while nodes represent intersections, terminals, or points of trip generation. This latter type of node is known as a *centroid*. Each zone in the system is represented by a centroid, and all trips into and out of the zone are assumed to be destined to or originate from the zone centroid. Centroids are connected to the network by *dummy links*, which, unlike regular links in the approximating network, do not represent any component of the real network.

Once a network is defined, it generally forms the basis of most system performance analyses. The links and nodes of the network can be aggregated to form *routes* or *paths* from point to point within the system. Analysis may well be performed on any of these network elements, depending on the problem focus. Examples include

Links. Analysis of a section of an urban freeway or of a transit line.

Nodes. Analysis of the flow through an intersection or of terminal operations.

Routes. Analysis of the operation of a transit route.

Networks. Analysis of network flows or of route interactions within the network.

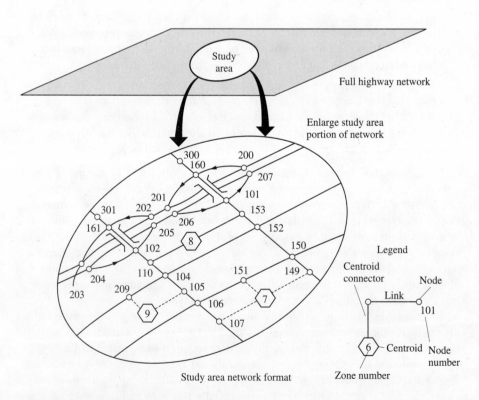

Figure 7.2   Example of a transportation network
SOURCE: Pedersen and Samdahl, 1982

An alternative approach to the use of networks is to represent the transportation system with a *continuum model* in which point-to-point network distances are approximated by the straight-line distances between such points. Such an approach eliminates the need for a detailed network representation and facilitates the analytical treatment of a number of transportation system performance measures (e.g., average walk time). The major disadvantage of such an approach is that it is not applicable to the analysis of specific links or components of the transportation network, particularly where link or system capacity is an important issue.

Several performance characteristics are fundamental to understanding the physical operations of transportation facilities. These include speed, volume, rate of flow, density, capacity, level of service, and headway. Most of these characteristics are related to one another and to the particular transportation technology in use. The

Typical format for display of network characteristics[1]

Link	Facility type	Lanes Orientation	Number	Length, mi	Traffic control	Other
205–206	Freeway	One-way (NB)	3	0.6	Grade separated	Industry
201–202	Freeway	One-way (SB)	3	0.6	Grade separated	Industry
102–110	Arterial	Two-way	4	0.3	Signals	Commercial
106–107	Arterial	Two-way	6	0.6	Grade separated	Industry
151–150	Arterial	Two-way	4	0.8	Signals	Residential

Continued

Node	Approach	Configuration[2]	Traffic control	Other
160	N	Ramp – 2 lane	Stop	One-way link
	S	–No south approach – 1 way SB –		
	E	3T, 1L	No stop	No right turn
	W	3T	No stop	No left turn
150	N	2T	Signal (g/c = 0.4)	
	S	2T	Signal (g/c = 0.4)	
	E	2T, 1L	Signal (g/c = 0.6)	
	W	3T	Signal (g/c = 0.6)	No left turn

Continued

[1]Refer to figure for diagram of network
[2]T = through lanes; L = left-turn lanes; R = right-turn lanes

Figure 7.2 Example of a transportation network (continued)

Highway Capacity Manual defines these terms as follows [Transportation Research Board, 2000]:

Speed: The rate of motion or distance traveled per unit time, generally expressed as miles or kilometers per hour. Several methods for calculating speed result in the following definitions.

- *Average running speed.* Length of a segment divided by average running time of vehicles to cover this distance. Delay time is not counted.

- *Average travel speed.* Length of a segment divided by average travel time of vehicles to cover this distance, including delay time.

- *Space mean speed.* Length of a segment divided by the average of the travel times of vehicles over this distance.

- *Time mean speed.* Average of vehicle speeds at a given point.

- *Free flow speed.* Average speed of vehicles under low-volume conditions.

Volume: The total number of vehicles passing a point on a transportation facility during a given time period, usually expressed as annual, daily, hourly, peak, or off-peak traffic volumes.

Rate of flow: The number of vehicles passing a point on a facility during some period of time. This is often expressed as the equivalent hourly rate at which vehicles pass by a given point during a given time interval, usually 15 minutes. For streets, the saturation flow rate is the vehicles per hour per lane that could pass through a signalized intersection if the signal were constantly green. The saturation flow rate is a critical input into the estimation of the capacity of signalized intersections.

The difference between volume and flow rate is important to understand. Suppose the following observations are made of the number of buses using a high-occupancy vehicle lane:

7:00–7:15	8 buses
7:15–7:30	10 buses
7:30–7:45	15 buses
7:45–8:00	13 buses

The bus volume during this hour is 46 buses, but the *rate* of flow for each 15-minute period is as follows:

7:00–7:15	8 buses/0.25 hour = 32 buses/hour
7:15–7:30	10 buses/0.25 hour = 40 buses/hour
7:30–7:45	15 buses/0.25 hour = 60 buses/hour
7:45–8:00	13 buses/0.25 hour = 52 buses/hour

These flow rates become important factors in determining facility capacity and flow control strategies (e.g., vehicle dispatching).

A peak-hour factor (*PHF*) is calculated to relate peak flow rates to hourly volumes. This relationship is estimated as

$$PHF = \frac{V}{4 \times V_{15}}$$ [7.1]

where
 PHF = peak hour factor
 V = peak hour volume (vehicles)
 V_{15} = volume for peak 15-minute period

For the above example, the peak hour factor would be calculated as

$$PHF = \frac{46 \text{ buses/hour}}{4 \times 15 \text{ buses/15-min period}} = 0.77$$

Density: The number of vehicles or pedestrians occupying a given length of a facility at a given point in time.

Capacity: The maximum rate of flow that can be accommodated on a facility segment under prevailing conditions. As shown in Table 3.5 on page 105, there are several factors that can influence facility and service capacity.

Level of service: A qualitative measure of the effects of a number of factors on the performance of a facility such as density, speed, travel time, traffic interruptions, safety, comfort, operating costs, and volume-to-capacity ratios. These qualitative measures are grouped into different levels to represent different facility or service conditions (see Sec. 3.1.5).

Headway: The amount of time between vehicle departures in a traffic stream measured at a given point. Headway is the inverse of vehicle frequency. Thus,

$$h = \frac{3600}{f}$$ [7.2]

where
 h = headway (seconds/vehicle)
 f = frequency (vehicles/hour)

Headways are used in analyzing vehicle departures at signalized intersections, as well as transit route performance.

For the special case of freeways or other higher levels of roads, the relationship between three of these characteristics—volume flow, average running speed, and density—is one of the fundamental relationships describing the conditions of vehicle flow. This relationship, expressed mathematically as

$$V = \bar{S} \times D$$ [7.3]

where
 V = volume flow through network element, vehicles per hour (or per kilometer)
 $\bar{S}$ = average speed of vehicles in network element, miles per hour (or kilometers per hour)
 D = density of vehicles within network element, vehicles per mile (or per kilometer)

is the basis for determining the capacity of a facility and its level of service. Additional relationships between flow and speed, flow and density, and speed and density limit the range of combinations possible among the three variables. These relationships, shown in Fig. 7.3, also demonstrate the relationships between the maximum rate of flow or capacity (V_m), critical density (D_o), and critical speed (S_o). The operating conditions identified in Fig. 7.3 as "oversaturated flow" are often experienced during peak hours when congested conditions result in lower speeds and stop-and-go conditions.

Note that Eq. 7.3 can be converted to passenger flow by multiplying V by the average number of persons per vehicle. Equation 7.3 is derived in the following section through the use of time–distance diagrams. At this point, it is important to note that all of the techniques discussed in Secs. 7.2.2 through 7.4 represent various ways of conceptualizing flows within transportation networks and of estimating system performance with respect to characteristics of the network and to the transportation system as a whole.

7.2.2   Performance Analysis Concepts

Most of the models and methods that analyze system performance are based on one or more ways of representing travel flows in a network. Before discussing specific applications, therefore, it will be useful to introduce the following concepts: (1) time–distance diagrams; (2) queuing theory; (3) fluid-flow approximations; (4) simulation; and (5) mathematical programming.

Time–Distance Diagrams   The simplest technique for analyzing transportation system performance is the time–distance or time–space diagram. The time–distance

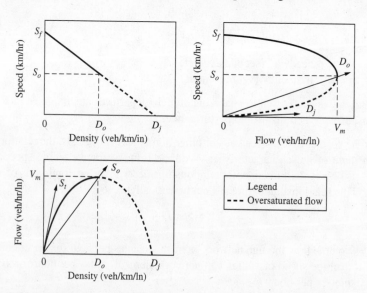

Figure 7.3 Relationship among flow, density, and speed for vehicle movement
SOURCE: TRB, 2000

diagram plots the trajectory of vehicles over the distance traveled along a route or link. Figure 7.4 illustrates the use of such a diagram. The diagram is constructed by choosing an arbitrary point in time and location along the route as the origin. The horizontal axis of the diagram is the time dimension; thus, the horizontal distance between the origin and any point in the diagram represents the time that has elapsed. The vertical axis of the diagram represents distance. The vertical difference on the trajectory between the origin and any point in the diagram represents the physical distance between the point on the route and that point on the route chosen as the origin. Most routes are not straight; they change directions any number of times. For the purposes of the time–distance diagram, however, it is assumed that one can "straighten out" the route and characterize it solely in terms of the single dimension of length or distance traveled. This is an easy and nonrestrictive assumption to make for the types of problems applicable to time–distance analysis.

The location of a given vehicle within the system can be plotted for each point in time as shown in Fig. 7.4. The locus of these location–time points (i.e., the line connecting these points) represents the time–distance trajectory of a vehicle traveling through the system. As indicated in Fig. 7.4, the instantaneous velocity of the vehicle at any point is defined by the tangent to the vehicle's trajectory at that point (dx/dt), while the average speed of the vehicle between any two locations and points in time is defined by the slope of the arc connecting these two points. It follows from these observations that a straight-line trajectory implies a vehicle traveling at constant speed and that a trajectory with increasing slope implies a vehicle that is accelerating. Conversely, a trajectory with decreasing slope implies deceleration, while a horizontal trajectory indicates a vehicle that is not moving.

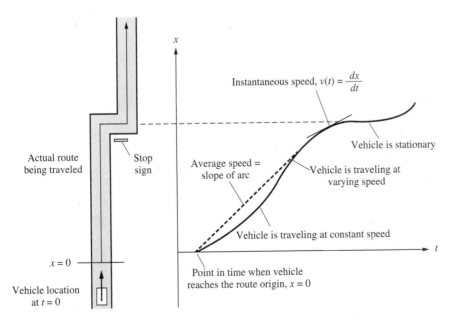

Figure 7.4   Example of time–distance diagram

Figure 7.5 illustrates the derivation of the fundamental equation of traffic flow using a time–distance approach. This figure depicts the flow of vehicles along a route. A *time–distance domain* can be drawn as shown in Fig. 7.5a for a particular segment of the route and for a particular observation period. If the length of the route segment is X feet and the length of the observation period is T seconds, then the "area" of the domain is

$$A = (X) \times (T) \text{ feet-seconds} \tag{7.4}$$

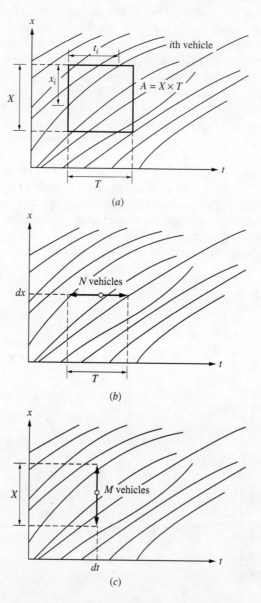

Figure 7.5   Derivation of the fundamental equation of traffic flow

Using the subscript i to denote individual vehicles that are within the route segment during the observation period, denote x_i as the distance traveled by vehicle i along the segment during the observation period and t_i as the amount of time vehicle i is within the route segment during the observation period. The total number of "vehicle-feet" of flow that occurs within the segment during the observation period is thus represented by $\Sigma_i \, x_i$. The area of domain A expresses the total number of "feet-seconds" used by this flow. The volume flow rate can thus be defined as the number of vehicle-feet that pass through the system per feet-second, or

$$V = \frac{\Sigma_i x_i}{A} \text{ vehicles per second} \qquad \textbf{[7.5]}$$

Alternatively, if one "shrinks" the route segment down to a single point in space with "width" dx as shown in Fig. 7.5b, the area of the time–distance domain is Tdx. If N vehicles pass through the domain, generating a total of Ndx vehicle-feet, the flow rate is

$$V = \frac{Ndx}{Tdx} = \frac{N}{T} \text{ vehicles per second} \qquad \textbf{[7.6]}$$

Similarly, $\Sigma_i \, t_i$ represents the total number of "vehicle-seconds" of flow within the route segment during the observation period. The density of vehicles within the domain shown in Fig. 7.5a can be defined as the number of vehicle-seconds that occur within the system per feet-second or,

$$D = \frac{\Sigma_i t_i}{A} \text{ vehicles per foot} \qquad \textbf{[7.7]}$$

Alternatively, shrinking the observation period to a single point in time of "length" dt yields a time–distance domain of area Xdt (see Fig. 7.5c). M vehicles within the route segment at the instant of time being considered generate a total of Mdt vehicle-seconds, and the density of flow is

$$D = \frac{Mdt}{Xdt} = \frac{M}{X} \text{ vehicles per foot} \qquad \textbf{[7.8]}$$

Finally, the average speed of flow through the domain is equal to the distance traveled by all vehicles within the domain divided by the total time taken by these vehicles within the domain

$$\bar{S} = \frac{\Sigma_i x_i}{\Sigma_i t_i} \text{ feet per second} \qquad \textbf{[7.9]}$$

Equation [7.9], however, is equivalent to

$$\bar{S} = \frac{V}{D} \text{ feet per second} \qquad \textbf{[7.10]}$$

which is the fundamental equation of traffic flow presented in Eq. [7.3].

One of the more important uses of time–distance diagrams is for analyzing the effects of access management strategies along major arterial roads and the spacing of traffic signals. The more interruptions to traffic flow along a road, the lower its capacity. Figure 7.6 shows a time–distance diagram that is used to determine traffic signal timings. The bandwidth in both directions in this figure represents the trajectories of vehicles that can move along this highway without stopping. As shown, traffic in this bandwidth will meet a green light at equally spaced traffic signals (1320 feet) and at a speed of progression of 30 miles per hour if the cycle length is 60 seconds [Parsonson, 1998]. In such a strategy, every other signal would be timed to turn green at the same instant. As indicated on the figure, the "offset" (lag in the start of green from one intersection to the next) is 50 percent of the cycle length, in this case 50 percent of 60, or 30 seconds. This type of analysis is commonly used to determine optimal traffic signal timing.

Time–distance diagrams are an extremely simple but useful method for representing the actual flow of vehicles within a system. They can be applied to the analysis of either exclusive or shared rights-of-way, of intersection signals or other route control systems, and of the interaction of conflicting flows. For simple problems, time–distance diagrams may well be all that an analyst needs to recommend improvements to system performance. The major limitation of the technique, however, is that it quickly becomes cumbersome to use as the complexity of the analysis increases, particularly as the number of vehicles in the system grows. In such cases, a time–distance diagram often serves as the basis of more complicated analysis techniques. Many simulation models, for example, are based on time–distance analyses, only applied at much higher levels of complexity. For further discussion of time–distance diagrams, see Morlok [1978]; Sheffi [1985]; and Daganzo [1997].

Queuing Theory   Queuing theory is used to analyze systems that can be characterized as having one or more *servers,* each performing a set of prescribed tasks for *customers.* As an example, an intersection might be considered to be a server performing the task of controlling the flow of vehicles from one street link to another. Another example would be a tollbooth. The "customers" in these cases are the vehicles wishing to pass through. The activities of the server are denoted in queuing theory as the *service process,* while the characteristics of customer arrival are determined by the *arrival process.* Because customer service takes a finite amount of time, it is often the case that other customers will arrive during the time taken to serve a customer. In such cases, the customers must either wait until the server is free and thus form a *queue,* or they must forgo the service and return at another time.

Queuing systems are characterized in terms of

1. The nature of the arrival process. The arrival process can be either *deterministic* (i.e., the time of each arrival is known with certainty) or *stochastic* (i.e., arrivals happen randomly, according to a known underlying probability distribution).

2. The number of servers available.

3. The nature of the service process. Again, this process can be either deterministic or stochastic in nature.

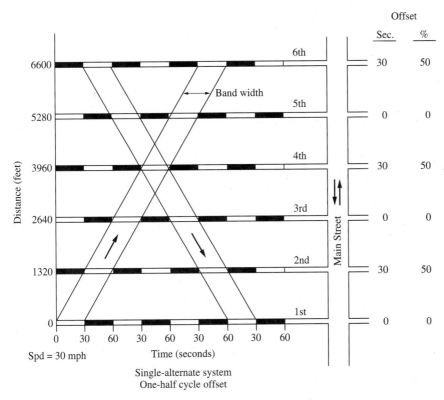

Figure 7.6 Time–distance diagrams used for determining signal timing
| SOURCE: [Parsonson, 1998]

4. The protocol for queue management. Are arrivals served on a first-come, first-served basis, or does a priority system exist for servicing customers? If multiple servers exist, how is the queue allocated among the servers (e.g., a separate queue for each server or one queue with the first member of the queue being served by the first available server)?

Many transportation system components can be thought of as queuing systems. For intersections, the service process is determined by the signal cycle time; the arrival process is determined by the volume flow on the link preceding the intersection; and arrivals are processed on a first-come, first-served basis. Other queuing phenomena include freeway sections under "forced-flow" conditions, tollbooths, terminal operations, and passenger arrivals at transit stops. If the travel flow and service characteristics satisfy certain assumptions, queuing theory can be very useful for deriving key descriptors of system performance such as

1. The expected time a customer spends in the system.
2. The expected delay experienced by a customer in the system.
3. The average queue length.

4. The expected maximum queue size.

5. The percent time a certain number of customers will be in the system.

A single server system is the simplest illustration of how queuing theory can be applied. Assume random arrival and service processes, both of which are described by a Poisson[1] distribution with an average arrival rate of λ customers per unit time and an average service rate of μ customers per unit time. Under steady-state conditions, the probability of n customers in the system at any given time is given by

$$P(N = n) = \rho^n(1 - \rho) \qquad \text{[7.11]}$$

where ρ, often referred to as the *service ratio* or *traffic intensity*, is equal to λ/μ. The important performance measures that can be estimated with these assumed conditions include

expected time T spent by customer in the system (where system is defined as the server and queue)

$$E(T) = \frac{1}{\mu - \lambda} \qquad \text{[7.12]}$$

expected number of customers in system N:

$$E(N) = \frac{\lambda}{(\mu - \lambda)} \qquad \text{[7.13]}$$

expected waiting time in the queue W:

$$E(W) = \frac{\lambda}{\mu(\mu - \lambda)} \qquad \text{[7.14]}$$

average length of queue L:

$$E(L) = \frac{\lambda^2}{\mu(\mu - \lambda)} \qquad \text{[7.15]}$$

Equation [7.12] is plotted in Fig. 7.7. As shown in the figure, expected time in the system increases as the service ratio ρ increases. As ρ approaches 1, that is, as the arrival rate equals the service rate, the queue length grows infinitely large, and the system is said to be *saturated* or at *capacity*.

To illustrate the use of Eqs. [7.11] through [7.15], consider a loading dock that on average can unload a truck in 30 minutes or, in other words, an average service rate μ of 2 trucks per hour. The average rate of arrival of trucks at the dock λ is 1 per hour. Thus ρ equals 0.5 for this system. Assuming that both truck arrivals and the unloading of trucks can be represented as Poisson processes, from Eqs. [7.14] and [7.15] the expected time spent by a truck waiting and being unloaded and the expected queue length are $1/(2 - 1) = 1$ hour and $1/(2 - 1) = 1$ truck, respectively.

---

[1] The Poisson distribution is used to describe the probability of n events occurring within a given period of time, given that the time between arrivals is a random number and that it is independent of the time of the previous arrivals. Any probability and statistics text will discuss the specifics of a Poisson process.

Similarly, the probability of 2 or more trucks being in the queue at any given time is

$$P(N \geq 2) = 1 - P(N = 0) - P(N = 1)$$

which, from Eq. [7.11], is

$$
\begin{aligned}
P(N \geq 2) &= 1 - \rho^0 (1 - \rho) - \rho^1 (1 - \rho) \\
&= 1 - 0.5^0 (1 - 0.5) - 0.5^1 (1 - 0.5) \\
&= 1 - 0.5 - 0.25 \\
&= 0.25
\end{aligned}
$$

This simple example serves to illustrate the importance of randomness in influencing system performance. Even though this system is far from saturated, the average queue length and service time vary considerably from the deterministic case, which would consist of a queue length of 0, a 0.5 hour service time, and the server being "idle" 50 percent of the time.

Queuing theory can be a very powerful analytical tool. Unfortunately, it cannot be used in transportation analysis as often as one might expect or hope. Reasons for this include

1. Arrival and service processes can rarely be described by convenient probability distributions. Thus, the number of problems that can be expressed in a tractable fashion is relatively small (although this number is increasing with time as useful approximations are developed for new problem applications).

2. Queuing theory deals with steady-state conditions. Transportation queuing phenomena are often transient in nature and hence not well suited to this form of analysis.

Nevertheless, the concept of the queue is an important starting point for the conceptualization and analysis of transportation systems. It is the basis for simple fluid-flow approximations as well as many complex simulation models where queuing theory is analytically intractable or otherwise cannot be applied.

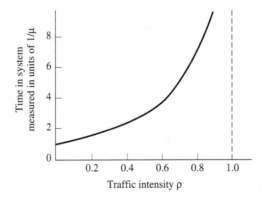

Figure 7.7   Wait time versus traffic intensity for Poisson queuing process

Fluid-Flow Approximations Figure 7.8a depicts a typical vehicle arrival process for a transportation facility, such as vehicles arriving at an intersection or passengers at a transit stop. Each step in the curve indicates a new arrival. The vertical height of the curve indicates the total number of arrivals that have occurred up to that point in time. However, in many transportation applications, the discrete event of the arrival and servicing of an individual customer is of little interest. What is of importance is the behavior of *flows* of customers through the system. In other words, a single customer is a negligibly small quantity within the system, and the uncertainty due to random fluctuations in the number of arrivals is small compared to the observed number of arrivals. In such cases, the discrete step function of Fig. 7.8a can be replaced by the "fluid-flow" approximation of Fig. 7.8b, that is, by a continuous flow of infinitesimally sized arrivals.

The cumulative number of customers served by the system can be plotted versus time and overlaid on the cumulative arrival curve. Figure 7.9 illustrates this for the case of an intersection. For simplicity of discussion, the representation of the intersection is idealized in that the yellow portion of the signal cycle is included in the green times shown. At the beginning of the process, the signal is green, no queue exists, and the arrival and departure curves coincide as customers are "served" the

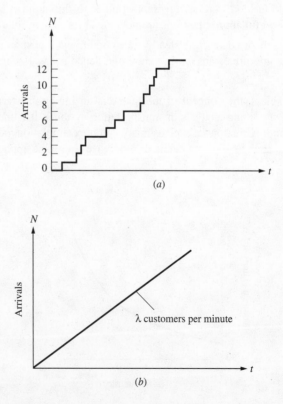

Figure 7.8   Discrete (a) and continuous (b) arrival processes

instant they arrive. When the light turns red, the departure curve becomes horizontal (no vehicles are departing), and a queue forms (indicated by the vertical gap between the arrival and departure curves). When the signal once again turns green, the queue begins to dissipate, and if the arrival rate (i.e., the slope of the arrival curve) is sufficiently lower than the departure or service rate (i.e., the slope of the departure curve), the queue ultimately vanishes. If, however, the arrival rate is large enough, the queue does not vanish during one signal cycle, and thus it will lengthen over time.

A number of important system descriptors can be obtained directly from diagrams such as Fig. 7.9. These include

1. The queue length n at any point in time is given by the vertical distance between the arrival and departure curves at that point.
2. The waiting time w experienced by any arrival is given by the horizontal distance between the arrival and departure curves.
3. The total delay experienced by all customers in the system is represented by the area A shown between the two curves.
4. The average queue length is given by the area between the two curves, divided by the time interval over which this area is calculated, or

$$\bar{n} = \frac{A}{T}$$

[7.16]

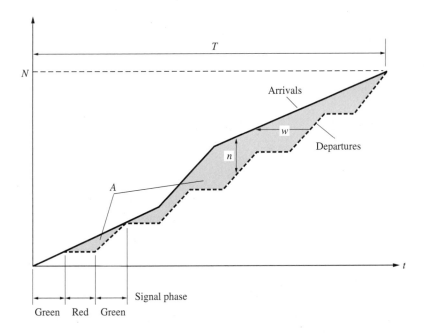

Figure 7.9 Fluid-flow approximations of a traffic signal's operation

5. The average wait time or delay is given by the area between the two curves, divided by the number of arrivals, or

$$\overline{w} = \frac{A}{N} \qquad\qquad [7.17]$$

Fluid-flow approximations indicate zero delay when the arrival rate is less than or equal to the service rate and no previously developed queue exists as indicated during the first green phase of Fig. 7.9. Finite delays occur when the arrival rate exceeds the service rate for finite periods of time. This can be contrasted with the results of queuing theory where delays occur when the service ratio is less than 1, and the delay grows infinitely large as the service ratio approaches unity. These seemingly contrary results highlight the differences between the fundamental assumptions underlying the two approaches. Queuing theory predicts steady-state results for a discrete system in which individual customers interact with individual servers. Because a server takes a finite amount of time to service a customer, customers arriving during this time period will experience delay, and this delay will tend to grow infinitely large as the arrival rate increases relative to the service rate.

Fluid-flow approximations, on the other hand, adopt a more macroscopic view of the system. This view treats the server as a "channel" through which customers flow. The service rate defines the channel capacity, while the arrival rate defines the volume flow through the channel. If this flow rate is less than the channel capacity, no delay occurs because microscopic interactions between the individual flow elements (i.e., customers and channel-server) are being ignored. If the flow rate is greater than the capacity or if the channel is temporarily obstructed (for example, by a signal turning red), delay occurs. If this represents the steady-state condition, the delay will be infinitely large. But, in general, the channel capacity is exceeded for only short periods of time, and the queue can eventually dissipate as shown by Fig. 7.9, resulting in finite delays.

Queuing theory and fluid-flow approximations complement one another in that queuing theory provides insight into how systems behave when operating in the region $\rho < 1$, while fluid-flow approximations describe their behavior in the region $\rho > 1$. Neither technique, however, appears to deal adequately with the case of $\rho \approx 1$, which may actually be the operating region of interest in many transportation planning problems. In such cases, simulation models may be required to examine system performance in detail.

The advantages of fluid-flow approximations are that they do not require steady-state assumptions and that they are capable of handling virtually any level of complexity in the arrival and service processes *provided* that arrivals and departures can be represented as continuous processes. For simple cases, but ones too complex for queuing theory, the analysis can often be performed directly by graphical methods. For more complicated cases, the techniques can be computerized for analytical or numerical solution. The graphical representation can be extremely useful in assisting the analyst to visualize and understand the processes at work. A major limitation of this approach is that it generally involves the analysis of the system's performance over a single time interval. If conditions during this time interval are not representative of conditions during other time intervals, the analysis must be repeated over an ensemble of intervals in order to derive average or steady-state values.

Simulation Simulation involves constructing a mathematical model of a system representing the cause-and-effect relationships that determine eventual outcomes. These relationships are assumed to remain stable over time. Thus, future outcomes such as system performance can be predicted by using estimates of future input variables. This relationship can be generalized with the following equation:

$$y(t + \Delta t) = f\left[x(t + \Delta t), y(t)\right] \qquad \text{[7.18]}$$

where

$y(t)$ = vector of variables that characterize system performance at time t

$x(t)$ = vector of exogenous variables that influence system performance at time t

$f(\bullet)$ = system of mathematical equations, computer algorithms, probability distributions, etc. that determines system performance at time $t + \Delta t$, given $x(t + \Delta t)$ and $y(t)$

Simulation models share the following characteristics:

1. They are explicitly dynamic; that is, they replicate system performance over time.

2. System performance at any point in time depends explicitly and generally in complex ways on system performance at previous points in time. That is, performance at time $t + \Delta t$ cannot be computed without knowing the performance at time t.

3. The "performance procedure" is sufficiently complex that it cannot be solved analytically. This complexity typically includes treating certain system processes probabilistically (e.g., random arrival and service rates).

If, as is typically the case, the simulation model is stochastic rather than deterministic, it must be applied many times so that the frequency distribution or expected values of the outcomes associated with these stochastic processes can be determined. This is an important aspect of simulation models that is often ignored in practice. Simulation modeling is thus a form of experimentation in which a series of trials of system performance is carried out under a range of operating conditions determined by the random processes embedded in the model. Given the complexity and size of simulation models and the need to perform a large number of model runs or replications in order to generate usable information, simulation modeling will generally be more time consuming than other approaches. This is particularly the case when no "off-the-shelf" model is available, in which case a new model must be developed. This would involve considerable time and cost to gather the required data, develop the computer programs, and make the programs operational.

An example of a multimodal simulation is the analysis of the passenger, pedestrian, and vehicle flow to and through Union Station, the rail passenger station in Los Angeles. Approximately 107 passenger and commuter train arrivals and departures were occurring at Union Station in 1995. By 2025, this was expected to increase to 380. Union Station also handled 280 subway arrivals, 1800 bus trips, and between 40,000 and 50,000 pedestrian trips each day, each of which was to increase significantly by 2025. The interaction among all of these modes was so complex that computer simulation models were used to determine a recommended program of improvements. The rail simulation assumed different levels of track and station improvements that would affect how many trains could use the station. Pedestrian movements were simulated by developing a person trip database that was linked to train arrivals and nontrain-related pedestrian activity. Internal and external pathways and waiting areas were incorporated into the model. Pedestrian facilities exceeding a criterion of 18 square feet per person were targeted for improvements, as were the rail facilities that resulted in an average delay per train of more than 1 minute in a 1-hour period [Kimley-Horn and Assocs., 1995]. A series of recommended facility improvements for both rail and pedestrian movements were identified based on the predicted levels of delay that resulted from the simulation models.

With advances in computer technology over the past decade, simulation modeling has become a more feasible analysis tool. The TRansportation ANalysis SIMulation System (TRANSIMS) effort supported by the U.S. DOT as part of its program for improving the state of the art in transportation modeling, has been one of the most extensive efforts to develop a regional simulation model. This model develops specific trip routing plans to satisfy the activity desires of each traveler [Barrett et al., 1995]. Each link in the road network is divided into sequential cells representing 7.5 meters of road space, the amount of space occupied by a typical car. For every simulated second, each cell in the network is examined to see if it is occupied by a vehicle. If it is, simple vehicle movement rules are used to move it to a next cell for the following second of simulated time. Using these vehicle movement rules, the vehicle "moves" across the network from its origin to destination as defined by the trip plan. Vehicles in the simulation accelerate, decelerate, turn, change lanes, and generally interact with other "vehicles" as real vehicles would. More recently, TRANSIMS has been enhanced to include a multimodal trip planner that develops trip plans across all modes of transportation and corresponding simulation of transit vehicles.

TRANSIMS was applied in Dallas to a 25-square mile area having approximately 200,000 vehicle trips over a 4-hour time period. The demonstration concluded that

1. The simulations successfully reproduced observed traffic patterns.
2. The effects of both infrastructure and operational improvements were evident.
3. Estimates of travel time reliability could be produced.
4. Equity impacts on population subgroups were readily available [Kimley-Horn and Assocs., 1998].

However, in order to produce these results, the model's coded network for the road system had to be extensively updated. Table 7.2 shows the type of data that had to be included as attribute information to the different components of the network. In order to implement this microsimulation modeling approach, the modeled road network had to represent the real world in much finer detail than it had previously.

Simulation modeling will likely become even more widely used in the future because there is often no other method available to analyze a given problem at the required level of detail and realism. In such cases, if the problem is to be analyzed at all, it must be analyzed using simulation.

Table 7.2 Network coding details for TRANSIMS application in Dallas

File	Information to be Coded
Node	• Location (X–Y coordinates)
Link	• End nodes
	• Length
	• Setback distance from center of intersection to stop line, by direction
	• Number of permanent lanes in each direction
	• Pocket lanes in each direction
	– Number to the left of the permanent lanes
	– Number to the right of the permanent lanes
	• Speed limit in each direction
Pocket lane	• Node into which the pocket lane leads
	• Link ID on which the pocket lane lies; lane number of the link
	• Type: turn, pull-out, or merge
	• Starting position and/or length
Parking places	• Location of traffic loading nodes/points
Unsignalized node	• Sign control on each incoming link
Lane connectivity	• Identification of all allowed movements through an intersection
Signalized node	• Seconds of offset from a base intersection
	• ID# of the timing plan
	• Start time for the timing plan
Phasing plan	• ID# of the timing plan
	• Incoming link, outgoing link, and turn protection indicator for each movement allowed in a particular phase number
Timing plan	• Length of green, yellow, and red clearance interval for each traffic signal phase number

SOURCE: Los Alamos, 1999

Mathematical Programming A *mathematical program* consists of an *objective function* that is to be maximized or minimized subject to a set of *constraints* (e.g., minimize the cost of operating a given service, subject to the constraint that all customers requesting service must be picked up and delivered). The objective function is a mathematical function of a set of *decision variables* (e.g., the number of buses to be assigned to each route in the system) that expresses how these decision variables are to be combined to yield the quantity to be maximized or minimized. The set of constraints is also a set of mathematical functions that define feasible ranges of values for these decision variables. A *linear program* is a special type of mathematical program in which the objective function and all the constraints are linear; an *integer program* is one in which the decision variables are constrained to take on integer values (e.g., one cannot buy 0.637 buses or carry 142.94 passengers).

Transportation system problems that can be addressed by mathematical programming procedures include

1. Scheduling of transportation services, vehicles, and drivers (typically through the use of heuristics).
2. Analysis of flows through networks.
3. Calibration of models typically by maximizing a likelihood function.
4. Expansion of survey samples to universe populations.

Figure 7.10 provides a classic example of a transportation application of mathematical programming (in fact, this formulation is often referred to as "the transportation problem"). In this system, goods are supplied at a set of origins, consumed at a set of destinations, and shipped along transportation routes that have costs attached to them. Thus, if we define the key variables in the following way,

O_i = quantity of goods produced at origin i, $i = 1, \ldots, n$

D_j = quantity of goods required at destination j, $j = 1, \ldots, m$

x_{ij} = quantity of goods shipped from origin i to destination j

c_{ij} = cost per unit quantity of shipping goods from i to j

then the mathematical programming formulation of the problem that will determine a set of flows (i.e., a set of x_{ij}'s) minimizing total shipping costs is

$$\min \sum_{i=1}^{n} \sum_{j=1}^{m} c_{ij} x_{ij} \qquad [7.19]$$

subject to
$$\sum_{j=1}^{m} x_{ij} \leq O_i \quad \text{for all } i, i = 1, \ldots, n \qquad [7.20]$$

$$\sum_{j=1}^{n} x_{ij} \geq D_j \quad \text{for all } j, j = 1, \ldots, m \qquad [7.21]$$

Equation [7.19] is the expression for the total shipping costs in the system for any given set of flows. The mathematical program determines a set of flows that minimizes this function, subject to the constraints that the total flow out of each origin zone does not exceed the amount produced in the zone (Eq. [7.20]) and that the total flow into each destination zone is at least sufficient to meet the consumption requirements of the zone (Eq. [7.21]). Note in Fig. 7.10 that the "transportation problem" has only origins and destination nodes. Physical transportation networks on the other hand have nodes that are neither origins nor destinations, such as intersections or transit stations. In network analysis, nodes that neither produce nor attract flows are called transshipment nodes. A network flow problem that includes transhipment nodes is referred to as a minimum cost flow problem.

When the problem is generalized to include transshipment nodes and multiple commodities, the resultant mathematical programming formulation is referred to as a multicommodity network flow problem. This type of problem formulation obviously

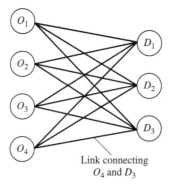

Figure 7.10 The "transportation" problem in mathematical programming

has great potential application to transportation system analysis. However, it should also be obvious that a comprehensive multicommodity network flow problem model of a regional urban transportation network would be a very complex formulation. Optimization of such a full-scale network problem represents a significant challenge.

Spreadsheets and software programs are available to solve linear programs involving large numbers of decision variables and constraints. These programs also provide "postoptimality" sensitivity analyses of these solutions; that is, how does the optimal solution change if key input parameters are changed? For example, if transportation costs increase by 10 percent, is the proposed solution still optimal?

If the objective function or constraints are defined as nonlinear or the optimal solution must be integer, then solving the program becomes more complex. The techniques for these types of problems vary widely and depend greatly on the nature of the particular problem. Specialized solution procedures must often be developed in order to exploit the special structure or characteristics of a problem. One class of integer programming problems known as combinatorial optimization problems is particularly challenging. These problems choose an optimal set of items (e.g., schedules of pickups and deliveries for a fleet of demand responsive transit vehicles) from among a very large set of feasible combinations of such items. In cases where optimal solutions cannot be found using formal mathematical programming procedures, it is often possible to develop *heuristic* procedures that are designed to find "good," but not necessarily optimal, solutions. Many of the procedures for determining transportation scheduling use heuristics rather than optimization procedures due to their complexity.

7.3 PERFORMANCE ANALYSIS METHODS FOR FACILITIES

The selection of which method to use in analyzing the performance of transportation facilities relates to several operational and design characteristics of the facilities themselves. Of these, it is most useful to distinguish facilities by whether vehicles have exclusive use of a guideway or whether vehicles share rights-of-way.

7.3.1   Exclusive Right-of-Way or Guideway Operations

The most common examples of dedicated right-of-way operations are heavy-rail transit (e.g., subways and commuter rail); light-rail transit operating on dedicated rights of way; "advanced" transit technologies, such as people movers; and reserved lanes on freeways for buses and other high-occupancy vehicles. Performance on such systems is dictated by the following:

1. Technological capabilities of the vehicles.
2. Systems used for controlling vehicle flow (e.g., track signal systems).
3. Spacing of stops along the right-of-way.
4. Amount of time spent at the stop, known as dwell time.
5. Operating policies (e.g., minimum vehicle following-distances).

The methods for analyzing performance on exclusive rights-of-way range from time–distance diagrams to simulation models. This section presents analytical models that estimate route or link capacity, headway, average travel time, and average speed.

Models of transit line performance often assume deterministic operational characteristics or, equivalently, that random variations in system conditions and performance (resulting from such things as interaction with other vehicles in the system or delays due to passenger movements), are negligible. The models described in the following are applicable to any service that can be assumed to be operating under such an assumption. Such a condition is more likely to hold under an exclusive right-of-way operation with its greater degree of control over vehicle flow than when the right-of-way is shared. If random variations in system conditions cannot be ignored, then more complicated analysis approaches, such as simulation, must be employed, regardless of the type of right-of-way.

Headway and Capacity   The design capacity (C_d) of a transit facility is directly related to (1) the frequency of service (f) on the facility; (2) the number of vehicles N_c operated together in a "transit unit" (TU) (e.g., a subway might operate five-car trains, a light-rail system might operate two-vehicle trains, or a busway accommodates single buses); and (3) the passenger capacity P_c of each vehicle. That is

$$C_d = f \times N_c \times P_c \text{ passengers per hour} \qquad [7.22]$$

or

$$C_d = \frac{3600 \times N_c \times P_c}{h} \qquad [7.23]$$

where h is the time between arrivals or service *headway* in seconds and is inversely proportional to the frequency; that is, $f = 3600/h$. Headway represents the time it takes for a transit unit to travel between stations, including the acceleration and deceleration time, the dwell time in the departure station, and the operations safety factors that keep a minimum distance between transit units. Figure 7.11 shows a time–distance plot of two consecutive trains and the corresponding variables that are important for determining headway. Equation [7.24], shown in Fig. 7.11, is the equation for headway.[2]

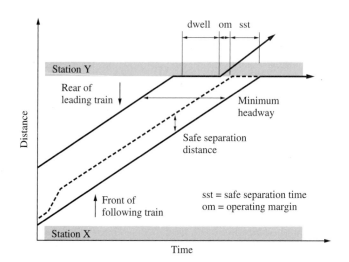

$$h = \sqrt{\frac{2(L+D)}{a_s}} + \frac{L_t}{v_a} + \left(\frac{100}{K} + B\right)\left(\frac{v_a}{2d_s}\right) + \frac{a_s t_{os}^2}{2v_a}\left(1 - \frac{v_a}{v_{max}}\right) + t_{os} + t_{jl} + t_{br} + t_d + t_{om} \qquad \textbf{[7.24]}$$

[typical values shown in square brackets]

where:

h	=	station headway (s);
L_t	=	length of the longest train; *[200 m or 660 feet]*
D	=	distance from front of stopped train to start of station exit block; *[10 m or 33 feet]*
v_a	=	station approach speed (m/s);
v_{max}	=	maximum line speed (m/s);
K	=	braking safety factor—worst case service braking is K% of specified normal rate—typically 75%; *[75]*
B	=	separation safety factor—equivalent to number of braking distances plus a margin (surrogate for blocks) that separate trains;
t_{os}	=	time for overspeed governor to operate; *[3 s]*
t_{jl}	=	time lost to braking jerk limitation; *[0.5 s]*
t_{br}	=	operator and brake system reaction time; *[1.5 s]*
t_d	=	dwell time; *[45 s]*
t_{om}	=	operating margin; *[20 s]*
a_s	=	initial service acceleration rate; *[1.3 m/s² or 4.3 ft/s²]* and
d_s	=	service deceleration rate; *[1.3 m/s² or 4.3 ft/s²]*

Figure 7.11   Distance–time plot of two consecutive trains
SOURCE: TRB, 1999

[2] The TRB *Transit Capacity and Quality of Service Manual* [TRB, 1999] provides modified equations for headway, assuming different signal control systems and vertical grades of the track alignment entering the station.

Dwell time (t_d) will vary by the number of passengers boarding and alighting at the maximum load station and the corresponding unit time for each. The equation for t_d is

$$t_d = (P_a \times t_a) + (P_b \times t_b) + t_{oc} \qquad \text{[7.25]}$$

where

t_d = dwell time (seconds)

P_a = passengers alighting through the busiest door during the peak 15 minutes (passengers)

t_a = alighting time per passenger (seconds/passenger)

P_b = passengers boarding through the busiest door during the peak 15 minutes (passengers)

t_b = boarding time per passenger (seconds/passenger)

t_{oc} = door opening and closing time (seconds)

The following example illustrates the use of these equations. A transit agency wants to determine the design capacity of a new rail line with a fixed block cab control system. At the maximum load point, 15 passengers board and 4 passengers depart through the busiest subway car door during the peak 15-minute period. Each passenger boarding and alighting takes 2.0 seconds, and the door opening and closing time is 4.0 seconds. The relevant data are as follows

L_t	=	200 meters	t_{os}	=	3.0 secs
D	=	10 meters	t_{jl}	=	0.5 secs
v_a	=	14.4 m/sec	t_{br}	=	1.5 secs
v_{max}	=	27.8 m/sec	a_s	=	1.3 m/sec^2
K	=	75 percent	d_s	=	1.3 m/sec^2
B	=	1.2	t_{om}	=	20 secs
N_c	=	4 cars/TU	P_c	=	100 pass/car

In order to use Eq. [7.24], the dwell time must be first calculated using Eq. [7.25].

$$\begin{aligned} t_d &= (20 \text{ passengers})(2 \text{ seconds/passenger}) + (4 \text{ passengers}) \\ &\quad (2 \text{ seconds/passenger}) + 4 \text{ seconds} \\ &= 52 \text{ seconds} \end{aligned}$$

The headway can now be calculated as

$$h = \sqrt{\frac{(2)(200+10)}{1.3}} + \frac{200}{14.4} + \left[\frac{100}{75} + 1.2\right]\left[\frac{14.4}{2 \times 1.3}\right] + \frac{(1.3)(3)^2}{2(14.4)}\left[1 - \frac{14.4}{27.8}\right].$$

$$+ 3.0 + 0.5 + 1.5 + 52 + 20$$

$$= (18.0 + 13.9 + 13.8 + 0.2 + 3.0 + 0.5 + 1.5 + 52 + 20) \text{ seconds}$$

$$= 122.9 \text{ seconds}$$

For purposes of analysis, an assumed headway of 120 seconds, or 2 minutes, would be appropriate. The design capacity is thus

$$C_d = \frac{3600 \text{ seconds/hour} \times 100 \text{ passengers/car} \times 4 \text{ cars/TU}}{120 \text{ seconds/TU}}$$

$$= 12,000 \text{ passengers/hour}$$

If a moving-block signaling system with a safety distance of 50 meters were used, the corresponding capacity using Eq. [3.5] found in [TRB, 1999] would be 14,545 passengers/hour. The impact of the signal technology on design capacity is significant.

The design capacity represents the theoretical level of ridership that could be handled given operating characteristics. The achievable capacity that reflects variations in demand is determined by

$$C_a = C_d \times PHF \qquad\qquad \text{[7.26]}$$

where

C_a = achievable capacity

C_d = design capacity

PHF = peak hour factor = $\dfrac{P}{4 \times P_{15}}$

P = peak hourly passenger volume

P_{15} = peak 15-minute passenger volume

In the preceding example, if the peak hourly volume at the maximum load point was 1600 passengers and the peak 15-minute passenger count was 500 passengers, then

$$PHF = \frac{1600}{4 \times 500} = 0.8$$

and the achievable capacity would be

$$12,000 \text{ passengers/hour} \times 0.80 = 9600 \text{ passengers/hour}$$

Figure 7.12 shows the typical maximum passenger capacities for grade-separated rail transit.

Travel Time and Average Speed Given a link i of length L_i feet or meters, the time T_i that elapses from the instant the transit unit comes to a stop at the first stop to the time it comes to rest at the second stop is estimated as

$$T_i = SM\left[\frac{v_{max}}{2}\left(\frac{1}{a} + \frac{1}{d} + t_{jl} + t_{br}\right) + \left(\frac{L_i + L_t}{v_{max}}\right)\right] + t_{di} + t_{om} \qquad \text{[7.27]}$$

where

SM = speed margin (ranges from 1.0 to 1.2)

v_{max} = maximum speed on link i, meters/second or feet/second

L_i = length of link i, meters or feet
L_t = length of train, meters or feet
a = acceleration rate, meters/second^2 or feet/second^2
d = deceleration rate, meters/second^2 or feet/second^2
t_{di} = dwell time for station i, seconds
t_{om} = operating margin, seconds

The average speed for this link is L_i/T_i. If there are m stops on the route, and if R is the turnaround time in seconds required at one end of the route, then the average speed over the entire route is

$$\bar{S} = \frac{3600 \times L}{(m)(SM)\left[\frac{v_{max}}{2}\left(\frac{1}{a} + \frac{1}{d} + t_{jl} + t_{br}\right)\right] + R + \sum_{i=1}^{m}\left[\left(\frac{L_i + L_t}{v_{max}}\right) + t_{di} + t_{om}\right]} \qquad \text{[7.28]}$$

where
$\bar{S}$ = average speed, kilometers/hour or miles/hour
L = one-way distance of route, kilometers or miles
R = turnaround time, seconds
m = number of stations
and other variables defined as in Eq. [7.27]

The denominator of this equation is simply the summation of Eq. [7.27] over all stops on the line, and the constant value of 3600 converts the expression from seconds to hours.

To illustrate the use of Eq. [7.28], consider a 10-kilometer light-rail transit route with stations uniformly spaced every kilometer. The maximum operating speed on

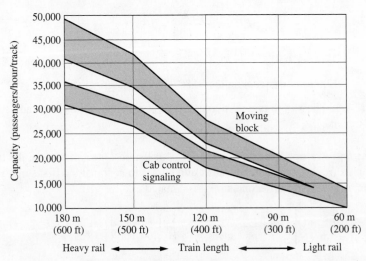

NOTE: Assumes peak hour average passenger loading of 5.0 p/m of length for light rail and 5.0 p/m of length for heavy rail. Capacity for one track of a grade-separated rail transit line. Operating margin ranges from 45 seconds (lower bound) to 70 seconds (upper bound).

Figure 7.12   Typical maximum passenger capacities of grade-separated rail transit
I SOURCE: TRB, 1999

the route is 30 meters/second. Assume acceleration and deceleration rates of 1.3 meters/second^2, station dwell times of 20 seconds per station, and a 5-minute turnaround time at the end of the line. Transit units consist of two-car trains with each car carrying a maximum of 80 passengers. Determine the design capacity for this light rail line and the average speed of this service. The data are as follows:

m = number of stops, 10 stops
L_i = length of the ith link, 1000 meters for all i
L_t = length of longest train, 60 meters
L = one-way length of line, 10 kilometers
v = cruise speed, 30 meters/second
R = turnaround time, 5 minutes = 300 seconds
a = acceleration, 1.3 meters/second^2
d = deceleration, 1.3 meters/second^2
t_{di} = dwell time at stop i, 20 seconds for all i
t_{jl} = time lost to braking jerk limitation, 0.5 seconds
t_{br} = operator and brake reaction time, 1.5 seconds
t_{om} = operating margin, 20 seconds

SM = 1.1

Substituting these values into Eq. [7.28] yields the following estimate for average speed:

$$\overline{S} = \frac{3600 \times 10 \text{ kilometers}}{10(1.1)\left[\frac{30}{2}\left(\frac{1}{1.3}+\frac{1}{1.3}+0.5+1.5\right)\right]+300+\sum_{i=1}^{10}\left[\frac{(1000+60)}{30}+20+20\right]} \quad [7.29]$$

= 22.8 kilometers/hour

7.3.2 Shared Right-of-Way Operations

Virtually all streets, highways, and sidewalks operate as shared rights-of-way. When determining the performance of shared right-of-way facilities, an important distinction is made between *controlled access facilities* (e.g., freeways) and *uncontrolled access facilities* (e.g., urban streets and sidewalks). Each of these is discussed below.

Controlled Access or Uninterrupted Flow Facilities A typical urban freeway or controlled access roadway consists of two or more lanes of travel in each direction, directional separation of traffic movements, no signalized control of operations, and access and egress to the facility limited to a small number of interchanges spaced relatively far apart. Under normal operating conditions, traffic on such facilities is generally free flowing and corresponds closely to the fundamental relationship of traffic flow (see Eq. [7.3]). As flow levels increase, however, interchanges typically begin to act as bottlenecks because of weaving maneuvers and other phenomena associated with vehicles exiting and entering the traffic flow. Large volumes of traffic also include many vehicle interactions that tend to slow down traffic.

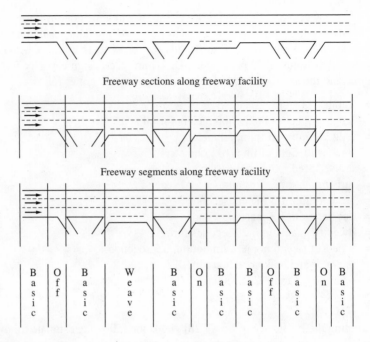

Freeway sections along freeway facility

Freeway segments along freeway facility

B a s i c	O f f	B a s i c	W e a v e	B a s i c	O n	B a s i c	B a s i c	O f f	B a s i c	O n	B a s i c

Figure 7.13   Freeway components for capacity analysis
I   SOURCE: TRB, 2000

For purposes of capacity analysis, freeways are assumed to consist of three segments—ramps, weaving sections, and basic freeway segments. Figure 7.13 shows how a freeway facility can be divided into these segments and its corresponding segment capacities determined over short time intervals, usually 15 minutes. The overall approach for estimating freeway facility performance is shown in Fig. 7.14. As can be seen, the calculation of segment capacities is critical to estimating overall performance. In each of the three cases—basic freeway segment, weaving, or ramp segments—the approach estimates segment capacity under ideal or base conditions and then reduces this estimate for the prevailing conditions that are not ideal. It is beyond the scope of this book to discuss this estimation process in detail. To illustrate the approach, however, calculating the capacity of a basic freeway segment is presented in the following paragraphs.

The base conditions for a freeway segment include a free-flow speed of 110 kilometers/hour or greater, 3.6 meters minimum lane width, 1.8 meter right shoulder lateral clearance, a traffic stream of only passenger cars, interchange spacing of greater than 3 kilometers, level terrain with grades no greater than 2 percent, and a driver population familiar with the facility. The level of service (LOS) measures for a basic freeway segment are shown in Table 7.3. The maximum density for LOS E represents the capacity of the segment. Figure 7.15 shows the relationship among speed, flow, and density for a basic freeway segment. The corresponding levels of service values for the density thresholds are also indicated.

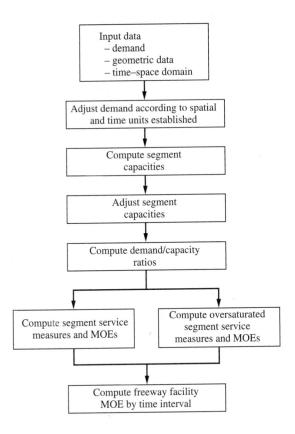

Figure 7.14   Approach to estimate freeway capacity
| SOURCE: TRB, 2000

The equivalent passenger car flow rate is given by

$$V_p = \frac{V}{PHF \times N \times f_{HV} \times f_p}$$ **[7.30]**

where

PHF = peak hour factor

V = hourly volume (vehicles/hour)

V_p = 15-minute passenger car equivalent flow rate (vehicle/hour/lane)

N = number of lanes

f_{HV} = heavy vehicle adjustment factor

f_p = driver population factor

The heavy vehicle factor (f_{HV}) adjusts the flow rate to account for nonpassenger cars in the traffic stream. The f_{HV} is calculated as

Table 7.3   Level of service measures for basic freeway segments

| Free-Flow Speed | Criteria | LOS | | | | |
		A	B	C	D	E
120 km/h	Max Density (pc/km/ln)	7	11	16	22	28
	Minimum Speed (km/h)	120.0	120.0	114.6	99.6	85.7
	Max v/c	0.35	0.55	0.77	0.92	1.00
	Max Service Flow Rate (pc/h/ln)	840	1320	1840	2200	2400
110 km/h	Max Density (pc/km/ln)	7	11	16	22	28
	Minimum Speed (km/h)	110.0	110.0	108.5	97.2	83.9
	Max v/c	0.33	0.51	0.74	0.91	1.00
	Max Service Flow Rate (pc/h/ln)	770	1210	1740	2135	2350
100 km/h	Max Density (pc/km/ln)	7	11	16	22	28
	Minimum Speed (km/h)	100.0	100.0	100.0	93.8	82.1
	Max v/c	0.30	0.48	0.70	0.90	1.00
	Max Service Flow Rate (pc/h/ln)	700	1100	1600	2065	2300
90 km/h	Max Density (pc/km/ln)	7	11	16	22	28
	Minimum Speed (km/h)	90.0	90.0	90.0	89.1	80.4
	Max v/c	0.28	0.44	0.64	0.87	1.00
	Max Service Flow Rate (pc/h/ln)	630	990	1440	1955	2250

Note: The exact mathematical relationship between density and v/c has not always been maintained at LOS boundaries because of the use of rounded values. Density is the primary determinant of LOS. Speed criterion lists speed at maximum density for a given LOS.
SOURCE: TRB, 2000

$$f_{HV} = \frac{1}{1 + P_T\left(E_T - 1\right) + P_R\left(E_R - 1\right)}$$ [7.31]

where

P_T = proportion of trucks in traffic stream

P_R = proportion of recreational vehicles (RVs) in traffic stream

E_T = passenger car equivalent of trucks

E_R = passenger car equivalent of recreational vehicles

The passenger car equivalent for trucks and RVs represents the number of cars that would use the same amount of freeway capacity as one truck or RV under prevailing conditions. These values vary from 1.2 and 1.5 for RVs and trucks, respectively, at road grades less than 2 percent, to 6.0 for RVs on a greater than 5 percent and longer than 0.8 kilometer upgrade, and 7.5 for trucks on a greater than 6 percent and longer than 6.4 kilometer downgrade. The driver population factor (f_p) represents the degree of driver familiarity with the facility. Values range from 1.0 for commuting traffic to 0.85 for recreational traffic. The values for different combinations of these input parameters are found in the *HCM 2000*.

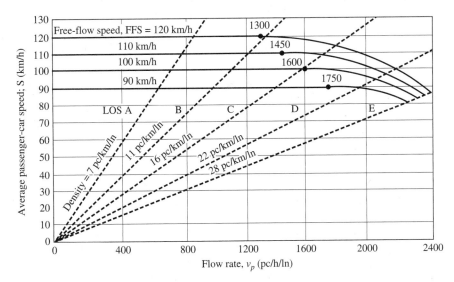

Note:
Capacity varies by free-flow speed. Capacity is 2400, 2300, and 2250 pc/h/ln at free flow-speeds of 120, 110, 100 and 90 km/h, respectively.

For $90 \leq \text{FFS} < 120$ and for flow rate (v_p)
   $(3100 - 15\,\text{FFS}) < v_p \leq (1800 + 5\,\text{FFS})$ then

$$S = \text{FFS} - \left[\frac{1}{28}(23\,\text{FFS} - 1800)\left(\frac{v_p + 15\,\text{FFS} - 3100}{20\,\text{FFS} - 1300} \right)^{2.6} \right]$$

For $90 \leq \text{FFS} < 120$ and
   $v_p \leq (3100 - 15\,\text{FFS})$ then
   $S = \text{FFS}$

Figure 7.15   Relationship among speed, flow, and density for basic freeway segment
| SOURCE: TRB, 2000

The free-flow speed for a basic freeway segment is estimated by

$$FFS = BFFS - f_{LW} - f_{LC} - f_N - f_{ID} \qquad \textbf{[7.32]}$$

where
   FFS = free-flow speed, kilometers/hour
   $BFFS$= base free-flow speed (110 kilometers/hour for urban)
   f_{LW} = adjustment of lane width (kilometers/hour)
   f_{LC} = adjustment for right shoulder lateral clearance (kilometers/hour)
   f_N = adjustment for number of lanes (kilometers/hour)
   f_{ID} = adjustment for interchange density (kilometers/hour)

The values for these different adjustments are shown in Table 7.4.
Basic freeway segment level of service is estimated by applying Eq. [7.3]. The flow rate in passenger cars/hour/lane is measured and the average passenger car speed is determined. The density can be calculated as

$$D = \frac{V_p}{S}$$

where the terms were defined in Eq. [7.3]. The calculated values of D and FFS are then compared to Table 7.3 to determine in which LOS range the facility is operating.

For example, suppose an engineer wants to determine the LOS of a six-lane suburban freeway. The specific data are as follows:

V = 4000 vehicles/hours in one direction   0.9 interchanges/kilometer
P_T = 15 percent, P_R = 3 percent   level terrain
PHF = 0.85   commuter traffic ($f_p = 1.0$)
3.6 meter lane width $BFFS$ = 110 kilometers/hour
1.8 meter lateral clearance for right shoulder $E_T = 1.5$ $E_R = 1.2$

The first step is to calculate the heavy vehicle factor

$$F_{HV} = \frac{1}{1 + (0.15)(1.5 - 1) + 0.03(1.2 - 1)} = 0.925$$

therefore,

$$V_p = \frac{4000 \text{ vehicles/hour}}{(0.85)(3)(0.925)(1.0)} = 1696 \text{ passenger cars/hour/lane}$$

Free-flow speed is calculated as

$$FFS = 120 - 0 - 0 - 4.8 - 8.1 = 107.1 \text{ kilometers/hour}$$

Using Table 7.3, the combination of a FFS of 107.1 kilometers/hour with a flow rate of 1696 passenger car/hour/lane results in a LOS C. Chapters 24 and 25 of the *HCM 2000* present the comparable analysis for freeway weaving and ramp freeway segments.

Table 7.4 Adjustments for level of service calculations for basic freeway segment

Adjustment for Lane Width	
Lane Width (m)	**Reduction in Free-Flow Speed (Km/Hour)**
3.6	0.0
3.5	1.0
3.4	2.1
3.3	3.1
3.2	5.6
3.1	8.1
3.0	10.6

(continues)

Table 7.4 Adjustments for level of service calculations for basic freeway segment (continued)

Adjustments for Right-Shoulder Lateral Clearance

Right-Shoulder Lateral Clearance (m)	Reduction in Free-Flow Speed f_{LC} (Km/Hour)			
	Lanes in One Direction			
	2	3	4	5
>1.8	0.0	0.0	0.0	0.0
1.5	1.0	0.7	0.3	0.2
1.2	1.9	1.3	0.7	0.4
0.9	2.9	1.9	1.0	0.6
0.6	3.9	2.6	1.3	0.8
0.3	4.8	3.2	1.6	1.1
0.0	5.8	3.9	1.9	1.3

Adjustments for Number of Lanes

Number of Lanes (One Direction)	Reduction in Free-Flow Speed f_N (Km/Hour)
>5	0.0
4	2.4
3	4.8
2	7.3

Adjustments for Interchange Density

Interchanges/Kilometer	Reduction in Free-Flow Speed f_{ID} (Km/Hour)
<0.3	0.0
0.4	1.1
0.5	2.1
0.6	3.9
0.7	5.0
0.8	6.0
0.9	8.1
1.0	9.2
1.1	10.2
1.2	12.1

SOURCE: TRB, 2000

Walkways and pedestrian paths can also be considered as uninterrupted flow facilities if pedestrians do not experience any delays due to other pedestrians or non-motorized modes of transportation. The relationships among speed, density, and flow are similar to that found for roadways. Figure 7.16 shows these relationships graphically. The only difference in definitions from previous applications is that density is measured as the average number of pedestrians per unit area (pedestrians/meter^2). For pedestrian facilities, the unit flow rate is calculated as

$$V_p = \frac{V_{15}}{15 \times W_E}$$

[7.33]

where

V_p = pedestrian unit flow rate, pedestrians/minute/meter
V_{15} = peak 15-minute flow rate, pedestrians/15 minutes
W_E = effective walkway width, meters

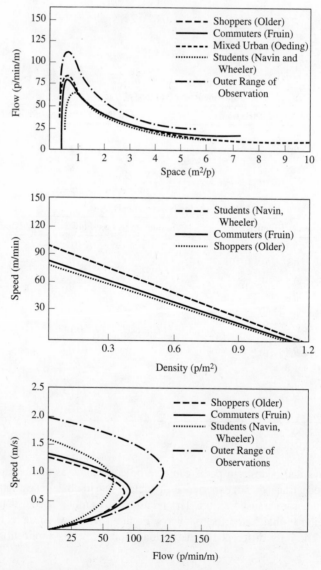

Figure 7.16   Speed, density, and flow relationships for pedestrian travel
| SOURCE: TRB, 2000

Similar to the analysis for freeways, flow rates for pedestrian facilities are adjusted for situations where geometric or operational conditions will result in reduced travel flow. This is the purpose of W_E in Eq. [7.33]. For analysis purposes, the total width of a walkway is reduced by an amount that represents the portion of a walkway that will not be used by pedestrians. Typical reductions are shown in Fig. 7.17. The level of service criteria for walkways and sidewalks are shown in Table 7.5. As for other facilities, the capacity flow rate will occur at the maximum flow rate value for LOS E, or 75 pedestrians/minute/meter. The *TRB 2000* presents different LOS criteria for situations that might alter pedestrian flow rates, such as shared pedestrian/bicycle operations, "bunching" of pedestrians as they move along a sidewalk, and pedestrian movements that cut across the main flow.

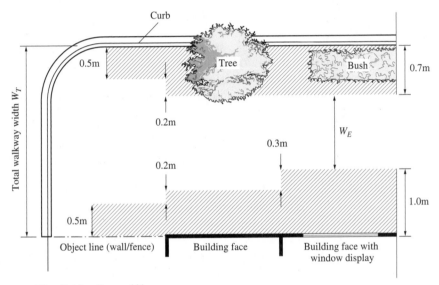

W_T = Total walkway width
W_E = Effective walkway width

Figure 7.17   Adjustments for walkway width
| SOURCE: TRB, 2000

Table 7.5   Level of service criteria for walkways and sidewalks

LOS	Space (m²/p)	Flow Rate (p/min/m)	Speed (m/s)	v/c Ratio
A	>5.6	≤16	>1.30	≤0.21
B	>3.7–5.6	>16–23	>1.27–1.30	>0.21–0.31
C	>2.2–3.7	>23–33	>1.22–1.27	>0.31–0.44
D	>1.4–2.2	>33–49	>1.14–1.22	>0.44–0.65
E	>0.75–1.4	>49–75	>0.75–1.14	>0.65–1.0
F	≤0.75	variable	≤0.75	variable

| SOURCE: TRB, 2000

Uncontrolled Access or Interrupted Flow Roads   The analysis of uncontrolled access facilities must take into account the movements of the vehicle or person using the facility and the corresponding interactions among the users that result in conflicts and reduced capacity. With a large variety of "vehicles" using the street (e.g., automobiles, trucks, buses, streetcars, motorcycles, scooters, bicycles, and pedestrians), the urban street is far more complicated to analyze than an urban freeway. The performance of urban streets with intersections will be strongly affected by the type of traffic control (e.g., signalization or signing) found at these points. One-way streets, parking restrictions, and pedestrian and bicycle movement will also affect street capacity.

The starting point for any model of urban street performance is that most of the vehicle delay occurs at intersections, not in midblock. For simple problems, the urban street can be thought of as a giant queuing system. Time–distance diagrams can be used for tracing the interaction between flows of vehicles and signal cycles and conflicting movements. Fluid-flow approximations are used when the departure process from one intersection defines the arrival process for the next intersection downstream. Simulation is used when the interactions among vehicles are so complex that simple mathematical relationships cannot replicate reality with any degree of accuracy.

The *HCM 2000* methodology for capacity and LOS analysis of urban streets is based on the estimate of average through-vehicle travel speed for the street. This analysis considers the different types of urban streets that would likely be found in a typical metropolitan area. For example, in some cases, many traffic signals will be located along a street segment, which could result in increased delays. Urban street LOS is thus characterized by urban street classification as shown in Table 7.6.

Average travel time on an urban street segment consists of two parts—running travel time and control delay at intersections. Running time can be influenced by numerous factors, ranging from parking on the side of the street to the number of driveways. The *HCM 2000* presents segment running time by urban street class, free-flow speed, and average segment length. The control delay at intersections is calculated with a series of equations reflecting arrival distribution, queue delays, the degree to which signals are coordinated, and other operational characteristics of intersections.

Stairways are another example of an interrupted flow facility. The TRB *Transit Capacity and Quality of Service Manual* describes the approaches used to calculate the LOS and capacity for such facilities. The basic concept is similar to that described earlier for pedestrian uninterrupted flow facilities. Interested readers are referred to this reference.

Transit Operations in Mixed Traffic   Two types of transit services operate in a mixed traffic environment—bus routes and light rail or streetcar operations on urban streets. Figure 7.18 shows the numerous factors that determine person-carrying capacity of bus routes. These factors reflect the technology characteristics of the transit vehicle such as vehicle type and size; the design of the infrastructure used by the vehicle such as loading area design or traffic signal timing; approaches for handling passengers such as fare payment; operational strategies such as skip-stop operations; and agency policies such as desired passenger loading standards. Each of these factors is incorporated into the determination of transit route capacity through calculation or by using look-up tables.

Table 7.6 Urban street level of service

Lanes	Service Volumes (veh/h)				
	A	B	C	D	E
			Class I		
1	N/A	740	920	1010	1110
2	N/A	1490	1780	1940	2120
3	N/A	2210	2580	2790	3040
4	N/A	2970	3440	3750	4060
			Class II		
1	N/A	N/A	620	820	860
2	N/A	N/A	1290	1590	1650
3	N/A	N/A	1920	2280	2370
4	N/A	N/A	2620	3070	3190
			Class III		
1	N/A	N/A	600	790	840
2	N/A	N/A	1250	1530	1610
3	N/A	N/A	1870	2220	2310
4	N/A	N/A	2580	2960	3080
			Class IV		
1	N/A	N/A	270	690	790
2	N/A	N/A	650	1440	1520
3	N/A	N/A	1070	2110	2180
4	N/A	N/A	1510	2820	2900

	Class			
	I	II	III	IV
Signal density	0.5 signal/km	2 sig/km	3 sig/km	6 sig/km
Free-flow speed	80 kph	65 kph	55 kph	45 kph
Cycle length	110 secs	90 secs	80 secs	70 secs
Effective green ratio	0.45	0.45	0.45	0.45
Adjusted sat. flow rate	1850	1800	1750	1700
Arrival type	3	4	4	5
Unit extension	3 secs	3 secs	3 secs	3 secs
Initial queue	0	0	0	0
Other delay	0	0	0	0
Peak hour factor	0.92	0.92	0.92	0.92
% lefts, % rights	10%	10%	10%	10%
Left turn bay	Yes	Yes	Yes	Yes

SOURCE: TRB, 2000

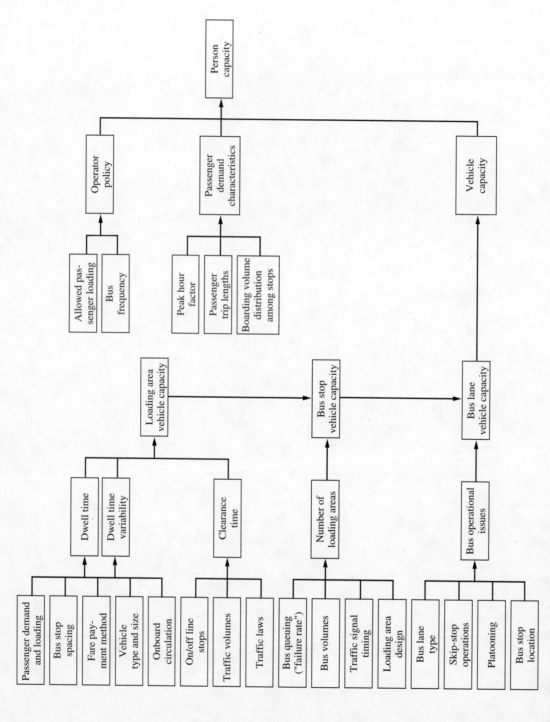

Figure 7.18   Factors influencing bus capacity
SOURCE: TRB, 1999

Vehicle capacity is usually calculated for three different locations along a route—loading areas at a station, bus stops, and bus lanes. Similar in concept to freeway capacity analysis, whichever one of these locations has the smallest estimated capacity is the controlling factor for the capacity of the entire route.

The maximum number of buses per *loading area* per hour is given by

$$B_{bb} = \frac{3600(g/C)}{t_c + (g/C)t_d + Z_\alpha c_v t_d} \qquad [7.34]$$

where

B_{bb} = maximum number of buses per loading area per hour
g/C = ratio of effective green time to total traffic signal cycle length
t_c = clearance time between successive buses, seconds
t_d = average (mean) dwell time, seconds
Z_α = one-tail normal variate corresponding to the probability that queues will not form behind the bus stop
c_v = coefficient of variation of dwell times

Note in this equation the variables c_v, Z_α, and g/C reflect the mixed traffic flow operating environment for buses. The coefficient of variation of dwell times (c_v), defined as the standard deviation of the dwell time distribution divided by the mean dwell time, ranges from 40 to 80 percent. The variable Z_α represents the probability that a queue of buses will not form behind a bus stop, known as failure rate.[3] This represents the uncertainty associated with traffic conditions along the route that could cause buses to "bunch" together. The value of Z_α for different failure rates is shown in Table 7.7. Capacity analysis occurs at a 25 percent failure rate. The g/C ratio reduces the capacity of the loading area recognizing that bus arrivals and departures could be delayed due to a nearby traffic signal. The average dwell time is calculated

Table 7.7 Values of $Z\alpha$ for capacity determination

Failure rate	$Z\alpha$
1.0%	2.330
2.5	1.960
5.0	1.645
7.5	1.440
10.0	1.280
15.0	1.040
20.0	0.840
25.0	0.675
30.0	0.525
50.0	0.000

SOURCE: TRB, 1999

[3] In statistical terms, $Z\alpha$ represents the area under one "tail" of a normal distribution beyond the acceptable levels of probability of a queue forming at a bus stop.

as before with Equation [7.25], and the clearance time (t_c) is the time it takes for a vehicle to leave a loading area or bus stop, usually between 10 and 15 seconds.

The capacity of *bus stops* is estimated as

$$B_s = N_{eb} \times B_{bb} = N_{eb} \times \frac{3600(g/C)}{t_c + (g/C)t_d + Z_\alpha c_v t_d} \qquad [7.35]$$

where

B_s = maximum number of buses per bus stop per hour
N_{eb} = number of effective loading areas

N_{eb} reflects the phenomenon that increasing the number of loading areas at a curb bus stop results in a nonproportional increase in bus-stop capacity due to vehicle interference. The values of N_{eb} are shown in Table 7.8 for different number of loading spaces. As seen, doubling the number of loading spaces from one to two results in only an 85 percent increase in capacity.

The capacity of *bus lanes* will vary by type of lane. Two lane types are of particular interest—exclusive arterial bus lanes and mixed traffic operations. The capacity of a non-skip-stop bus service on an arterial bus lane is calculated as:

$$B = B_{bb} \times N_{eb} \times f_r \qquad [7.36]$$

where

B = bus lane capacity, buses/hour
B_{bb} = bus loading area vehicle capacity at the critical bus stop
N_{eb} = number of effective loading areas at the critical bus stop
f_r = capacity adjustment factor for right turns at the critical bus stop

The value of f_r is estimated with the following equation:

$$f_r = 1 - f_l\,[v_r/c_r] \qquad [7.37]$$

where f_l is taken from Table 7.9 and v_r and c_r are the volume of right turns and the capacity of the right-turn lane at the specific intersection, respectively.

On-street light rail or streetcar operations can be treated similarly to bus operations. Thus, Eqs. [7.34] and [7.35] can be used to calculate line capacity. However, vehicles often turn in front of a train, resulting in delay. The following equation is used to determine minimum headway for trains operating on a street:

$$h_{os} = \max \left\{ \begin{array}{l} \dfrac{t_c + (g/C)t_d + Z_\alpha c_v t_d}{(g/C)} \\ 2C_{max} \end{array} \right\} \qquad [7.38]$$

where

h_{os} = minimum on-street section train headway(s)
g = effective green time(s), reflecting the reductive effects of on-street parking and pedestrian movements (mixed traffic operation only), as well as any impacts of traffic signal preemption
C = cycle length(s) at the stop with the highest dwell time
C_{max} = longest cycle length(s) in the line's on-street section
t_d = dwell time(s) at the critical stop

Table 7.8   Effective loading areas for bus stop capacity

Loading area #	On-Line Loading Areas		Off-Line Loading Areas	
	Efficiency Percent	# of Cumulative Effective Loading Areas	Efficiency Percent	# of Cumulative Effective Loading Areas
1	100	1.00	100	1.00
2	85	1.85	85	1.85
3	60	2.45	75	2.60
4	20	2.65	65	3.25
5	5	2.70	50	3.75

| SOURCE: TRB, 1999

Table 7.9   Bus-stop location factor, f_l, for bus lane type

Bus Stop Location	Type 1: Exclusive Lane Does Not Use Adjacent Lane	Type 2: Exclusive Lane Has Partial Use of Adjacent Lane	Type 3: Exclusive Lane Has Full Use of Adjacent Lane
Near side	1.0	0.9	0.0
Mid block	0.9	0.7	0.0
Far side	0.8	0.5	0.0

| SOURCE: TRB, 1999

t_c = clearance time between trains(s), defined as the sum of the minimum clear spacing between trains (typically 15–20 seconds or the signal cycle time) and the time for the cars of a train to clear a station (typically 5 seconds per car);

Z_α = one-tail normal variate corresponding to the probability that queues of trains will not form

c_v = coefficient of variation of dwell times (typically 40 percent for light rail, while streetcars running in mixed traffic have c_v values similar to buses)

Once the headway is calculated, capacity can be estimated as before.

7.3.3   Analysis of Multimodal Corridor Performance

Most transportation corridors consist of more than one facility or transportation service. Thus, the performance of the transportation corridor as perceived by users must include all of the modal facilities and services if a true picture is to be obtained of system performance. The *Highway Capacity Manual* defines a corridor as "a set of essentially parallel and competing facilities and modes with cross connectors that serve trips going between two designated points" [TRB, 2000]. Figure 7.19 shows how a typical corridor can be divided into its constituent components. The basic concepts in determining multimodal person and freight capacity include the following

- The person capacity of a link or node is calculated independently of freight capacity unless joint-use facilities exist. Then, a portion of joint-use facility capacity may be allocated to person or goods transport with each based on vehicle mix. A possible allocation is to assign all capacity to either person or goods movement.

- The one-directional capacity of a path is determined by the smallest capacity of any component link or node.

- The two-directional capacity of a link or corridor is equal to the capacity in one direction plus the capacity in the opposite direction.

- Assuming a corridor is composed of parallel paths, the total capacity of the corridor is equal to the sum of the capacities of each path [Cambridge Systematics et al., 1998].

Because freight and person trips often take place on different facilities or use the same facilities only at different times, capacity can be calculated separately for each.

The discussion so far has focused on corridor capacity. In most urban corridors, however, the opportunity for travelers to use alternate routes or to substitute one mode for another creates a challenge in determining corridor-level performance. The use of individual corridor facilities or services and the resulting level of service on each could vary from one day to another. However, the maximum allowable flow rate through the corridor, that is, the capacity, will stay the same until improvements are made. Determining multimodal corridor capacity consists of three steps [Cambridge Systematics et al., 1998].

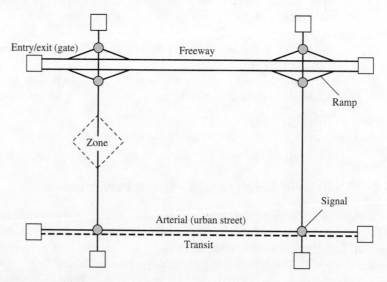

Figure 7.19   Constituent components of a corridor
I SOURCE: TRB, 2000

Step 1: Identify and characterize the links and nodes. Each modal network should be defined with mode-specific links. Even in those cases where a transportation facility is shared (e.g., a road that handles cars and transit buses), the road network and the transit network would each include the road link. In this case, the road network would define this link in road capacity, whereas it would be defined with transit capacity in the transit network. In some cases, a "composite mode" would be used on this link to represent the operational impact of both modes (e.g., a vehicle length that is the weighted average of all the vehicle types using the link).

Step 2: Calculate the modal capacity of each component. This step uses all of the estimation methods discussed in previous sections for the types of facilities in the corridor. Freight capacity methods are found in [Cambridge Systematics et al., 1998].

Step 3: Calculate the multimodal capacity of the corridor. The multimodal capacity of a transportation corridor is the sum of the capacities of the most constrained segment of all the individual facilities in the corridor. The following example illustrates this concept. A corridor is 10 kilometers long and has a highway, bicycle lane, and a light rail line. The two-way bicycle lane begins at kilometer 2 and goes to kilometer 7. The highway has three lanes in each direction from kilometer 0 to kilometer 5 and two lanes in each direction from kilometer 5 to kilometer 10. The light rail line is a dual-track operation running the entire length of the corridor. What is the total person-carrying capacity of the corridor? Figure 7.20 illustrates the physical components of the corridor. As shown, the maximum rate of person flow that can be handled in the corridor is limited by the segment from kilometer 7 to kilometer 10. Thus, the multimodal corridor capacity is 46,000 passengers/hour.

Multimodal capacity is useful information because it provides a sense of the mobility opportunities available in a corridor. However, it does not represent what travelers are likely to do. For example, in Fig. 7.20, if 46,000 persons per hour actually wanted to use this corridor, it is not likely that they would be modally distributed as suggested by the respective capacities. Measures that reflect actual corridor performance are thus of great interest to transportation planners. Several common corridor performance measures relate to the intensity, duration, extent, and variability of congestion and corridor accessibility. Readers are referred to [TRB, 2000] for a discussion on how these measures are estimated.

7.4 NETWORK MODELS

The preceding sections have focused on the performance of individual facilities and their components. Many analysis problems, however, occur at the level of the network as a whole. Such problems include (1) roadway/freeway/subarea network optimization; (2) coordination of signal cycle times and offsets over road network segments; (3) determination of equilibrium flows for every link in a network, given an origin–destination demand pattern; and (4) a range of logistical or scheduling problems. Each of these is discussed briefly in the following.

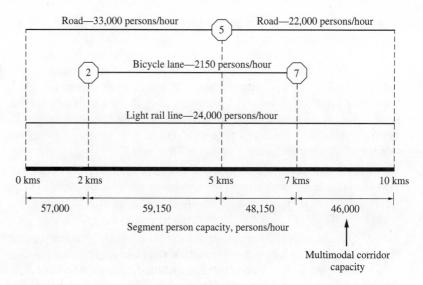

Figure 7.20   Illustration of multimodal capacity analysis

7.4.1   Roadway/Freeway/Subarea Network Optimization

Several large-scale simulation and optimization models can be used to estimate the impact of different transportation management strategies on such measures as total delay, total number of stops, and systemwide fuel consumption and emissions. The FHWA has supported the development of an integrated set of simulation models called TRAF that are used to analyze network operations [FHWA, 1999]. The TRAF system includes the following component models:

- NETSIM, a microscopic stochastic simulation model of urban traffic.
- FRESIM, a microscopic stochastic simulation model of freeway traffic.
- NETFLO (1 and 2), a macroscopic simulation model of urban traffic.
- FREFLO, a macroscopic simulation model of freeway traffic.

   The NETSIM and FRESIM models have been combined in a program called CORSIM (for corridor simulation) and the NETFLO and FREFLO models have been integrated into a program called CORFLO for macro corridor flow analysis. Both NETSIM and FRESIM are good examples of how such models work.

   NETSIM simulates vehicle movement by treating a vehicle as a distinct object and updating its position, speed, and acceleration at each time step. This is referred to as time-scan simulation. Up to nine vehicle classifications can be defined, such as auto, car pool, or truck, each having different operating and performance characteristics. Turn movements, free-flow speeds, and queue discharge headways are assigned stochastically. Traffic control systems such as traffic signals are also updated for each simulated time step. Vehicles are moved according to a car-following logic embedded in the simulation.

Similar to NETSIM, FRESIM treats each vehicle as a separate entity that interacts with the freeway geometry and other vehicles. Car-following procedures are incorporated into the simulation to guide vehicle paths through the network.

Table 7.10 shows the size limitations of the network characteristics for a CORSIM simulation. Of particular interest in this table are the definition of network components and thus the types of strategies or actions that can be analyzed. For example, FRESIM will assess the impacts of roadway incidents, the effectiveness of ramp metering, and the implications of traffic management strategies on bus operations. These types of issues are so interconnected with the operating environment that simulation models are the best tool for analyzing the resulting impacts.

Table 7.10   Size limitations of CORSIM network characteristics

Characteristic	NETSIM	FRESIM
Nodes	500	500
Links	1000	1000
Vehicles	20,000	20,000
Buses	256	200
Bus stations	99	X
Bus routes	100	100
Actuated traffic controllers	100	X
Detectors	300	300
Detector data stations	X	70
Events	200	X
Incidents	X	20
Disjointed freeway segments	X	20
Entries per segment	X	35
Through lanes	7	5
Auxiliary lanes	X	3
Through lanes per ramp link	X	3
Links per segment	X	200
Lane adds/drops per link	X	3
Detectors for a metered signal	X	10
Exits per segment	X	35
Ramp metering signals	X	150

SOURCE: [FHWA, 1999]

7.4.2   Signal Optimization

One of the earliest network model applications was determining the optimal settings for traffic signals in a road network. Given the interdependencies of one intersection's performance with those that are nearby, optimization models became a good method for developing the "best" traffic control strategy. As for the models described previously, these signal optimization models are designed to achieve user-defined objectives, such as minimize total user delay on the network. Models commonly used include TRANSYT-7F, PASSER II, and MAXBAND [FHWA, 1991]. Major characteristics of TRANSYT-7F, which are typical of this class of model, include the following:

1. It is a macroscopic, deterministic simulation model.

2. It represents vehicles in terms of platoon shapes (i.e., closely bunched groups of vehicles that change as vehicles proceed through the system).

3. The arterial is represented as a set of nodes (intersections) connected by a series of unidirectional links.

4. Model inputs include flows, free speeds, saturation flows, link lengths for every link in the system, and signal timing parameters at each node.

5. Model outputs by link include total travel time, total distance traveled, uniform and random delay encountered, number of stops made, maximum back of queue size, the degree of saturation, and fuel consumption.

The system is optimized by defining a performance index (PI) and employing a search procedure to find the set of decision variables that minimizes the PI (if the PI is to be minimized, the PI is then referred to as a disutility index or DI). The PI could be (1) a linear combination of delay and stops, fuel consumption, and maximum length of queue; (2) excess operating cost; or (3) a traffic progression function. For example, in version 7F of TRANSYT, one possible objective function is to minimize:

$$DI = \sum_{i=1}^{n} \left\{ \left(w_{di}d_i + Kw_{si}s_i\right) + U_i\left(w_{d(i-1)}d_{i-1} + Kw_{s(i-1)}s_{i-1}\right) + QB_i\left[w_q\left(q_i - c_i\right)^2\right] \right. \qquad \textbf{[7.39]}$$

where

DI = disutility index

d_i = delay on link i in vehicle hours

K = weight attached to stops relative to delay

s_i = stops on link i in stops/second

w_{xi} = link specific weighting factors for delay and stops

U_i = binary variable set to "1" if link-to-link weighting exists

Q = binary variable set to "1" if back of queue penalty in force

B_i = binary variable set to "1" if maximum back of queue exceeds storage

w_q = networkwide penalty applied to the excess queue spillover

q_i = computed maximum back of queue on link i

c_i = maximum back of queue capacity for link i

This objective function is typical of the types of performance measures found in such simulation models—delay, number of stops, and length of queue. Note also that the model user can assign weights that reflect the importance of one variable over another.

The decision variables that can be adjusted in order to minimize DI are the traffic signal splits (i.e., the fraction of the signal cycle allocated to the green light) and the signal offsets (i.e., the staggering of signal cycles as one moves from signal to signal along the route).

7.4.3 Network Flow Analysis

Network flow analysis problems include

1. Determining the maximum flow through a network with links of finite capacity.
2. Determining the equilibrium flows through a network, given a known set of origin–destination patterns.
3. Determining the shortest paths between nodes within the network.

Equilibrium flow assignment techniques have already been discussed in Sec. 5.4.4. Fundamental to virtually all network analyses, however, is the calculation of the shortest or minimum paths between selected points or nodes in the network. Equilibrium assignment techniques, for example, typically require the calculation of minimum paths from all origin centroids to all destination centroids at each iteration of the algorithm. Thus, the need exists for efficient algorithms that can quickly compute shortest paths for large numbers of origins and destinations. One example of such an algorithm follows.

Given the following notation:

N = set of all nodes in the network
M = set of labeled nodes (defined in the following)
$\overline{M}$ = set of the unlabeled nodes $(M \cup \overline{M} = N)$
n_i = ith node
$c(n_i, n_j)$ = "cost" (travel time, distance, etc.) of traveling on link from n_i to n_j; $[c(n_i, n_j) \geq 0$ for all $i, j]$
$C(n_i)$ = total cost of travel between "home" node (defined in the following) and node n_i

A typical shortest path algorithm can be defined by the following steps:

1. Set $k = 1$. Choose a node as a home node (i.e., the origin from which minimum paths are to be computed). Label this node $[- , 0]$, where the minus sign indicates that there is no predecessor node, and the zero indicates that there is zero cost associated with reaching this node. Set M equal to $\{n_1\}$, where n_1 is equal to the node number for the chosen home node.
2. Set $k = 2$. Find an unlabeled node n_2 such that $c(n_1, n_2)$ is a minimum. Assign n_2 the label $[n_1, c(n_1, n_2)]$. That is, n_1 is its predecessor node (i.e., it is linked to n_1), and the cost of reaching n_2 from the home node is $c(n_1, n_2)$. Add n_2 to the set of labeled nodes (i.e., M now equals $\{n_1, n_2\}$).
3. Set $k = k + 1$
4. Find an unlabeled node n_k, such that

$$C(n_k) = \min_{n_x \in M} \left\{ C(n_x) + \min_{n_y \in \overline{M}} \left[(n_x, n_y) \right] \right\}$$

$$= C(n_i) + c(n_i, n_k) \text{ for some } n_i \in M, n_k \in \overline{M}$$

Assign n_k the label $[n_i, C(n_k)]$ and add n_k to M. Eq. [7.39] searches over all currently labeled nodes, that is, nodes already connected to the home node that possess links with unlabeled nodes (i.e., nodes that are not yet connected to the home node). It seeks a new node to connect to the home node (i.e., to add to the labeled set), such that the path to the new node has the lowest cost of travel from the home node.

5. If all nodes have been labeled, that is, all nodes have been connected to the home node, stop. Otherwise go to step 3.

The final product of this algorithm is a *shortest path tree* rooted at node n_1. It is a tree that identifies the minimum paths between the home node n_1 and all other nodes in the network. Figure 7.21 illustrates the use of this algorithm for the case of a simple six-node network.

Determining the shortest path through a network depends on how "cost" is defined. As noted in Chap. 5, a generalized cost function can be used to represent all of the factors that influence a traveler's decision. One of the critical factors is trip travel time, or from a network perspective, the sum of individual link times over a vehicle's path. A speed-flow curve developed by the predecessor agency to the FHWA, the Bureau of Public Roads (BPR), has been used in many models to determine link travel speeds and thus link travel time. These "BPR curves," which have been incorporated into models as link performance functions, are of the form

$$s = \frac{s_f}{1 + a(v/c)^b} \qquad [7.40]$$

where

s = predicted mean speed
s_f = free-flow speed
v = volume
c = capacity
a and b = parameters defined through model calibration

Recent research has recommended the following values of a and b [Dowling et al., 1997]:

$$a\begin{cases} = 0.05 \text{ for signalized facilities} \\ = 0.20 \text{ for all other facilities} \end{cases} \qquad [7.41]$$

$$b = 10$$

The free-flow speed can be estimated with a variety of techniques. Posted speed limits, look-up tables, *HCM* free-flow speed methods, and speed counts have all been used to determine values for s_f. Figure 7.22 shows the updated BPR curve compared to the original BPR curve and to computer simulation results of a freeway and arterial in California. As noted in the research report that recommended the updated BPR curve, once accurate estimates of free-flow speed and capacity are known, "the updated BPR curve can estimate speeds for both arterials and freeways with accuracy approaching those of the HCM and simulation models" [Dowling et al., 1997].

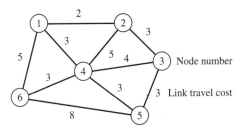

Objective: Build a shortest path tree from node 6.

Step 1: $n_1 = 6$ Label = (–,0)
          $M = \{6\}$ $\bar{M} = \{1,2,3,4,5\}$

Step 2: $n_2 = 4$ Label = (6,3)
          $M = \{6, 4\}$ $\bar{M} = \{1,2,3,5\}$

Step 3: $C(6) + c(6,1) = 5$ $C(4) + c(4,1) = 3 + 3 = 6$
          $C(6) + c(6,5) = 8$ $C(4) + c(4,2) = 3 + 5 = 8$
                                             $C(4) + c(4,3) = 3 + 4 = 7$
                                             $C(4) + c(4,5) = 3 + 3 = 6$
          $n_3 = 1$ Label = (6,5)
          $M = \{6,4,1\}$ $\bar{M} = \{2,3,5\}$

Step 4: $C(6) + c(6,5) = 8$ $C(4) + c(4,2) = 8$ $C(1) + c(1,2) = 5 + 2 = 7$
                                             $C(4) + c(4,3) = 7$
                                             $C(4) + c(4,5) = 6$
          $n_4 = 5$ Label (4,6)
          $M = \{6,4,1,5\}$ $\bar{M} = \{2,3\}$

Step 5: $C(4) + c(4,2) = 8$ $C(1) + c(1,2) = 7$ $C(5) + c(5,3) = 6 + 3 = 9$
          $C(4) + c(4,3) = 7$
          $n_5 = 3$ Label = (4,7)
          $M = \{6,4,1,5,3\}$ $\bar{M} = \{2\}$

Step 6: $C(4) + c(4,2) = 8$ $C(1) + c(1,2) = 7$
          $n_6 = 2$ Label = (1,7)
          $M = \{6,4,1,5,3,2\}$ $\bar{M} = \{\phi\}$

Result: The shortest path tree is

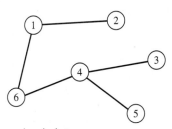

Figure 7.21   Example of shortest path calculation

7.4.4   Out-of-Vehicle Performance Measures

Demand modeling focuses on the entire trip, from origin to destination. This means
that every component of the trip, including the time for the line-haul movement as

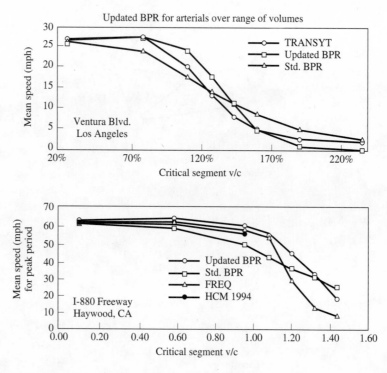

Figure 7.22 BPR curve compared to other speed-flow curves
| SOURCE: Dowling et al., 1997

well as the access and egress time and the existence of a transfer, become a critical consideration. In fact, studies have indicated that access and waiting times and a transfer are considered much more onerous to a traveler than the time in transit. For example, a study in Boston indicated that the existence of a transfer on a transit trip was considered equivalent by transit riders to another 12 to 15 minutes of transit in-vehicle time [CTPS, 1997]. The most important out-of-vehicle performance measures are access and egress *walk* time; *wait* times associated with vehicle arrivals; and *transfer* times if one or more transfers are needed between vehicles or routes. For modeling automobile trips, the only out-of-vehicle measure considered is the walk time, which occurs at the destination end of the trip. Access or "home-based" walk times, for example walking from the doorstep to the garage, are generally assumed to be negligibly small. Walk, wait, and transfer times, however, can be important variables in modeling transit trips.

Walk Time Estimating expected walk time is critical to the determination of expected transit ridership because it defines the likely market for transit users. That is, riders are assumed willing to walk for only so long to reach the transit station or a destination. The industry rule of thumb has been the amount of time it takes to walk a distance of 1/4 mile to or from a transit stop. However, studies in Boston and Atlanta have shown that transit users are willing to walk longer distances. In Boston, for example, over 40 percent of transit trips involved a walk of more than 1/4 mile, with over 10 percent having a walk of more than 1/2 mile [CTPS, 1997]. In Atlanta, between 35 and 40 percent of the subway riders walked more than 1/4 mile from the station [Shurbajji, 1993].

The typical walk analysis problem estimates average zonal walk times, either between two randomly located points within a zone or to a specified point in the zone, such as a shopping center. If trips are assumed to occur uniformly over the zone, then

1. The expected walk time between two randomly located points in a rectangular zone of dimensions x and y (see Fig. 7.23a) in which people must walk in directions parallel to the zone sides (i.e., "right-angled" or "Manhattan" distances corresponding to paths through a rectangular grid network) is

$$E(WT) = \frac{x + y}{3w} \qquad \text{[7.42]}$$

where $E(WT)$ is expected walk time and w is walking speed.

2. The expected right-angle walk time between a randomly located point within a rectangular zone and a corner of the zone is (see Fig. 7.23b)

$$E(WT) = \frac{x + y}{2w} \qquad \text{[7.43]}$$

3. The expected right-angle walk time between a randomly located point and a specified point within a rectangular zone is (see Fig. 7.23c)

$$E(WT) = \frac{\left[\Sigma (x_i + y_i) a_i\right]}{2A \times w} \qquad \text{[7.44]}$$

where

$x_i, y_i =$ dimensions of ith zone section, $i = 1, \ldots, m$
$a_i \quad =$ area of ith zone section $= x_i \times y_i$
$A \quad =$ area of zone $= \Sigma a_i$

Similar results can be obtained for circular zones or zones of more complex shapes and for nonuniform probability distributions, although the computational complexity involved typically increases considerably. Ultimately, a point is reached where the analysis is no longer analytically tractable, and alternative techniques such as simulation must be employed.

Geographic information systems (GIS) have made such calculations much easier. With spatially located employment and housing sites and with household-specific information on travel patterns, the GIS can calculate average walking times from any location in a zone to any other. Figure 7.24 shows the calculation of walking distances for prospective light rail stations in Portland, Oregon. By analyzing the typical distances people are willing to walk, planners can estimate the number of destinations that could be served by transit.

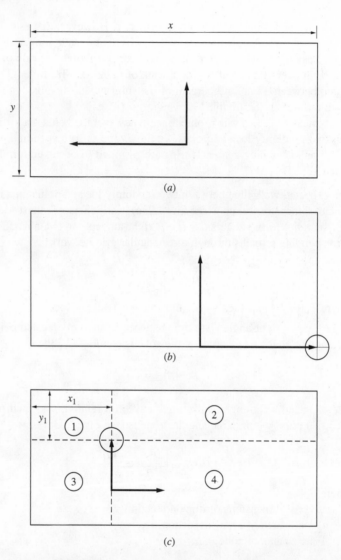

Figure 7.23   Average walk time calculations: (a) average distance between two random points; (b) average distance to a zone corner; (c) average distance to an interior point

Wait Time   Passenger wait times at transit stops are a function of the scheduled headway of the transit service, schedule adherence, and the extent of the passengers' knowledge of the service schedule. For simple analyses, it is generally assumed that

1. Passengers arrive randomly at the transit stop, with a uniform arrival rate for services with relatively short headways (e.g., 15 minutes or less).

2. Transit vehicle arrivals are exactly on schedule, as determined by their design headway.

Walk Area Analysis
I-5/Portland Blvd.
Five Minute Walk

Figure 7.24 Example of a GIS application for determining walk times and transit market area

| SOURCE: Portland METRO, 1995

3. Passengers will have some information about the scheduled vehicle arrival times and plan their own arrivals accordingly for services with relatively long headways (e.g., greater than 15 minutes).

If a vehicle cannot accommodate the number of passengers waiting to board, some passengers will have to wait for the next vehicle, thus increasing their expected delay. In such cases, it becomes necessary to treat the phenomenon as a more complicated queuing system. Passenger arrivals are assumed to follow some underlying probability distribution, bus arrivals are a function of scheduled headways and stochastic roadway conditions, and waiting times at any given stop are dependent upon the "upstream" stops on the route (e.g., heavy passenger ons or offs at an upstream stop will generally increase the delays experienced by passengers waiting at downstream stops). While queuing theory or fluid-flow approximations might be applied to simple formulations of this problem, simulation is the only feasible approach to the problem in its general form.

The traditional method of estimating average wait time is to assume that it equals one-half the headway. Given the assumption that vehicles arrive on schedule and given the likelihood of this not happening in mixed traffic flow during peak hours, some planning efforts have assumed a higher proportion of the scheduled headway (60 to 70 percent is used in Seattle) [CTPS, 1997].

Transfer time   Transfers between routes can be treated in a manner similar to wait times if it can be assumed that the arrival times of the passengers on the first route and the vehicle on the second route to which they are transferring are randomly related to each other. In such cases, the expected transfer time can be taken as half the headway of the second route. If coordinated transfers are part of the schedule, the scheduled amount of time for the transfer can generally be used as the average transfer time. As with wait times, if more detailed analysis is needed, methods such as simulation are likely to be required.

From a user's perspective, transfer wait time is considered much more negatively than the initial wait to catch a bus or train. The Boston study mentioned previously determined that the user perceives the transfer wait as being between 1.5 and 1.8 times more onerous than the initial wait time and between two and three times more onerous than in-vehicle time.

7.4.5   Representing Access In a Transportation Network

Access to the transportation system is a critical component of a traveler's perception of how desirable one mode is over another. As previously mentioned, out-of-vehicle travel time (i.e., wait and transfer times), as well as the existence of a transfer itself, become important determinants of mode choice. The disutility associated with these trip components is incorporated into the demand model specification through variables and weighting factors. Several methods have been used to represent access times and costs in the zonal system and network model [SG Assocs., 1998].

The simplest approach is to provide one or two centroid connector links for each zone that is "connected" to a transit network (note: automobile access is already accounted for by the definition of the model's road network). The performance function or access impedance for this link would be the amount of time it takes on average to walk or bicycle to the station from the zone. Another approach is to divide the traffic analysis zone into "submarkets" that each represent homogeneous access characteristics (Fig. 7.25). However, in typical urban situations, multiple means of accessing a transit station are available, and given that demand models predict trips by mode, these multiple paths must be represented in the network.

Figure 7.26 shows how a multiple path scenario can be modeled. This example of network coding from Sacramento is particularly complex in that not only was walk-to-rail a feasible option (represented by the links to the dummy node 714–803 and 713–803), but other access markets had to be represented as well. These included drive to park-and-ride lot and walk to the rail station (e.g., link path 5355–903–703), walk to bus and transfer to rail for which the option of walking directly to the station existed (e.g., link path 714–5344–5366–703), and transferring from a bus to rail for which the option of walking directly to the station did not exist (e.g., link path 5399–803–703). As noted previously, the distance people are assumed to walk to take transit is a significant variable. If evidence suggests that transit riders are willing to walk longer distances for rail service than for bus, a phenomenon that was observed

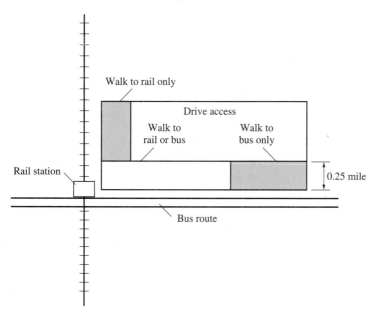

Figure 7.25   Zone with multiple access markets
I SOURCE: [SG Assocs., 1998]

in the Boston and Atlanta studies mentioned previously, then the market "overlap," the portion of a zone that has the potential for walk to bus or rail as shown in Fig. 7.25, is large. This makes the ridership prediction effort more complex and places great emphasis on network and zone definition.

7.5   IMPACT MODELS

Determining the impacts of transportation system performance is another important role for supply analysis. Analysis methods of varying scales of application and degrees of sophistication have been developed in particular for air quality, noise, and fuel consumption impacts. Each of these impacts will be discussed in the following sections.

7.5.1   Air-Quality Impact

The relationship between air quality and transportation system performance has been a concern of transportation planners for many years. Models have been developed that estimate the levels of pollutant emissions resulting from different system or facility performance characteristics. One set of models converts information on driving conditions, vehicle and driver behavior, and environmental factors into estimates of motor vehicle emissions. Known as emission models, these models can be

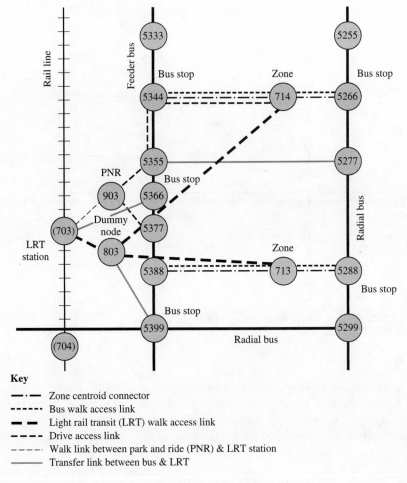

Figure 7.26   Example of transit network coding for access modes, Sacramento
I SOURCE: SG Assocs., 1998

used with facility performance models to estimate the total level of pollutant emissions associated with facility performance. Emission models are based on relationships between vehicle activities and vehicle emissions. For example, the following vehicle activities result in emissions that must be accounted for in a model:

Vehicle Activity	Type of Emission
Vehicle miles traveled	Emissions during movement (running exhaust)
	Emissions from evaporation during movement (running evaporative)
Cold engine starts	Higher running exhaust at beginning of trip
Warm engine starts	Higher running exhaust following start

Vehicle Activity	Type of Emission
Vehicle sitting after use (hot soaks)	Evaporative
Engine idling	Running exhaust and elevated evaporative
Exposure to temperature cycles	Evaporative
Vehicle refueling	Evaporative
Changing engine behavior (modal behavior)	Elevated running exhaust

Several factors can influence vehicle emission rates (Table 7.11). In order to account for these factors, a general modeling approach has been developed that (1) tests representative vehicles on a standard federal test procedure (FTP), which reflects the different speeds and acceleration rates typically found in an urban environment; (2) develops baseline emission rates (grams/mile or grams/hour) for various vehicle groups on the FTP; and (3) adjusts these baseline emission rates for conditions that differ from the FTP [Guensler, 1997].

The U.S. EPA has developed a computer model called MOBILE to estimate motor vehicle emissions in the United States. This model computes HC, CO, and NO_x emissions for different vehicle types in three major geographic regions of the United States (low altitude, high altitude, and California) [U.S. Environmental Protection Agency, 1994]. A planner using this model is able to input data on vehicle mix in the traffic flow, annual mileage rates and vehicle registration distributions for each vehicle type, basic emission rates for vehicle types, the existence of a vehicle inspection and/or maintenance program, average trips and miles traveled per day for each vehicle type, fuel type, and specific characteristics of the region in which the highway facility is located. Fig. 7.27a shows how the MOBILE model can be used to predict air quality.

Table 7.11 Factors affecting vehicle emission rates

Vehicle parameters	Vehicle classification
	Model and year (weight, engine size, etc.)
	Accrued vehicle mileage
	Fuel delivery system
	Emission control system
	Onboard computer control system
	Control system tampering
	Inspection and maintenance history
Fuel parameters	Fuel type
	Oxygen content of fuel
	Fuel volatility
	Sulfur content (continued)

Table 7.11   Factors affecting vehicle emission rates (continued)

Fuel parameters (continued)	Benzene content
	Olefin and aromatic content
	Lead and metals content
	Trace sulfur-catalyst effect
Environmental factors	Altitude
	Humidity
	Ambient temperature
	Diurnal temperature range
	Road grade
Vehicle operating conditions	Cold and hot start engine mode
	Average vehicle speed
	Engine modal activities—enriched conditions
	Load (e.g., towing and air conditioning)
	Trip length and number of trips per day
	Driver behavior

SOURCE: Guensler, 1997

In order to estimate the total emissions in an urban area, known as an emissions inventory, a model like MOBILE is used to determine emissions rates (Fig. 7.28). These rates are then multiplied by the corresponding level of activity (e.g., vehicle miles traveled) that are predicted by a travel demand model. Put simply,

Activity-specific emissions rates × Emission-producing vehicle activities
= Emissions inventory

A second category of models, called dispersion models, uses data on emissions, meteorological conditions, and topographic characteristics to compute the dispersion of pollutants in the atmosphere. The model then predicts the concentrations of pollutants at sensitive receptor locations over specified time periods. Dispersion models are more complex than emission models in that they must account for the transport of pollutant emissions over some distance from the source. Several models have been developed based on alternative theories of how this transport phenomenon occurs. One of the earliest models assumed that pollutant concentrations are uniform with a rectangular "box" defining the air-quality space over an emissions source. The steady-state concentration of pollutants is then related to emission rates and to other factors affecting the concentration of pollutants within this space.

The most common dispersion modeling approach is called Gaussian plume modeling. This approach represents the concentration of pollutants as a function of the emission rate, wind speed and direction, length of line source, and coefficients that reflect the assumed dispersion process that is based on the Gaussian or normal distribution function. For a more detailed discussion of this modeling approach, see Rau and Wooten [1980] and Horowitz [1982].

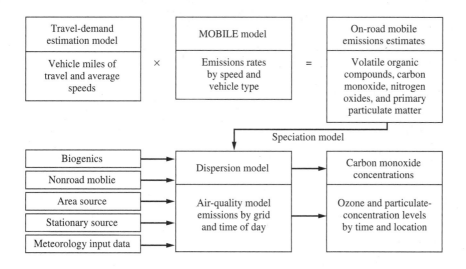

(a) Air-quality estimates

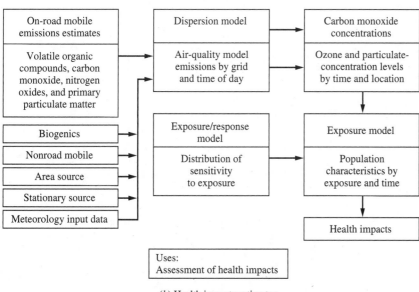

(b) Health impacts estimates

Figure 7.27 Uses of MOBILE in air-quality planning

SOURCE: NRC, 2000

```
MOBILE5a (26-Mar-93)
VOC HC emission factors include all evaporative HC emission factors, except for refueling emissions.

Emission factors are as of July 1st of the indicated calendar year.
Cal. Year: 1990    I/M Program: No         Ambient Temp: 75.0 / 75.0 / 75.0  (F)    Region: low
                   Anti-tam. Program: No   Operating Mode: 20.6 / 27.3 / 20.6       Altitude: 500. ft.
                   Reformulated Gas: Yes   ASTM Class: C
Metropolis 1990                            Minimum Temp: 60. (F)                     Maximum Temp: 90. (F)
       Period 1 RVP: 11.5                  Period 2 RVP: 8.7
  Ether Blend Market Share: 0.650       Alcohol Blend Market Share: 0.350
  Ether Blend Oxygen Content: 0.020     Alcohol Blend Oxygen Content: 0.025
                                         Alcohol Blend RVP Waiver: Yes
```

Veh. Type:	LDGV	LDGT1	LDGT2	LDGT	HDGV	LDDV	LDDT	HDDV	MC	All Veh
Veh. Speeds:	19.6	19.6	19.6	19.6	19.6	19.6	19.6	19.6	19.6	
VMT Mix:	0.653	0.164	0.082	0.031	0.008	0.002	0.053	0.008	0.008	
Composite Emission Factors(gm/Mile)										
VOC HC:	4.54	5.25	7.83	6.12	15.87	0.73	1.08	3.30	7.27	5.191
Exhaust HC:	2.16	2.83	4.21	3.29	7.36	0.73	1.08	3.30	2.26	2.644
Evaporat HC:	1.13	1.42	1.95	1.60	5.56	4.66	1.338			
Refuel L HC:	0.00	0.00	0.00	0.00	0.00	0.000				
Runing L HC:	1.17	0.93	1.61	1.16	2.83	1.136				
Rsting L HC:	0.08	0.07	0.07	0.07	0.12	0.074				
Exhaust CO:	26.58	31.48	42.71	35.23	114.25	1.68	2.03	13.71	17.71	30.392
Exhaust NOX:	1.83	2.13	2.82	2.36	6.29	1.65	1.97	20.96	0.85	3.092

```
Evaporative Emissions by Component        Weathered RVP: 11.6        Hot Soak Temp: 75.0  (F)
(Hot Soak: g/trip, Diurnals: g, Crankcase: g, Refuel: g/gal, Resting: g/hr)  Running Loss Temp: 75.0  (F)
                                                                             Resting Loss Temp: 75.0  (F)
```

	LDGV	LDGT1	LDGT2	LDGT	HDGV	MC
Hot soak	4.14	4.48	6.71	5.19	12.67	8.03
WtDiurnal	17.11	23.97	35.96	27.76	90.19	35.67
Multiple	35.31	46.93	57.86	50.39	91.59	
Crankcase	0.02	0.04	0.15	0.08	0.37	0.00
Refuel	0.00	0.00	0.00	0.00	0.00	
Resting	0.09	0.08	0.09	0.08	0.12	0.14

Figure 7.28 Example of MOBILE output

CALINE-4 is a good example of a dispersion model. As shown in Fig. 7.29, input variables include *site characteristics* like wind speed (u), wind direction relative to the y axis (BRG), a measure of atmospheric stability ($CLAS$), the mixing height of pollutant concentrations ($MIXH$), the averaging time for concentration formation ($ATIM$), a measure of roughness of the surrounding land surface (ZO), the velocity of deposition and settling (VD) and (VS), and ambient air quality; *link characteristics* such as coordinates, use (vehicles per hour), emission factor (EF), height (H), width (W), and type; and the *location of sensitive receptors* where pollutant concentrations are to be estimated. CALINE-4 can also be used to estimate carbon monoxide concentrations from queued vehicles at a series of intersections. Given these inputs, the model produces an estimate of the pollutant concentrations in parts per million.

In using both emission and dispersion models, planners often generate worst case scenarios consisting of the highest levels of traffic volumes, worst meteorological conditions, and in the case of development impact studies, the largest amount of development (the build-out scenario). Air-quality impacts are estimated for several target years—the current year or the latest year for which data are available, the year in which the project is to open, and a future year in which maximum utilization is expected.

Fig. 7.27b shows the next level of application of MOBILE-type models, assessing the health impacts of mobile emissions. This relationship, very seldom modeled as part of the transportation planning process, is a good example of the distinction between performance measurement of output versus outcome discussed in Chap. 4. The emissions output leads to health-related outcomes. Presumably, public officials would want to know such impacts as part of their consideration of transportation investment options, especially for those actions that will have a significant impact on travel behavior, such as metropolitan transportation investment programs. In addition, as specified in Fig. 7.27, there is a direct relationship between emissions modeling and transportation modeling. Fig. 7.30 shows the "scale" effect of this relationship. At the regional level, current models use aggregate measures such as average speed and regional VMT as the activity factors and have a high level of temporal aggregation. At the vehicle level, the data become more disaggregate and can be obtained much more frequently, but often at greater cost.

Computer models like MOBILE and CALINE provide planners with tools that can quickly estimate the impact of transportation facility performance on air quality over a period of years. Although such models represent the current approach to air-quality impact modeling, advancements in computer technology and an increased understanding of the underlying phenomena of air pollution will likely result in the development of more accurate models in the future.

Current research on motor vehicle emissions, for example, has focused on the different pollutant-emitting characteristics of engine operation. As shown in Fig. 7.31, emission levels vary by what the engine is doing or its mode of operation. Starting the car after it has been sitting for more than 1 hour results in a substantial release of emissions (2 to 4 grams of HC, 1 to 3 grams of NO_x, and 30 to 50 grams of CO). For a typical 20-mile (32.8 kilometer) commute trip in a 1994 vehicle, about 30 percent of the CO and 10 percent of the NO_x and HC for the entire trip is

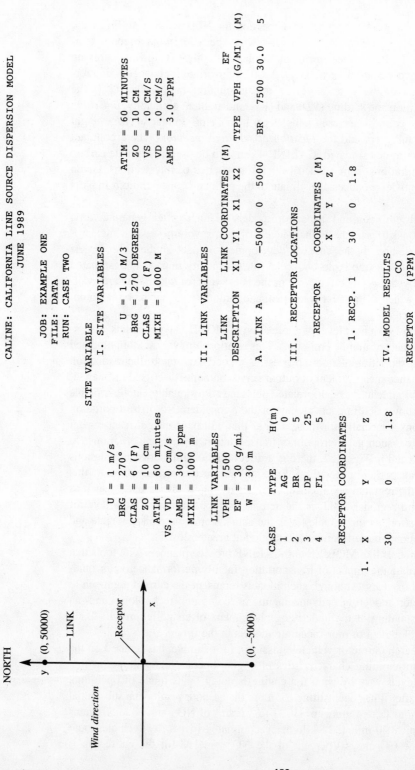

Figure 7.29 Example of pollutant dispersion model—CALINE-4

452

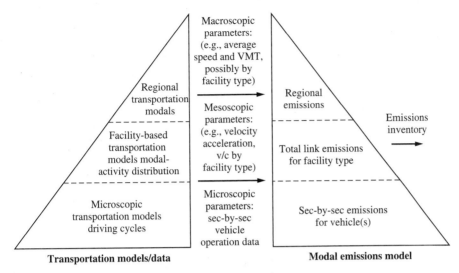

Figure 7.30   Transportation and emissions model interface
| SOURCE: NRC, 2000

associated with this cold start [Guensler, 2000]. Figure 7.31 also shows emission spikes that are attributed to vehicle acceleration or heavy power load on the engine. These spikes, so-called enrichment events, occur when more fuel is provided to the engine to provide power, and this fuel is not completely combusted. CO emission rates (grams/second) can be as high as 2500 times greater in an enrichment condition as compared to normal operation; HC can be higher by a factor of 100. New computer models that take this phenomenon into account are being developed to better estimate vehicle emissions under a variety of conditions. Even with this capability at the individual vehicle level, most modeling approaches group similar types of vehicles based on their class, technology, and model year. At the most aggregate level, vehicles could be classified as either passenger cars or trucks [NRC, 2000].

Air-quality analysis has recently been combined with geographic information systems to provide a very powerful tool for illustrating the relationships between pollution and transportation. Figure 7.32, for example, shows the level of CO emissions for different time periods in Atlanta. With such tools, planners can determine both the temporal and spatial characteristics of the transportation-related air-quality problem [Bachman, 1998].

7.5.2   Noise Impact

Traffic noise is measured in sound level units known as decibels, weighted with a standard "A" filter (dBA). The dBA measure has frequency response characteristics that correlate to human impressions of loudness. The A filter reduces the intensity of low and very high noise frequencies at which the human ear has lower sensitivity. It also gives additional weight to the more annoying high frequencies in the frequency spectrum.

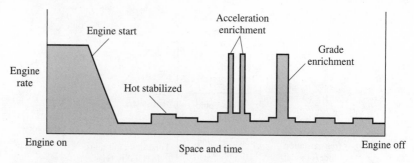

Figure 7.31   Vehicle emissions as a function of engine mode
| SOURCE: Bachman, 1998

A widely used measure of noise is the percent of time certain noise levels are exceeded during a specified time interval. Common levels include

L_{90} = noise level exceeded 90 percent of the time
L_{50} = noise level exceeded 50 percent of the time
L_{10} = noise level exceeded 10 percent of the time
L_{dn} = noise level averaged over 24 hours
L_{max} = highest sound level measured during a given time period

The most common measure is called the *equivalent sound level,* denoted by L_{eq}, which represents the average energy level reaching an observer during a specified period of time. A key advantage of L_{eq} over other measures is that it is not as dependent on the characteristics of traffic flow such as peaking factors and the assumption of uniform vehicle spacing. It also provides good estimates of noise levels in low volume situations, whereas the L_{90}, L_{50}, L_{10} measures can be very sensitive to traffic flow characteristics.

The methods for estimating noise impact range from look-up tables to simulation models. A noise model developed by the FHWA is the most common method for noise analysis in the United States. This model predicts noise levels by making adjustments to a reference energy mean noise emission level for different vehicle types. The adjustments represent differences due to flow characteristics (e.g., volume and speed), distance between the roadway and receiver, the length of roadway, and the effect of noise shielding or ground effects. When the distance from the centerline of traffic to the observer is greater than 15 meters, the equation for this method is [Barry and Reagan, 1978; Bowlby, 1981]

$$L_{eq}(h)_i = \left(L_o\right)_{Ei} \qquad \text{reference energy mean emission level} \qquad \textbf{[7.45]}$$

$$+ 10\log_{10}\left(\frac{N_i \times \pi \times D_0}{S_i \times T}\right) \qquad \text{traffic-flow adjustment}$$

$$+ 10\log_{10}\left[\frac{D_0}{D}\right]^{1+\alpha} \qquad \text{distance adjustment}$$

$$+ \Delta_s - 13 \qquad \text{shielding adjustment}$$

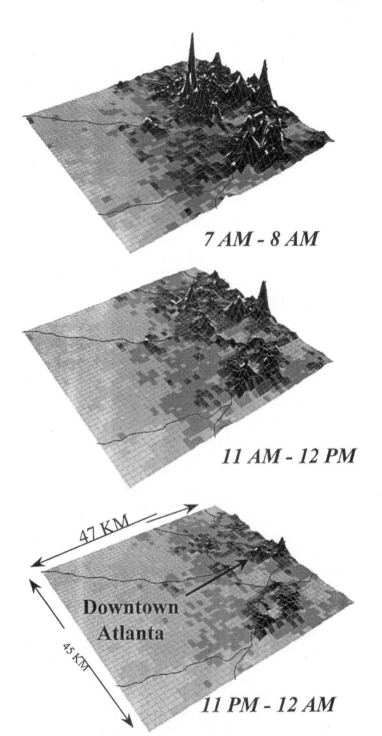

7 AM - 8 AM

11 AM - 12 PM

47 KM

**Downtown
Atlanta**

45 KM

11 PM - 12 AM

Figure 7.32 CO emissions spatially and temporally allocated in Atlanta
I SOURCE: Bachman, 1998

where

$L_{eq}(h)_i$ = hourly equivalent sound level for the ith vehicle type

L_{oEi} = reference energy mean emission level for vehicle type i

N_i = number of class i vehicles passing a specified point during time T

S_i = average speed for the ith vehicle class, kilometers per hour

T = period for which L_{eq} is desired, hours

D = receptor location, meters from centerline of traffic lane

α = site condition parameter

Ψ = an adjustment for finite length roadways

Δ_s = attenuation provided by shielding such as barriers, rows of houses, or densely wooded area, dBA

Each component of Eq. [7.45] will be discussed in the following. More detailed information on each measure can be found in [FHWA, 1995, 1997, 1998; Lee et al., 1996; Lee et al., 1998; Cohn et al., 1999].

Three classes of vehicles are incorporated into the FHWA model: auto (A), medium truck (MT), and heavy truck (HT). The reference energy mean emission level for each vehicle class is represented by the following equations:

$$(L_o)_{EA} = 38.1 \log_{10}(S_A) - 2.4 \qquad \text{automobiles} \qquad \textbf{[7.46]}$$

$$(L_o)_{EMT} = 33.9 \log_{10}(S_{MT}) + 16.4 \qquad \text{medium trucks} \qquad \textbf{[7.47]}$$

$$(L_o)_{EHT} = 24.6 \log_{10}(S_{HT}) + 38.5 \qquad \text{heavy trucks} \qquad \textbf{[7.48]}$$

where S is the average vehicle speed in kilometers per hour of each vehicle type.

Because vehicle characteristics can vary from one region to another, Eqs. [7.46] to [7.48] might not be appropriate for all areas. Also, these equations are based on assumed cruise conditions between 50 and 100 kilometers per hour on a flat roadway. Local areas might therefore need to develop their own reference levels.

The first adjustment to $(L_o)_{Ei}$ reflects the fact that noise impacts are caused by traffic flows, rather than from a single vehicle. Taking into consideration constants, the second term in Eq. [7.45] can be simplified to $10 \log_{10}(N_i D_o/S_i) - 25$. From this equation, it can be seen that if the total number of vehicles (N_i) is held constant, the adjustment factor results in a 3 dBA reduction per doubling of the speed. Likewise, holding average speed constant and doubling the volume produces a 3 dBA increase.

The second adjustment to $(L_o)_{Ei}$ takes into account the impact of measuring noise at distances of greater than 15 meters. This term also considers the acoustic nature of the surface between the roadway and the observer. When the surface is such that sound is reflected, the surface is called "hard," and α in the adjustment term is equal to zero. With such a surface, a doubling of distance results in a decrease of 3 dBA. When the ground is absorptive or "soft," α is approximately equal to 0.5 and the drop off in dBA's is 4.5 dBA per doubling of distance.

The third adjustment to $(L_o)_{Ei}$ recognizes that it is often necessary to divide a road into segments of finite length to account for changes in topography, traffic flows, and area exposed to the observer. This adjustment factor becomes complicated by the fact that not only do length adjustments have to be made, but the nature of the terrain must also be included. The easiest way of explaining how this adjustment term is calculated is through Figs. 7.33 and 7.34. Figure 7.33 shows three possible locations of a road segment and the resulting angle measurements with regard to an observer's location. Note that angles measured to the left of a perpendicular line to the segment always have negative values (ϕ_1 is always the angle connecting the observer and the leftmost end of the segment). The value of this adjustment term can then be found for both hard and soft sites by using the graphs shown in Fig. 7.34, where $\Delta\phi = \phi_2 - \phi_1$.

The final adjustment is made for cases where an object located between the road and observer interferes with the propagation of sound waves. Certain types of objects in these locations, such as dense woods, rows of houses, or noise barriers, result in lowered noise levels. For example, dense woods 30 meters deep, with trees extending at least 5 meters above the line of sight, can reduce the equivalent sound level by 5 dBA. In those instances where barriers are constructed to reduce noise, the level of reduction depends on the shape of the barrier, material of construction, and how much of the roadway is shielded from the observer.

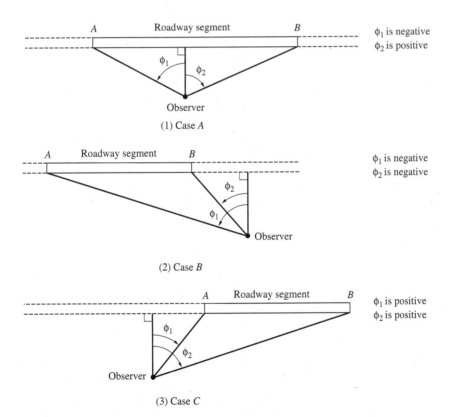

Figure 7.33   Angle identification of roadway segments for noise assessments
I SOURCE: Barry and Reagan, 1978

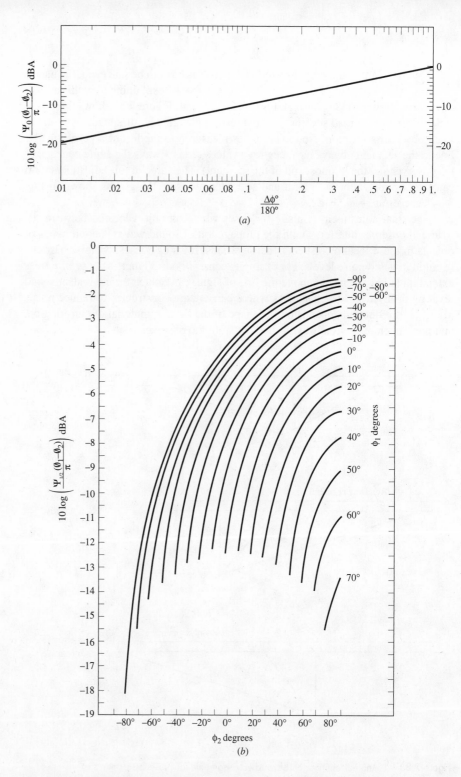

Figure 7.34 Adjustment factors for finite length roadways for hard and soft sites
I SOURCE: Barry and Reagan, 1978

458

Because sound can come from different vehicle types and from different distances (e.g., one traffic lane versus another), some means of combining different noise levels into an equivalent sound measure is necessary. Equation 7.49 shows how this is done.

$$L_{eq}(h) = 10\log\left[10^{\frac{Leq(h)_A}{10}} + 10^{\frac{Leq(h)_{MT}}{10}} + 10^{\frac{Leq(h)_{HT}}{10}}\right] \qquad [7.49]$$

The same equation can be used for calculating equivalent sound levels from different distances. In this case, the $L_{eq}(h)$ values would be sound levels at a particular distance rather than vehicle sound levels.

The following example illustrates the application of the preceding equations. Assume that the eastbound lane of a two-lane highway carries an hourly volume of 400 automobiles, 30 medium trucks, and 40 heavy trucks, whereas the westbound lane carries 350 automobiles, 20 medium trucks, and 30 heavy trucks. Average speeds are 60 kilometers per hour for trucks and 85 kilometers per hour for automobiles. The lane width is 3.6 meters. The observer is 50 meters south of the centerline of the eastbound lane, with ϕ_1 equal to $-45°$ and ϕ_2 equal to $+45°$. The ground surface between the road and the observer is reflective, with no obstructions. What is the total hourly equivalent sound level?

The reference energy mean emission levels for the three vehicle types are determined from Eqs. [7.46] to [7.48], which yield 71.1 dBA, 76.7 dBA, and 82.2 dBA for automobiles, medium trucks, and heavy trucks, respectively. Starting with 71.1 dBA, the eastbound lane traffic flow adjustment is $10\log(400 \times 15/85) - 25 = -6.5$ dBA. Similarly, the westbound traffic flow adjustment is -7.1 dBA. The distance adjustment for the eastbound lane is $10\log[(15/(50)^{1+0} = -5.2$ dBA. For the westbound lane, the adjustment is $10\log[15/(50 + 3.66)]^{1+0} = -5.5$ dBA. From Fig. 7.34, the adjustment for finite road length where $\Delta\phi = 90°$ is -3 dBA. There is no shielding adjustment in this case. The $L_{eq}(h)$ for automobiles for the eastbound lane is thus $71.1 - 6.5 - 5.2 - 3 = 56.4$ dBA. Similar calculations can be made for trucks, resulting in $L_{eq}(h)_{MT} = 52.3$ dBA and $L_{eq}(h)_{HT} = 59$ dBA. The total hourly equivalent sound from the eastbound lane is $L_{eq}(h)_{EB} = 10\log(10^{5.64} + 10^{5.23} + 10^{5.9}) = 61.5$ dBA. The $L_{eq}(h)$ for the westbound lane is 60.1 dBA. The total noise heard by the observer is thus $L_{eq}(h) = 10\log(10^{6.15} + 10^{6.01}) = 63.9$ dBA.

Computer models have been developed to estimate noise impacts resulting from the implementation of planned projects or programs. A model called SNAP allows the user to predict traffic noise for multilane highways, four vehicle types, different roadway types, and varying vehicle speeds [Bowlby, 1981]. Another set of models called STAMINA and OPTIMA also provide noise predictions, with the OPTIMA package calculating cost-effectiveness ratios for alternative designs of noise barriers [Bowlby et al., 1982].

The FHWA has developed a new model called the Traffic Noise Model ® (TNM) that is based on the same principles as the earlier model, but which has updated and expanded parameters and a graphical interface that allows the user to more easily portray the analysis results [Pritchett, 1999]. The TNM uses an iterative process to calculate sound levels at each receiver. The model subdivides the road into segments and calculates the sound levels at each receiver for five vehicle types that originate from each segment. These sound levels are then aggregated at the receptor site to obtain a total sound measure. This model has just been released and is beginning to be used in practice.

7.5.3   Fuel Consumption

Estimating fuel consumption is not as involved as that for air quality and noise impact. The most common method is to estimate the change in number of vehicle miles associated with a particular project and multiply by a fuel consumption factor that reflects the average amount of fuel consumed per vehicle mile by vehicle type and model year. Fuel consumption rates can be obtained from a series of reports published by the U.S. Department of Energy (see, for example, [Davis, 1999]).

The following example from a light rail study in Portland, Oregon, illustrates this approach.

For highway vehicles

Step 1: Categorize vehicles into 10 categories, including light-duty gasoline automobiles, light-duty gasoline trucks, medium-duty gasoline trucks, heavy-duty gasoline trucks, light-duty diesel automobiles, light-duty diesel trucks, heavy-duty diesel trucks, standard buses, articulated buses, and motorcycles.

Step 2: Estimate the percent total vehicle miles traveled for each vehicle type.

Step 3: Multiply the percent VMT by total VMT to calculate the daily VMT for each vehicle type.

Step 4: Divide daily VMT by average fuel consumption in miles per gallon for each vehicle type to determine the daily fuel consumption in gallons for each vehicle type.

Step 5: Multiply daily fuel consumption by a constant BTUs per gallon of gas or diesel to give the daily vehicle energy consumption for each vehicle type.

For light rail system

Step 1: Multiply the number of LRT car miles by the average electrical energy consumption factor in kilowatt-hours of electrical use.

Step 2: Multiply total kilowatt-hours by a BTU conversion factor to determine energy consumption in BTUs for LRT [METRO, 1996].

An example of energy impact analysis at the regional plan level is shown in Table 7.12. The MPO in the San Francisco Bay Area used the changes in mode split, travel speeds, and daily VMT predicted from the regional model to estimate transportation energy use. Many of the projects in the RTP were not well defined, and thus a range of energy values was provided to reflect this uncertainty.

One of the first improvements to the common method previously described has been to develop relationships between fuel consumption and vehicle average speed.

Table 7.12   Energy consumption estimates for San Francisco Bay Area

Energy consumed in BTUs (Billions) 1990, No Project and Project Alternatives

Alternative	Construction	On-Road Vehicle use	Transit Use	Total Energy
Existing (1990)	49	1351	22	1422
No Project (2020)	31	1908	24	1963
Project (2020)	75	1919	25	2019

Energy consumed in BTU (Billions)

Project Category	Total	Yearly Average	Daily Average
Bay Area share of state operations and maintenance	5000–18,100	2270–8810	6.21–22.41
Streets and roads maintenance	2150–7440	990–3500	2.68–9.58
Nonpavement maintenance and local bridges	3690–13,290	1680–6040	4.60–16.65
Highway projects	7550–15,140	3440–6880	9.42–18.84
Toll bridge replacements	3930–4210	1790–1910	4.90–5.23
Toll bridge seismic retrofits	2430–3640	1110–1650	3.01–4.52
Rail transit construction	15,000–18,000	6820–8180	18.68–22.40
All Bay Area	262,700–638,200	18,080–36,280	49.48–99.53

SOURCE: MTC, 1998

Such relationships have been developed for different vehicle types for use in identifying appropriate fuel consumption rates based on average vehicle speed (Fig. 7.35). Because fuel consumption depends on engine technology, an analysis of fuel consumption impacts based on this approach must take into account the mix of vehicle types in the traffic flow, changing fuel efficiency, and the use of alternative fuels [Leiby and Rubin, 1999].

Two other approaches for estimating fuel consumption impact have also been developed. The first uses linear regression techniques that relate fuel consumption rates to vehicle speeds and road grades. Such an approach was originally used to estimate the fuel consumption impact of traffic engineering improvements [Hall, 1980]. Based on the results of several hundred vehicle test runs, an equation was formulated to predict vehicle fuel consumption rates, which were then used to estimate the total amount of fuel used for specific traffic control strategies.

The second approach uses simulation to model vehicle engine performance together with the factors affecting this performance to estimate the impact of alternative vehicle control strategies on fuel consumption. The major purpose of the simulation model is to calculate the propulsive and resistive forces that act on a vehicle as it proceeds along a given route. Sims and Miller [1982], for example, used a computer simulation model to assess the energy efficiency of alternative fixed-route transit modes. In this model, the resistive force consisted of the sum of rolling friction, aerodynamic drag, and forces due to the grade and curvature of the way.

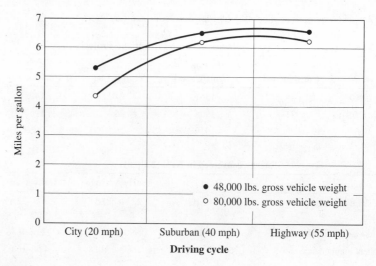

Figure 7.35   Fuel consumption as a function of vehicle speed
| SOURCE: As reported in TRB, 1995

Propulsive forces were directly related to the engine technology employed, with the user being able to specify engine performance curves for different modes. This simulation model was used to investigate the effect on vehicle fuel consumption of vehicle speed, stop spacing, grade, curvature, head wind, stop dwell times, and acceleration rates.

7.6   COST MODELS

Estimating the construction and operation costs of transportation facilities and services and the costs associated with the use of the private automobile are the two most common cost estimation tasks in transportation planning. User costs for the automobile are estimated simply by calculating the vehicle operating costs per mile, parking charges and tolls, and the "sunk" costs of insurance, depreciation and licensing fees (see *www.apta.com/stats/trvauto/drivcost.htm*). The societal costs of automobile use are not considered in these types of models but should be incorporated into the overall evaluation process (discussed in Chap. 8). The more complex cost estimation process relates to facility construction and operations/maintenance/management costs, which is discussed in this section.

7.6.1   Basic concepts

System costs are considered as either *fixed* or *variable*. Fixed costs, which are generally equivalent to *capital* costs, are those associated with the construction of a facility and the purchase of equipment. That is, they are the costs of putting an operational system into place. Variable costs, which are generally equivalent to *operat-*

ing costs, are the costs associated with the actual operation of the system. Variable costs can be further categorized as being *direct, indirect,* or *joint.* Direct costs are attributable to specific components such as labor, fuel, and maintenance associated with the operation of a given facility or service. Indirect costs cannot be identified with specific components except in rather arbitrary ways or with difficult and costly experiments (e.g., heating and lighting costs of maintenance facilities). Finally, joint costs are those that are shared by two or more services, such as streetcar routes drawing power from the same substation or a commuter rail line sharing track maintenance costs with intercity passenger and freight lines.

Fixed costs do not vary with the level of day-to-day operations, whereas variable costs do. *Average* or *unit* costs can be determined by dividing cost by the level of output in order to obtain an average cost per unit of output. Figure 7.36 illustrates these concepts by showing "typical" fixed, variable, and total (fixed plus variable) cost curves, as well as by showing how these translate into average unit cost curves.

A typical way of classifying capital costs, especially for transit projects, is to consider them as fixed costs, systemwide costs, and dependent costs. Table 7.13 shows an example of this classification for a light rail project in Portland, Oregon. The dependent costs include those items whose estimates are considered as a percentage of total project cost. Contingencies will be discussed later in this section.

Another type of unit cost is *marginal* cost. The marginal cost of production is the cost associated with the production of the last unit of output, or mathematically, the marginal cost at any output level is given by the slope of the total cost curve at

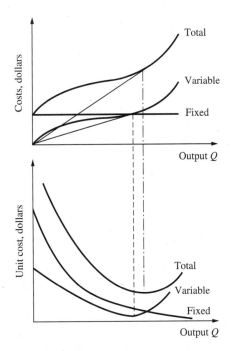

Figure 7.36 Fixed, variable, and unit costs

Table 7.13   Capital cost categories for a light rail project

Fixed Costs	Systemwide Costs	Dependent Costs
Right-of-way	Traction electrification system	Contingency
Utility relocation	Signals	Engineering and administration
Street construction	Communications	
Light rail grade construction	Light rail vehicles	
Structures	Operations and maintenance	
Track installation and materials	Facility	
Road crossings		
Stations		
Park-and-ride		
Fare collection		
Special conditions		

SOURCE: METRO, 1997

that output point. Figure 7.37 illustrates this concept and provides a comparison between typical marginal cost and average cost curves. As indicated by this figure, it is typically the case that marginal costs are minimized at a lower level of output than average costs.

For the automobile user, fixed costs correspond to such things as depreciation, insurance, and residential parking charges. Variable costs represent operating costs such as gasoline, oil, tires, maintenance, and parking charges at trip destinations. It is generally assumed that automobile users rarely consider their total costs of travel, except perhaps when making vehicle purchases, at which time they may roughly calculate the expected total costs associated with the purchase and use of the vehicle. Rather, travel patterns are affected by perceived costs, which are considered to consist of the *out-of-pocket costs* such as gasoline and parking charges and maybe some allocation for "normal" maintenance costs associated with the trip. Even when people consider the full range of variable costs associated with a trip, their perceptions of these costs tend to vary considerably in relatively unpredictable ways from the "true" values [Adiv, 1982; Clark, 1982].

7.6.2   Capital Cost Models

Good capital cost estimates are important to the transportation planning process for two major reasons. First, the evaluation of plan and project cost is a critical criterion for comparing alternatives. For example, by its very definition, cost-effectiveness, one type of evaluation, must have estimates of capital costs in order to be used. Second, transportation planning in the United States must result in a transportation program that realistically reflects the amount of available funding (known as financially constrained planning). In order for officials to satisfy as many needs as pos-

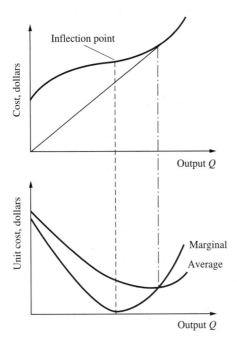

Figure 7.37 Average and marginal costs

sible, they must know project costs so that the constrained program can use all available resources.

In most cases, costs in the planning stage of project development are estimates based on general concepts of the mode, alignments, and design standards. As the project development process progresses toward plans and specifications, cost estimates become more specific and reflect project requirements that come from detailed engineering drawings.

Because transportation projects are designed with many repetitive features, for example, lengths of road, track, or sidewalk that are very similar, estimating capital costs is a straightforward process. A "typical section or segment" is defined for these lengths, a unit cost per length for this section is determined, and this cost per length is then multiplied by the total length of similar sections in the project. To this would be added special costs that might not be able to be defined in this way, such as cost for utilities, environmental mitigation, maintenance facilities, and communications systems.

Figure 7.38 shows an example of this typical section approach as applied to a proposed light rail line. A composite unit cost per section is developed (in this case, $1,640 per lineal foot) based on all of the components of which the section consists. This composite unit cost is then multiplied by the total length of this type of section in the project. The unit cost values usually reflect the local history for cost of construction. In those situations where a facility or service new to an area is being considered, cost estimates are often based on comparable projects elsewhere in the United States, adjusted to reflect local rates and conditions.

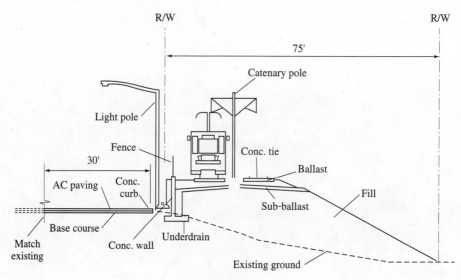

Proposed

Item no.	Description	Qty	Unit	Unit cost	Total cost of civil construction per L.F.	Total cost of track materials and installation per L.F.	Total cost of section per L.F.
1	Clear and guide	100.00	S.F.	0.16	16.00		
2	Pavement removal	5.58	S.Y.	10.00	56.00		
3	Excavation	0.78	C.Y.	7.00	5.00		
4	Backfill	22.30	C.Y.	20.00	446.00		
5	Excavate trench, common earth	0.30	C.Y.	4.00	1.00		
6	Underdrain	1.00	L.F.	17.00	17.00		
7	Landscaping	30.00	S.F.	0.50	15.00		
8	Fence	1.00	L.F.	24.00	24.00		
9	Base course, 1 1/2" stone, 12" deep	2.22	TON	27.00	60.00		
10	AC paving, 6" deep	1.11	TON	56.00	61.00		
11	Sub-ballast	1.58	TON	42.00		68.00	
12	Ballast	2.37	TON	48.00		114.00	
13	Rail	2.00	T.F.	103.00		386.00	
14	Concrete footings	0.37	C.Y.	226.00	84.00		
15	Concrete walls	0.56	C.Y.	450.00	250.00		
16	Concrete curbs	1.00	L.F.	17.00	17.00		
17	Lighting	1.00	L.F.	20.00	20.00		
18							
19							
20							
	Totals				$1072.00	$568.00	$1640.00

Figure 7.38   Composite unit cost for a light rail section
| SOURCE: Portland METRO, 1997b

Recognizing the uncertainties associated with cost estimates is an important part of the estimation process. Indeed, many of the criticisms of transportation investment (especially for transit) are aimed at much higher than expected construction costs to complete projects [Pickrell, 1989]. However, it is quite common during project development for there to be changes in project scope and design standards, for unexpected construction problems to be encountered, and for errors in cost estimates and quantities to become evident. One way of incorporating uncertainty into the cost estimation process is through the use of contingencies. Figure 7.39 shows typical contingencies for different cost categories and for different project phases. As shown, as the project gets closer to construction, the contingency values decrease.

7.6.3   Operations, Maintenance, and Management Costs

Many transportation projects have significant operations, maintenance, and management (OMM) costs associated with their lifetime operation. For example, traffic management components of control centers or HOV lanes, transit services, and traffic signal improvement programs can include substantial labor costs. OMM costs are particularly important for transit projects where they can constitute as much as 80 percent of annual expenditures to operate a service.

OMM cost models are based on an accurate estimate of service characteristics and a detailed recent annual budget. Similar to capital cost models, unit costs are applied against quantities of service, resulting in a total cost estimate. Two common approaches for OMM modeling are cost-allocation models and resource build-up models. Cost-allocation models assign every budget item to one of several service variables. The costs for each variable are summed and then divided by the total amount of service to obtain an aggregate unit cost. Typical model forms include [NTI et al., 1996]

$$\text{Roads:} \quad \text{OMM} = [C_m \times \text{(lane miles)}] + [C_p \times \text{(pavement area)}] \quad \text{[7.50]}$$
$$+ [C_s \times \text{(major structures)}]$$

$$\text{Transit:} \quad \text{OMM} = [C_m \times \text{(vehicle miles)}] + [C_n \times \text{(vehicle hours)}] \quad \text{[7.51]}$$
$$+ [C_v \times \text{(peak vehicles)}] + [C_p \times \text{(passengers)}]$$

where C's are unit costs per variable. For example, C_m is dollars per lane mile.

If a new transit service is to be implemented and the additional vehicle miles traveled, vehicle hours in operation, number of additional vehicles in the peak period, and the number of passengers are known, Eq. [7.51] could be used to estimate the incremental costs associated with this new service. A finer level of detail could be built into these models such as disaggregating lane miles or vehicle miles traveled by facility or service type.

Because this model is based on a previous year's data, it is often difficult to use for predicting costs affiliated with service conditions that are different from the base year. In addition, if unusual events occurred during the base year, such as a labor strike, the unit costs might not be a good measure of future costs. Choosing the base budget year thus becomes an important basis for developing a valid cost model. Table 7.14 shows how a cost-allocation model is developed.

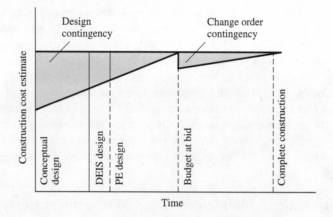

Fixed costs	Contingency	Systemwide costs	Contingency
Right-of-way	20%	Traction electrification system	20%
Utility relocation	30	Signals	20
Street construction	25	Communications	20
Light rail grade construction	25	Light rail vehicles	10
Structures	25	Operations/maintenance facility	25
Track installation and materials	25		
Road crossings	20		
Stations	20		
Park-and-ride	20		
Fare collection	15		
Special conditions	25		

Project phase	Design contingency percentage
Conceptual definition	35%
Conceptual and detailed	25
Detailed definition	22.5 (approximate)
Preliminary engineering	20 (approximate)
100 percent final design	0

Figure 7.39   Variation of costs through project development and contingency values
⏐ SOURCE: METRO, Portland, 1997

Resource build-up models provide more accurate estimates of OMM costs by computing the costs of labor and material needed for a given level of service provision. This type of model takes the form

OMM = (unit of service) × (resources per unit of service) × (resources unit cost)

Table 7.15 shows an example of a resource build-up model. Note that the productivity ratios are key to the modeling process. These ratios can take several forms:

- *Continuously variable* such that the rate stays the same over time as service units are added (e.g., vehicle maintenance per revenue vehicle mile).

Table 7.14 Example of cost-allocation model for hypothetical transit agency

Fleet size	60 buses
Vehicle usage	54 buses peak; 37 buses midday
Number of routes	11
Annual revenue miles	2,448,200
Annual revenue hours	183,700
Total unlinked passenger trips	7,195,000
Total employees	104

Function and Expense Object	Annual Expense	Basis for Assignment		
		Revenue Hours	Revenue Miles	Peak Vehicles
Labor				
Vehicle operations	2,691,695	100%		
Vehicle maintenance	578,945		100%	
Nonvehicle maintenance	0	100%		
General Administration	384,245			100%
Fringe Benefits				
Vehicle operations	892,650	100%		
Vehicle maintenance	189,365		100%	
Nonvehicle maintenance	0	100%		
General administration	77,235			100%
Services	160,180			100%
Materials/Supplies				
Vehicle operations	537,820		100%	
Vehicle maintenance	344,365		100%	
Nonvehicle maintenance	0		100%	
General administration	110,845			100%
Liability	301,500		100%	
Taxes				
Vehicle operations	20,470		100%	
Vehicle maintenance	0		100%	
Nonvehicle maintenance	0	100%		
General administration	21,170			100%
Purchased Transportation	16,895			100%
Miscellaneous Expenses	84,145			100%

The expenses for each of the three categories divided by the amount of service provision for that category are

Dollar per revenue hour:	$3,584,345/183,700 revenue hours = $19.51/revenue hour
Dollar per revenue mile:	$1,670,965/2,488,200 revenue miles = $0.67/revenue mile
Dollar per peak vehicle:	$1,476,375/54 peak vehicles = $27,340/peak vehicle

The OMM cost model is thus $OMM = (\$19.51) \times (\text{revenue hours}) + (\$0.67) \times (\text{revenue miles}) + (\$27,340) \times (\text{peak vehicles})$

Table 7.15   Example of resource build-up model

Cost Item	Cost Type	Unit Cost	Driving Variable	Number of Employees	Total Cost 1990 $	Assumed Productivity
Station Maintenance and Operations						
Janitorial	CONTR	$4920	STATION		$39,360	
Grounds	CONTR	4200	STATION		33,600	
Utilities	CONTR	2000	STATION		16,000	
Parking lot maintenance	CONTR	2080	STATION		16,640	
Building maintenance	CONTR	14880	STATION		119,040	
Administration	ADMIN	5769	STATION		46,154	
Car Cleaning						
Cleaner foreman	LABOR	45,360	# Cleaners	1	45,360	
Cleaners	LABOR	30,240	PKTOTCAR	32	967,683	
Cleaning materials	MATL	100	TOTCAR		7000	
Security/Fare Collection						
Security	CONTR	100,000	Lump sum		100,000	
Fare machinery service	LABOR	38,559	STATION	4	154,234	2 station/emp.
Fare machinery maintenance	LABOR	50,057	STATION	2	100,115	4 station/emp.
Station agent	LABOR	50,057	STATION	2	100,115	
Other Costs						
Administration	ADMIN	40	TRIPS		754,800	$51.23 admin/trip
Marketing	ADMIN	0.10	PASS		377,400	
Other transit marketing	ADMIN	0.05	PASS			$0.12 mktg/pass
Audit	ADMIN	1.0%	RR COST			
		Total cost			22,913,212	

Cost Summary

Cost per train hour:	$971
Cost per train mile:	$30.36
Cost per car mile:	$5.06
Cost per passenger:	$6.07

SOURCE: Seattle Metro, 1991

- *Step-wise variable* that change in specified amounts with changes in service conditions (e.g., adding new space to a maintenance facility will add utility and staff costs).
- *Fixed,* whose marginal cost is zero over the expected range of system variables (e.g., administrative and legal support).

Another approach to estimating costs is to use multivariate statistical techniques in which a number of observations relating total or operating costs to operating variables are used to determine a multiple regression equation. This equation specifies the relationship between operating costs and operating variables. Typically, data from a number of different transit systems are used to provide observations, although observations from one system over time can also be used. The multivariate approach is often difficult to apply in that

1. Costs often vary dramatically from system to system because of differences in labor agreements, equipment, and scale of operations, rendering the pooling of data from several systems suspect.

2. Within a given system there may be insufficient variation in operating characteristics over the observation period to generate reliable regression estimators.

The unit cost approach is therefore the more commonly adopted of the two approaches.

Because operating, maintenance, and management costs represent long-term commitments to operating agencies, the cost estimates for each alternative are of great interest to state and local governments. This usually results in different ways of portraying the level of commitment and resulting benefits associated with this commitment. Figure 7.40 illustrates this concept for a transit analysis in Seattle. The OMM costs were shown in relation to capital costs, cost per trip, cost to each county, and annual trips served.

7.7   CHAPTER SUMMARY

1. The *supply* of transportation services can be characterized in terms of the *performance* of the transportation system (e.g., travel times, headways, and capacities), the *impacts* this system has on the environment (which includes nonusers of the system), and the *costs* incurred in building, maintaining, and using the system.

2. Six major components of the transportation system that are generally addressable by supply-related policies are the transportation *infrastructure* (rights-of-way, signal systems, other fixed facilities), the *vehicles* that operate within this infrastructure, the *routes* and *schedules* for the system, the *drivers* of the vehicles, the *procedures* or rules for operating the system, and the *costs* borne by operators and users of the system.

3. Types of supply-related policies or actions include *construction* of transportation infrastructure, *design* and *manufacture* of transportation equipment (most notably, vehicles and signal systems), *provision* of transportation services, *maintenance* of transportation infrastructure and vehicles, *regulation* of transportation services and systems, *enforcement* of transportation regulations, the *financing* of transportation infrastructure and vehicles, *subsidization* of transportation system operating costs and/or users of the system, *taxation* of transportation services and equipment, and *pricing* of transportation services.

4. Major decision makers or actors involved in supply-related policy making include *elected representatives, regulators, operators, implementers, contractors,* and *law enforcement agencies.*

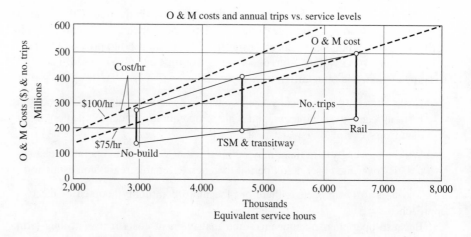

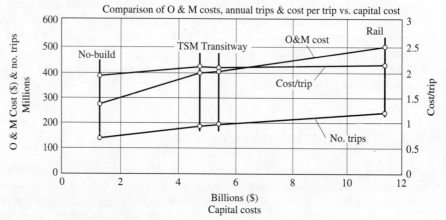

Figure 7.40   O & M cost-effectiveness measures for transit alternatives in Seattle

5. Supply analysis interacts with the planning process in three major ways. First, performance and cost characteristics are important determinants of the demand for transportation, while demand levels in turn affect system performance. Second, measures of system performance, impacts, and costs are important evaluation criteria. Third, supply analysis contributes toward the monitoring and diagnosis planning functions.

6. *Network* definition, consisting of the representation of the real or physical network by an "approximating network" composed of a connected set of *links* and *nodes,* is perhaps the single most important step in the entire analysis process in that it determines the maximum level of accuracy and detail of analysis obtainable, the quantity and quality of data required to represent and analyze the system, and the type(s) of analysis techniques that can be applied.

7. The *fundamental equation of traffic flow* $V = \bar{S} \times D$ summarizes the fundamental relationship between *volume V,* average *speed S,* and *density D* for flows on transportation links.

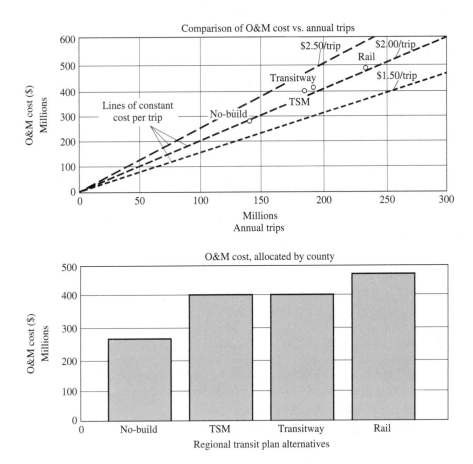

Figure 7.40 (continued) O & M cost-effectiveness measures for transit alternatives in Seattle
I SOURCE: METRO, Seattle, 1992

8. *Time–distance diagrams* are used to plot vehicle trajectories through time and space (i.e., to identify vehicle locations along a route at any point in time). They can be applied to a range of simple performance analysis problems involving either shared or exclusive rights-of-way, the analysis of intersection signal systems, the analysis of conflicting flows, and so on.

9. *Queuing theory* is used to analyze systems in which customers must queue up in order to receive service (e.g., cars waiting at an intersection for the light to turn green). While transportation queuing theory applications are somewhat limited because of the nature of most transportation problems, the concept of the queue is a most important one and forms the basis for many analysis techniques (e.g., fluid-flow approximations and many simulation models).

10. *Fluid-flow approximations* are used to analyze transportation queuing problems in which customer arrivals and servicing can be treated in terms of continuous *flows* of customers (e.g., flows of vehicles on a street), rather than in terms of discrete, individual customers.

11. Simulation involves constructing a mathematical model of a system and then operating or exercising this model over time. Simulation models are *dynamic* in that they replicate system performance over time. That is, system performance at time $t + \Delta t$ cannot be computed without knowing the performance at time t. In addition, simulation models typically include stochastic or probabilistic elements that represent events or processes that cannot be modeled deterministically.

12. *Mathematical programming* techniques are used to find optimal values for a set of *decision variables,* where optimality is defined in terms of an *objective function*, which is maximized (or minimized) subject to a set of *constraints* that define feasible ranges for the decision variables.

13. The choice of an appropriate technique to analyze transportation system performance depends on the level of detail required, data availability, the time and budget available for the study, and so on. It also depends on the nature of the system being analyzed. In particular, techniques vary according to whether one is analyzing exclusive or shared rights-of-way, whether access to the shared right-of-way is controlled or uncontrolled, and whether one is analyzing a component of a network (e.g., a link or a route) or the network as a whole.

14. Transportation *impacts* include *air quality* impacts, *noise* impacts, and *fuel consumption* impacts. In general, transportation impacts can be analyzed by a variety of techniques, ranging from the use of simple nomographs or other "handbook" techniques to the use of complex computer simulation models of the system's performance as it relates to the generation of the impacts of interest.

15. Transportation costs are either *fixed* or *variable* in nature. "Ballpark" estimation of fixed or capital costs can be done on a unit cost basis. That is, the cost per unit (e.g., cost per vehicles or cost per mile of right-of-way) is determined and then multiplied by the number of units required. Variable or operating costs are modeled through the use of *cost-allocation* techniques or the use of multivariate statistical techniques (usually regression analysis). Cost allocation involves assigning each cost component (e.g., driver wages) on logical or empirical grounds to a system performance characteristic (e.g., vehicle hours of operation). Average unit costs for each operating variable can then be determined (e.g., dollars per vehicle hour of operation), and total operating costs for any given service can be computed by multiplying these unit costs by the relevant operating variables and summing over all such terms.

QUESTIONS

1. Consider a section of one-way roadway that at time $t = 0$ is empty, with a vehicle about to enter (i.e., the location of the leading edge of the vehicle is $x = 0$ at $t = 0$) and with a stream of vehicles uniformly spaced behind the lead vehicle at a density of 100 vehicles per mile. All vehicles (including the lead vehicle) are moving at a uniform speed of 30 miles per hour. At $t = 1$ second, a traf-

fic signal turns red at location $x = 500$ feet. Assuming that the first vehicle stops instantaneously when it reaches the stoplight location (i.e., $x = 500$) and that each succeeding vehicle stops instantaneously such that its leading edge is exactly 15 feet behind the leading edge of the vehicle in front of it, draw a time–distance diagram that shows the trajectories of the leading edges of the first five vehicles entering the section. Given this diagram,

(a) Connect the points representing the time–distance locations at which each vehicle comes to rest. What shape is the resulting curve?

(b) This line is known as a *shock wave*. Physically, what does it represent?

(c) Compute the slope of this line. Physically, what does it represent?

At $t = 31$ seconds, the light turns green. Assuming that the vehicles instantaneously accelerate to 30 miles per hour once they start moving and assuming that there is a 1/2-second response time between the time when the first vehicle and the second vehicle (the second and the third, etc.) start moving, do the following:

(d) Draw the time–distance diagram showing the initialization of movement of the first five vehicles in the queue.

(e) Draw on this diagram the shock wave associated with the initialization of movement through the queue.

(f) Compute the slope of this line.

(g) Explain the physical interpretation of these results.

Given the results of your analysis, draw a new time–distance diagram at a larger scale that will enable you to calculate graphically the total number of vehicles stopped by the red light.

2. For the system analyzed in Question 1, use a fluid-flow diagram to calculate

(a) The total number of vehicles stopped by the light.

(b) The maximim queue length of vehicles stopped at the light.

(c) The average delay experienced by vehicles because of the red light.

3. Consider a simple intersection of two one-way streets, one of which is a major "through" street and the other is a minor street controlled by a stop sign. Vehicles arrive along the minor approach according to a Poisson process at an average rate of two vehicles per hour. In order for a vehicle on the minor street to cross the major street, a *gap* in the flow of vehicles along the major street of sufficient size for the vehicle to cross safely must occur. Assume that such gaps in the major street flow arrive at the intersection according to a Poisson process at an average rate of μ gaps per hour (actually, gap arrivals tend to be determined by more complicated processes than the Poisson). Using the queuing theory results presented in this chapter, plot average wait times and queue lengths for vehicles on the minor street versus λ for the cases of μ equal to 200, 100, and 50. What criteria would you use to determine the flow levels at which it might be worthwhile to signalize the intersection? What additional information would you require to make this determination?

4. One approach to obtaining a more detailed analysis of the intersection examined in Question 3 is to use a simulation model. Develop a flowchart of the major tasks and information required to construct such a simulation model. In developing the flowchart, discuss

(a) How stochastic elements of the system are to be treated.

(b) Data requirements of the model.

(c) System performance measures required and how these will be computed.

5. Consider a typical transit passenger stop that experiences average arrival rates of passengers over time as given in Table P7.5a. Buses with a design capacity of 80 passengers (seated and standing) arrive at the stop at the times shown in Table P7.5b with the loadings shown in the same table. Assuming that the buses load passengers instantaneously at the stop (up to their maximum of 80), draw a fluid-flow diagram of operations at this stop for the time period 9 A.M. to 10 A.M., given that at 9 A.M. there are no passengers waiting at the stop. Calculate for this system the average wait time experienced, the maximum wait time, the average number of passengers waiting, the largest number of passengers ever waiting, and the average number of passengers on board the bus as it leaves the stop.

Table P7.5a   Average passenger arrival rates

Time Period	Arrival Rate, Passengers per Minute
9:00 to 9:15	1
9:15 to 9:30	2
9:30 to 10:00	1

Table P7.5b   Bus arrival times and loadings

Bus Number	Arrival Time	Number of Passengers on Board
1	9:10	40
2	9:20	65
3	9:25	30
4	9:44	75
5	9:47	70
6	9:50	60
7	10:00	50

6. Compute minimum path trees from nodes 2 and 5 for the network shown in Fig. P7.6.

7. Given a total cost curve of the form

$$TC = a + bQ + cQ^2$$

where Q is the quantity produced, derive the corresponding average and marginal cost curves. Plot these curves given $a = 100$, $b = 1$, and $c = 0.001$. Indicate the points of minimum average and minimum marginal costs. Why are these points not the same?

8. Assume you have been asked to calculate the person-carrying capacity of a bus route. You know that at the maximum load point there are 15 people boarding a bus (with each taking 3 seconds) and 5 people alighting (with each taking 2 seconds). A nearby traffic signal has a cycle length of 60 seconds and a green time of 30 seconds. The clearance time for the bus is 15 seconds, the coefficient of variation of the dwell time is 0.6, and you are seeking a failure rate of not greater than 10 percent in queue formation. If there is room for only one bus at this stop at one time, what is the person-carrying capacity of this route? (Note the dwell time will be the maximum time it takes for passengers to either board *or* alight.)

9. The transit agency is designing a new transit station. The general manager is interested in knowing the expected performance of the ticketing booth during the peak hour because he wants to make sure there are not any complaints about poor service. As luck would have it, the behavior of people arriving at a ticket booth can be viewed as a queuing phenomenon. Assume that the passenger arrivals can be represented by a Poisson distribution and the amount of time it takes to serve each customer can be represented by a negative exponential distribution with service rate μ. If 400 passengers per hour are expected to arrive at the ticket booth and it takes approximately 5 seconds to serve each passenger, calculate the following:

(a) Average number of people in the system.

(b) Average time in the system.

(c) Average time waiting in the queue.

(d) Percent time the booth is not serving a passenger.

(e) Probability that there are 3 people in the system.

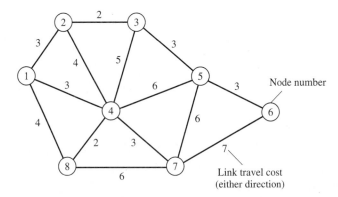

Figure P7.6 Network for minimum path calculations

10. You have been asked to design the pedestrian walkway in the new airport. Because of the crush load expected during the peak period when several airplanes discharge their passengers during the same 15-minute period, the designer has been asked to determine the width of the terminal passageway for a design capacity level of service. Given the following characteristics, how wide should the passageway be?

    (a) Peak 15-minute flow rate = 2200 pedestrians per 15-minute period. On either side of the passageway, seats are provided for airline passengers to wait for boarding. Given the luggage and assorted activities of such passengers, you have determined that for calculations this "border" on both sides of the passageway can be treated like building faces with window displays.

    (b) Assume the same characteristics as in (a). How wide should the passageway be if you wanted to design for LOS B?

REFERENCES

Adiv, A. 1982. *Perception of travel cost by automobile to work: Empirical study in San Fransico Bay Area.* Presented at the 61st annual meeting of the Transportation Research Board. Washington, D.C.

Bachman, W. 1998. *A GIS-based modal model of automobile exhaust emissions.* Final Report. EPA-600/R-98-097. Washington, D.C.: Office of Research and Development. Aug.

Barrett, C. et al. 1995. *An operational description of TRANSIMS.* Report LA-UR-95-2393. Travel Improvement Program. Los Alamos, NM. June.

Barry, T. M. and J. A. Reagan. 1978. *FHWA highway traffic noise prediction model.* Report FHWA-RD-77-108. Washington, D.C.: Federal Highway Administration. Dec.

Bell, M. and Y. Iida. 1997. *Transportation network analysis.* New York: Wiley.

Benson, P. E. 1979. *CALINE-3—a versatile dispersion model for predicting air pollutant levels near highways and arterial streets.* California Department of Transportation Report FWHA/CA/TL-79/23. Sacramento, CA. November.

Bowlby, W. (ed.) 1981. *Found procedures for measuring highway noise: final report* Federal Highway Administration Report FHWA-DP-45-lR. U.S. Department of Transportation. Washington, D.C. August.

Bowlby, W., J. Higgins, and J. Reagan. 1982. *Noise barrier cost reduction procedure STAMINA 2.0/OPTIMA: User's manual.* Report FHWA-DP-57-1. Washington, D.C. April.

Cambridge Systematics et al. 1998. Multimodal corridor and capacity analysis manual. *NCHRP Report 399.* Washington, D.C.: National Academy Press.

Central Transportation Planning Staff. 1997. *Transfer penalties in urban mode choice modeling.* Travel Model Improvement Program. Washington, D.C.: U.S. Department of Transportation. Jan.

Clark, J. E. 1982. Modeling travelers' perception of travel time. Presented at the 61st annual meeting of fraud time. Presented at the 61st annual meeting at the Transportation Research Board. Washington, D.C. Jan.

Clarke, A. and R. Disney. 1998. *Probability and random processes for engineers and scientists.* New York: Wiley.

Cohn, L., R. Harris, and P. Lederer. 1999. Environmental and energy considerations. In J. Edwards (ed.) *Transportation planning handbook.* Washington, D.C.: Institute of Transportation Engineers.

Daganzo, C. F. 1997. *Fundamentals of transportation and traffic operations.* New York: Elsevier Scientific.

Davis, S. 1999. *Transportation energy data book: Edition 1999.* Report ORNL-6958. Oak Ridge, TN. Sept.

Dowling, R., W. Kittelson, J. Zegeer, and A. Skabardonis. 1997. Planning techniques to estimate speeds and service volumes for planning applications. *NCHRP Report 387.* Washington, D.C.: National Academy Press.

Dueker, K. et al. 1998. Strategies to attract auto users to public transportation. *TCRP 40.* Washington, D.C.: National Academy Press.

Environmental Protection Agency. 1994. *Users guide to MOBILE 5.* Report EPA-AA-AQAB-94-91. Ann Arbor, MI: Office of Mobile Sources and Office of Air and Radiation. May.

Federal Highway Administration. 1991. *The MOST reference manual,* Vol. 1. Washington, D.C.: U.S. Department of Transportation. Dec.

_____. 1995. *Highway traffic noise analysis and abatement policy and guidance.* Washington, D.C.: Office of Environment and Planning. June.

_____. 1999. *Traffic software integrated system users guide. Version 4.3.* Washington, D.C. June.

Guensler, R. 2000. *Understanding the air quality impacts of urban transportation.* Continuing education course notes. Georgia Institute of Technology. Sept.

Hall, J. 1980. *Traffic engineering improvement priorities for energy conservation.* NMEI Report No. 78-1128. Albuquerque, NM: New Mexico Energy Institute.

Handy, S. and D. Niemeier. 1997. Measuring accessibility: An exploration of issues and alternatives. *Environment and Planning A.* Vol. 29.

Horowitz, J. L. 1982. *Air quality analysis for urban transportation planning.* Cambridge, MA: MIT Press.

Kimley-Horn and Assocs. 1995. *Long-range capacity and access study, Los Angeles Union passenger terminal.* Los Angeles, CA: Southern California Association of Governments. Jan.

Larson, R. C. and A. R. Odoni. 1981. *Urban operations research.* Englewood Cliffs, NJ: Prentice-Hall.

Lee, C. and G. Fleming. 1996. *Measurement of highway-related noise.* Report FHWA-PD-96-046. Washington, D.C.: Office of Environment and Planning. March.

_____ and J. Burstein. 1998. *FHWA traffic noise model, Version 1.0. Look-up tables.* Report DOT-VNTSC-FHWA-98-5. Washington, D.C.: Federal Highway Administration. Office of Environment and Planning. July.

Leiby, P. and J. Rubin. 1999. Sustainable transportation: Analyzing the transition to alternative travel vehicles. *Transportation Research Record 1492.* Washington, D.C.: National Academy Press.

Los Alamos National Laboratory. 1998. *Transportation analysis simulation system (TRANSIMS): The Dallas case study.* Travel Model Improvement Program. Washington, D.C.: U.S. Department of Transportation and U.S. Environmental Protection Agency. Jan.

Manheim, M. 1979. *Fundamentals of transportation system analysis.* Cambridge, MA: MIT Press.

May, A. 1990. *Traffic flow fundamentals.* Englewood Cliffs, NJ: Prentice-Hall.

METRO, Portland. 1994. *Light rail transit representative alternatives, Conceptual design and order of magnitude cost estimate. South–North Transit Corridor Study.* Portland, OR. May 27.

_____. 1995. *Design option narrowing technical summary report.* Walk isochrone compendium. Portland, OR. June.

_____.1996. *Social, economic, and environmental methods report.* South–North Transit Corridor Study. Draft Environmental Impact Statement. Portland, OR. May 20.

_____. 1997a. *Capital costs methods report.* South–North Transit Corridor Study. Portland, OR. March 20.

_____.1997b. *Capital costs results report.* South–North Transit Corridor Study. Portland, OR. Sept. 3.

METRO, Seattle. 1991. *Regional transit project: Capital cost methodology report.* Seattle, WA: Nov. 20.

_____. 1992. *Operating and maintenance cost estimate results report.* Seattle, WA. Aug.17.

Metropolitan Transportation Commission. 1998. *1998 Regional transportation plan: Draft environmental impact report.* Oakland, CA. Aug.

Morlok, E. 1978. *Introduction to transportation engineering and planning.* New York: McGraw-Hill.

National Research Council. 2000. *Modeling mobile source emissions.* Washington, D.C.: National Academy Press.

National Transit Institute and Parsons Brinckerhoff. 1996. *MIS desk reference.* Prepared for the Federal Highway Administration and the Federal Transit Administration. Washington, D.C. Aug.

Newell, G. F. 1971. *Applications of queuing theory.* London: Chapman and Hall.

_____. 1980 *Traffic flow on transportation networks.* Cambridge, MA: MIT Press.

Oppenheim, N. 1995. *Urban travel demand modeling.* New York: Wiley Interscience.

Ortuzar, J. and L. Willumsen. 1994. *Modelling transport.* 2d ed. New York: Wiley.

Parsonson, P. 1998. *Influence of signal spacing on arterial-traffic progression.* Paper presented at Third National Conference on Access Management. Fort Lauderdale, FL. Oct. 4–7.

Pedersen, N. and D. Samdahl.1982. Highway traffic data for urbanized area project planning and design. *NCHRP Report 255.* Washington, D.C.: National Academy Press.

Pickrell, D. 1989. *Urban rail transit projects forecast versus actual ridership and costs.* Washington, D.C.: Urban Mass Transportation Administration. Oct.

Pine, R. et al. 1998. Transit scheduling: Basic and advanced manuals. *TCRP Report 30.* Washington, D.C.: National Academy Press.

Pritchett, J. 1999. *The old FHWA noise model and TNM (traffic noise model): A comparison.* Unpublished paper. School of Civil and Environmental Engineering. Georgia Institute of Technology, Atlanta, GA.

Rau, J. G. and D. C. Wooten (eds.). 1980. *Environmental impact analysis handbook.* New York: McGraw-Hill.

SG Assocs. 1998. *Guidelines for network representation of transit access: State-of-practice summary.* Report DOT-T-99-05. Washington, D.C.: U.S. DOT. June.

Sheffi, Y. 1985. *Urban transportation networks.* Englewood Cliffs, NJ: Prentice-Hall.

Shurbajji, M. 1993. *Use of spatially-defined travel characteristics in transit service planning.* Unpublished Ph.D. dissertation. School of Civil and Environmental Engineering. Georgia Institute of Technology, Atlanta, GA.

Sims, D. and E. Miller. 1982. Energy consumption of alternative fixed-route transit modes. *TRAC Forum.* Vol. 5. No. 1. Toronto.

Taylor, M., W. Young, and P. Bonsall. 1996. *Understanding traffic systems: Data, analysis and presentation.* Brookfield, VT: Ashgate.

Transportation Research Board. 1995. Expanding metropolitan highways. *Special Report 245.* Washington, D.C.: National Academy Press.

_____. 1999. *Transit capacity and quality of service manual.* Washington, D.C.: National Academy Press.

_____. 2000. *Highway capacity manual.* Washington, D.C.: National Academy Press.

Vuchic, V. R. 1981. *Urban public transportation systems and technology.* Englewood Cliffs, NJ: Prentice-Hall.

chapter

8

Transportation System and
Project Evaluation

8.0   INTRODUCTION

The preceding chapters have identified numerous analysis tools that can be used to predict travel demand and to estimate the impacts of transportation alternatives. Decision makers, however, often require information on the consequences of alternative projects and plans in addition to quantitative estimates of use. For example, information on the implementation process and on alternative financing strategies can be critical considerations in the decision-making process. Perhaps the most important information is how the alternatives compare to one another. This chapter discusses how the system and project evaluation process provides the information needed by decision makers to select among alternatives.

Two distinctions are made in this chapter that are often not included in other texts. First, evaluation *techniques* (e.g., benefit/cost analysis) are used within an evaluation *process* (e.g., the interaction among key participants in planning). Transportation planning is as much a political process as it is a technical one. Thus, understanding the process of evaluation is as important as knowing how the techniques work. Second, a distinction is made between *a priori* evaluation, which considers the evaluation of plan or project alternatives yet to be implemented; and *ex post* evaluation, which considers projects or programs after implementation. With a growing interest in service-oriented transportation actions, the latter approach to evaluation is increasingly being used by transportation planners. A careful consideration of the impacts of previously implemented projects can provide transportation officials with useful information on how to change existing services to better meet customer desires or on the desirability of implementing similar projects elsewhere in the metropolitan area.

The next section defines evaluation and provides guidance on the general principles for successful evaluation efforts. The key questions that form the basis of any type of evaluation are presented. Section 8.2 describes the general characteristics of benefit and cost measurement, which are used within an evaluation framework discussed in Sec. 8.3. Sections 8.4 to 8.6 present several techniques for assessing the relative worth of alternatives, including cost-effectiveness, economic evaluation techniques, and multiobjective evaluation. Section 8.7 presents several case studies of evaluation for regional, corridor, and freight planning. Ex post evaluation, including both the underlying concepts and alternative experimental designs, is discussed in Sec. 8.8.

8.1   WHAT IS EVALUATION?

Evaluation is the process of determining the desirability of different courses of action and of presenting this information to decision makers in a comprehensive and useful form. Determining the desirability of an alternative requires (1) defining how value is to be measured; (2) estimating the source and timing of the benefits and costs of the proposed actions; and (3) comparing these benefits and costs to determine a level of effectiveness for that alternative. Evaluation thus provides information to decision makers on the estimated impacts, trade-offs, and major areas of uncertainty associated with the analysis of alternatives. Not only does the magnitude of the impact have to be determined, but those who are positively or negatively affected should also be identified.

The function of evaluation in the transportation studies of the 1950s and early 1960s was primarily to estimate in quantitative terms the impacts for each alternative and to assign monetary values to project benefits and costs. The value of an individual project was thus determined by comparing these monetary benefits and costs. The alternative that maximized the return on investment was recommended for implementation. Beginning in the late 1960s, however, transportation planners became increasingly interested in the consequences of transportation investment that could not be easily measured in monetary terms. In part because of legal requirements, issues such as air quality, community cohesion, energy consumption, equitable distribution of resources, and economic development were incorporated into the evaluation process. With this change in the focus of evaluation, questions other than, "Which alternative maximizes the monetary benefits returned for the costs incurred?" became important in the evaluation process. Evaluation began to include questions relating to a broad definition of effectiveness, the efficiency of resource allocation, the equitable distribution of resources, and the feasibility of implementation (Table 8.1).

Multimodal transportation planning presents a special challenge to the evaluation process. As noted in a review of innovative practices, "multimodal transportation planning is best carried out when all modes are analyzed simultaneously and interactions among the modes are accounted for" [Transmanagement, Inc., 1998]. Being able to compare modal alternatives in a way that does not bias the results toward one or another of the modes is a significant goal of the evaluation process.

Although defining alternatives occurs early in the process, evaluation procedures are often used to help determine the suitability of proposed alternatives. Table 8.2, for example, shows the evaluation criteria used in Denver to define, refine, and evaluate major investment alternatives during different steps of the planning process. The initial step in this process, called fatal flaw analysis, identifies those impacts of the alternative so severe or constraining that the alternative is clearly infeasible, even without obtaining further information. The criteria for detailed evaluation are divided into two categories that are usually found in planning studies—system performance and system impacts.

Table 8.1   Questions that form the basis of evaluation

Appropriateness

What information on impacts and trade-offs is required for the decisions that need to be made?

Do the objectives attained by the alternative reflect previously specified community goals and objectives?

Equity

What is the distribution of benefits and costs among members of the community?

Do any groups pay shares of the costs that are disproportionate to the benefits they receive?

Effectiveness

Is the alternative likely to produce the desired results?

To what extent are planning and community goals attained through the implementation of the alternative?

Adequacy

Does the alternative correspond to the scale of the problem and to the level of expectation of problem solution?

Are there other alternatives that might be considered?

Efficiency

Does the alternative provide sufficient benefits to justify the costs?

In comparison with other alternatives, are the additional benefits provided (or foregone) worth the extra cost (or cost savings)?

Implementation Feasibility

Will the funds be available to implement the alternative on schedule?

Are there any administrative or legal barriers to alternative implementation?

Does the organizational capability (e.g., staff and expertise) exist to implement the alternative?

Are there groups who are likely to oppose the alternative?

Sensitivity Analysis

How are the predicted impacts modified when analysis assumptions are changed?

What is the likelihood of these changes occurring?

The Federal Transit Administration (FTA) has suggested six principles for the development of a set of alternatives:

1. Alternatives should be defined in terms of their design concept and scope.

2. Alternatives should respond directly to a clear statement of the purpose of and need for transportation improvements . . . and to a potentially broader set of considerations outlined in the goals and objectives for the corridor.

3. Alternatives should be developed through a process that considers all reasonable options and then uses appropriate analyses to narrow the focus on the most competitive alternatives.

4. The set of alternatives should be structured to provide a range of options to decision makers that illustrates the trade-offs among costs, transportation benefits, and other impacts.

5. Each alternative should be defined to make it as competitive as possible and then refined in light of the information developed on its performance.

6. The alternatives should be identified and refined in an open, well-documented process that obtains appropriate input and reviews from all participants [National Transit Institute et al., 1996].

Table 8.2   Evaluation criteria for major investment studies in Denver

Prescreening Criteria (Unsuitability/Fatal Flaw Analysis)

- Is the alternative consistent with regional goals and objectives?
- Is the alternative affordable?
- Does the alternative have an irresolvable environmental impact?
- Does the alternative have irresolvable community or agency opposition?
- Is the technology proven in revenue service?

Screening Criteria

- How consistent is the alternative with regional goals/policies?
- How affordable is the alternative?
- What are the primary environmental impacts of the alternative?
- How well does the alternative address the corridor's mobility problems?

Detailed Level Evaluation

Performance Criteria	*Impact Criteria*
• Project person-carrying capacity	• Wetlands
• Potential person-carrying capacity	• Parks, historic properties, wildlife refuges
• Maximum link utilization	• Air quality
• Number of users	• Endangered species
• System utilization (regional basis)	• Environmental justice
• Corridor congestion	• Displacements
• Travel times	• Neighborhood disruption/community cohesion
• Regional delay	
• Travel time reliability	• Hazardous materials
• Impact to goods movement	

SOURCE: As reported in Smith, 1999

In the United States, the definition of alternatives for corridor- and project-level studies where federal dollars might be spent must include a no-build and a transportation system management (TSM) alternative. The TSM alternative includes low-cost improvements such as traffic engineering, transit service, and travel demand management strategies that are designed to enhance the efficiency of the existing transportation system. The evaluation process thus needs to assess not only the impacts of large-scale infrastructure improvements, but also the likely effects of system operation and service changes.

Once the alternatives are defined, evaluation should have the following characteristics in order to be effective within a decision-oriented planning process:

1. *Evaluation should focus on the decisions being faced by decision makers.* The selection of the evaluation criteria, the planning time horizon, the scope of the analysis (i.e., the factors to be included), and the scale of analysis will all depend on the nature of the decision(s) to be made.

2. *Evaluation should relate the consequences of alternatives to goals and objectives.* Similar to the first point, the information provided by the evaluation process must be directly tied to the stated goals and objectives of the decision-making and planning processes. Because these objectives can relate to a wide-ranging set of issues, the evaluation process must be able to deal with both quantitative and qualitative information and to potentially give as much attention to social, economic, and environmental impacts as it does to transportation system performance.

3. *Evaluation should determine how different groups are affected by transportation proposals.* Changes to the transportation system can affect different groups in a variety of ways. An improvement to a transportation facility, for example, would provide benefits to the users of the facility but, at the same time, could create "costs" to nearby residents in terms of higher noise levels, decreased air quality, and increased congestion on local streets.

4. *Evaluation should be sensitive to the time frame in which project impacts are likely to occur.* By its very nature, transportation investment involves trading present expenditures for future net benefits. Evaluation should identify the time stream of benefits and costs as they are expected to occur over the useful life of the project. Because it is often difficult to predict future benefits and costs, any uncertainty associated with such predictions should be clearly documented.

5. *Evaluation should, in the case of regional transportation planning, produce information on the likely impacts of alternatives at a level of aggregation that permits varying levels of assessment.* Because many of the significant impacts of transportation actions occur at the local level, the evaluation process must be able to assess impacts at this level. A subarea impact assessment is especially important in determining the distributional consequences of transportation investments.

6. *Evaluation should analyze the implementation requirements of each alternative.* Implementation feasibility can include funding, labor requirements, construction capability, engineering and design expertise, and public outreach. The evaluation process must explicitly consider these factors so that the plans and projects emerging from the planning process are indeed feasible.

7. *Evaluation should assess the financial feasibility of the actions recommended in the plan.* ISTEA in 1991 required for the first time that transportation plans and programs be financially constrained. This means that the sources of funding for all of the proposed projects should be identified. Such sources include not only the traditional fuel tax or general fund revenues, but also public/private sector funding arrangements such as donations, tolls, and impact fees.

8. *Evaluation should provide information to decision makers on the value of alternatives in a readily understandable form and in a timely fashion.* Because most decision makers are not experts in the field of transportation planning, the information produced by the evaluation process should be presented in a format that is easily understood and that highlights the trade-offs among the alternatives. Evaluation should also occur throughout the planning process to structure the information flow that takes place among the participants in the planning process.

These eight characteristics suggest an evaluation process that involves interested parties, that summarizes in understandable terms the key issues to be considered by decision makers, and that guides much of the technical analysis activity during the planning process. Not only does evaluation identify the type of information that must be produced (and, hence, the data that must be collected), it also determines the procedures to be used and the schedule that must be followed to produce timely input into decision making. Above all, to be useful to planners, an evaluation framework must be adaptable to different problem contexts.

A definition of benefits and costs is basic to evaluation at any level of application. In recent years, a great deal of attention has been given to how both are defined and measured. The next section presents an overview of the key concepts associated with these important steps in the evaluation process.

8.2   CHARACTERISTICS OF BENEFIT AND COST MEASUREMENT

The definition of benefits and costs, and the relative weighting of each, are dependent upon the groups likely to experience them. Historically, evaluation focused on the benefits and costs relating to transportation system users, with only those readily expressed in dollar terms being included in the evaluation process. As has been noted throughout this book, however, the transportation system is strongly linked to other socioeconomic and environmental systems. Changes to the transportation system can have both direct and indirect impacts on the social and economic activities in a metropolitan area. The challenge in evaluation is to understand these complex relationships and to identify a set of benefits and costs that accounts for the diverse impacts associated with any particular action.

8.2.1   General Characteristics of Benefits and Costs

Several characteristics of benefits and costs affect how they are used in the evaluation process. First, there is a difference between *real* and *pecuniary* impacts. Real benefits are those realized by the final consumers of a project or that add to a community's overall welfare. Pecuniary benefits are gained at the expense of other individuals or groups (i.e., a redistribution of income). The increase in land values resulting from improved transportation accessibility is a good example of a pecuniary benefit. Although the owners of the land will benefit monetarily from changes to the transportation system, consumers of the land will ultimately have to pay these costs in terms of increased purchase prices or rents. From a societal viewpoint, there will be no net welfare gain for the economy. In general, strictly pecuniary effects should not be included in an evaluation unless a redistributional impact among income groups is a major objective of the investment [Musgrave and Musgrave, 1976]. From a political perspective, however, knowing such pecuniary impacts might be important for identifying those groups that would support or oppose specific alternatives.

Second, benefits and costs can be *direct* or *indirect*. Direct benefits and costs are related specifically to the objectives of the investment, while indirect benefits and costs are, in some sense, by-products. For example, a transit improvement can provide direct benefits to users in terms of reduced travel times and indirect impacts to the community of increased demand for housing near a rail station and a resulting change in community character [Louis Berger and Assocs., 1998]. Another indirect impact of such an improvement might be the short-term travel time savings realized by users of other facilities due to the diversion of travelers who abandoned their previous mode to use the new service. In many cases, the distinction between direct and indirect effects is difficult to determine. For this reason, a clear statement of planning goals and objectives, along with a statement of how the alternatives under consideration relate to these goals, is important to the evaluation. Table 8.3 shows the possible direct and indirect effects of transportation projects on ecosystems.

The real costs of an investment can also be both direct and indirect. The direct costs of an alternative include its initial capital costs as well as the costs associated with operation, maintenance, and rehabilitation. Indirect costs include expenditures required of other government agencies (e.g., additional costs for police agencies to enforce speed limits and parking restrictions or to provide protection at transit terminals); and the societal costs of additional air and noise pollution, increased congestion, and any adverse impacts on the viability of alternative transportation services (such as declining transit ridership due to highway investment).

Table 8.3   Some possible effects on ecosystems from transportation projects

Direct Effect	Indirect Effect	Some Manifestations	Possible Consequences (From Individual Effects or Combinations of Effects)
Physical alteration—habitat destruction	Habitat fragmentation	• Creation of smaller patches • Creation of barriers • Creation of more edges • Draining or ponding	• Local extinction of wide-ranging species • Loss of interior or area-sensitive species • Direct mortality impacts • Erosion of genetic diversity and amplification of inbreeding • Increased probability of local extinction from small population sizes and reduced likelihood of re-establishment • Increased abundance of weedy species • Reduced biological diversity
Introduction of pollutants—toxicity and behavioral effects	Degradation of habitat	• Changes in reproductive behavior and rates • Changes in food sources	• Changes in community structure • Changes in ecosystem structure and function
Alteration of natural presses—e.g., hydrology, species interactions	Altered energy flows	• Changes in population sizes from effects on births, deaths, immigration, and emigration • Vegetative changes	• Change in ecosystem ability to support life

SOURCE: Louis Berger and Assocs., 1998

Third, the differences in the degree to which benefits and costs can be "measured" result in *tangible* or *intangible* impacts. Tangible benefits and costs can be assigned monetary values, with benefits being measured by the price a service would command in the marketplace and costs being measured by the price of the inputs needed to deliver the service; or they can be measured in quantifiable, nonmonetary terms. Intangible benefits and costs conversely are those that cannot be easily measured or associated with prices in the marketplace (e.g., the aesthetic value of a bridge design). Intangible benefits and costs can nevertheless play an important role in the decision-making process and should thus be included in some form in alternatives assessment. They should be described as explicitly as possible and, where feasible, use subjective terms that indicate relative worth (e.g., high, medium, or low).

Fourth, benefits and costs can be defined as being either *internal* or *external* to the study area. Major improvements to urban transportation facilities are often paid for by local municipalities (e.g., those belonging to a regional transit district), while the benefits of such improvements, both direct and indirect, can accrue to groups beyond the jurisdictional boundaries of these municipalities. For example, improvements to a major metropolitan transportation facility can benefit both short- and long-distance travelers, thus making a determination of exactly who will benefit somewhat difficult. Similarly, an improvement in air quality due to transportation improvements in one region could benefit another region downwind.

Fifth, a distinction between *user* and *nonuser* costs is especially relevant to transportation planning. For many years, the most commonly used evaluation criteria focused on user benefits and costs that were usually measured in monetary terms. The monetary value of user travel time savings was an important benefit incorporated into evaluation and was often the single largest contributor to a project's overall benefits. In recent years, the definition of benefits and costs has been expanded to include many nonuser impacts, including the dislocation of businesses and homes, environmental degradation, and impacts on land-use patterns.

Sixth, a distinction is also made between *total* and *incremental* costs and benefits. An estimate of total costs, for instance, includes the total outlay of dollars used to construct and operate an alternative. Incremental costs are those that represent the additional costs associated with the proposed changes to the existing system. By using the incremental cost method, planners can (1) gain a better understanding of the interrelationship among transportation system components; (2) clearly identify the separable costs associated with project implementation so that decision makers are aware of the highest cost components and the subsequent demands on different revenue sources needed to cover additional costs; and (3) obtain information on the sensitivity of certain components of the proposed change. For example, those components having high costs associated with them, and yet whose impacts are uncertain, are possible targets for more detailed analysis. In order to use an incremental cost approach for evaluation, the planner must also have the capability to perform a marginal or incremental assessment of benefits.

One final observation needs to be made with respect to the concept of benefit and cost accounting. There is a potential that some impacts will be overlooked while others may be double counted (Table 8.4). Double counting means that the positive or negative effects are already being attained indirectly and thus should not be considered explicitly in the analysis. For example, as noted earlier, increases in the value of land due to transportation investment will be paid for by those who buy or rent the land and are not a net economic benefit to society. The measurement of benefits and costs should be based on theoretically sound and tested procedures. Perhaps most important, the evaluation framework must be applied consistently across all alternatives to provide a valid basis for comparative evaluation. This consistency is especially important in the application of the same definitions of costs and benefits to all alternatives under consideration.

Table 8.4   Items in evaluation to avoid double counting

Cost Item	Omit	Include
Fares	Always	
Tolls	If used for highway construction, finance, or operations and maintenance	If used for costs not covered elsewhere in the cost estimate
Parking charges	If used to cover capital or operations and maintenance costs that are covered elsewhere in cost estimates	If used for costs not covered elsewhere in cost estimates
Fuel taxes or other highway user fees	Always (it is a transfer payment; not a measure of resources consumed)	
Insurance costs	Portion attributable to accident costs (usually assume all of it is attributable)	
Land value increase	Always	

SOURCE: Cohen, Stowers, and Petersilia, 1978

8.2.2   Estimates of Economic Benefits and Costs

The six characteristics of benefits and costs previously described—real versus pecuniary, direct versus indirect, tangible versus intangible, internal versus external, user versus nonuser, and total versus incremental—are important for assessing the value of transportation actions. The definition of what constitutes a benefit or cost, however, is central to the entire notion of evaluation. Of the two, the definition of a benefit has been the more difficult.

Benefits   In simple terms, benefits are the desirable effects of an investment, where "desirable" suggests some positive impact on user, community or decision-maker goals and objectives (see [Beimborn et al., 1993] and [Cambridge Systematics et al., 1996] for a discussion of the wide range of benefits associated with transit investment). Much of the literature on the definition of benefits focuses on the net increase to economic welfare. This focus has occurred for macro-level benefits (i.e., at the level of a state or regional economy) and at the micro level (i.e., the benefits accruing to individual travelers). At the macro level, the economic objectives of many transportation investments can relate to the distribution of economic activity and/or to productivity gains. As discussed in Sec. 3.2.4, productivity gains do represent a major source of economic benefit to a region. However, impacts such as job creation most often are simply a redistribution of economic activity from elsewhere and thus do not contribute any net economic benefit. Lewis [1991] notes,

> While productivity gains alone can often justify the economic costs of transportation investments, this is rarely (if ever) the case with the employment, income and other targets of regional redistribution. . . . Transportation executives need to emphasize productivity and growth over the redistribution of economic activity as the principal objectives of transportation policies and investment programs.

As noted in Chap. 3, macroeconomic impacts fall into one of three categories—generative, redistributive, and financial. Numerous methods can be used to measure each of these impacts. As shown in Table 8.5, these methods can be used for both a priori and ex post evaluations.

Table 8.5   Methods for measuring economic impacts of investments

Methods for Generative Impacts						
Impacts	**Regional Transportation Models**	**Benefit/Cost Analysis**	**Input/Output Models**	**Forecasting/ Simulation Models**	**Multiple Regression/ Econometric Models**	**Nonstatistical and Statistical Comparisons**
User benefits	Predictive	Predictive				
Employment and income growth			Predictive	Predictive		
Agglomeration/ urbanization benefits	Predictive			Predictive	Evaluative	Evaluative
External benefits	Predictive				Evaluative	
Social benefits	Predictive				Evaluative	Evaluative
Reduced development costs					Evaluative	Evaluative

Methods for Redistributive Impacts								
Impacts	**Compare Cases**	**Interviews, Focus Groups, Surveys**	**Physical Condition Analysis**	**Real-Estate Market Analysis**	**Fiscal Impact Analysis**	**Development Support Analysis**	**Regression Models**	**Statistical and Nonstatistical Comparisons**
Land development	Predictive	Predictive, evaluative	Predictive, evaluative	Predictive		Predictive	Predictive, evaluative	Evaluative
Employment and income shifts	Predictive	Evaluative				Predictive		Evaluative
Increased economic activity	Predictive	Evaluative						

(continues)

Table 8.5   Methods for measuring economic impacts of investments (continued)

				Methods for Financial Transfer Impacts			
Impacts	Compare Cases	Interviews, Focus Groups, Surveys	Physical Condition Analysis	Real-Estate Market Analysis	Fiscal Impact Analysis	Development Support Analysis	Multipliers from Input/ Output Models
Employment and income growth							Predictive
Tax impacts		Evaluative			Predictive	Predictive	
Joint development	Predictive	Predictive, evaluative		Predictive			

SOURCE: Cambridge Systematics et al., 1998

At the micro level, the definition of benefits relates to economic interpretations of how a change in price will affect consumer welfare or, in other words, the value to a consumer of a change in the price of a good (i.e., the "willingness to pay"). The quantity to be measured in this case is the amount the consumer would pay or would need to be paid in order to be as well off after a price change as before the change. The two economic measures that represent this benefit accurately are compensating variation and equivalent variation measures, which can be estimated through the use of disaggregate travel demand models. In practice, however, most measurements of economic benefit are based on the concept of consumer surplus.

As shown in Fig. 8.1, consumer surplus is the area under the market demand curve that represents the total user benefit minus the cost to users. At the original price of travel P_1, which includes out-of-pocket costs, travel time, and other factors that influence travel behavior, travelers would be willing to pay amounts to the left of point D for use of the facility. The total benefit to the number of travelers (V_1) would be equal to the area under the demand curve to the left of point D (area $OCDV_1$). The net benefit or "consumer surplus" to users of the facility at price P_1 is this area $OCDV_1$ minus area OP_1DV_1, or area P_1CD. A change in total net user benefit stemming from a price change would thus be the change in consumer surplus, which in Fig. 8.1 would be area P_2CE minus P_1CD, or the shaded area P_2P_1DE. This total net user benefit consists of two types of benefit: that gained by the original V_1 travelers (represented by area P_2P_1DA) and that realized by the new users of the facility induced to travel because of a "lower" travel price (represented by area ADE). Note that the "change in price" that produces a benefit to users can consist of many components, and great care must be used in defining this change. From a theoretical perspective, the dynamics of this change over time (i.e., how and when a new equilibrium is established) are significant considerations.

Based on this concept of total net user benefits, numerous equations and theoretical formulations have been developed to define benefits according to different underlying assumptions. Through simple geometry, the change in consumers' surplus can be approximated as

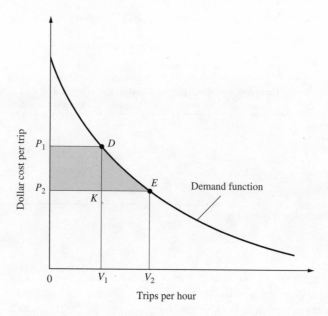

Figure 8.1   The measurement of consumer surplus

$$NB = 1/2\left(P_1 - P_2\right)\left(V_1 + V_2\right)$$ [8.1]

where

    NB = net benefits to users
    P_1 = original price of travel
    P_2 = reduced price of travel
    V_1 = volume of travel at P_1
    V_2 = volume of travel at P_2

The evaluation of user benefits is primarily a process of determining how great a reduction in costs will occur if an improvement is made (e.g., the reduction in the price of travel due to an improvement). From the perspective of the user, reductions in the number of accidents, trip costs, and travel time comprise the most direct benefits of transportation projects. The most significant of these is the reduction in user travel times (e.g., between 72 percent and 81 percent of the benefits of the U.S. interstate highway system can be attributed to travel time savings [Fallon et al., 1970]). These types of benefits are usually incorporated into the evaluation process by applying a unit monetary value to each measure and then multiplying this value by activity factors that represent the total amount of travel on the system. For example, vehicle operating cost savings of $0.05 per mile x 100,000 vehicle miles saved equals $5000 of vehicle operating benefits. In each case, however, the determination of appropriate unit values has created considerable controversy among researchers and practitioners, a good illustration being the appropriate measure for "value of time."

Benefits from Travel Time Savings   The underlying basis for assigning a monetary value to travel time is that time not spent in travel can be used for other activities having economic value. In the case of work travel, a reasonable estimate of the value of time for work trips can be related to the traveler's wage, about 50 percent of the before tax wage rate [Small, 1999]. For other trip purposes, the value of travel time becomes less obvious [Henscher, 1989; Waters, 1996]. In addition, the value of time has been found to be sensitive to trip purpose, a traveler's income level, and the length of the trip. Table 8.6, for example, shows the results of a survey sent to 2500 residents in a highway corridor in southern California. The purpose of the survey was to determine the values of time and of travel time reliability in congested conditions. The standard deviation of travel time was used to measure reliability. For the average length of trip and for median household income, the value of $12.60 per hour of standard deviation was estimated for travel time variability as compared to a value of travel time of $5.30 per hour for normal travel time. This indicates that travelers place a much greater value on travel time reliability [Small et al., 1999]. A similar result was found in a study of value of time associated with travel time variability due to freeway incidents [Cohen and Southworth, 1999]. Value of time measures can be found in tabular or graphical form in many evaluation handbooks. A 1997 study of value of time recommended the values shown in Table 8.7 [FTA, 1997]. An excellent overview of the use of value of time in transportation studies is found in [Henscher, 1997].

Benefit from Reduction in Crashes   Another benefit attributable to facility improvement is the reduction in crashes and thus the value of reduced fatalities and injuries. As was the case with the value of travel time, estimating the cost of traffic crashes is a difficult task, as the costs can vary from one urban area to another and can differ by type of crash. In the case of a fatal crash, one faces the difficult problem of

Table 8.6   Value of travel time and reliability of travel time

Household Income (000s)	Value of Travel Time ($/Hour)
15	$2.64
35	3.99
75	5.34
95	6.70
	8.05

Trip Type and Income	Value of Reliability ($ Per Minute of Standard Deviation)
Work trip, higher income	$0.26
Work trip, lower income	0.22
Nonwork trip, higher income	0.21
Nonwork trip, lower income	0.17

SOURCE: Small et al., 1999

Table 8.7   Ranges for hourly values of time

Category	1995 Dollars Per Person Hour
Local Travel	
Personal (auto and transit)	$6.00–$10.20
Business (auto and transit)	15.00–22.60
All purposes (auto and transit)	6.40–10.70
Walk, waiting, and transfer time, personal	17.00
Walk, waiting, and transfer time, business	18.80
Truck driver	16.50
Intercity Travel	
Personal (auto and transit)	$10.20–$15.30
Business (auto and transit)	15.00–22.60
All purposes (auto and transit)	10.40–15.70
Truck driver	16.50

SOURCE: Smith, 1999

estimating the value of a human life. A crucial step in estimating crash costs is determining what elements should be included. For example, should *net* future earnings or *total* future earnings of crash victims be used? Should nonmonetary items such as loss to family or community service be included? And should a value be assigned to pain and suffering?

From an economic perspective, the value of safety improvements relates directly to the aggregate willingness to pay for a reduction in the risk of a crash. Small [1999], for example, notes in a review of safety studies that this willingness to pay is approximately $5 million for saving one life. In an earlier study, Small [1992] concluded that the willingness to pay to reduce the risk of a serious, but nonfatal injury was about 10 percent of the willingness to pay to reduce the risk of a fatality. Given the large number of injury crashes, he concluded that this adds significantly to the estimated benefits of safety improvements. The willingness-to-pay concept as used by the U.S. DOT includes surveying individuals to determine their willingness to pay for increased health and safety as well as estimating

- The wage differential paid to induce people to take risky jobs.
- The values implied by product demand and price in markets for safety-related products such as safer automobiles, smoke detectors, houses in areas with little air pollution, or cigarettes as more has become known about their health effects.
- The trade-offs people make between time, money, comfort, and safety through their speed choice, use of pedestrian tunnels, safety belt use, and purchase and use of child safety seats and motorcycle helmets [FHWA, 1999].

A representative estimate (1995 dollars) of societal benefit for crash or death costs averted is shown in Table 8.8. These values should be viewed with caution because of the variation in crash costs among urban areas, by crash type, and according to the socioeconomic status of those injured. A good overview of the societal costs of transportation crashes is found in [Miller, 1997].

Benefit from Reduced Cost of Vehicle Operation The final user benefit usually considered in transportation project evaluation is the change in costs of vehicle operation. Included in this category of costs are fuel, oil, maintenance and repairs, and depreciation associated with vehicle wear. Typical unit values of operating costs for vehicles on freeways and arterial streets can be found in many handbooks and automobile industry studies and are also calculated from highway cost allocation studies (see, for example, [U.S. DOT, 1997]). These costs depend on the characteristics of the vehicles involved, the vehicle mix, roadway design, traffic levels, drivers and trip-making behavior, and the respective costs in a particular metropolitan area.

Table 8.8   Crash costs (1995 dollars)

Urban Functional System	Death	Cost/Nonfatal Injury	Property Damage Cost/Crash
Interstate		$27,047	$5148
Other freeway/expressway		35,002	6435
Other principal arterial	$2.6 million	28,638	6435
Minor arterial		39,775	6435
Collector		31,820	5148

SOURCE: FHWA, 1999

These three measures of user benefit—value of time, accident costs, and vehicle operating costs—are commonly used in the comparative assessment of project and system alternatives. Due to difficulties of measurement, these measures are often considered with average unit values. However, some economists would argue that marginal unit values, for example, the cost of adding one additional unit to the traffic stream, would be the most appropriate cost measure. Given that these measures are so important in evaluation, planners should (1) review with decision makers the unit values to be used in the assessment and (2) conduct tests on how sensitive the evaluation results are to the values used.

Costs The determination of costs is generally considered to be a much easier task than the assessment of benefits. However, depending on how "costs" are defined, this perception might be inappropriate. Some of the misconceptions that lead to this perception include (1) little distinction is typically made between dollar expenditures and total costs, the latter including some representation of the social costs and opportunity costs of investment; (2) nonquantifiable cost measures are customarily dismissed as noncost considerations or as negative benefits; and (3) benefits are considered to occur over longer periods of time while most costs are assumed to occur only early in project implementation [Quade, 1975]. As transportation planning becomes more integrated with sustainable development goals, the estimation of "costs" could become as challenging as that for benefits.

Just as different definitions and classification schemes apply to benefits, there are also different ways of defining and aggregating costs. The cost classification scheme chosen for an evaluation depends on the objectives of the planning effort, the amount of detailed data needed to determine costs, and the format requirements of government regulations. For example, costs could be determined on the basis of who must bear them (e.g., agencies, system users, or system nonusers), the components or commodities purchased (e.g., in the case of user costs, vehicle depreciation, fuel and oil costs, maintenance costs, insurance fees, time costs, and fares), or the activities with which they are associated (e.g., research and development, planning and design, right-of-way acquisition and construction, finance, operating, maintenance, and management). No single cost scheme will be appropriate for all purposes. At the project level, cost estimating entails defining typical cross sections of the facilities being considered, multiplying them by a unit cost, and summing over all such sections in the project. For a metropolitan plan evaluation, however, the costs will be greater than simply the engineering costs of building a facility.

8.2.3   Social Costs of Transportation

The previous section focused on those benefits and costs that have been historically included in evaluation. Ever since the early 1990s, however, numerous studies have examined the much broader social costs of transportation, in particular, the question of whether motor vehicle users pay the true costs they are imposing upon society. These costs include not only those associated with providing the transportation infrastructure, but also the costs associated with such things as personal illness due to motor vehicle pollution and the costs of delay imposed upon other users of a facility when a vehicle is added to a traffic flow. Delucchi [1997] provides the following classification of motor vehicle costs that would be considered in a social cost analysis:

1. *Personal nonmonetary costs of using motor vehicles, such as,*

   uncompensated nonwork travel time.

   personal time working on motor vehicles.

   noise inflicted on oneself.

   air pollution inflicted on oneself.

2. *Explicitly priced private sector motor vehicle goods and services, net of producer surplus, taxes and fees, such as,*

   annualized cost of the entire car and truck fleet, excluding taxes.

   automobile insurance.

   parking away from the residence.

   overhead expenses of business, commercial, and government fleets.

3. *"Bundled" private sector goods (implicitly priced), such as,*

   annualized cost of nonresidential off-street parking included in price of goods and services or offered as an employee benefit.

   annualized cost of home garages and other residential parking included in the price of housing.

4. *Government services charged partly to motor vehicle users, such as,*

   annualized cost of public highways and highway maintenance.

   highway law enforcement.

   environmental regulation, protection, and clean up.

   annualized cost of municipal off-street parking.

5. *Monetary externalities (unpriced), such as,*

   costs of travel delays caused by others.

   accident costs not paid for by responsible party.

   price effect of using petroleum fuels for motor vehicles.

6. *Nonmonetary externalities (unpriced), such as,*

   air pollution inflicted on others.

   accidents; pain, suffering, and death not paid by responsible party.

   extra uncompensated travel time due to delays caused by others.

   global warming.

   noise and water pollution inflicted on others.

   habitat and species damage.

   aesthetics of infrastructure and vehicle intrusion into visual environment.

All of these costs, and many more not in the preceding list, can be incorporated into a more complete picture of what the true costs are of motor vehicle uses. Several studies have in fact attempted to do this. However, the results vary significantly based on the assumptions made concerning how costs should be accounted for. Gomez-Ibanez [1997], for example, reviewed five studies that examined the underpayment of motor vehicle users to society (or alternatively, society's subsidy to motor vehicle users). Table 8.9 summarizes the results of these studies. As shown, the estimated underpayment ranges from 3.4 cents per passenger mile to 55.3 cents per passenger mile. Gomez-Ibanez noted that some of the variation is due to the different types of trips considered; however, "much of the variation is due to different estimating methods, including disagreements as to whether certain costs should be classified as externalities or not." He concluded that the results of these studies—that motor vehicle users do not pay their total societal cost—is probably true but that the amount of this underpayment is not as much as indicated by these studies.

Social cost accounting, or more likely a cost accounting system that takes into account sustainability factors such as ecosystem health, will likely continue to be of interest to transportation planners (see, for example, [National Academy of Engineering, 1996; World Bank, 1996; National Research Council, 1997]). The primary benefit of such cost accounting is that it provides decision makers with yet another source of information on which to gauge the relative worth of alternatives. As long as the assumptions and applications are consistent across all alternatives, the measured differences should provide useful insights (for a thorough review of the social cost of motor vehicle use in the United States see [Murphy and Delucchi, 1998]).

Table 8.9   Estimates of social costs and societal subsidy for motor vehicle use (cents per passenger mile)

	World Resources Institute	Transport and Environment	National Resources Defense Council	Conservation Law Foundation	Litman
Government Facilities and Services					
Capital	0.8		0.4	2.8	2.2
O&M	0.9		2.4	0.3	3.5
Other government	1.8		0.3–0.9	1.2	1.4
Subtotal	3.4		3.1–3.7	4.3	7.1
Externalities					
Congestion			0.4		15.5
Air pollution	1.0	3.82	4.0–7.0	6.6	7.5
(climate change portion)	(0.7)	(0.9)	(2.2–4.6)		(1.1)
Noise pollution	0.1		0.1–0.2	0.1	0.9
Water pollution			0.1		1.2
Solid waste					0.2
Crashes	1.4	2.74	3.3	0.6	3.2
Energy	0.7		1.5–5.0	1.9	2.6
Parking	2.7		0.8–3.2	5.2	10.9
Other			0.01		8.2
Subtotal	5.9	6.8	10.2–19.2	14.4	50.2
User Payments					
Fares, tolls			0.0	0.0	0.0
Taxes		3.4	0.7	3.0	2.0
Subtotal	1.0	3.4	0.7	3.0	2.0
Net Subsidy	8.3	3.4	12.6–22.2	15.7	55.3

SOURCE: Gomez-Ibanez, 1997; as found in Kageson, 1993; Litman, 1994; MacKenzie et al., 1992; Apogee, 1994; Miller and Moffet, 1993

8.2.4   Distributional Impacts

As noted in previous sections, the distributional impacts of transportation projects are an important part of the evaluation process. GIS-based evaluation will provide important information on such impacts to the transportation planning process. The types of questions that will be part of this effort include the following [Marchese, 1999]:

What are the travel activity patterns of different income and racial groups?

     Concentrations of minority and low-income populations within metropolitan area.

     Use of transportation modes by race and income.

Population by race and income within accessible distance to transportation facilities.

Car ownership by race and income.

Do low-income and minority populations shoulder a disproportionate share of the burdens of transportation facilities?

Comparison of carbon monoxide exposure by race and income.

Relocation of homes and businesses due to highway or transit construction by race and income of owners.

Comparison of locations of bus depots by race and income levels of communities.

Comparison of reduction or elimination of green space by highway or transit construction in communities of different races or income levels.

Demographics of location of current or planned air pollution monitors.

Do low-income and minority populations receive a proportionate share of transportation benefits?

Access to jobs by race and income.

Access to other quality of life destinations by race and income.

Number of destinations available by transit to communities of different races and incomes.

Commute times by race and income by mode of transportation.

Frequency of transit service by race and income of community being served.

Ratio of transit seat-miles to total number of passengers by race and income of population being served.

Cost of travel compared by race and income.

Number of bike and pedestrian accidents occurring in communities of different races and income levels.

Customer satisfaction.

Where are transportation investments being spent with respect to populations of different races and income levels?

Comparison of financial investments in transportation by mode to the use by race and income.

Comparison of financial investments in transportation by location to race or income level of community being served.

Analysis such as that shown in Fig. 8.2 will likely be used to show how proposed transportation investments relate to concentrations of low-income and minority populations. The important lesson emerging from these experiences is that changes to the transportation system—and, in particular, changes to the funding of services—can have substantial equity impacts that adversely affect those least able to afford such changes.

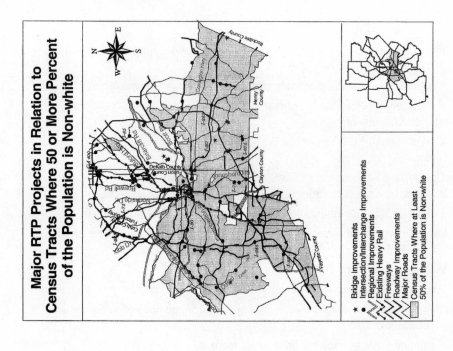

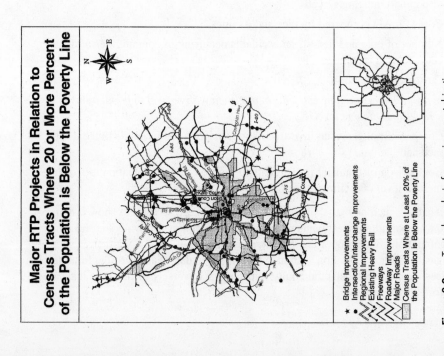

Figure 8.2   Typical analysis for environmental justice.
¡ SOURCE: Atlanta Regional Commission, 2000

8.3 A FRAMEWORK FOR EVALUATION

Figure 8.3 shows a hierarchy of benefits and costs commonly considered as part of a transportation planning process. In this case, the hierarchy is related to the evaluation of transit investments, but certainly it could be used to evaluate any form of transportation system change [Cambridge Systematics, Inc. et al., 1996]. A typical way of presenting the information that results from this evaluation to decision makers is an information tableau or matrix as shown in Fig. 8.4. Such a matrix relates the consequences of each alternative to the attainment of goals and objectives that are represented by measures of effectiveness (MOEs). The overall effectiveness of each alternative is then a determination by decision makers of how one alternative's values for all MOEs compare to those for the other alternatives. The significance of this approach to evaluation is its recognition that the final authority and responsibility for choice lie with decision makers. The evaluation methodology does not produce the "correct" decision, rather, the approach produces a structured set of information that can be used by decision makers to choose the preferred alternative.

Section 4.3.2 described how MOEs can be used in a transportation planning study; thus, this discussion will not be repeated here. However, one aspect of the definition of MOEs that needs reemphasis is that they can be targeted on affected population groups or geographic areas. Every program or project will have *some* distributional consequence, either in the benefits that accrue to or in the costs that

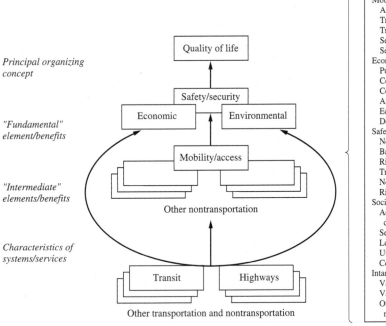

Figure 8.3 Hierarchy for impact measurement and valuation
| SOURCE: Cambridge Systematics, et al., 1996

	Alt. 1	Alt. 2	Alt. 3	Alt. 4	Alt. 5
MOE 1					
MOE 2					
MOE 3					
MOE 4					
MOE 5					
MOE 6					
Economic evaluation	B/C or NPV	B/C or NPV	B/C or NPV	B/C or NPV	B/C or NPV

Goals and objectives →

Data that measures impact and /or cost-effectiveness information

B/C = Benefit/cost ratios; for "best" alternative one must conduct incremental B/C analysis
NPV = Net present value

Figure 8.4   Evaluation matrix

will be borne by groups in the community [Starling, 1979]. These consequences can include the equitable distribution of the transportation services provided, the incidence and distribution of the externalities associated with the project, and/or the distribution of the financial costs of building and operating a transportation facility or service. A determination of which of the distributional impacts are desirable, or at least tolerable, as opposed to those that are clearly unacceptable is a value judgment that is generally made as part of the political decision-making process (influenced by laws and regulations).

The most common way of portraying distributional consequences is an impact–incidence matrix. Such a matrix contains information concerning the impacts on groups directly or indirectly affected (Fig. 8.5). By seeing the impact information displayed in this way, decision makers can identify which groups are adversely affected. The level of aggregation of the information provided in such a matrix depends on the stage of the planning process in which it is to be used. Less detail is required in the early alternatives screening phase, while more disaggregate information is needed as the process nears design evaluation. Geographic information systems (GISs) are very useful for providing this type of information in that they can identify the socioeconomic characteristics of the households that are geo-

			Impacts				
			MOE 1	MOE 2	MOE 3	MOE 4	MOE n
Groups impacted	Direct	1					
		2					
		3					
	Indirect	1					
		2					
		3					
		4					

Figure 8.5   The impact-incidence matrix

Employment density

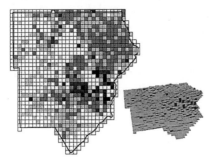

Housing unit density

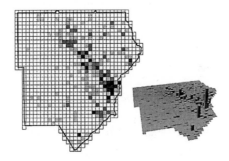

Proposed light rail lines

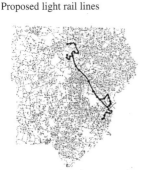

1/2 mile buffer of proposed light rail station
and surrounding land use

What potential station locations optimize accessibility?
How many employees are within walking distance of a potential station?
How many homes are within walking distance of a potential station?
What is the maximum allowable future accessibility?

Figure 8.6 GIS applications for evaluation purposes

graphically located within a certain distance of a transportation facility or service (Fig. 8.6).

As can be seen in Fig. 8.4, the evaluation matrix not only includes estimates of an alternative's benefits and costs as they relate to MOEs, but it can (and should) provide some sense of the trade-offs between level of effectiveness and cost. Perhaps most importantly, evaluation should provide information on the relative worth of alternatives when compared to one another. Transportation planners use several methods to provide this type of information: cost-effectiveness evaluation, economic evaluation, and multiobjective rating schemes.

8.4 COST-EFFECTIVENESS EVALUATION

Cost-effectiveness evaluation estimates the level of goals and objectives attainment per dollar of net expenditure [Hudson, Wachs, and Schofer, 1974; Campbell and

Humphrey, 1978]. It is a useful tool for illustrating possible trade-offs between the level of effectiveness that can be achieved for the costs incurred. Effectiveness, in this approach, is a single measure that captures as much as possible all of the benefits associated with a project. For example, the Federal Transit Administration (FTA) has developed a cost-effectiveness index of "cost per new transit rider" to evaluate new transit proposals. The measure of effectiveness in this case—new transit riders—is a surrogate for all of the benefits associated with a project alternative. Such a measure would capture such benefits as reduced highway congestion (assuming some of the new riders would divert from single-occupant vehicles) and the corresponding reductions in environmental impacts. Other types of measures that could be similarly considered include travel time savings for existing transit riders or highway users and total travel hours saved. Such measures should be used in cost-effectiveness evaluation only if they represent a large portion of the benefits associated with the alternatives or if the study is so focused on one measure that it will provide the information desired.

One of the limitations of the effectiveness measures discussed previously is that they are mode-specific. However, multimodal transportation planning should evaluate alternatives in a way that shows no bias for one mode or another. One way of doing this is to use a composite measure of direct user benefits. Mode choice models produce an overall impedance measure that reflects the characteristics of all modes available for travel (see Sec. 5.4.3). Changes in this impedance value due to changes in transportation system characteristics could represent the benefits or disbenefits that could be used in a cost effectiveness index.

The cost values in a cost effectiveness index relate to the capital, operating, maintenance, and rehabilitation costs associated with an alternative. These are often referred to as "operator or owner" costs. Section 8.2.2 noted that changes in user costs such as fares, motor vehicle costs, and travel time are usually considered as changes in benefits (positive or negative). Societal costs can also be considered as changes in benefits and thus incorporated into the effectiveness component of the cost-effectiveness index or simply portrayed as impacts in the evaluation matrix. Figure 8.7 shows an example of cost-effectiveness evaluation. In this example, seven alternatives are being considered, each exhibiting different levels of effectiveness in relation to the costs incurred. The alternatives not dominated by other alternatives lie on an "efficiency frontier." In this example, alternatives 1, 2, 4, 5, and 7 lie on such a frontier. Alternative 3 is dominated by alternative 4 because additional benefits will occur for the same costs as alternative 3. Likewise, alternative 6 is dominated by alternative 5 because the same benefits can be obtained as alternative 6 for much lower costs. Choosing one alternative over another is inherently a value-laden one and depends on the willingness of the decision makers to trade-off level of effectiveness for cost. Thus, cost-effectiveness evaluation focuses attention on the *differences* between alternatives. In Fig. 8.7, the ratio of additional benefits for costs incurred for alternative 2 compared to the no-build alternative is the highest of any comparison between alternatives (that is, the line joining the two has the greatest slope). More benefits will occur for additional costs by implementing alternative 4 and so on along the efficiency frontier. However, as shown, the incremental increase in effectiveness in relation to additional costs declines (once again represented by the slope of the line joining two alternatives). Decision makers must judge whether this additional benefit is worth the added costs. The role of cost-effectiveness evaluation is thus one of providing "useful information to decision makers in a well-thought-out discussion of these differences, including explanations of why the differences arise" [National Transit Institute et al., 1996].

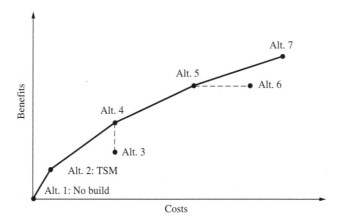

Figure 8.7   Example of cost-effectiveness evaluation

Figure 8.8 shows an example of cost-effectiveness evaluation for a narrowly defined problem—how to reduce salt infiltration into a reservoir bordering a state highway that had been subject to snow and ice removal through the use of salt. The question facing the highway agency was how to meet a water-quality objective while still keeping the road safe during winter months. The measure of effectiveness was defined as cost per long term reduction in sodium concentration (milligram/liter) in the reservoir. The strategies considered resulted in different levels of cost-effectiveness. The most cost-effective strategies were those to the left of the $10,000 per milligram/liter slope in that alternatives 3a and 3b dominated all others to the right of this boundary. The alternative selected was chosen because of the reduction of sodium concentration desired and the level of budget available to solve this problem.

Although providing useful information to decision makers, the cost-effectiveness approach does not indicate which alternative is the "best." Such information can be extremely important to decision makers. Two approaches for doing this include (1) economic evaluation techniques (e.g., benefit/cost ratios, net present value, annual worth, and rate of return) and (2) the use of weighting schemes that produce a "score" for each alternative. Economic evaluation techniques estimate the monetary value of the benefits and costs for individual plans/projects and their comparative worth. In these approaches, the costs and benefits for each year of the useful life of a project are estimated, discounted to a base year, and then compared on the basis of some decision rule; for example, the ratio of benefits to costs must be greater than 1.0. Being able to compare different streams of monetary benefits and costs over time requires that there be some method for a fair comparison. This is found in the concepts of discounting and capital recovery.

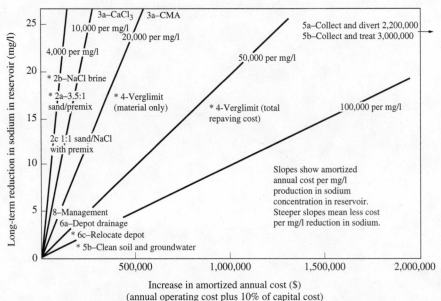

Figure 8.8   Cost-effectiveness evaluation of snow removal strategies
I SOURCE: Massachusetts DPW, 1988

8.5 ECONOMIC CONCEPTS OF DISCOUNTING AND CAPITAL RECOVERY

The worth of an individual alternative can be determined by estimating the monetary return on an initial capital investment. A concept critical to doing this is the changing value of money over time, commonly known as the *real value of money*. Put simply, if one invests $10,000 today with an interest rate of 10 percent for a 1-year period, at the end of 1 year, the investment will nominally be worth

$$\$10,000 + (\$10,000) \times (0.10) = \$11,000$$

Thus, $10,000 now is equivalent to a future sum of $11,000 at the end of year 1 at 10 percent interest. The $10,000 is the *present* or *discounted* value of the future $11,000. In this case, the interest rate is the rate of return paid to the investor. As a general convention, the term *interest rate* will be used to denote the process of determining future values; the term *discount rate* will be used to denote the process of discounting benefits and costs to an earlier time.

The equation used to compare sums of money that exist at two distinct times is

$$F = P \times (1 + i)^n \tag{8.2}$$

where

F = future amount of money
P = present amount of money
i = interest rate
n = periods of repayment or project life

Using the previous example, the future value of $10,000 after 2 years would be

$$F = \$10{,}000 \times (1 + 0.10)^2 = \$12{,}100$$

Alternatively, the present value of future cost or revenues can be determined by rearranging Eq. [8.2] and where r is the discount rate.

$$P = F/(1+r)^n \tag{8.3}$$

Again using the previous example, the present value of a future sum of $12,100 with a 10 percent discount rate is

$$P = \frac{\$12{,}000}{(1+0.10)^2} = \$10{,}000$$

In Eq. [8.3], the factor $1/(1 + r)^n$ is called the *present worth factor* and can be found in tabular form with different discount rates in numerous engineering economy texts. Table 8.10 gives the present worth factors for discount rates of 5 percent, 8 percent, 10 percent, 12 percent, and 15 percent for different time periods. Figure 8.9 shows that discounting does not occur linearly over time. When determining present values for an entire time stream of costs, therefore, one should not use measures of costs and benefits at a single forecast year, a mistake commonly made.

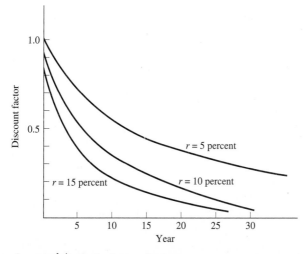

Figure 8.9   Impact of discount rate over time

An important investment consideration is the present value of a constant stream of equal payments over several periods. This form of payment is common to most home and auto mortgages. Based on Eq. [8.2], it can be shown that the relationship between an initial investment and the annual payment over n years needed to repay the initial investment is

$$A = P\left[\frac{i(1+i)^n}{(1+i)^n - 1}\right]$$

[8.4]

where A is the uniform payments required over n periods.

The term $i(1 + i)^n / [(1 + i)^n - 1]$ is commonly called the *capital recovery factor* and represents the proportion of an initial investment that has to be recouped as benefits in each of n periods in order to return the same value as was invested (see Table 8.10). Using the previous example, the annual payment required over 10 years to return the value of a present amount of $10,000 is

$$A = \frac{(\$10,000)\left[(0.1)(1.1)^{10}\right]}{(1.10)^{10} - 1} = \$1627$$

The value of money concept and associated equations are extremely important components of plan and project evaluation. For comparative purposes, the benefits and costs of alternatives are represented as occurring at a single point in time. In most cases, the time streams of benefits and costs are expressed in terms of present values (that is, one discounts to the present time) or as equivalent annual costs. An example of how these methods can be used is included later in this chapter.

As shown in Fig. 8.9, the selection of a discount rate can have a significant impact on the level of discounted benefits and costs. The options available for selecting a discount rate, which is expressed as net of inflation, include

1. The percentage rate of return that the investment would otherwise provide in the private sector.

2. The government borrowing rate for capital.

3. The "social discount rate," which recognizes the additional societal value of investment in public services and infrastructure.

4. A discount rate explicitly chosen to reflect the risk associated with an alternative.

5. A discount rate mandated by a government agency for investments using federal dollars (for example, in the United States, the Office of Management and Budget stipulates such a discount rate for federally aided projects).

It is beyond the scope of this book to outline the advantages and disadvantages of each approach in selecting a discount rate. However, the reader should realize that the selection of a discount rate for evaluation, in essence a relative weighting of present costs versus future benefits, is a value judgment that should be related specifically to the context of the alternatives being considered. The discount rate can significantly influence whether one project is more desirable than another. Accordingly, the evaluation should include a sensitivity analysis with different discount rates to determine how the relative assessment of the alternatives is affected by

Table 8.10   Present worth and capital recovery factors

Present Worth Factors					
Year	5%	8%	10%	12%	15%
1	0.9524	0.9259	0.9091	0.8929	0.8696
2	0.9070	0.8573	0.8264	0.7972	0.7561
3	0.8638	0.7938	0.7513	0.7118	0.6575
4	0.8227	0.7350	0.6830	0.6355	0.5718
5	0.7835	0.6806	0.6209	0.5674	0.4972
10	0.6139	0.4632	0.3855	0.3220	0.2472
15	0.4810	0.3152	0.2394	0.1827	0.1229
20	0.3769	0.2145	0.1486	0.1037	0.0611
50	0.0872	0.0213	0.0085	0.0035	0.0009

Capital Recovery Factors					
Year	5%	8%	10%	12%	15%
1	1.0500	1.0800	1.1000	1.1200	1.1500
2	0.5378	0.5607	0.5762	0.5917	0.6151
3	0.3672	0.3880	0.4021	0.4163	0.4380
4	0.2820	0.3019	0.3155	0.3292	0.3503
5	0.2310	0.2505	0.2638	0.2774	0.2983
10	0.1295	0.1490	0.1627	0.1770	0.1993
15	0.0963	0.1168	0.1315	0.1468	0.1710
20	0.0802	0.1019	0.1175	0.1339	0.1598
50	0.0548	0.0817	0.1009	0.1204	0.1501

changes in this discounting value. One manual has suggested that the discount rates included in this sensitivity analysis should range from a low rate that reflects the current government borrowing rate for capital (a rate that many economists feel is artificially low because of tax exemptions and the reduction of risk attributable to government backing of such borrowing) to a high rate that represents an expected private-sector return on capital [Cohen et al., 1978].

By carefully considering the choice of discount rate and by recognizing the judgmental nature of this rate through sensitivity analysis, the assessment of project benefits and costs can generate information useful to decision makers. The under-lying assumptions and consequences of discount rate selection should also be presented to decision makers so that they too can fully understand the explicit and implicit values incorporated into this selection.

8.6   COMPARATIVE ASSESSMENT METHODS

Two types of comparative assessment methods are discussed in this section: a method that focuses on a single objective of maximizing net economic benefit and another that incorporates multiple objectives.

8.6.1   Single-Objective Assessment Methods

Four methods have been used in transportation planning to determine an alternative's net economic benefit:

1. *Present worth method.* Discount the costs and benefits for each alternative to its equivalent present value and then compare these present values to get a net present worth.

2. *Annual worth method.* Determine the discounted annual equivalent benefits and costs for each alternative and then compare these annualized values to get an annual worth.

3. *Benefit/cost method.* Separating costs from benefits, discount the cash flows to their equivalent annual (or present) values and compare the equivalent benefits to the equivalent cost for each alternative. A benefit to cost ratio is determined for each alternative.

4. *Return-on-investment method.* Find the interest rate that balances present and future cash flows and compare it to a minimum attractive return rate specified before the evaluation process begins.

The use of these methods can be illustrated with the example time stream of benefits and costs shown in Fig. 8.10. The costs represent the capital costs for construction, continuing costs of operation and maintenance, and the future costs associated with major rehabilitation. The benefits represent changes in user and nonuser benefits that can be measured in dollar terms.

For the present worth method, all benefits and costs are discounted to the present time. This is represented mathematically by summing the value of the costs (or benefits) multiplied by the present worth factor appropriate for the year the costs (or

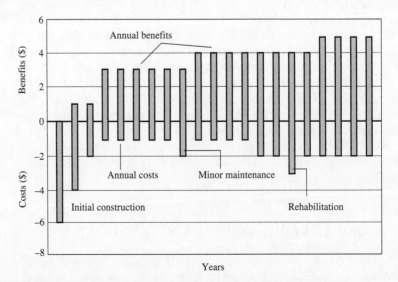

Figure 8.10   Time stream of benefits and costs for a hypothetical project

benefits) occur. The net present value of an alternative then equals the difference between the present value of the benefits and the present value of the costs. In mathematical terms,

$$NPV_{y,r} = \sum \left(pwf_{r,t}\right)\left(benefits_{y,t}\right) - \sum \left(pwf_{r,t}\right)\left(costs_{y,t}\right) \qquad [8.5]$$

where

$NPV_{y,r}$ = net present value of project y with discount rate r
$pwf_{r,t}$ = present worth factor with discount rate r and time t
$benefits_{y,t}$ = benefits of project y in time period t
$costs_{y,t}$ = costs of project y in time period t
n = economic life of project y

If this difference is positive, the alternative is feasible from an economic efficiency perspective.

By applying Eq. [8.5] to the time stream of benefits and costs shown in Fig. 8.9, it can be shown that the net present value of the project with a discount rate of 10 percent is $26.149 − $21.208 = + $4.941 million.

In the annual worth method, the discounted costs are summed and multiplied by the appropriate capital recovery factor. In the case above, the summation of the discounted costs is equal to $21.208 million, of the discounted benefits $26.149 million, and the capital recovery factor for a 20-year time period at a discount rate of 10 percent is 0.1175. The equivalent annual worth for the project is thus calculated as

Equivalent annual costs: ($21.208) (0.1175) = $2.492 million

Equivalent annual benefits: ($26.149) (0.1175) = $3.073 million

Equivalent annual worth = $3.073 – $2.492 = +$0.581 million

The benefit/cost (B/C) method develops a ratio of discounted benefits to discounted costs. If the alternative has a B/C ratio of less than 1.0, it is a likely candidate for rejection. (Once again, the ultimate decision for rejection rests with the decision makers; the B/C measure is simply an indication of efficient use of resources.) In the preceding case, the total discounted benefits were equal to $26.149 million, and the total discounted costs equaled $21.208 million. The ratio of benefits to costs for this alternative is thus $26.149/$21.208 = 1.23. At this point in the analysis, this alternative is economically feasible (although not necessarily the best choice).

Finally, the rate of return method determines the discount rate at which the present value of both the present and future costs will equal the present value of both the present and future benefits. In mathematical terms, we are trying to determine the discount rate r in which

$$\sum \left(pwf_{r,t}\right)\left(benefits_{y,t}\right) = \sum \left(pwf_{r,t}\right)\left(costs_{y,t}\right) \qquad [8.6]$$

where the terms are the same as in Eq. [8.5]. When the discount rate is identified, it can be compared to a predefined acceptable rate of return. If the calculated rate of return from the alternative is greater than this acceptable rate, the project can be considered economically feasible. In the preceding example, the rate of return for the project is just over 15 percent, which means that if the required rate of return had been set to less than or equal to 15 percent by decision makers, this project would be a feasible candidate.

In the previous examples, the four methods were used to evaluate the benefits and costs for one project. In practice, planners must use these methods to determine the worth of alternatives relative to one another. For the net present value and the annual worth methods, the alternative having the largest net positive worth will always be the best alternative from an economic efficiency perspective. The net present value and rate of return methods will produce the same recommended alternative if the same project life cycle is assumed for each. One of the key advantages of the annual worth method is that it can be used to compare alternatives having unequal project lives. For the net present value and benefit/cost methods (where benefits and costs are discounted to the present time), projects having unequal lives must first be expanded to a common life span and then the respective benefits and costs discounted to a common time.

The benefit/cost method requires steps beyond the initial calculation of B/C ratios for each alternative. An incremental benefit/cost approach must be used with successive pairs of alternatives such that the B/C assessment between alternatives a and b is

$$[B/C]_{b-a} = \frac{B_b - B_a}{C_b - C_a} \qquad \text{[8.7]}$$

which must be greater than 1.0 if alternative b is preferred over alternative a. To avoid inconsistencies in this assessment, the higher-cost alternative should always be compared to the lower-cost alternative (i.e., the higher-cost alternative in Eq. [8.7] is alternative b). The "best" choice is the highest-cost alternative whose B/C ratios with all lower-cost alternatives are greater than 1.0.

The use of Eq. [8.7] is shown in Fig. 8.11. The first part of this figure shows the present value of all costs (in millions) associated with five alternatives. In this simple case, the user costs of the do-nothing alternative are quite high because of increased time delays that would result if no improvements are made. The second part of the figure presents the first benefit/cost assessment, which indicates that all of the alternatives when compared to the do-nothing alternative are considered economically feasible. It now becomes necessary to compare each alternative with the others to determine which one provides the greatest return for the dollars expended. Such an assessment is shown in the third part of Fig. 8.11 in which each alternative is compared with all other lower capital cost alternatives. Alternative 3 is the "best" alternative based on the benefit/cost analysis. Notice that alternative 3 could have been chosen quickly by calculating the net present value of the alternatives; that is, $150 - 17 = 133$ for alternative 3 is the largest net equivalent worth of all the alternatives.

Although the benefit/cost method has seen extensive use in transportation planning, it has significant disadvantages (see, for example, [Wilbur Smith and Assocs., 1993; Apogee, 1994b; Kaliski et al., 2000]). In many cases, the benefit/cost ratio is ambiguous and can generate information that is based on arbitrary definitions of cost and benefits. Also, it is difficult to judge the significance of two benefit/cost comparisons that produce ratios marginally different from one another.

Present value of costs, $ millions

Alternative	User cost	Operating and maintenance	Capital
0 do nothing	250	150	3
1	150	170	10
2	200	175	12
3	125	125	20
4	110	130	35

(a)

B/C comparison with "do nothing"

Alternative	Change in user, O&M costs	Change in capital costs	B/C ratio
0			
1	80	7	11.4
2	25	9	2.8
3	150	17	8.8
4	160	32	5.0

(b)

			B/C ratios in comparing alternatives		
Alternative	User + O&M costs	Cap. costs	Alt. 2	Alt. 3	Alt. 4
1	320	10	−$55/2 = −27.5 (no benefit)	$70/10 = 7.0	$80/25 = 3.2
2	375	12	. . .	$125/8 = 15.6	$135/23 = 5.9
3	250	20	. . .	. . .	$10/15 = 0.7
4	240	35			

(c)

Alternative 2 is not justified because it shows no benefit over alternative 1. Alternative 3 is possibly justified because it dominates alternatives 1 and 2. Alternative 4 also dominates alternatives 1 and 2 but has a B/C ratio with alternative 3 of less than 1. Therefore, alternative 3 is the best alternative based on economic efficiency. Note that the alternative with the highest B/C ratio when compared with the do-nothing alternative did not result in the best alternative.

Figure 8.11   Example of application of benefit/cost analysis

Each of the preceding comparative assessment methods is in use today. Even though one might be more appropriate than another in a given situation, in general, the net present value method provides the "correct" answer to decision makers with the easiest calculations and is thus the recommended method for economic efficiency evaluation. Of course, each of these methods suffers from the problem of assigning a monetary value to the benefits and cost of the alternatives under consideration. For this reason, *use of cost-effectiveness or economic efficiency methods must take place within the overall context of the evaluation matrix described earlier.*

8.6.2   Multiobjective Assessment Methods

The methods discussed in the preceding section have one major characteristic in common: the many dimensions of a transportation problem are reduced to dollar terms to maximize net economic benefits. While the methods that seek to maximize net benefits can provide useful information to the planning process, they do not systematically incorporate multiple, often conflicting, objectives into the assessment process. Several assessment methods have been developed to do this. One of these methods, similar to the evaluation framework in Fig. 8.4, is called a goals–achievement matrix [Hill, 1973]. The goals–achievement matrix assigns relative weights to each objective. These weights are then multiplied by the score given for each impact and then summed across all objectives. The weights assigned to the objectives or evaluation criteria are usually determined by panels of experts or representatives of community groups. The preferred plan is the one with the largest score.

A modification to this goals–achievement matrix involves use of an assessment scale to determine whether goal attainment is enhanced (+ 1), decreased (− 1), or if there is no effect (0). The weights of individual objectives and their incidence can be introduced and an overall index for each plan determined. Another approach uses scores that reflect the relative importance attached to specific characteristics or projects (for example, on a scale from 1 to 5, rank the following . . .). Figure 8.12 shows an example of this approach as applied in Toronto.

The two advantages of these approaches are that (1) community objectives can be weighted to reflect the preference of decision makers and (2) impacts need not be expressed in monetary terms [Hill, 1973; Cohen et al., 1978]. The goals–achievement matrix can also include categories targeted at how particular community groups might be impacted. In addition, a goals–achievement approach permits a determination of the extent to which threshold values or standards could be met:

> By determining how various objectives will be affected by proposed plans the goals–achievement matrix can determine the extent to which certain specified standards are being met. Is the transportation plan likely to meet minimum accessibility requirements and minimum standards of comfort and convenience? Are the levels of air pollution and noise likely to exceed specified standards? . . . These are the types of questions that the goals achievement matrix is designed to answer [Hill, 1973].

Although offered as an improvement over the economic evaluation approach, the rating approach exhibits as many, if not more, problems. For example, subjective weighting procedures raise the question of whose values are being applied in the assessment. Also, the rating approach does not provide useful information on

	Transportation criteria				Socioeconomic criteria				Cost criteria				
Generic opportunities	Pass. Cap. increase/traffic flow improv.	Choice increase	Demand reduction	Subtotal	Emmissions control	Economic impact	Public acceptance	Subtotal	Reduce goods movement costs	Capital cost	Operating cost	Subtotal	Overall score
Demand management													
Land use mix and density (compact urban form)	1	2	3	6	3	2	1	6	2	3	3	8	20
Parking pricing/ management policies	2	1	2	5	2	2	2	6	2	3	3	8	19
Public info. on environment/ energy/tradeoffs	2	2	2	6	2	1	3	6	2	3	2	7	19
Ride-sharing programs	2	1	2	5	2	2	3	7	1	3	2	6	18
Flexible/staggered work hours	2	1	2	5	2	1	2	5	2	3	3	8	18
Transit fare integration/schedule coord.	2	2	1	5	2	2	3	7	1	3	2	6	18
Road pricing/tolls	3	2	3	8	2	2	1	5	2	1	2	5	18
Truck backhaul matching service	2	1	2	5	2	2	2	6	2	2	2	6	17
Reduced off-peak transit fares	1	0	2	3	2	2	1	5	1	3	3	7	15
Truck road use pricing/VWD regs	2	1	2	5	1	0	3	4	2	2	2	6	15
Nighttime truck deliveries	2	0	2	4	2	1	2	5	0	3	1	4	13
CBD vehicle restrictions	1	0	3	4	1	1	0	2	2	3	2	7	13
Supply Management													
Improved commuter rail	3	3	1	7	3	3	3	9	2	1	1	4	20
Rapid transit improvements	3	3	1	7	3	3	3	9	2	0	1	3	19
Improved real-time user info.	3	2	1	6	2	2	3	7	2	2	2	6	19
Express bus extensions to gateway	2	3	2	7	2	2	2	6	2	2	1	5	18
HOV lanes/transit priority	2	2	2	6	2	2	2	6	2	1	2	5	17
More one-way arterial streets	2	1	1	4	2	2	1	5	2	3	3	8	17
New/improved arterials and expressways	3	2	1	6	1	3	2	6	3	0	2	5	17

Figure 8.12 Toronto mobility study screening of improvement opportunities (continues)

SOURCE: Ministry of Transport, 1990

Generic opportunities	Transportation criteria				Socioeconomic criteria				Cost criteria				Overall score
	Pass. Cap. increase/ traffic flow improv.	Choice increase	Demand reduction	Subtotal	Emmis-sions control	Economic impact	Public accept-ance	Subtotal	Reduce goods movement costs	Capital cost	Operating cost	Subtotal	
Supply Management (continued)													
Computerized traffic management systems	2	2	1	5	2	2	3	7	2	1	2	5	17
Improved traffic/transit operations and control	2	1	1	4	2	2	3	7	2	2	2	6	17
Expanded off-street loading facil./curb mgmt.	3	1	1	5	2	2	2	6	2	1	2	5	16
New by-pass highway (414)	3	2	0	5	2	2	2	6	3	0	2	5	16
Designated truck lanes/routes	2	1	2	5	2	2	1	5	2	1	2	5	15
Signed hospital access routes	1	1	1	3	1	2	2	5	1	2	3	6	14
Funding/implementation													
Private sector funding rail transit	2	2	1	5	2	3	2	7	2	3	2	7	19
New road taxes dedicated to transit improv.	2	2	1	5	2	3	2	7	1	3	2	6	18
Employer tax break for subsid. transit passes	1	1	2	4	2	1	3	6	2	3	2	7	17
Increased traffic enforcement	2	1	1	4	2	2	2	6	2	3	2	7	17
Employer tax to fund transit	2	1	1	4	2	0	2	4	2	3	2	7	15
Parking tax	1	0	2	3	2	2	0	4	2	3	2	7	14

Rating scale: 0 = Unfavorable
1 = Neutral
2 = Favorable
3 = Highly favorable

Figure 8.12   Toronto mobility study screening of improvement opportunities (continued)
1 SOURCE: Ministry of Transport, 1990

whether the costs of alternatives are justified by the benefits expected. The major drawback of this assessment approach, however, is that even though the weights are supposed to be objectively determined, there is no unambiguous way of doing this. If the weights do not reflect the true preference of the community, this approach is not very helpful.

Another type of multiobjective assessment method, based on expected utility theory, is the most formal methodology for analysis of decisions in an uncertain environment [Keeney and Raiffa, 1993; Raiffa, 1997; Hammond and Keeney, 1998]. This method uses a multiattribute utility function that assigns a unique value to any combination of impacts. The concept of a multiattribute utility function is similar to the utility function discussed in Chap. 5 with regard to demand analysis. The utility for an individual alternative is a function of the utilities of the various attribute levels associated with the alternative. For example, the utility of an impact could be expressed in the following form:

$$U(x) = \sum K_i U_i(x_i)$$ [8.8]

where
  $U(x)$ = overall utility for impact x
  K_i = scaling factors
  $U_i(x_i)$ = single dimensional utility over attribute x_i

Such an approach was used to evaluate a new airport for Mexico City [Keeney and Raiffa, 1993]. In this case, six objectives were identified as being important to the analysis: (1) minimize total construction and maintenance costs; (2) provide adequate capacity to meet air-traffic demands; (3) minimize access time to the airport; (4) maximize safety of the system; (5) minimize social disruption of a new airport; and (6) minimize effects of noise pollution due to air traffic. Measures of effectiveness were developed for each objective, and interviews with key decision makers were used to identify characteristics of the utility function associated with each measure (e.g., at what point were decision makers indifferent as to the direction and magnitude of the measure of effectiveness?). After determining the parameters of the overall utility function (which was found to be the product of the individual utility functions), the expected utility for specified alternatives was determined, and a list was developed that showed the best solutions based on the decision makers' own judgment of how important the evaluation criteria were.

8.6.3 Treatment of Uncertainty in Evaluation

No aspect of evaluation is as pervasive to the process, and yet as often ignored, as uncertainty. For example, transportation officials rarely incorporate uncertainty or risk in any quantitative way into decision making [Lewis, 1991; Mehndiratta et al., 2000]. Uncertainty is present in all facets of transportation planning, from estimating the amount of travel demand for a facility to predicting the economic and technological factors that will influence this travel behavior. Uncertainty can be considered from the perspective of where in the planning process it occurs. Quade [1975] notes

Uncertainty might be (a) *conceptual*: what precisely is the problem? (b) *factual*: what are the relevant facts associated with the alternatives and the current situation? (c) *predictive*: what changes in the situation are likely to occur before any decision can take effect? and what are the likely consequences and reactions to the alternatives between which a choice must be made? (d) *strategic*: what counteractions may be expected to be taken by opposing interests? (e) *ethical*: what should the goals be and which of the potential outcomes would be preferable in the light of those goals?

To answer these questions, and thereby deal with the uncertainties they represent, planners have adopted several different approaches. They can do one of the following:

1.   Assume the useful life of the project or system under design to be less than its economic life. By doing this, the initial capital outlay will be expected to be recouped over a reduced period of time, or the project will not be undertaken.

2.   Add a "risk premium" to the discount rate used in evaluation. Although quite arbitrary in nature, increasing the discount rate reduces the expected value of net benefits and requires larger expected future benefits for the project to be chosen.

3.   Use scenario planning approaches to identify alternative futures and transportation needs, given different future circumstances. This approach usually requires the input of experts in different areas (e.g., energy, economic development, social and community values) to identify alternative scenarios of the future, defined along dimensions determined to be critical for influencing future transportation needs (see Sec. 4.3.1 on the use of land-use scenarios in transportation planning).

4.   Undertake sensitivity analyses of the important variables in the evaluation (e.g., discount rate, value of time, and other uncertain costs or benefits) to judge how the results of evaluation vary with changes in important input parameters.

5.   Use decision theory techniques (e.g., decision flow diagrams, expected monetary value approaches, and game simulation) that employ probability distributions of events occurring to incorporate uncertainty into risk analysis [Raiffa, 1997].

6.   For cost estimates, use contingency factors that reflect unexpected costs. Such factors are applied at different stages of project development. In addition, ranges of cost estimates are often provided (usually low, medium, and high) that reflect varying assumptions in what might happen to influential factors.

7.   Build flexibility into the design of the system or facility. Projects can be staged over time so that the completion of one phase of construction initiates a reexamination of the future and consideration of alternative strategies for completing the planned facility.

This last approach, incorporating flexibility into the transportation program and/or plan, requires a systematic investigation of alternative strategies for system development over the long term and consideration of short-term actions that do not foreclose future options. A good example of this approach was a policy developed by the U.S. DOT in 1976 to guide investment in major urban mass transportation projects [Urban Mass Transportation Administration, 1976]. In this policy statement, federal officials encouraged the concept of "incremental development," whereby initial segments of system improvement would occur only in those corridors in which the need was justified in the short term. Other corridors were to receive improvements appropriate to their needs, with the level of service being progressively upgraded as demand developed. The purpose of this approach was "to ensure that high priority corridors receive initial attention; that appropriate balance is maintained between the transportation requirements of the entire region and those of local communities within the region, and between long range and short range needs for transportation improvements; that flexibility is preserved to respond to changing technology, land use patterns and growth objectives; and that the fiscal burden is spread over a long period of time." An example of this incremental development approach might be the construction of a fixed guideway rail project in corridor 1, a detailed study of rail transit and implementation of exclusive bus lanes in corridor 2, express bus on freeway in corridor 3, and minor bus service changes in corridor 4. As demand increased in these other corridors, additional infrastructure or services would be added.

One of the more interesting developments in the treatment of uncertainty has been the application to public sector decisions of risk management concepts as applied in private sector investment decisions. A survey of transportation decision makers showed that they were quite aware of the risks associated with the projects being considered [Mehndiratta et al., 2000]. These risks included financial risk (receiving less money than needed); short-term political risk associated with changing politics; long-term political risk associated with changing societal values and their impact on transportation projects; the risk of local opposition; technological risks; and the risks of a changing market. However, very little analysis had been incorporated into the planning process to address these risks.

Brand et al. [2000] have suggested a way of doing this. They propose adopting a *real options approach* to investment that has been used in private investment for some time. The basic concepts in this approach are as follows:

- Transportation investment opportunities have potential benefits, but there is a risk that they will turn out differently than expected.

- Transportation investment opportunities are decision options that once made are *irreversible*.

- Deferring an investment decision to await new information has value because it can lower risks. The value of the decision to wait for new information is an *opportunity cost* whose lost value should be included as a cost in the net present value calculation.

- The *optimal timing* of transportation investment changes could be to defer investments [Chu and Polzin, 1997].

- The alternatives set should include flexible options that can be *staged over time* and that have the potential to capitalize on future favorable conditions.

The risk management options associated with different types of risks are shown in Table 8.11. These risks were identified through interviews with MPO planners and senior decision makers; thus, they can be considered representative of the types of risks present in metropolitan transportation decision making.

Urban Transportation Planning

Table 8.11   Sources of risk and risk management options

Risk Category	Source	Risk Management Options
Before a Project Is Implemented		
Political: Short term	• Change in political party, reducing or eliminating support for project • Impacted neighborhoods oppose project, forcing design changes • Natural or economic events occur, changing spending priorities	• Incremental planning, building coalitions, staged implementation
Political: Long term	• Public opinion and values change • Laws governing project design change, thus changing costs • Promises made to project opposition have unintended consequences	• Incremental planning, building coalitions, staged implementation
Forecasting (market)	• Validity of input assumptions regarding future • Validity of assumed relationships • Validity of modeling techniques	• Incremental planning, sensitivity analysis, staged implementation • Same, plus data collection and model • Methodological research
Funding (financial)	• Expected stream and form of funding does not appear • Reductions in funding levels • Constrained funding	• Building coalitions, develop plans with smaller discrete phases
Litigation	• Possible litigation risks at the planning stage from laws and regulations • Lowered ridership and revenue forecasts and higher cost estimates	• Additional consideration and forecasts of impacts required by relevant statutes, building coalitions • Legislation to limit or clarify liability
Cost (including time)	• Delays, cost overruns • Litigation by losing bidders	• Additional engineering and cost estimation, appropriate contract language, turnkey contracts • Transparency in bidding process
Technology	• Innovations in technology that make project prematurely obsolete	• Staged implementation

Table 8.11   Sources of risk and risk management options (continued)

Risk Category	Source	Risk Management Options
After a Project Is Implemented		
Market	• Low ridership/travel volumes	• Staged implementation, marketing and advertising, pricing flexibility
Operational performance	• System does not perform as planned • Travel volumes lead to congestion	• Additional prototype testing, engineering development, flexible procurement, performance-based contracts • Incremental planning, staged implementation
Operating and maintenance costs	• Higher costs than expected	• Same as above
Institutional	• Possibilities of labor disputes • Inability to manage project	• Long-term contracts
Political: Long term	• Special interest legislation • Change in regulatory structure	• Staged implementation
Financial	• Lower levels of subsidy	• Staged implementation, long-term contracts
Political: Short term	• Change in public attitudes on issues such as toll roads, pricing, etc.	• Staged implementation, building coalitions
Liability	• Litigation over project outcomes	• Legislation to limit/clarify liability, incremental planning, staged implementation

SOURCE: Brand et al., 2000

8.7   EXAMPLES OF EVALUATION IN TRANSPORTATION PLANNING

Because evaluation is so closely tied to the context of the decisions being faced, there is no common approach to conducting an evaluation process. The examples discussed in the following show a range of applications and methodologies that were considered appropriate for the studies at the time. Those aspects of effective evaluation presented in this chapter and that are illustrated in each case study are highlighted.

8.7.1 Evaluation of Alternative Regional Plans—The Southeastern Wisconsin Example

In 1994, the Southeastern Wisconsin Regional Planning Commission (SEWRPC) developed a transportation plan that was based on seven principles. Transportation systems planning was to (1) be regional in scope, focusing on a single integrated transportation system; (2) be conducted concurrently with land-use planning; (3) plan highway and transit systems together; (4) plan transportation facilities and management measures as an integrated system; (5) recognize the existence of a limited natural resource base; (6) recognize the role of transportation in the achievement of personal and community goals; and (7) recognize the importance of properly relating the regional transportation system to the state and national systems [SEWRPC, 1994].

A 33-person advisory committee, established to oversee the development of the plan, recommended nine transportation system objectives to guide plan development. These objectives and corresponding standards were used to both identify problems and evaluate alternative plans. Three plans resulted from a preliminary screening of alternatives: the no-build alternative and two alternative regional transportation system plans that were differentiated by two important characteristics. One alternative assumed the cost of automobile use was doubled from 5.5 cents per mile to 11 cents per mile (equivalent to a $1.10 increase in the gas tax), and transit service (revenue miles) was increased 125 percent. The second alternative increased public transit service by 53 percent but did not assume any increase in the cost of automobile use. Both alternatives included such actions as a freeway traffic management system, restricted parking on major arterials, widespread application of traffic operational improvements, transportation demand management actions, land-use strategies to promote transit ridership, and the possibility of light or commuter rail transit that would be subject of further studies.

Table 8.12 shows the results of the evaluation of these three alternatives. The benefit/cost ratios (under objective 2) were based on the estimated costs of building, operating, and maintaining the proposed systems, while the benefits were associated with reduced user costs. The value of time for automobile and transit travel was assumed to be $3.00 per hour, $9.00 per hour for light commercial truck trips, and $15 per hour for heavy-duty truck operations.

The advisory committee chose alternative 3 to be advanced for further approval, partly based on its perception that significant increases in automobile pricing would not occur as assumed in alternative 1. However, alternative 3 also had a substantial funding shortfall. Accordingly, the HOV/busway lanes and potential rail guideway projects were kept in the plan but only as "potential" future facilities. A large section of the plan was devoted to identifying the scope of major investment studies that would be conducted for these facility investments.

Observations Several characteristics of this process reflect the approach toward evaluation discussed earlier.

Table 8.12 Evaluation of transportation system plans, Southeastern Wisconsin

	Year 2010 Alternative Transportation System Plans					
Objectives and Standards	**No-Build**		**Alternative Plan 1**		**Alternative Plan 3**	
Obj. 1: Effectively serve regional land-use pattern	*Highway*	*Transit*	*Highway*	*Transit*	*Highway*	*Transit*
• Percent of population within:						
30 mins. of 40% of jobs	93.6	7.1	95.2	54.4	95.2	25.8
35 mins. of three major centers	92.9	3.6	88.1	22.7	88.1	6.6
40 mins. of major medical center or 30 mins. of hospital/clinic	100	49.3	100	65.6	100	54.8
40 mins. of major rec. center	100	44.5	100	74.6	100	50.4
40 mins. of higher ed. facility	100	58.1	100	89.6	100	72.7
60 mins. of airport	100	22.1	100	58.9	100	44.9
Obj. 2: Minimize costs						
• Sum of capital and oper. costs	$360 million annually		$546 million annually		$520 million annually	
• Benefit/cost ratio			1.68		1.18	
Obj. 3: Provide flexible, balanced transportation system						
• Residents served by transit	1,436,500		1,493,900		1,469,700	
• Jobs served by rapid transit	146,700		412,900		408,300	
• Local peak transit headways	10 to 45 mins.		10 to 30 mins.		10 to 45 mins.	
• Rapid transit peak headways	15 to 60 mins.		5 to 30 mins.		5 to 30 mins.	
• Percent CBD trips by transit	10		15		11	
Obj. 4: Minimize disruption						
• Residential/nonresidential structures to be dislocated	Residential: 39 Nonresidential: 14		Residential: 651 Nonresidential: 163		Residential: 720 Nonresidential: 165	
• Value of land used for trans.	$4.7 million		$145.4 million		$170.7 million	
• Amount of property tax reduction	$3.7 million		$129.1 million		$152.7 million	
Obj. 5: Protect natural environment						
• Acres of prime environ. land lost	15		432		475	
• Tons of air pollution per day						
VOC	29.8		25.6		28.7	
NO_x	81.4		71.3		79.9	
CO	248.4		211.5		238.2	
• Acres of prime farmland lost	8		1039		1644	

(continues)

Table 8.12   Evaluation of transportation system plans, Southeastern Wisconsin (continued)

Objectives and Standards	Year 2010 Alternative Transportation System Plans		
	No-Build	Alternative Plan 1	Alternative Plan 3
Obj. 6: Facilities traffic flow			
• Ave. weekday pass hours of travel			
Arterial and highway	1.8 million	1.5 million	1.7 million
Transit	50,000	72,000	51,000
• Ave. weekday vehicle-hours travel			
Arterial and highway	1.2 million	1 million	1.1 million
Transit	5700	9200	6700
• Ave. weekday VMT			
Arterial and highway	44.5 million	38.4 million	42.9 million
Transit	66,000	142,000	97,000
• Percent of arterial mileage under design capacity	79.5	96.4	95.3
Obj. 7: Reduce accident exposure			
• Percent of pass miles on facilities with lowest accident exposure	39.5	38.4	38.2
Obj. 8: Minimize energy consumed			
• Gallons of fuel consumed annually	468.1 million	412.0 million	454.7 million
Obj. 9: Facilitate linked trip making			
• Ave. wait time at transit stops	11.6 mins.	9.1 mins.	10.4 mins.

SOURCE: SEWRPC, 1994

- The evaluation process was linked directly to community goals and objectives. To ensure that community and decision-maker concerns were reflected in the evaluation process, numerous public information meetings, public hearings, information dissemination techniques, and an advisory committee were used.

- Scenarios served as the basis of alternatives definition. For example, alternative 1 assumed a doubling of automobile costs, something that eventually was considered unlikely by the advisory committee. Alternative transit and highway network configurations were also tested during the development of the alternatives.

- Benefit and cost assessment was used within an overall evaluation framework. The assessment was simply one more piece of information for decision makers. In addition, a sensitivity analysis was conducted on the discount rate to gauge how the results would change given a change in this key assumption.

- The overall approach to the planning process was one of providing information to decision makers. Decision-maker concerns were explicitly incorporated into the evaluation criteria, and special analyses were undertaken in response to questions that decision makers had during the process. The result of planning was a set of information that could be used by decision makers to determine the best alternative from their perspective of maximizing community welfare.

Such a decision-oriented planning process at the regional level is not undertaken effectively without some cost. The process is dependent upon an analysis capability that can permit a broad examination of numerous issues. In the SEWRPC case, for example, many different alternative system configurations had to be simulated and the performance results interpreted. The process is also dependent on a close interaction between planners and decision makers that requires substantial staff time for meetings and presentations. Even with these costs, the results of the planning process, because of this close relationship to the decision-making process, become an important source of information to decision makers.

8.7.2 Portland's (Oregon) Transportation Plan Implementing a 2040 Vision

In 1991, Portland METRO, the metropolitan planning organization for the Portland region, adopted a set of regional urban growth goals and objectives in response to state planning requirements. As part of these goals and objectives, METRO created a vision of what the metropolitan area should look like in 2040 that was to guide the regional planning and decision-making processes. In 1997, METRO adopted a Regional Framework Plan, a comprehensive set of policies aimed at integrating land use, transportation, water, parks, and open spaces in support of the 2040 vision. The 2040 growth concept directed most development to urban centers and along existing transportation corridors and provided increased emphasis on transit and non-motorized transportation. In 2000, METRO developed a regional transportation plan that identified a 2020 preferred transportation system, which was viewed as an important stepping stone for implementing the 2040 vision.

An extensive public involvement process was used to identify desired transportation projects. In addition, a congestion management system, a process of monitoring and evaluating transportation system performance in order to identify alternative actions, provided input in several areas—regional transportation demand strategies such as parking pricing and transit subsidies; system management strategies including ITS, access management and transit preferential treatment; high-occupancy vehicle treatments; regional transit, bicycle, and pedestrian system improvements; and land-use policies. The environmental impacts of proposed projects and actions, in particular air quality and endangered species, were evaluated in some detail. In this latter case, more than 150 road culverts needed repairs to allow fish on the endangered species list to pass under the roads.

A financial analysis of the Preferred System Plan indicated that the cost of desired improvements was more than four times the revenue projections. As a result, a 2020 Strategic System Plan was developed aimed at implementing the most critical improvements to achieve the 2040 vision. The Preferred System Plan included 820 projects, whereas the Strategic System included 615. As noted in the latter plan, "while the 2020 Preferred System is a full statement of need, the 2020 Strategic

System is a statement of the highest priority need, given current transportation funding constraints" [Portland METRO, 1999]. In addition, the Strategic System Plan delayed the need for major road expansion projects in some cases and carefully phased in other improvements to get the most use out of the existing system.

The principles for identifying the projects in the Strategic System Plan included the following:

Vision for consistency with the 2040 Growth Concept
- Implements the transportation needs for the most significant primary land-use components of 2040.

- Addresses many transportation needs for secondary land-use components.

- Addresses some needs for other 2040 Growth Concept land-use components.

- Substantially preserves a regional highways function.

Structure for consistency with the 2040 Growth Concept
- Central city and most regional centers served by light rail transit should have direct access to the regional highway system and contain a mix of arterial street, pedestrian, and bicycle systems improvements.

- Most industrial areas should have strong connections to the regional highway system and intermodal facilities.

- Most town centers, corridors, and main streets should be served by regional transit and contain a mix of arterial street, pedestrian, and bicycle systems improvements.

- Many neighborhoods and employment areas should be served by community transit, arterial capacity improvements, and some improvements to the pedestrian and bicycle systems.

2020 Strategic System performance
- Meets modal performance targets.

- Meets most regional motor vehicle performance measures.

- Meets state planning requirements.

- Maintains current regional operations, maintenance, and preservation needs.

- Meets 20-year benchmarks for 2040 Growth Concept implementation.

The evaluation data presented as part of the Strategic System Plan alternative are shown in Table 8.13. In addition to the systems-level data shown in this table, the evaluation was conducted for targeted corridors and transit lines to indicate levels of performance at a scale the decision makers could understand.

Observations   The Portland Strategic System evaluation illustrates several characteristics of good practice.

- Plan evaluation was directly tied to a community vision that focused on desired development and infrastructure characteristics 40 years into the future. The 20-

Table 8.13   Transportation system plan performance evaluation in Portland, Oregon

	1994	2020 Preferred System	2020 Strategic System	Difference 1994–2020 Strategic
Average weekday person trips	4,864,738	7,534,953	7,548,706	+55%
Average weekday work trips	939,578	1,547,213	1,549,214	+65%
Average weekday nonwork trips	3,925,162	6,036,811	6,046,674	+54%
Average home-based work trip length (miles)	6.45	6.62	6.52	+3%
Average weekday VMT	16,112,462	24,061,990	23,929,950	+48.5%
Average weekday VMT/person	14.10	14.44	14.36	+1.8%
Average weekday VMT/employee	20.36	18.12	18.02	−11.5%
Average motor vehicle speed	25 mph	22 mph	21 mph	−16%
Average motor vehicle travel time	11 minutes	13 minutes	13 minutes	+18%
% freeway miles congested	14.9%	28.6%	26.6%	+78%
% arterial miles congested	6.0%	15.3%	16.3%	+172%
Total motor vehicle hours delayed	7,509	34,280	37,690	+402%
Freeway hours of delay	2,441	10,182	10,984	+350%
Arterial hours of delay	5,068	24,098	26,706	+427%
Walk trips	5.18%	6.81%	6.82%	+32%
Bike trips	0.97%	1.25%	1.22%	+26%
Transit trips	3.55%	7.32%	6.92%	+95%
Average weekday transit trips	172,464	551,757	52,700	+203%
% households w/in 1/4 mile transit	78%	83%	83%	+6.4%
% jobs w/in 1/4 mile transit	86%	88%	88%	+2.9%
Average weekday truck trips	54,598	72,118	72,118	+32%
Average weekday average trip length (miles)	22.64	23.90	23.91	+5%
Two-hour peak truck vehicle hours of delay	130	732	809	+522%
Two-hour peak average truck travel time (minutes)	36.53	43.28	43.98	+20%

SOURCE: Portland METRO, 1999

year transportation plan was thus considered an important step to meeting the vision's goals.

- Constrained revenues resulted in a preferred plan that was beyond the region's capabilities. The evaluation process thus focused on identifying those projects and actions that would meet the most important transportation needs of the region.

- Numerous measures of effectiveness were used to identify the likely consequences of this reduced investment strategy. These measures included multiple modal performance indicators, land-use accessibility by transit, and freight mobility.

- The regional plan included an evaluation of impacts at the subregional and corridor levels to better define the consequences of proposed actions.

8.7.3   Albany (New York) New Visions Planning

As part of a regional transportation planning process, the Albany MPO conducted an extensive public participation effort to identify a new vision for the region (see Sec. 4.3.1). This vision became the foundation upon which much of the subsequent planning in the region was based. In particular, a set of core and supplemental performance measures identified through this process was used to guide every plan and project evaluation that followed. The core measures were divided into three categories: transportation service, resource requirements, and external effects. Supplemental measures were defined in four areas—transit futures, goods movement, infrastructure, and special transportation needs—when it was thought that additional evaluation information was needed. Those impacts that could be converted to monetary values fit into a system cost category. Other impacts were defined subjectively. No effort was made to represent one type of measure as being more important than another. As noted in the MPO report, "trade-offs between monetary and nonmonetary performance measures can be made by policy makers, without reducing the decision to a single monetary value"[CDTC, 1995a]. Two examples of how the core and supplemental measures were used in evaluation include a fixed guideway transit evaluation and a highway major investment study.

Fixed Guideway Transit Study   One of the results of the visioning effort was an increased public interest in the potential for nonsingle-occupant automobile transportation, especially various forms of transit. Decision makers asked questions in five areas relating to potential transit investment that became the focus of an effort to assess this potential:

- Is the region of a size and configuration suited to fixed guideway service?
- Are there available technologies that meet the region's needs?
- What could be accomplished through fixed guideway implementation? Would such investment significantly alter our expectations of future congestion, land use, access to transportation options, resource requirements and other performance indicators?
- How dependent is fixed guideway investment upon other policy choices related to land use, highway, and parking pricing?
- How much would specific applications cost, and do they appear to be worth the investment [CDTC, 1995b]?

Four fixed guideway transit alternatives were evaluated: (1) light rail service in a freeway median leading into downtown Albany; (2) commuter rail service into downtown Albany; (3) light rail service in an arterial corridor with increased development densities around transit stations; and (4) circulator guideway service linking major shopping, office, and other trip generators in the central part of the region. The technical evaluation compared each fixed guideway alternative with a TSM or "best bus" alternative. By so doing, the evaluation process not only provided useful information on the feasibility of fixed guideway investment, but also provided an understanding of the potential for bus-related actions, land-use strategies, and pricing programs. The analysis of these options was based on a mode choice model that was sensitive to land-use patterns, local site design, household characteristics, and system performance. Figure 8.13 shows the results of this evaluation. In keeping with the evaluation principle presented earlier that the purpose of evaluation is to structure the information presented to decision makers, it is interesting to note the use of symbols as a means of conveying information. This is particularly striking given that quantitative estimates for many of these measures did exist; they were simply not used to represent the impacts. Also important to note is that each of the alternatives evaluated assumed that a parking pricing program was implemented in downtown Albany that would support transit use.

Major Investment Study   The second example from Albany is a major investment study that examined the feasibility of building a major arterial connection between an interstate highway and a technology park/community college. Figure 8.14 shows the process that was followed in this planning study. Sixteen alternatives were evaluated. The "major flaw analysis" identified 10 alternatives that did not meet minimum levels of acceptability. A second-level screening ("comparative analysis") applied 33 criteria to determine the level to which each alternative met the following project objectives:

- Improve access, reduce travel time, and reduce forecast congestion to promote economic development in the study area.
- Improve mobility and accessibility for pedestrians, bicyclists, and transit users by linking residential neighborhoods and commercial activities in the corridor.
- Preserve the long-term corridor function through access management.
- Support the land-use management visions, goals, and master plans of local communities in the corridor.
- Establish an ITS laboratory facility that could be used to test new ideas and concepts in regard to traffic management.

Each of the alternatives included transportation demand management and congestion management actions that were common to each, including enhanced ride sharing, improved traffic signal timing, parking supply reductions, enhanced transit service, and mixed-use land-use development.

The results of the evaluation were presented to decision makers in a variety of ways. An evaluation matrix as defined in this chapter was used to list all of the impacts relevant to the decision-making process. This matrix is shown in Table 8.14. Notice the variety of information found in this table, including numerical (e.g., "1160 vehicles per hour"), relative (e.g., "land-use compatibility level C versus D"), rankings (e.g., "fourth most favorably considered by public comment"), and

Core Measures	Applic. 1	Appl. 2 urban	Applic. 2	Applic. 3	Applic. 4
		Applic. 1 with			
	Albany-Schen.	urban	Northway	Circulator	
	LRT/Bus	reinvestment	LRT/Busway	LRT/AGT	Commuter Rail
Transportation service					
Access — Availability of reasonable alternatives	✔✔✔	✔✔✔	✔✔✔	✔✔	✔✔
Provision of alt. with time advantage	✔✔✔✔	✔✔✔✔	✔✔✔✔	✔✔✔✔	✔✔✔✔
Modal alternatives for freight					
Accessibility — Travel time by best mode	✔	✔	✔	✔	✔
Congestion — Excess hours of delay	✔	✔✔	✔	✔	✔
Flexibility — Reserve capacity	✔	✔	✔	✔	✔
Non-highway emergency capacity	✔	✔	✔	✔	✔
Corridor alternatives during disruption	✔	✔	✔	✔	✔
Fixed capacity risk	✘✘✘✘	✘✘✘✘	✘✘✘✘	✘✘✘✘	✘✘
Resource requirements					
Safety — Societal costs of accidents	✔	✔	✔	✔	✔
Energy — Total energy consumption	✔	✔	✔	✔	✔
Economic cost — Total marginal user, gov't and societal costs	✘	✔	✘	✘	✘
External effects					
Air quality — Daily emissions	✔	✔	✔	✔	✔
Attainment status	✔	✔	✔	✔	✔
Land use — Amount of open space	✔	✔			
Disruption of residences and businesses	✘	✘			
Highway/land use compatibility index	✔	✔	✔	✔	
Support of community character	✔	✔	✔	✔	✔
Environmental — Sensitive areas impacted					
Exposure to undesirable noise levels					
Economic — Overall support for economic health	✔	✔	✔	✔	✔
Supplemental measures					
Transit system usage	✔✔✔	✔✔✔	✔✔✔	✔✔✔	✔✔✔
Public cost per person trip served (transit)	✘✘✘✘	✘✘✘✘	✘✘✘✘	✘✘✘✘	✘✘✘
Public cost per person trip served (auto)	✔	✔	✔	✔	✔
Marginal cost per new rider served	$9.00	no net cost	$14.00	$8.50	$9.65
Overall government transit costs	✘✘✘✘	✘✘✘✘	✘✘✘✘	✘✘✘✘	✘✘✘
Benefit/cost measuring only $	0.6	2.8	0.2	0.5	0.2
Rider "friendliness"	✔	✔	✔	✔	✔

✔✔✔✔	Positive impact greater than 50%, relative to the null (at current service levels)
✔✔✔	Positive impact between 20% and 50%
✔✔	Positive impact between 10 and 20%
✔	Positive impact less than 10% or cannot be estimated.
	Negligible impact expected.
✘	Negative impact less than 10% or cannot be estimated.
✘✘	Negative impact between 10 and 20%
✘✘✘	Negative impact between 20 and 50%
✘✘✘✘	Negative impact greater than 50%, relative to the null.

Figure 8.13 Impact of performance measures on fixed guideway investment, Albany, New York

I SOURCE: CDTC, 1995b

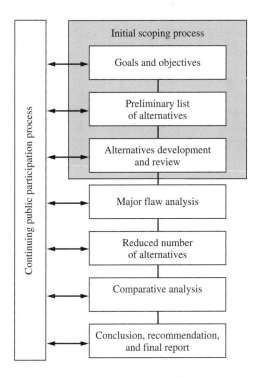

Figure 8.14   Major investment planning process in Albany, New York

I SOURCE: CDTC, 2000a

subjective (e.g., "mostly consistent with local plans"). The evaluation information was further summarized as shown in Fig. 8.15. In this figure, a solid circle meant that the alternative met and exceeded the established criterion, a half circle that it met the criterion, and an open circle that it did not satisfy the criterion.

Finally, a trade-off analysis was conducted that showed decision makers the strengths and weaknesses of each alternative (Table 8.15). The benefits common to all alternatives were also listed.

Based on this evaluation, two alternatives that were hybrids of the six were defined for the next stage of the process, which was to conduct an environmental impact analysis. The best project elements from alternatives C and D were chosen for this next step.

• Performance measures played a strong role in this evaluation. However, some of these measures were different from those used elsewhere. Monetary values were attached only to those impacts that involved direct or indirect monetary consequences and that were not distributional in nature. Travel time, for example, did not have a value of time associated with it, except for commercial and on-the-job travelers (this is very different from other cases). The remaining changes in travel time were incorporated into an accessibility measure. The safety measure used a societal cost of crashes and the economic cost measures used a marginal cost approach.

Criteria	No-build	Alternative A-1	Alternative A-2	Alternative A-3	Alternative B	Alternative C	Alternative D
1. Reduce travel time and/or travel distances		●	●	●	●	●	●
2. Improve LOS		●	●	●	●	●	●
3. Provide or improve system flexibility		●	●	●	●	●	●
4. Improve access to jobs		●	●	●	◐	●	●
5. Reduce nonauto travel times and/or distances		●	●	●	●	●	●
6. Improve transportation performance on Route 4		●	●	●	●	●	●
7. Compatibility with local plans		●	●	◐	◐	○	●
8. Public acceptability		◐	◐	◐	○	○	◐
9. Minimize adverse socioeconomic, cultural, and environmental impacts		◐	◐	◐	◐	◐	◐
10. Provide ITS testbed features		●	●	●	●	●	●
11. Cost-effectiveness		●	◐	◐	○	◐	◐

● Meets or exceeds criteria
◐ Meets criteria with mitigation
○ Does not meet criteria

Figure 8.15 Summary of evaluation information for Albany MIS

SOURCE: CDTC, 2000b

Table 8.14 Evaluation matrix for MIS in Albany, New York

	No-Build	Alt A-1	Alt A-2	Alt A-3	Alt B	Alt C	Alt D
Objective 1: Improve Access, Reduce Travel Times, Reduce Forecast Congestion to Promote Economy							
Criterion 1: Reduce Peak Period							
Travel time (minutes) during P.M. peaks between							
• Community college and Exit 8	15.4	8.0	7.3	6.9	6.2	8.1	7.3
• Tech park and Exit 8	11.5	5.4	4.9	4.9	5.5	4.8	4.9
Reduction in vehicle hours during peak hours		802	910	851	992	891	910
Criterion 2: Improve Peak Period Levels of Service on Roadways, Interchanges, and Intersections							
Reduction in daily recurring excess delay		1167	1251	1197	1315	1294	1251
Reduction in number of lane miles during P.M. peak with less than LOS D		8.5	7.8	8.2	3.7	5.1	7.8
Number of intersections that improve to from LOS F to LOS D or better		3	3	3	3	3	3
Criterion 3: Provide or Improve Flexibility of the Transportation System							
Reserve capacity (peak vehicles per hour)							
• NB traffic between Jordan Rd. and Winter St.	None	1160	1080	1100	1060	1100	1080
• SB traffic between Valley and Blooming Gr.	None	750	820	810	810	850	820
Provision for other modes during closure	No	Yes	Yes	Yes	Yes	Yes	Yes
Criterion 4: Improve Accessibility to Jobs							
Number of P.M. peak-hour work trips within 0.25 mile of interchanges/ intersections		2721	2721	2721	1617	2721	2721
Objective 2: Improve Mobility and Accessibility for Nonauto Users							
Criterion 5: Reduce Travel Time and/or Distances for Nonauto Modes Between Residences and Activity Centers							
Capability of improving nonauto transport	No	Yes	Yes	Yes	Yes	Yes	Yes
Objective 3: Preserve Long-Term Route 4 Corridor Function Through Access Management							
Criterion 6: Improve Transportation Performance Compared to Forecast Future No-Build Condition							
Average travel speed on segments (P.M. peak)							
• NB from Winter St. to Williams Rd.	17 mph	31 mph	31 mph	31 mph	30 mph	30 mph	31 mph
• SB from Blooming Grove to Valley View Rd.	10 mph	32 mph	27 mph	27 mph	29 mph	27 mph	27 mph
Land-use/transportation compatibility index							
• Route 43/Route 4 to Winter St.	C	B	B	B	B	C	B
• Winter St. to Mazoway Ave.	C	B	B	B	C	C	B

(continues)

Table 8.14 Evaluation matrix for MIS in Albany, New York

	No-Build	Alt A-1	Alt A-2	Alt A-3	Alt B	Alt C	Alt D
Objective 4: Support Land-Use Management Visions, Goals, and Master Plans of Local Communities							
Criterion 7: Alternative Should be Institutionally Feasible							
Consistent with adopted or approved local land-use management plan							
• Town of North Greenbush	No	Partial	Mostly	Mostly	Partial	Partial	Mostly
• CDTC (transit agency)		Yes	Yes	Yes	Yes	Yes	Yes
• Rensselaer County		Yes	Yes	Yes	Yes	Yes	Yes
Criterion 8: Generally Acceptable to the Public							
Community quality of life/impact assessment				See following table			
Ranking based on public comments	7th	1st	4th	3rd	5th	6th	2nd
Criterion 9: Minimize Adverse Social, Economic, and Environmental Consequences							
Number of residences/businesses displaced	0	1	2	2	2	1	3
Land-use/transportation compatibility index							
• Washington Ave.: I-90 to Rock Cut Rd.	C	C	C	C	C	C	C
• Washington Ave.: Rock Cut Rd. to Route 4	E	D	D	D	E	D	D
• Route 4: Route 4/Route 43 to Winter St.	D	C	C	C	C	C	C
• Route 4: Winter St. to Mazoway Ave.	F	E	E	E	E	E	E
Number of Section 4(f) resources affected	0	0	0	0	0	0	0
Approximate area of wetlands affected (square meters)							
• State wetlands	0	0	0	0	0	0	0
• Potential federal wetlands	0	835	885	885	935	1415	885
Acreage of protected habitat taken	0	0	0	0	0	0	0
Reduction in equivalent gallons of fuel per day	0	200	591	591	545	438	615
Objective 5: Establish ITS Laboratory Facility							
Criterion 10: ITS Test Bed Should be Feasible							
Relative ability to implement ITS test bed		Good	Good	Good	Good	Good	Good
Achieve Preceding Objectives in a Cost-Effective Manner							
Capital cost (in millions)		$25	$35	$38	$54	$38	$37
Benefit/cost ratio		5.62	4.92	4.35	3.56	4.42	4.67
Community Quality of Life Assessment							
Community college	F	B	B	B	B	C	B
Glenmore/Glenwood	A	D	D	D	E	E	C
Route 4/Tech Park business and residents	F	A	B	B	C	B	B
Van Alstyne residents	F	A	B	B	D	B	B
DANA	C	A	B	B	B	B	B
Washington Avenue	B	B	D	D	D	D	D

NOTE: A. Excellent/least disruptive   F: Failing or most disruptive
SOURCE: CDTC, 2000b

Table 8.15   Trade-off analysis for MIS in Albany, New York

Advantages	Disadvantages
No-Build	
No disruption	Traffic on Route 4
No capital cost	Discourages economic development
Alternative A-1	
Most cost-effective	Lower safety benefits
Least disruptive of the build alternatives	Lower fuel benefits
Most traffic removed from Route 4	North-end community impacts
Least community impact at south end	
High expressed public acceptability	
Alternative A-2	
Improved traffic operations at the south end	Lesser access to/from south end
Improved travel time	Cost of south-end interchange
	North-end community impacts
Alternative A-3	
Improved traffic operation at the south end	Lesser access to/from south end
Improved travel time	Cost of south-end interchange
Improved community college egress	Visual impacts of bridge at community college
	North-end community impacts
	Cost of bridge at community college
Alternative B	
Best through travel times	Less flexibility
Best safety	Highest cost
Improves south end through traffic operations	Least consistent with community goals
	Least cost-effective
	Less diversion from Route 4
	Lesser accessibility
	Greatest environmental effects
Alternative C	
Improved travel operations at the south end	Lesser access to/from south end
Improved travel time	Cost of south-end interchange
Best system connectivity	Affects greatest number of properties
Improves Blooming Grove/Williams/Route 136 intersections	Most north-end community impacts
	Adds signal to Connector
Removes signal on Route 4	One of the least publicly acceptable
Reduces armory traffic on Glenmore	Takes least traffic off Route 4
	Least land-use compatibility index
Alternative D	
Improved traffic operations at the south end	Lesser access to/from south end
Improved travel time	Cost of south-end interchange
Furthest removed from Glenwood	Highest potential building displacements
High expressed public acceptability	Greatest impact on natural ravines
	Potential impacts to South Troy neighborhood

SOURCE: CDTC, 2000b

- The highway/land-use relationship was considered explicitly. Compatibility was a measure of transportation intrusion into residential areas defined as daily traffic divided by average residential driveway spacing. The community character measure combined population and employment changes with measures of mobility, real-estate patterns, and cultural factors.

- Cost-effectiveness measures relating to the public cost per person trip and the marginal cost per new rider were used, as was a benefit/cost measure that reflected only monetary terms (although there did not seem to be any incremental benefit/cost analysis conducted).

- For transit, the study used symbols to represent the level to which the alternatives met the measures of effectiveness.

8.7.4   Urban Corridor Analysis in Salt Lake City

The I-15 corridor south of Salt Lake City, a major transportation artery serving the metropolitan area, was experiencing serious congestion levels that were only expected to get worse. In order to improve mobility in and through this corridor, the state DOT and MPO initiated an alternatives analysis that examined 12 alternatives, including no-build, TSM, bus service, additional general purpose freeway lanes, combinations of freeway and HOV lanes, light rail in two alignments, and a combination of light rail and general purpose freeway lanes [Green and Assocs., 1987]. Interestingly, this was one of the first corridor studies that included trade-offs between transit and freeway infrastructure. The I-15 study was thus one of the first multimodal transportation studies conducted in the United States.

An example of the evaluation results is shown in Table 8.16 where five of the 12 alternatives are presented [Rutherford, 1994].

Table 8.16   Urban Corridor Analysis in Salt Lake City

Evaluation Measure	Alternative 1* (No-Build)	Alternative 2 Rehab I-15/ Best Bus	Alternative 3 1 Lane I-15/ Best Bus	Alternative 4 2 Lanes I-15/ Best Bus	Alternative 9 1 Lane I-15 UPRR LRT
I. Cost					
A. Total Capital Cost					
1. *Total Capital Cost*					
a. 1987 Dollar (Millions)	$57	$283	$437	$526	$574
b. Current Dollars (Millions)	69	392	575	711	729
2. *I-15 Improvements (1987 $ Millions)*					
a. TSM Improvements	$10.07	$10.07	$10.07	$10.07	$10.07
b. SR 201/I-15/I-80 Interchange Improvements	—	23.11	92.95	95.96	92.95
c. Freeway Mainline Improvements (New)	—	—	54.61	112.58	54.61

Table 8.16 Urban Corridor Analysis in Salt Lake City (continued)

Evaluation Measure	Alternative 1* (No-Build)	Alternative 2 Rehab I-15/ Best Bus	Alternative 3 1 Lane I-15/ Best Bus	Alternative 4 2 Lanes I-15/ Best Bus	Alternative 9 1 Lane I-15 UPRR LRT
d. Structure Replacement	—	103.83	92.15	93.29	92.15
e. Pavement Replacement	—	56.04	51.13	48.18	51.13
f. Improve Existing Interchange	—	—	17.53	46.33	17.53
g. New Interchanges	—	—	29.89	30.90	29.89
h. Special I-215 HOV Access Ramps	—	—	—	—	—
Total	$10.07	$193.05`	$348.33	$437.31	$348.33
3. *Transit Improvements (1987 $ Millions)*					
a. SRTP Improvements	$39.13	$39.13	$39.13	$39.13	$39.13
b. LRT Construction, ROW and Mitigation Allowance	—	—	—	—	101.69
c. Special Park-and-Ride Lots	—	—	—	—	12.88
d. Light Rail Transit Vehicles	—	—	—	—	24.20
e. Light Rail Transit Maintenance Facility	—	—	—	—	8.62
f. Standard Transit Buses	6.13	38.68	37.80	37.80	29.40
g. Transit Buses Maintenance Facility	1.96	12.38	12.10	12.10	9.41
Total	$47.22	$90.19	$89.03	$89.03	$225.33
4. *Total Equivalent Annual Capital Cost (1987 $ Millions)*					
a. Total	$8.04	$35.35	$53.34	$63.76	$67.45
b. I-15	1.18	22.59	40.75	51.17	40.75
c. Transit	6.86	12.75	12.59	12.59	26.70
B. Annual O & M Cost (1987 $ Millions)					
a. Total	$29.26	$41.86	$41.92	$41.90	$45.20
b. I-15	1.38	1.38	1.68	1.85	1.68
c. Transit	27.88	40.48	40.24	40.05	43.52
C. Total Annualized Cost (Capital and O & M) (1987 $ Millions)					
a. Total	$37.30	$77.20	$95.26	$105.66	$112.65
b. I-15	2.56	23.97	42.43	53.02	42.43
c. Transit	34.74	53.23	52.83	52.64	70.22
D. Annual Time Cost Savings to Transit Riders (Compared to Alternative 3) (2010) ($ Millions)	N/A	—	0	—	$3.03

(continues)

Table 8.16   Urban Corridor Analysis in Salt Lake City (continued)

Evaluation Measure	Alternative 1* (No-Build)	Alternative 2 Rehab I-15/ Best Bus	Alternative 3 1 Lane I-15/ Best Bus	Alternative 4 2 Lanes I-15/ Best Bus	Alternative 9 1 Lane I-15 UPRR LRT
E. Annual Time Cost Savings to HOV Users (Compared to Alternative 3) (2010) ($ Millions)	—	—	0	—	—
F. Annual Time Cost Savings to Highway Users (Compared to Alternative 2) (2010) ($ Millions)	—	0	$0.34	$0.73	$0.56
II. Effectiveness (Transportation System Performance)					
A. Utilization by mode					
1. *Daily Transit Person Trips (2010) (Linked) (Thousands)*	87.77	100.10	99.68	99.79	105.80
2. *Daily Work Trips by Mode (2010) (Millions)*					
a. Transit (linked)	47.41	55.20	54.78	54.89	58.29
b. HOV (3+ persons)	59.56	58.73	59.05	59.06	58.82
c. Auto (1 and 2 persons)	822.7	815.8	815.8	815.8	812.6
3. *Annual Transit Trips (2010) (Millions)*					
a. Linked	23.91	27.22	27.11	27.14	28.77
b. Unlinked	29.88	34.03	33.89	33.93	37.37
4. *Daily "Guideway" Passengers (2010)*					
a. Rail	—	—	—	—	23,400
b. Express Bus and HOV on I-15	3100	4200	4100	4100	—
5. *Mode Split for Work Trips (2010)*					
a. % Transit	5.10%	5.94%	5.89%	5.90%	6.27%
b. % HOV (3+ persons)	6.41	6.31	6.35	6.35	6.33
c. % Auto (1 and 2 persons)	88.49	87.75	87.76	87.75	87.40
6. *Mode Split to Downtown SLC (2010) (Work Trips)*					
a. % Transit	21.9%	23.9%	23.8%	23.8%	23.9%
b. % HOV (3+ persons)	7.5	7.5	7.5	7.5	7.5
c. % Auto (1 and 2 persons)	70.6	68.6	68.7	68.7	68.6
B. Level of Service (LOS)					
1. *I-15 Volumes, V/C Ratio, LOS and A.M. Peak Period Speeds at Selected Locations for the General Purpose Lanes*					
a. 7200 South–9000 South					
Volume	5726	5726	6736	7432	6736

Table 8.16 Urban Corridor Analysis in Salt Lake City (continued)

Evaluation Measure	Alternative 1* (No-Build)	Alternative 2 Rehab I-15/ Best Bus	Alternative 3 1 Lane I-15/ Best Bus	Alternative 4 2 Lanes I-15/ Best Bus	Alternative 9 1 Lane I-15 UPRR LRT
V/C	1.08	1.08	0.95	0.84	0.95
LOS	F	F	E	D	E
Speed (mph)	<30	<30	43	51	43
b. 3300 South–4500 South					
Volume	5655	5655	7244	8455	7244
V/C	1.07	1.07	1.03	0.96	1.03
LOS	F	F	F	E	F
Speed (mph)	<30	<30	<30	40	<30
c. 1300 South–2100 South					
Volume	4818	4818	5767	6195	5767
V/C	0.68	0.68	0.65	0.70	0.65
LOS	C	C	C	C	C
Speed (mph)	56	56	56	55	56
2. *Automobile Travel Times in A.M. Peak (Minutes)*					
a. Sandy to CBD	31	31	30	29	30
b. West Jordan to Fashion Place Mall	11	11	9	9	9
c. Sandy to South Salt Lake	23	23	23	23	23
3. *Transit Travel Times (Minutes)*					
a. Sandy to CBD	55	57	57	56	51
b. West Jordan to Fashion Place Mall	33	33	33	33	36
c. Sandy to South Salt Lake	53	51	51	51	50
4. *Vehicle Miles Traveled (VMT) Per Day on Congested Roadways (V/C>.9): I-15 Northbound (A.M. Peak Hour) (Thousands of Miles)*	56.7	56.7	63.7	67.2	63.7
5. *Total Miles of Congested Roadway (V/C>.9) (A.M. Peak Hour) (Corridor Area)*					
a. Total Miles	25.15	25.15	22.18	21.27	22.18
b. I-15 Miles	10.88	10.88	9.35	9.06	9.35

6. *LOS for Key Intersections*	*Alternatives 1,2*		*Alternatives 3,9*	*Alternative 4*
a. North Temple and I-15 (New Interchange)	Not included		F for Alternative I Design D for Alternative II Design	E for Alternative I Design D for Alternative II Design
b. CBD Intersections	9 Intersections: 2 at LOS A, 2 at LOSB, 2 at LOS C, 1 at LOS D, 2 at LOS E		Compared with 1, 2, 7, 8: • 7 are same LOS • 1 improves from D to C • 1 worsens from E to F	Compared with 1, 2, 7, 8: All are same LOS

(continues)

Table 8.16   Urban Corridor Analysis in Salt Lake City (continued)

Evaluation Measure	Alternative 1* (No-Build)	Alternative 2 Rehab I-15/ Best Bus	Alternative 3 1 Lane I-15/ Best Bus	Alternative 4 2 Lanes I-15/ Best Bus	Alternative 9 1 Lane I-15 UPRR LRT
6. *LOS for Key Intersections* (cont.)	*Alternatives 1,2*		*Alternatives 3,9*		*Alternative 4*
c. Local Street to Local Street Intersections	31 Intersections were selected for comparison: 1 at LOS A, 3 at LOS B, 14 at LOS C, 8 at LOS D, 2 at LOS E, 3 at LOS F		Compared with 1, 2, 7, 8: • 21 are same as LOS • 5 improve • 5 worsen Of the 31 intersections: 2 at LOS A, 3 at LOS B, 14 at LOS C, 4 at LOS D, 5 at LOS E, 3 at LOS F		Compared with 1, 2, 7, 8: • 18 are same LOS • 4 improve • 9 worsen Of the 31 intersections: 2 at LOS A, 3 at LOS B, 10 at LOS C, 9 at LOS D, 3 at LOS E, 4 at LOS F
d. Local Street to I-15 Interchanges	6 Interchanges	6 Interchanges	Compared with Alternative 2:	Compared with Alternative 2:	Compared with Alternative 2:
Existing Interchanges: 3300 South, 4500 South, 5300 South, 7200 South, 9000 South, 10600 South	1 at LOS D, 5 at LOS F	2 at LOS D, 1 at LOS E, 3 at LOS F	Of the 6 existing interchanges: • 2 improve • 4 the same	Of the 6 existing interchanges: • 5 improve • 1 the same	Of the 6 existing interchanges: • 3 improve • 3 the same
New Interchanges: (i) North Temple (ii) 11400 South			For the 2 New Interchanges: (i) at LOS D (ii) at LOS C	For the 2 New Interchanges: (i) at LOS D (ii) at LOS C	For the 2 New Interchanges: (i) at LOS D (ii) at LOS C
	Overall: • 1 at LOS D • 5 at LOS F	*Overall:* • 2 at LOS D • 1 at LOS E • 3 at LOS F	*Overall:* • 4 at LOS C • 1 at LOS D • 1 at LOS E • 2 at LOS F	*Overall:* • 3 at LOS C • 5 at LOS C	*Overall:* • 4 at LOS C • 1 at LOS D • 1 at LOS E • 2 at LOS F

III. Impacts to Natural and Socioeconomic Environments

A. Natural Environment

	Alternative 1*	Alternative 2	Alternative 3	Alternative 4	Alternative 9
1. *Geologic Hazards*	Seismic activity in the area will affect all alternatives similarly				
2. *Natural Resources/Water Quality/Vegetation/Wildlife*	No Impact	All alternatives would involve possible removal of mature trees and landscaping. Disrupted wildlife would return to corridor on their own accord after construction phase. Water quality and floodplains are not significantly affected.			
3. *Soils and Agriculture*	No Impact		Removes 2 acres of prime agricultural soil		Same as Alternative 7
4. *Wetlands*	No Impact		Will potentially displace or disrupt 18.6 acres of wetlands		
5. *Air Quality*	No reduction in regional pollutant burden		All build alternatives will reduce regional pollutant burden by a minor amount		
6. *Noise*	No Impact		I-15 alignment potentially impacts 38 noise-sensitive sites	Same as Alternative 3	UPRR and I-15 alignment potentially impacts 65 sites

Table 8.16 Urban Corridor Analysis in Salt Lake City (continued)

Evaluation Measure	Alternative 1* (No-Build)	Alternative 2 Rehab I-15/ Best Bus	Alternative 3 1 Lane I-15/ Best Bus	Alternative 4 2 Lanes I-15/ Best Bus	Alternative 9 1 Lane I-15 UPRR LRT
7. *Energy*	No reduction in energy consumption or saving travel costs	Minor reduction in energy consumption and saving travel costs	Daily Savings 263 barrels of oil $44,000 travel costs	Daily Savings 317 barrels of oil $44,500 travel costs	Daily Savings 333 barrels of oil $60,000 travel costs
B. Socioeconomic Environment					
1. *Land Use and Planning*	Does not conform with regional and local transportation plans	Complies only slightly with regional and local plans for improving	No significant impact to local planning		
2. *Displacement Residences/Business*	No Impact		Will displace: 4 acres 2 residences 0 mobile homes 0 businesses	Same as Alternative 3	Will displace: 49 acres 8 residences 1 mobile home 9 businesses
3. *Economics and Development*	No change for existing development along I-15 trends would continue		Minor enhancement of development along I-15 specifically near interchanges		
4. *Joint Development Potential*	—	—	—	—	2 sites
5. *Employment Impact (Employees) (due to transit)*					
a. Short-Term (during construction)	—	—	—	—	530
b. Permanent	1000	1500	1500	1500	1500
6. *Net Fiscal Impact*					
a. Construction related	—	—	—	—	$14 million/year
b. Ongoing	—	—	—	—	$11 million/year
c. Property Tax Base: Effects due to Light Rail (incremental annual revenues in $ millions)	—	—	—	—	$1.02 to $1.06
7. *Local Traffic Impact*	*Alternatives 1, 2, 7, 8*		*Alternatives 3, 4, 5, 6, 9, 10, 11, 12*		
a. North Temple Interchange	Interchange not included		• Better access to CBD. Without interchange, all interchanges north of North Temple would be negatively impacted. • Traffic in lower avenues impacted. Traffic on 2nd Avenue expected to increase 7% • Capitol Hill area would benefit by a reduction of approximately 20% in overall traffic		
8. *Visual*	No Impact		New interchanges at 11400 South and North Temple represent an intrusion into the visual environment		
9. *Parklands*	No Impact				Ball field

(continues)

Table 8.16   Urban Corridor Analysis in Salt Lake City (continued)

Evaluation Measure	Alternative 1* (No-Build)	Alternative 2 Rehab I-15/ Best Bus	Alternative 3 1 Lane I-15/ Best Bus	Alternative 4 2 Lanes I-15/ Best Bus	Alternative 9 1 Lane I-15 UPRR LRT
10. *Cultural Resource/ Historic Sites*	No Impact		Displaces two residences potentially eligible for National Register		8 residences, 3 businesses
11. *Construction (Temporary)*	No Impact	All build alternatives would experience similar temporary construction impacts. Disruption and reduced patronage to business adjacent alignments. Short-term economic gains due to influx of workers and purchase of supplies. Increase to truck traffic in the local area. Other impacts would include increased dust, noise, and traffic conflicts. Restricted access due to detours and construction activities. Increased energy consumption.			
IV. Financial and Institutional Feasibility					
A. Sources of Revenues for Capital Improvements					
1. *Forecast of Revenue to 2010 from Existing Sources (UMTA Section 3 at 50%) (Current $ Millions)*					
a. Total	$69	$374	$557	$563	$634
b. I-15	11	255	438	444	438
c. Transit	58	119	119	119	196
2. *Potential Deficit to 2010 (comparison of capital costs to revenues) (UMTA Section 3 at 50%) (Current $ Millions)*					
a. Total	$0	$18	$18	$148	$95
b. I-15	0	0	0	130	0
c. Transit	0	18	18	18	95
3. *Potential Deficit as a Percentage of Capital costs (2010) (UMTA Section 3 at 50%)*					
a. Total	0%	4.6%	3.1%	20.8%	13.0%
b. I-15	0	0	0	22.6	0
c. Transit	0	13.1	13.1	13.1	32.6
4. *Forecast of Revenues from Potential New Sources, by Mode*	Information not available				
B. Source of Revenue, for O & M					
1. *Forecast of Revenues from Existing Sources (Current $ Millions)*					
a. I-15—Annual O & M Cost in 2009–2010	$3.84	$3.84	$4.68	$5.16	$4.68
b. Transit—Total O & M Revenues through 2010	1284	1326	1325	1325	1352

Table 8.16   Urban Corridor Analysis in Salt Lake City (continued)

Evaluation Measure	Alternative 1* (No-Build)	Alternative 2 Rehab I-15/ Best Bus	Alternative 3 1 Lane I-15/ Best Bus	Alternative 4 2 Lanes I-15/ Best Bus	Alternative 9 1 Lane I-15 UPRR LRT
2. Potential Deficit (comparison of O & M costs to revenues) (2010) Current $ Millions)					
b. I-15	0	$0	$0	$0	$0
c. Transit (%)	(205) surplus	90	86	81	200
3. Potential Deficit as a Percentage of O & M Costs (2010)					
b. I-15	0%	0%	0%	0%	0%
c. Transit (%)	(19.0) surplus	6.4	6.1	5.8	12.7
4. Forecast of Revenues from Potential New Sources, by Mode	Information not available				
C. Equity Benefit and Burden					
1. Incidence of Financing Burden, by Population Subgroup and/or Area	IV A2 and IV B2 provide the projected shortfalls for all alternatives. There is no shortfall for Alternative 1, and all other shortfalls could be funded by one or more of the sources. This table provides equity issues associated with each revenue sources.				
2. Incidence of Natural and Socioeconomic Impact by Population Subgroup and/or Area					
a. I-15	No Impact	All alternatives that include a new freeway interchange at North Temple potentially impact three adjacent neighborhoods: Jackson, Guadaloupe, and Euclid. Traffic in the lower avenues, especially 2nd Avenue, will increase while traffic in the Capitol Hill area will be reduced.			
b. Transit	No Impact				
V. Cost-Effectiveness					
A. UMTA-Required Indices					
1. Federal Cost-Effectiveness Index (2010) ($ per new rider)	N/A	N/A	0	N/A	$4.40
2. Total Cost-Effectiveness Index (2010) ($ per new rider)	N/A	N/A	0	N/A	$8.65
B. Capital Cost-Effectiveness Comparison					
1. Capital Cost/Passenger (Transit and HOV) ($ per passenger)	$ .29	$ .47	$ .46	$ .46	$ .93
C. O & M Cost-Effectiveness Comparison					
1. O & M Cost/Passenger (Transit and HOV) (2010) ($ per Passenger)	$1.17	$1.49	$1.48	$1.48	$1.51

*Evaluation included 12 alternatives
SOURCE: Rutherford, 1994

- A no-build alternative and a TSM alternative (best bus) are part of the alternatives evaluated. Notice that the no-build alternative has costs attached to it—$10 million for operational improvements and $4 million for transit service. This is typical; the no-build alternative should include those costs associated with keeping the current system operating at a reasonable level.

- Evaluation results were presented in five major performance measure categories—costs, effectiveness in terms of transportation systems performance, impacts to the natural and socioeconomic environments, financial and institutional feasibility, and cost-effectiveness. Generally, the degree of quantification declines (or, alternatively, the degree of subjective evaluation increases) as one moves down the list.

- Costs were presented as total costs and equivalent annual costs. Time cost savings to three groups—transit riders, HOV users and highway users—were identified.

- System performance was measured as number of trips carried by each mode, mode split, level of service, travel times, and miles of congested roadway. In several cases, locations were used as part of the performance measure to add specificity to the result. For example, mode split *to downtown*; I-15 volumes, V/C ratio, LOS, and speeds *at 7200 South–9000 South*; travel times *from Sandy to the CBD*; and LOS for *local street to local street intersections*.

- Many of the natural and socioeconomic environmental impacts were measured qualitatively. Notice the range of impacts considered for the socioeconomic environment—from impact on regional and local planning to intrusions into the visual environment.

- Financial feasibility examined sources and prospective deficits of funds to implement each alternative. Equity considerations focused on the incidence of finance burden and of impacts on target population groups.

- Cost-effectiveness indexes presented information on the cost per new rider, the capital cost per passenger, and the operations and maintenance cost per passenger. The first index was required by the federal government; the latter two were desired by local decision makers.

A case study of the evaluation process concluded that "each decision maker made a decision on which alternative to select based on a subjective assessment of the information and the relative importance of specific measures . . . the most important criteria dealt with the performance of the individual modes" [Rutherford, 1994]. Not surprisingly, given the highway and transit agencies involved, the recommended alternative included elements of both modes.

8.7.5   New York Cross-Harbor Freight Major Investment Study

The New York region has one of the most extensive freight networks in the world. In an effort to strategically plan for the future of this network, several of the public agencies in the region undertook a series of studies focusing on the infrastructure necessary to handle future demands. One of these studies examined alternatives for freight movement across New York Harbor [Cambridge Systematics, 2000].

Four alternatives were selected for detailed analysis and evaluation.

No-build: Improvement in regional freight rail, but no cross-harbor rail service.

TSM/Expand float: Low-cost capital and operating improvements to existing and expanded cross-harbor float (i.e., barge) services.

Rail tunnel: Major capital investment in a new rail tunnel connecting to major intermodal freight yards.

Rail tunnel with port: Same as rail tunnel option, only with additional service to a major new container port.

The potential of diverting commodities from truck to rail was determined through a stated preference survey of New York shippers. From this survey, diversion forecasts of tons by alternative for each commodity type and origin/destination pair were developed. The impact of such truck diversions on the region's highway network was estimated by removing the forecasted number of trucks from the regional highway network travel demand forecasts and then determining the level of service on these highways. Because of the influence of travel cost on transportation modal choice, a sensitivity analysis was conducted by doubling truck tolls and by assuming two different charges per container or per carload for the rail tunnel. Not surprisingly, charging to use the rail tunnel caused almost a 10 percent drop in truck-related diversion to the rail mode.

The New York metropolitan area was divided into 14 macro regions to analyze general commodity flows. At this level, a total of about 725,000 daily truck trips were projected in 2020. This macro-level assessment was then followed by a detailed network analysis using all 2147 analysis zones from the regional model. By so doing, link-by-link results were obtained. The approach used to estimate the impacts of freight diversion on the region's highway network was based on the differences between the forecasted 2020 conditions with and without truck diversions. This is called "pivot point analysis" because the predicted changes in volumes, travel times, and vehicle miles and hours traveled "pivot" off of an original estimate.

A model called the Surface Transportation Efficiency Analysis Method (STEAM) was then used to estimate benefits. This model incorporates a value of travel time, emission costs, and accident costs to determine monetary benefits of the changes in transportation system performance. For example, the following are some of the unit costs used in this application:

Variable	Auto	Truck
In vehicle value of time ($/person hours of travel)	11.39	21.12
Out of vehicle value of time ($/PHT)	21.76	21.12
Cost per gallon of fuel	1.30	1.23
Noise damage costs by highway class ranging from	0.001	0.018
($/VMT) to	0.001	0.057
Highway maintenance costs ($/VMT)	0.003	0.128

Variable (continued)

Emission costs ($/ton)

HC	3856
CO	4040
NO_x	7609
PM_{10}	10,709

Cost per accident ($/accident)

Fatal	2,317,398
Injury	50,760
Property damage	2824

Cost of greenhouse gas emissions ($/ton)   3.56

These unit costs were then multiplied by the differences in estimates of vehicle miles traveled (VMT) and person hours traveled (PHT). All units were a function of these two factors; for example, tons of emissions were a function of grams per VMT, and accidents were a function of injuries per 100 million VMT.

Direct user benefits were calculated for three major groups: (1) existing rail users who would benefit from improved rail service; (2) existing truck users who would divert to the rail mode (and presumably faster shipping times and hence lower costs); and (3) highway users who would benefit from the reduction in truck traffic. Figure 8.16 shows the estimated user benefits for each alternative. These benefits were used in comparison to costs to determine the economic feasibility of the alternatives.

Observations   This study illustrates two key characteristics of effective evaluation.

- A sensitivity analysis was conducted on key variables such as shipping costs. Shipper sensitivity to rail costs led to the conclusion that substantial increases in truck costs (e.g., tolls) would have to exist in order to encourage modal diversions.

- Although the analysis focused on user benefits instead of a broader societal perspective, the study does illustrate the concept of assessing the consequences of change on specific groups.

8.7.6   Benefit/Cost Analysis of Light Rail in Portland (Oregon)

Portland has been a national leader in implementing light rail transit in an urban environment. In 1988, the metropolitan area was considering an extension of its Westside line, a decision that was to be based partially on an evaluation of the benefits associated with the project. A benefit/cost analysis approach was adopted for determining the worth of the project. The results of this analysis are shown in Table 8.17 and reported in [Cambridge Systematics et al., 1998]. The benefits and costs were discounted to their present value over a 30-year time frame. Benefits included

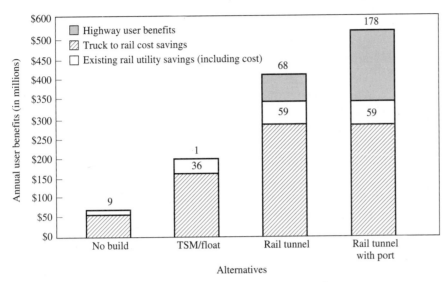

Figure 8.16 Comparison of alternatives in New York cross-harbor study
| SOURCE: Cambridge Systematics, 2000

user time savings for transit riders, motorists, and truck drivers; operating cost savings to those now taking transit; reduced parking, insurance, and vehicle ownership costs; and the cost savings associated with foregone expenditures for additional highway improvements that are no longer needed. The calculations that underlie this type of analysis can be seen in the following example [Cambridge Systematics et al., 1998].

Table 8.17   Benefit analysis for light rail transit in Portland, Oregon

	Annual	**Cumulative**	**Present Worth (000's)**
Time savings			
—Diverted motorists	−$2.138	−$64.14	−$25.656
—Transit users	0.728	21.84	8.736
—Continuing auto commuters	2.48	74.55	29.82
—Goods movement	0.115	3.45	1.38
Operating cost savings	0.28	8.4	3.36
Parking cost savings	9.084	272.52	109,008
Insurance cost savings	0.306	9.18	3.672
Second-car ownership savings	3.471	104.13	41.652
Infrastructure cost savings			524,172
Total benefit			$696,172

| SOURCE: Cambridge Systematics et al., 1998

$$\text{Benefit} = \left[S_w \times \frac{V_w}{60} \times A_w \right] + \left[S_{nw} \times \frac{V_{nw}}{60} \times A_{nw} \right] \qquad \text{[8.9]}$$

where

S_w = Daily time savings (minutes)
V_w = Value of time for work trips ($ per hour)
S_{nw} = Daily time savings of nonwork trips ($4.00 per hour, 1988 dollars)
V_{nw} = Value of time for nonwork trips ($2.00 per hour, 1988 dollars)
A_w = Annual conversion factor (250 average workdays in year)
A_{nw} = Annual conversion factor (300 average nonwork travel days in year)

As an example, to calculate the value of the time savings for those who continue to use their autos for work trips, the following data that came from the regional traffic forecasting model can be substituted:

$$\text{Benefit} = \left[149,081 \times \frac{(\$4.00)}{60} \times 250 \right] = \$2.485 \text{ million}$$

As can be seen in Table 8.17, the discounted benefits were valued at almost $700 million in 1988 dollars. The discounted present value of costs was $300 million.

Observations   The Portland light rail benefit/cost analysis illustrates some common characteristics of this approach to evaluation.

- Some of the largest contributions to user benefit comes from time and parking-cost savings. When small time savings over large numbers of travelers are summed, a large number will usually result. This is quite common in transportation benefit/cost analyses. Notice that there is a negative benefit for diverted motorists because their travel time on transit will be longer than previously.

- This analysis incorporated savings to individuals of not having to own or use a second car because of the new transit service. Such savings will very much depend on the level of transit service provided at the regional level and the socioeconomic characteristics of the traveler. A car left at home could be used by other members of the household for purposes not readily handled by transit.

- The Portland analysis also included the cost savings of not having to build new road infrastructure to handle demand. This is not found very often in benefit/cost analysis.

8.8   EVALUATION OF IMPLEMENTED PROGRAMS AND PROJECTS (EX POST EVALUATION)

The discussion up to this point has focused on the evaluation of prospective projects. Another type of evaluation that has become increasingly important in recent years is the assessment of programs and projects after they have been implemented. The purpose of such ex post evaluation is to answer four basic questions: What changes were made to the transportation system? What were the impacts of these changes? Why did these impacts occur? How successful were the actions taken to improve the transportation system? Not only does ex post evaluation show the level to which programs and projects achieve their objectives, the evaluation results also provide guidance to decision makers on the effectiveness of the implementation strategy used.

The most important concern to the planner in ex post evaluation is to correctly establish the causal factors for the changes that occurred. In order to establish causality, evaluators often develop an experimental design or evaluation strategy to collect the information needed to determine causality. The experimental design can take many forms, ranging from simple case studies to before-and-after studies of both impacted population groups and a control group.

The impacts of system changes can be measured by many different strategies for comparing the state of the system before and after a change. There are three strategies, however, that are most commonly used: comparison at different points in time, comparison of different geographic regions or population groups, and comparison between real and hypothetical systems [Billheimer and Lave, 1975]. These three comparison strategies are illustrated in Fig. 8.17. Because each of the strategies has particular strengths and weaknesses, a combination of the three approaches might be necessary. The experimental design must take into account the strengths and weaknesses of comparison strategies so that the selected strategy can address the problems likely to arise in establishing causality.

Perhaps the best discussion of experimental designs can be found in Campbell and Stanley [1963], where a number of such designs are related to experimental conditions that "threaten" the validity of experiments. Internal and external validity are defined as [Campbell and Stanley, 1966]:

- *Internal validity* is the basic minimum without which any experiment is uninterpretable: Did in fact the experiment treatments make a difference in this specific experimental instance?
- *External validity* relates to the ability to generalize. To what populations, settings, treatment variables, and measurement variables can this effect be generalized?

Seven major threats to internal validity were identified by Stanley and Campbell. The threats, along with transportation examples, are as follows:

1. *Exogenous influences.* Changes in the economy or social and political structure could occur at the same time as changes made to the transportation system. Thus, an explanation for the resulting impacts might be better related to these economic and sociopolitical forces than to the transportation change. For example, if transit service is being significantly improved at a time when the price of gasoline is rising rapidly, it might be difficult to distinguish between the two when trying to explain the response of ridership to new service. Other examples of exogenous influences include labor strikes, gasoline shortages, economic recession, and weather.

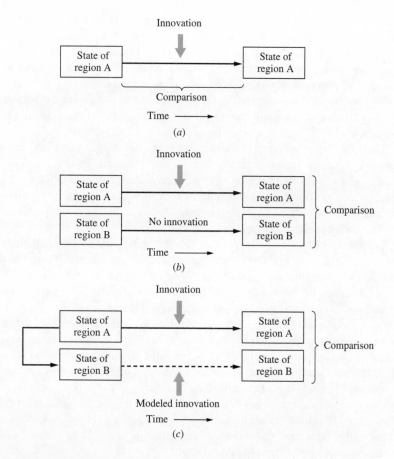

Figure 8.17   Alternative ex post evaluation comparison strategies: (a) before and after approach; (b) control region approach; (c) modeling approach
| SOURCE: Billheimer and Lave, 1975

2. *Maturation.* The natural evolution of the physical and economic characteristics of a metropolitan area might account for observed changes. For example, a deteriorating building stock near a rail station might be the cause of new building construction or renovation, as much as the existence of a new transit station. Long-term trends in transit ridership, accident rates, and gasoline consumption could explain events after a change as much as the change itself.

3. *Testing.* The process of being interviewed or of answering a questionnaire might change an individual's behavior simply because of the awareness of being the subject of evaluation. A subsequent interview or questionnaire with such a person would identify a change in behavior but not the reason for it.

4. *Instrumentation.* A change in the survey instrument, interviewer, and coders between data collection efforts could result in different types of data being collected. For example, some interviewees might ask for clarification of ambiguous

questions. This clarification, if not consistent across all interviewers, could result in different interpretations of the question from one group to the next.

5. *Statistical regression*. Sampling on the basis of one or two behavioral characteristics determined from previous surveys can result in wrong information. For example, on the day of a survey, one respondent reported taking a bus trip, an unusual event made necessary only by a malfunctioning car. If a subsequent survey is made, selected on the basis of bus riders, this respondent is likely to be reported as switching to a bus, when in fact the transit rider was always an automobile driver.

6. *Selection*. The criteria used for selecting a sample population could introduce bias into the results. For example, a survey sample based on telephone numbers biases the results in that those not having telephones (e.g., those with low income) or those having unlisted numbers would not be represented.

7. *Sample mortality*. In those cases where different population groups are used in the evaluation, the differential loss of respondents from these groups could invalidate the results. For example, in those neighborhoods having a high percentage of renters, any survey that is to be made over a period of years is likely to face a significant problem of previous respondents having moved away.

The major threats to external validity are similar to those in the preceding list but are related to the transferability of results to other situations. For example, one such threat is called "reactive testing" and is defined as a population group reacting differently once it had been surveyed.

Table 8.18 shows the type of experimental designs available and their impact on the threats to internal validity. The 0's in this table represent an observation or data-gathering effort for population group A or B. The X's represent a change introduced into the system. The R's mean a random sample is taken of that particular observation effort. A "+" indicates that the design addresses, to some extent, the threat to internal validity; a "−" indicates vulnerability; a "?" means that the effect depends on the application of the design; and a blank means the factor is not relevant.

It is beyond the scope of this text to discuss in detail the advantages and disadvantages of each experimental design. Such discussion can be found elsewhere [Charles River Assoc., 1972]. Indeed, the most effective experimental design depends on the specifics of the program or project being evaluated, the time available, and the level of financial resources that can be allocated to the evaluation task. In most transportation cases, the before-and-after study approach seems to be the most utilized (designs g, h, and i in Table 8.18.)

Because the evaluation of implemented programs and projects is so susceptible to factors not under the control of evaluators, one should carefully plan the approach to be followed. This might require the development of an evaluation or impact assessment plan in which the objectives, experimental approach, data required, data analysis strategy, and limitations are clearly specified.

As mentioned earlier, another purpose of ex post evaluation is to provide information on the implementation strategy used. Clearly, the institutional arrangements in an urban area, the coordination (or lack thereof) among agencies, and the timing of implementation can influence the results of any change to the transportation system. As noted by Williams as early as the mid 1970s, "the lack of concern for

Table 8.18 Experimental designs and impact on threats to internal validity

	Exogenous Influences	Maturation	Testing	Instrumentation	Statistical Regression	Selection	Sample Mortality	Interactions of Sources
a. One-shot case study $\times\ O$	−	−				−	−	−
b. One-group pretest–posttest design $O_1 \times O_2$	−	?	+	+	−	−	−	−
c. Static-group comparison $\times\ O_A$ O_B	−	−	−	−	−	−	−	−
d. Pretest–posttest control group $R\ O_{A1} \times O_{A2}$ $R\ O_{B1}\quad O_{B2}$	+	+	+	+	+	+	+	+
e. Posttest-only control group $R \times O_A$ $R\quad O_B$	+	+	+	+	+	+	+	+
f. Time series $O_1\ O_2\ O_3\ O_4 \times O_5\ O_6\ O_7\ O_8$	−	−	−	−	−	−	−	−
g. Before-and-after with control group $O_{A1} \times O_{A2}$ $O_{B1}\quad O_{B2}$	−	−	+	+	−	−	−	−
h. Before-and-after user study $R\ O \times$ $R\quad \times O$	−	−	+	?	+	+	+	+
i. Randomized before-and-after user study with control (A) $\dfrac{R\ O \times}{R\quad \times O}$ (B) $R\ O$ $\quad R\quad O$	+	+	+	?	+	+	+	+

−	Design is vulnerable to threat
+	Design accounts for threat
?	Design affects threat depending on application
blank	Threat not relevant

SOURCE: Charles River Associates 1972

implementation is currently the crucial impediment to improving complex operating programs, policy analysis, and experimentation" [Williams, 1976]. An examination of the implementation of an auto-restricted zone in downtown Boston illustrated the importance of an implementation strategy in successfully developing a transportation project [Lloyd and Meyer, 1984]. This examination concluded that the context for policy, program, and project implementation is usually in constant flux, with political variables changing, major actors leaving and then later reappearing in the process with new demands, and key decision makers not making decisions until the consequences of such actions are clearly spelled out. The following identifies several important characteristics of an "implementation" perspective in evaluation:

1. Successful implementation of a program or policy requires an individual or group of individuals who are committed to the project and who are able to orchestrate the often innumerable interactions necessary to overcome implementation obstacles. In this regard, there is a need for professionals who are comfortable serving multiple objectives, who are able to operate in complex political environments, who can provide expertise in a politically acceptable way, and who can operate at different levels or problem scales in response to different constituencies.

2. Because it is often difficult to predict with any certainty what the implementation characteristics of innovative or potentially controversial projects will be, project advocates must maintain a flexible approach with respect to how implementation will occur. Any implementation analysis that occurs prior to project adoption must remain flexible in the face of uncertainty and provide opportunities for compromise, if necessary.

3. The most important characteristic of successful implementation (where implementation here is considered as obtaining the desired project outcome) is the development of a constituency that can support the project in the adoption stages and then have enough at stake in the project to continue this support when project advocates begin new assignments. The more influential these constituencies, the easier it will be to implement and maintain the project.

4. Consistent communication and feedback are critical elements of successful implementation, required to gauge the response of constituent groups and modify strategy as appropriate. The media, which often report their own interpretations of project implementation, have an important role to play in shaping public perceptions and in determining the positions of key decision makers. In extremely controversial situations, the media might act as the only communication link between the major antagonists. Project proponents must develop and maintain credibility with the media and work to keep them informed about the project.

5. The success of the auto-restricted zone was attributed to, among other factors, the fact that the project proponents had strong professional values and goals that meshed closely with the political objectives of the chief decision maker—in this case, the mayor. The planners' desires to enhance the amenities of the downtown area and to limit the use of the automobile corresponded quite closely with the mayor's objectives of improving the business climate of the downtown area and

maintaining the city for its citizens, not for those from the suburbs who commuted in every morning and left the city behind every night. The "marriage" between professional goals and political power proved to be a significant factor in the success of the auto-restricted zone.

The implementation process of system change is thus a critical part of the success or failure of that change. As such, ex post evaluation should explicitly consider this process and identify where problems occurred.

With a new emphasis on operational strategies and ITS applications, a more complex implementation process can be encountered. Not only does one have the challenge of estimating behavioral change to such strategies, but understanding the implementation process also means knowing the institutional, legal, and technical aspects of new technology. A proposed set of evaluation criteria for ITS actions can be found in [Brand, 1998; Lippin et al., 1998; Transcore, 1998]. For these types of actions, the time frame of when the costs and impacts occur and the identification of specific user groups become an important part of the evaluation process.

In summary, ex post evaluation can be an important part of transportation planning. Not only does it provide useful information to decision makers on the achievement of previously implemented programs or projects, but it can also give them some idea of the feasibility of such programs or projects in other situations. The major challenge to evaluators is establishing the causality between project implementation and the resulting effect. To do this, careful consideration must be given to an experimental design or evaluation approach that takes into account possible threats to validity. The development of an evaluation plan before the program or project is implemented is necessary to ensure a successful evaluation effort.

8.9   CHAPTER SUMMARY

1.   *Evaluation* is the process of determining the relative value of individual alternatives and the desirability of one alternative over another. Evaluation provides information to decision makers on impacts, likely trade-offs, and major areas of uncertainty.

2.   An evaluation matrix that provides information on impacts and cost-effectiveness or economic efficiency is a good way of presenting the results to decision makers. This approach includes an assessment of alternatives based on *measures of effectiveness*, along with results from comparative assessment methods. An important characteristic of measures of effectiveness is that they can be specified for affected interests or geographic areas. The most common way of portraying these distributional consequences is through the use of an *impact-incidence matrix*.

3.   There are several characteristics of benefits and costs that must be considered in evaluation. These include the distinctions between *real* and *pecuniary*, *direct* and *indirect*, *tangible* and *intangible*, *internal* and *external*, *user* and *nonuser*, and *total* and *incremental*. Most measures of benefit are based on the concept of *consumer surplus*. The evaluation of user benefits is primarily a process of

determining how great a reduction in negative effects will occur if an improvement is made. User benefits can include the value to users of a reduction in travel time, a reduction in accidents, and decreased vehicle operating costs.

4. The *costs* used in comparative assessment methods include the dollar outlays necessary to construct, operate, and maintain a facility. Social costs of transportation systems have become an important issue over the past decade and will likely continue to be of concern to planners in the future.

5. Because costs and benefits of projects occur over time, a discounting procedure is needed to represent these impacts at a single point in time. The choice of a *discount rate* is extremely important because such a rate can significantly influence whether one project is more desirable than another.

6. Uncertainty is found in almost every aspect of the transportation planning process. Planners can account for uncertainty by adjusting project life cycles, doing a sensitivity analysis with discount rates, incorporating contingency factors into cost estimates, using scenarios, and adopting risk management approaches.

7. There are two major types of comparative assessment methods. The first type focuses on maximizing net economic benefits and includes such methods as *present worth, annual worth, benefit/cost,* and *return on investment.* The second type incorporates multiple objectives into the assessment method and includes such methods as goals achievement matrices and the use of multiutility functions.

8. *Ex post evaluation* serves to answer three basic questions: What changes were made to the transportation system? What were the impacts of these changes? Why did these impacts occur? Such evaluation also provides information to decision makers on the effectiveness of the implementation strategy used. To answer the questions, the evaluation process must be based on a valid experimental design.

QUESTIONS

1. Using the questions for evaluation found in Table 8.1, outline the type of information, and the likely sources of such information, needed to answer these questions.

2. For a transportation project with which you are familiar, develop a cost-effectiveness framework that you think (1) meets the needs of the decision-making process and (2) falls within the budget established for evaluation. Be specific with the MOEs to be defined and the comparative assessment techniques to be used.

3. A four-lane highway that runs through a densely populated urban area provides the major east–west connection across the southern part of a metropolitan area. Because of increased commercial and residential construction in the surrounding community, this highway has experienced increased levels of congestion in recent years. Several options for improving the situation have already been suggested by local officials, including expansion to six lanes, addition of a fifth

lane, traffic engineering improvements, and preferential treatment for high-occupancy vehicles. Outline an evaluation plan for a corridor planning study designed to address this problem. Be sure to specify the type of information needed and the techniques to be used in assessment. You are especially interested in the impact of prospective projects on existing travel flows and on the surrounding community. How would your evaluation plan change if the candidate highway were located in the city center and paralleled a rapid-transit line?

4. Given the benefits and costs schedule shown below for projects A, B, and C, determine which project is best from an economic efficiency point of view. Conduct the assessment for discount rates of 8 percent, 10 percent, and 12 percent. How does the discount rate affect the choice of the "best" project?

Year	Expected yearly cost			Expected yearly benefit		
	A	B	C	A	B	C
1	$10	$30	$10	$0	$0	$0
2	15	10	10	5	20	15
3	30	10	10	10	20	15
4	15	5	10	15	10	10
5	10	5	10	15	10	5
6	5	5	5	10	10	5
7	5	5	5	10	10	5
8	5	5	5	10	5	5
9	5	5	5	5	5	5
10	5	5	5	5	5	5

5. Select one of the approaches for dealing with uncertainty presented on pages 519 to 523. Discuss in detail the use of this approach and its advantages and disadvantages.

6. For the costs shown below, and assuming that benefits can be defined in terms of savings in costs relative to the do-nothing alternative, do a benefit/cost assessment to determine which alternative is preferred.

	Present value costs	
Alternative	User and Operating Costs	Capital
0 (do nothing)	$400	$5
1	200	40
2	300	20
3	350	25
4	250	30
5	320	15

7. Figure 8.15 illustrates how information can be presented to decision makers. For some of the categories listed, quantitative data were available that indicated the level of impact. Why do you think the information was presented as shown when quantitative data were available?

8. A new express bus service is about to be introduced in one corridor of a metropolitan area. Because of financial constraints, agency officials are interested to know what impact this service is likely to have on travel behavior in the corridor. Outline an experimental design structured to determine this impact.

9. To evaluate many innovative projects, planners are often asked to produce an evaluation plan that must be approved by sponsoring agency decision makers. Describe, in outline form, the contents of such a plan as it relates to an ITS strategy. Be sure to examine both the technical and institutional components of such an evaluation.

REFERENCES

Apogee Research, Inc. 1994a. *The costs of transportation: Final report.* Report prepared for the Conservation Law Foundation. Washington, D.C. March.

_____. 1994b. *Cost benefit analysis of highway improvement projects.* Report prepared for FHWA. Washington, D.C. October.

Atlanta Regional Commission 2000. *Regional transportation plan.* Atlanta.

Beimborn, E. et al. 1993. *Measurement of transit benefits.* Report DOT-T-93-33. Washington, D.C: Urban Mass Transportation Administration. June.

Bell, M. and T. McGuire. 1997. Macroeconomic analysis of the linkages between transportation investments and economic performance. *NCHRP Report 389.* Washington, D.C.: National Academy Press.

Berger, L. and Assocs. 1998. Guidance for estimating the indirect effects of proposed transportation projects. *NCHRP Report 403.* Washington, D.C.: National Academy Press.

Billheimer, J. and R. Lave. 1975. *Evaluation Plan for the Santa Monica Freeway Preferential Lane Project.* Contract No. DOT-TSC-1084. Submitted to the Transportation Systems Center. Cambridge, MA. November.

Billheimer, J. and R. Trexler. 1980. *Evaluation Handbook for Transportation Impact Assessment.* U.S. Department of Transportation Report UMTA-IT-06-0203-81-1. Washington, D.C. December.

Brand, D. 1997. Criteria and methods for evaluating intelligent transportation system plans and operational tests. *Transportation Research Record 1453.* Washington, D.C.: National Academy Press.

_____. 1998. Applying benefit/cost analysis to identify and measure the benefits of intelligent transportation systems. *Transportation Research Record 1651.* Washington, D.C.: National Academy Press.

_____. S. Mehndiratta and T. Parody. 2000. The options approach to risk analysis in transportation planning. *Transportation Research Record.* Washington, D.C.: National Academy Press.

Cambridge Systematics, Inc. and Apogee Research, Inc. 1996. Measuring and valuing transit benefits and disbenefits. *TCRP Report 20.* Washington, D.C.: National Academy Press.

————, R. Cervero, and D. Ashauer. 1998. Economic impact analysis of transit investment: Guidebook for practitioners. *TCRP Report 35.* Washington, D.C.: National Academy Press.

————. 2000. *Cross-harbor freight movement major investment study. Task 11—transportation impact and benefit methodology and analysis.* Cambridge, MA: Cambridge Systematics. January.

Campbell, B. and T. Humphrey. 1978. *Methods of cost-effectiveness analysis for highway projects.* NCHRP Synthesis 142. Washington, D.C.: National Academy Press.

Campbell, D. and J. Stanley. 1963. *Experimental and quasi-experimental designs for research.* Chicago: Rand McNally.

Capital District Transportation Committee. 1995a. *New visions workbook.* Albany, NY: CDTC. December.

Capital District Transportation Committee. 1995b. *Transit futures report.* Albany, NY: CDTC. October.

————. 2000a. *Interstate I-90 exit 8, Phase 2 connector, ITS test bed. Executive summary.* Albany, NY.

————. 2000b. *Draft expanded project proposal/major investment study, Interstate I-90 exit 8, Phase 2, connector, ITS test bed.* Albany, NY.

Charles River Associates. 1972. *Measurement of the effects of transportation changes.* Report CRA-166-2. Boston, MA: Charles River Assocs. August.

Chu, X. and S. Polzin. 1997. *Timing of major transportation investments.* Tampa, FL: Center for Urban Transportation Research. University of South Florida. August.

Cohen, H., J. Stowers, and M. Petersilia. 1978. *Evaluating urban transportation system alternatives.* Report DOT-P-30-78-44. Washington, D.C. November.

Cohen, H. and F. Southworth. 1999. On the measurement and valuation of travel time variability due to incidents on freeways. *Journal of Transportation and Statistics.* Vol. 2. No. 2. Washington, D.C.: Bureau of Transportation and Statistics. December.

Delucchi, M. 1997. The annualized social cost of motor-vehicle use in the U.S. based on 1990–1991 data: Summary of theory, data, methods, and results. In D. Greene et al. (eds.) *The full costs and benefits of transportation.* New York: Springer.

Dixit, A. and R. Pindyck. 1994. *Investment under uncertainty.* Princeton, NJ: Princeton University Press.

Fallon, G. et al. 1970. *Benefits of interstate highways.* Washington, D.C.: U.S. Government Printing Office Committee Print (91-41).

Federal Transit Administration. 1997. *Technical guidance for section 5309 new starts criteria.* Washington, D.C.: U.S. Department of Transportation. September.

Gomez-Ibanez, J. 1997. Estimating whether transport users pay their way: The state of the art. In D. Greene et al. (eds.) *The full costs and benefits of transportation.* New York: Springer.

Green, S. and Assocs. 1987. *I-15/State St. corridor alternatives analysis and environmental study: Evaluation results report.* Salt Lake City, UT: Wasatch Frontier Regional Council of Governments. November.

Hammond, J. and R. Keeney. 1998. *Smart choices.* Cambridge, MA: Harvard Business School Press. September.

Henscher, D. 1989. Behavioural and resource values of travel time savings: A bicentennial update. *Australian Road Research* 19. September.

_____. 1997. Behavioural value of time savings in personal and commercial automobile travel. In D. Greene et al. (eds.) *The full costs and benefits of transportation.* New York: Springer.

Hill, M. 1973. *Planning for multiple objectives.* Regional Science Research Institute Monograph Series No. 5. Amherst, MA.

Hudson, B., M. Wachs, and J. Schofer. 1974. Local impact evaluation in the design of large-scale urban systems. *Journal of the American Institute of Planners.* Vol. 40. No. 4. July.

Kageson, P. 1993. *Getting the prices right: A European scheme for making transport pay its true costs.* Report T&E 93/6. Brussels, BEL: European Federation for Transport and Environment. May.

Kaliski, J., S. Smith, and G. Weisbrod. 2000. *Major corridor investment–benefit analysis system.* Paper presented at the 79th annual meeting of the Transportation Research Board. Washington, D.C. January.

Keeney, R. and H. Raiffa. 1993. *Decisions with multiple objectives.* Cambridge, UK: Cambridge University Press. July.

Lewis. D. 1991. Primer on transportation, productivity, and economic development. *NCHRP Report 342.* Washington, D.C.: National Academy Press.

Lippin, J. et al. 1998. *Marketing ITS infrastructure in the public interest.* Washington, D.C.: Federal Highway Administration. May.

Litman, T. 1994. *Transportation cost analysis: Techniques, estimates, and implications.* Victoria, BC, Canada: Litman. December.

Lloyd, E. and M. Meyer. 1984. Strategies for overcoming opposition to project implementation: The Boston ARZ. *Transport Policy and Decision Making.* Vol 2. New York: Nijhoff Publishers.

MacKenzie, J. et al. 1992. *The going rate: What it really costs to drive.* New York: World Resources Institute.

Massachusetts Department of Public Works. 1988. *Massachusetts type II noise attenuation study.* Final report. Boston. March.

Mehndiratta, S., D. Brand, and T. Parody. 2000. How transportation planners and decision makers address risk and uncertainty. *Transportation Research Record.* Washington, D.C.: National Academy Press.

Miller, P. and J. Moffet. 1993. *The price of mobility: Uncovering the hidden cost of transportation.* New York: National Resources Defense Council.

Miller, T. 1997. Societal costs of transportation crashes. In D. Greene, et al. (eds.) *The full costs and benefits of transportation.* New York: Springer.

Ministry of Transport. 1990. *Opportunities for enhancing mobility in the GTA.* Ontario, Canada. September.

Murphy, J. and M. Delucchi. 1998. A review of the literature on the social cost of motor vehicle use in the U.S. *Journal of Transportation and Statistics*. Vol. 1. No. 1. Washington, D.C.: Bureau of Transportation Statistics. January.

Musgrave, R. and P. Musgrave. 1976. *Public finance in theory and practice.* New York: McGraw-Hill.

National Academy of Engineering. 1996. *Engineering within ecological constraints.* Washington, D.C.: National Academy Press.

National Research Council. 1997. *Building a foundation for sound environmental decisions*. Washington, D.C.: National Academy Press.

National Transit Institute and Parsons, Brinckerhoff, Quade, and Douglas, Inc. 1996. *MIS desk reference.* Prepared for the Federal Highway Administration and Federal Transit Administration. Washington, D.C. August.

Portland METRO. 1999. *Regional transportation plan.* Portland, OR.

Quade, E. 1975. *Analysis for public decisions.* New York: Elsevier.

Raiffa, H. 1997. *Decision analysis.* New York: McGraw-Hill.

Rutherford, S. 1994. Multimodal evaluation of passenger transportation. *NCHRP Synthesis 201.* Washington, D.C.: National Academy Press.

Small, K. 1992. Urban transportation economics. Vol. 51 of *Fundamentals of pure and applied economics.* Chur, Switzerland: Harwood Academic Publishers.

_____. 1999. Project evaluation. In J. Gomez-Ibanez, W. Tye, and C. Winston (eds.) *Essays in transportation economics and policy.* Washington, D.C.: Brookings Institution.

_____, R. Noland, X. Chu, and D. Lewis. 1999. Valuation of travel time savings and predictability in congested conditions for highway user-cost estimation. *NCHRP Report 431.* Washington, D.C.: National Academy Press.

Smith, S. 1999. Guidebook for transportation corridor studies: A process for effective decision-making. *NCHRP Report 435.* Washington, D.C.: National Academy Press.

Southeastern Wisconsin Regional Planning Commission. 1994. *A regional transportation systems plan for southeastern Wisconsin: 2010.* Planning Report No. 41. Milwaukee, WI. December.

Starling, G. 1979. *The politics and economics of public policy.* Homewood, IL: Dorsey.

Thomas, E. and J. Schofer. 1970. Strategies for the evaluation of alternative transportation plans. *NCHRP Report 96.* Washington, D.C: National Academy Press.

Transcore. 1998. *Integrating intelligent transportation systems within the transportation planning process: An interim handbook.* Report FHWA-SA-989-048. Washington, D.C.: Federal Highway Administration, January.

Transmanagement, Inc., M. Coogan, and M. Meyer. 1998. Innovative practices for multimodal transportation planning for freight and passengers. *NCHRP Report 404.* Washington, D.C.: National Academy Press.

United States Department of Transportation. 1997. *Federal highway cost allocation study*. Washington, D.C. July.

_____. 1999. *Highway economic requirements system, technical report.* Washington, D.C: Federal Highway Administration. March 29.

Urban Mass Transportation Administration. 1976. Major urban mass transportation investments. *Federal Register.* September 22.

Waters, W.G. 1996. Values of travel time savings in road transport project evaluation. In Proceedings of world conference on transport research, transport policy. *World Transport Research.* Vol. 3. New York: Elsevier.

Wilbur Smith and Assocs. 1993. *Guide to the economic evaluation of highway projects.* Prepared for the Iowa DOT. Ames, IA.

Williams, W. 1976. Implementation analysis and assessment. In Williams and Elmore (eds.) *Social program implementation.* New York: Academic Press.

World Bank. 1996. *Sustainable transport.* Washington, D.C.: World Bank. May.

chapter

9

Program and Project Implementation

9.0 INTRODUCTION

This chapter examines the project programming process and alternative schemes for establishing project priorities. The first section discusses the general characteristics of a programming process and a transportation program. Section 9.2 presents examples of how projects can be prioritized, while Sec. 9.3 describes the process of estimating future funding. The final section discusses implementation considerations.

9.1 CHARACTERISTICS OF A PROGRAMMING PROCESS

Programming is defined as "the matching of available projects with available funds to accomplish the goals of a given period" [Campbell and Humphrey, 1978]. The programming process considers three things: (1) *resource availability,* the amount and source of resources that will be available to fund a program over the investment period it covers (the organizational resources needed to implement and monitor the projects as they progress through the project development process is also an important part of resource commitment); (2) *resource distribution,* the absolute level and relative distribution of funds among modes, functional systems, political jurisdictions, and specific project types; and (3) the *staging of projects over time,* which takes into account interdependencies among projects, both geographically and temporally (i.e., the future implications of near-term decisions).

Programming is subject to the politics inherent in any public decision to expend funds. As noted in a synthesis of capital programming experiences in the United States, "the process of considering choices and trade-offs begins with an emphasis on technical information, but ultimately reflects many policy and political factors as final choices are made" [Neumann, 1997]. This technical information may be perceived to be both useful and desirable, but when forced to make a commitment, decision makers will often trust their political intuition over analysis.

An interesting example of this effect was found in suburban Chicago, where a priority rating scheme was developed to help decision makers in project selection [Wilson and Schofer, 1978]. Prior to this, highway programming decisions had been based on which project was proposed first or which was ready for immediate implementation. Twelve members of a decision-making committee identified seven project priority measures that would be considered in their decisions: (1) change in peak-hour travel time; (2) change in equivalent property damage (EDP) rate of accidents; (3) change in average daily congestion (volume/capacity ratio); (4) change in off-peak daily travel time; (5) change in noise pollution; (6) change in air pollution; and (7) number of dwelling units taken. A linear weighting scheme produced a project score by adding the product of weights and project values for each priority measure. The weights associated with each measure were determined by the decision makers, while the values for the measures were obtained from existing data sources and through the use of standard forecasting techniques. Cost-effectiveness indexes were then derived by dividing the project scores by capital costs. The most cost-effective program was defined by ranking the projects by decreasing cost-effectiveness, then choosing the highest-rated projects until the budget was expended.

As shown in Table 9.1, the program identified by the cost-effectiveness method was different from the one ultimately chosen by the decision makers. Observers of the process noted that "the difference between the rankings of the decision makers and those produced by the evaluation process is that the political process is still in control of investment decisions in the public sector and thus is still in a position to respond to the unique characteristics of the needs of individuals and groups" [Wilson and Schofer, 1978]. In other words, attempts to analytically structure the priority-setting process are a useful exercise for both planners and decision makers, but the final decision will still be based on political judgment.

The programming process will likely differ from one metropolitan area to another. In some cases, decision makers may be interested in specific projects and their chances of being programmed. In others, decision makers may be more con-

Table 9.1   Effect of project priority guidelines on programming process decisions

Projects Selected By Ranking Methodology			Projects Selected By Decision Makers		
Project	Cost-Effectiveness	Cost ($000,000)	Project	Cost-Effectiveness	Cost ($000,000)
A	62.6	0.385	A	62.6	0.385
B	28.6	0.300	B	Not chosen	
C	16.1	0.912	C	16.1	0.912
D	14.2	0.610	D	Not chosen	
E	7.9	0.600	E	Not chosen	
F	7.4	3.300	F	7.4	3.300
G	Not chosen		G	6.0	1.065
H	5.7	0.172	H	Not chosen	
I	3.7	0.601	I	3.7	0.601
J	Not chosen		J	2.9	0.570
	Total	6.88		Total	6.833

SOURCE: Wilson and Schofer, 1978

cerned with overall funding levels and the need to either maximize the expenditure of funds (i.e., show a constituency that progress is being made) or minimize such expenditure (i.e., show a constituency that government is responding to fiscal austerity). The programming objective in both cases is to determine which projects will be selected and when they will be constructed, while meeting the goals of the agency or of the regional transportation plan. Developing a transportation program, therefore, requires both an awareness of the impacts of alternative projects on the community *and* an understanding of the many local political agendas.

A *transportation program* lists all of the projects to be implemented in a specified investment period and the agency responsible for each. Additional information is often presented, such as each project's relationship to community goals, its priority ranking, and a summary of how the programmed funds are distributed. Given that the program usually includes projects for which commitments have been made in previous years, it does not represent a completely new investment strategy each time it is developed.[1] Projects can be found in the program that are in all stages of development, from initial planning studies to construction. The funds that support these projects can be restricted for specific uses, but in recent years, in the United States, flexibility has increased in transferring funds from one category to another or in making projects eligible for different categories.

Although the programming process will likely differ from one context to the next, some commonality among metropolitan areas results from U.S. DOT requirements for minimum characteristics of the programming process and the final document. Federal planning regulations in 1975 required each urbanized area receiving federal funds to have an MPO-endorsed transportation improvement program (TIP). The TIP was defined as a staged, 3- to 5-year program of transportation improvements that includes realistic estimates of total costs and revenues for the investment period. The projects in the TIP are to be drawn from the transportation plan. More recent federal regulations have provided the following requirements

- The TIP must be updated at least every 2 years.
- In air-quality nonattainment and maintenance areas for transportation-related pollutants, the FHWA, FTA, and MPO must make a conformity determination on any new or amended TIP. Projects are to be specified in sufficient detail to permit air-quality conformity analysis for individual projects.
- There must be reasonable opportunity for public comment.
- The TIP must cover a period of not less than 3 years but may cover a longer period if it identifies priorities and financial information for the additional years.
- The TIP must be financially constrained by year and must include a financial plan that demonstrates which projects can be implemented using current revenue sources and which projects are to be implemented using proposed revenue sources.

---

[1] This backlog of projects has been a major cause of the slow pace in changing metropolitan transportation priorities in response to changes in federal transportation policies.

- The TIP must include all regionally significant projects to be implemented with federal, state, or local funds consistent with the transportation plan [NCTCOG, 2000].

Given the dynamic character of transportation decision making in a metropolitan area, the TIP must also be a document that can be amended to reflect changing priorities and project status.

Several factors should be kept in mind when creating or updating a transportation program [Campbell and Humphrey, 1978]. The programming process must be closely linked to the planning that precedes it. The projects programmed for implementation should have already been subjected to analysis and evaluation. This is important to ensure consistency between transportation policy and planning goals as *articulated in the plan* and the actual policy as *implemented in the program.* The transportation program should thus be considered as one step toward the realization of an adopted transportation system plan and as one component of managing an agency's investment program. In addition to helping decision makers identify which projects are to be implemented, the transportation program provides a way to monitor the progress of previously programmed projects, the degree to which the projects being implemented conform to regional policies, and the specific obstacles that must be overcome to implement programmed projects.

Because the programming process represents an important stage in the effort to implement projects, opportunities should be provided for the involvement of interested parties, including elected officials, representatives of implementing and planning agencies, and the general public. An "open" programming process will help ensure consistency between the transportation program and the community goals outlined in the transportation plan or regional transportation policy.

Perhaps the most important consideration in the programming process, and certainly in the political bargaining process that accompanies it, is the amount of funding available for transportation investment. For the programming process to be effective and credible, the level of funds identified in the program must be a realistic estimate of what can be expected. This requires a good understanding of the many sources of funds that are used to finance transportation projects. In the United States, Congress placed an important restriction on transportation plans and programs in 1991, when ISTEA mandated that

> MPO long-range plans and programs must include financial plans that demonstrate how the transportation improvement program can be implemented, indicate resources from public and private sources that are reasonably expected to be made available to carry out the plan, and recommend any innovative financing techniques to finance needed projects and programs, including such techniques as value capture, tolls, and congestion pricing [U.S. Congress, 1991].

Case studies of transportation planning since the passage of ISTEA have shown that this mandate, known as "financially constrained planning and programming" or "fiscal constraint," has affected transportation decision making in several ways.

- Financial constraint has caused decision makers to consider investment decisions more carefully. Prior to this requirement, many transportation plans and

TIPs often included project lists that totaled between 40 to 100 percent more than the level of resources available. Such overprogramming caused many to question the credibility of the planning process.

- Financial constraint has discouraged the "wish list." Unfunded projects were removed from the transportation plan or became the subject of further study. More attention was given to formal criteria and analysis that would indicate the most worthy projects.

- Financial constraint has increased the demand for funds that were not restricted to any one mode. Having the flexibility of allocating funds to more than one modal project has become an important means of developing a regional consensus on a transportation program.

- Financial constraints have been imposed late in the planning process. In order to encourage innovation and creative thinking (and discouragement at the outset due to constrained resources), plans were initially developed without financial constraints. Once the plans were developed, projects were then prioritized within the expected budget levels [Transmanagement et al., 1998].

Fiscal constraint reflects the reality that there are seldom enough funds to satisfy all of the transportation needs in a metropolitan area. Consequently, some means of establishing priorities among the many projects identified during the transportation planning process is necessary.

Table 9.2 presents the key elements of a capital programming process. Although the table relates specifically to a state DOT, most of the elements shown are also relevant to a metropolitan programming process. Two elements in particular—setting priorities for project selection and determining the availability of funds—play a critical role in such a process. Each of these is discussed in the following.

Table 9.2   Key elements of the capital programming process

Key Element	Purpose/Main Activity
Setting program goals and objectives	• Establish clear and measurable statements of what the transportation agency wants to accomplish to meet its policy goals consistent with the state transportation plan
Establishing program performance measures	• Set criteria to enable the agency to measure the progress of program implementation and to evaluate the results of its program in terms of system performance, costs, and benefits
Assessing needs and identifying projects	• Identify and measure deficiencies, problems, and needs
	• Identify alternative solutions to address these needs
	• Evaluate proposed projects according to consistent criteria
Project evaluation	• Evaluate proposed projects according to consistent criteria (continued)

Table 9.2   Key elements of the capital programming process (continued)

Key Element	Purpose/Main Activity
Priority setting and program development	• Organize the agency's work into program areas reflecting distinct objectives and/or types of work • Identify priorities for each program area consistent with agency goals and objectives • Set priorities for projects within (or across) each program area, using criteria that reflect agency goals and objectives • Develop fiscally constrained candidate programs reflecting realistic project budgets and schedules
Program trade-offs	• Evaluate what the proposed program will achieve • Evaluate trade-offs for shifting resources among program areas or project types • Determine levels of resource allocation across program areas based on agency priorities, including the results of needs analysis
Budgeting	• Develop expenditure plan based on available resources and project and program costs
Program implementation and monitoring	• Implement program • Monitor progress in program delivery • Track system conditions and performance over time • Evaluate results based on established performance measures

SOURCE Neumann, 1997

9.2   SETTING PRIORITIES FOR PROJECT SELECTION

Every programming process sets priorities for project implementation. Even in cases where no formal process of prioritizing exists, the allocation of organizational and financial resources for the development of some projects over others demonstrates an implicit prioritization. The financial and organizational resources available for transportation investment are limited and may in some instances even be decreasing. Priorities must be established for project implementation to determine which projects should receive funding and which ones should be postponed or completely removed from consideration.

Five approaches have been used to provide information on project priorities: goal achievement, numerical ratings, priority indexes, programming evaluation matrices, and multiobjective systems analysis techniques.

9.2.1 Goals Achievement

The goals achievement approach identifies objectives that are important to a region and subjectively links project priorities to goals achievement. For example, the East–West Gateway Coordinating Council in St. Louis identified seven issues that regionally significant projects should address: system preservation, safety, congestion, access to opportunity, goods movement, sustainable development, and resource conservation. Using a subjective assessment of how each project related to these issues, planners determined cost-effectiveness measures of "high, medium, low" and applied them to prospective projects (Fig. 9.1). As noted in the transportation plan, "the prioritization process . . . accounts for other [than preservation] priorities through a value added approach. The value added approach accounts for the additional benefit related to every priority area addressed by a particular improvement" [East–West Gateway Coordinating Council, 1999].

The Puget Sound Regional Commission (PSRC) in Seattle developed a "regional policy focus" consisting of nine policy objectives that were to act as "a guide to help identify and focus the region's near-term investment and funding decisions" [PSRC, 1999]. Examples of these policy objectives included

- Support/complement core high-occupancy vehicle lane network.
- Support/complement development of urban centers.
- Support preservation of major regional transportation facilities.
- Support/complement freight-related transportation investment.
- Mitigate ground transportation impacts from Sea-Tac Airport.

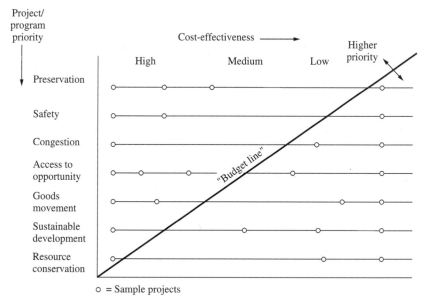

o = Sample projects

Figure 9.1   Project prioritization in St. Louis
SOURCE: East-West Gateway Coordinating Council, 1999

Note in these policy objectives the linkage to land use (i.e., development of centers) and to freight transportation issues.

A similar approach used in Sacramento included goals relating to transportation services for disadvantaged individuals, strategies for multimodal corridor improvements, funding for an ITS corridor project, and the construction of bicycle/pedestrian improvements [SACOG, 1999]. Additional criteria for regional priority projects were as follows

1. The project should include and/or affect more than one jurisdiction.
2. The project should be a true regional priority that will benefit the region.
3. The project should be one where the MPO Board, staff, or member jurisdiction have some control or influence.
4. The effort identified in the project should have some element that can be accomplished within 1 year.
5. There should be a potential benefit for making this project a priority.
6. There should be a realistic expectation that funding will be available.
7. There should be support of the member agencies where the project is located.

9.2.2   Numerical Ratings

A second method for assigning priorities uses the evaluation techniques discussed in Chap. 8 to determine the "best" projects. Benefit/cost analysis, net present worth, or cost-effectiveness methods can be used to assess the relative worth of projects over their complete life cycle. In some cases, these methods are used for certain types of projects, but not for others. For example, benefit/cost analyses tend to be used for safety projects more than any other project type [Neumann, 1997]. Cost-effectiveness, on the other hand, has been used for all types of projects.

Sufficiency/deficiency ratings are another common way to assign priorities, especially for bridge and pavement projects. These ratings are usually based on an assessment of the physical condition of the road or bridge. For pavements, the condition rating is based on the condition of pavement as well as the types of pavement distress that are present. These condition ratings are usually used as part of a pavement management system that predicts future deficiencies and assigns priorities for pavement resurfacing projects [Zimmerman et al., 1995]. For bridges, a sufficiency rating is used to determine priority for replacement and rehabilitation. This is a numerical rating based on the bridge's structural adequacy and safety, importance for public use, and serviceability and functional obsolescence [U.S. DOT, 1997]. These ratings usually range from 0 to 100 (in some cases 0 to 9), with those receiving below a 50 becoming eligible for replacement or rehabilitation. Those facilities having the worst ratings are given the top priority. Because of a bridge collapse in 1967 that resulted in 46 deaths, Congress required the U.S. DOT to develop and maintain a national bridge inventory that is based on bridge inspection data submitted by the state DOTs at regular intervals not to exceed 2 years.

9.2.3 Priority Indexes

The most commonly used approach for assigning priority among projects is the priority index. These indexes are based on measures of user benefit, environmental impacts, safety, and current condition of the facility. Maximum scores are assigned to each evaluation category, and every project is given points for each of these categories. The projects with the highest number of points are programmed until the budget constraint is reached. Different forms of this approach are found in the following examples from Dallas–Fort Worth, Massachusetts, northern New Jersey, and Columbus, Ohio.

Dallas–Ft. Worth   Because of the flexibility in using ISTEA funds, roadway, transit, and other transportation projects in the Dallas–Fort Worth metropolitan area were evaluated for the 1993 and 1995 TIPs with a single set of criteria [NCTCOG, 2000]. Surveys were sent to transportation professionals and local elected officials asking them to allocate a total of 100 points among 21 evaluation criteria that had been identified from ISTEA, the Clean Air Act, and the Americans With Disabilities Act. Based on the results of the survey, this initial set was reduced to five criteria with weights attached to each. Table 9.3 shows the criteria and associated weights for three types of projects—roadway, congestion/air quality, and transit/multimodal capacity improvements. A different weighting scheme was used for each project type because the federal funds that supported these programs had different goals. Cost-effectiveness was defined in the following way

Roadways Annualized dollar cost of making an improvement per estimated monetary travel time savings of motorists.

Transit and TDM Annualized dollar cost of making an improvement per monetary value of person hours of travel removed from the main traffic stream.

Bicycle facilities Annualized dollar cost of making an improvement per monetary value of person hours of travel removed from the main traffic stream by area type and congestion level. If the project was in an uncongested area, it was not scored for cost-effectiveness because of presumed lack of mobility benefits.

Table 9.3   Project priority criteria, Dallas–Ft. Worth

Criteria	Possible Points
Roadways: Total of 100 Points	
Current cost-effectiveness (1995)	25
Future cost-effectiveness (2020)	20
Air quality/energy conservation	20
Project commitment/local cost participation	20
Intermodal/multimodal/social mobility	15

(continued)

Table 9.3   Project priority criteria, Dallas–Ft. Worth (continued)

Criteria	Possible Points
Congestion Mitigation/Air Quality: Total of 100 Points	
Current cost-effectiveness, benefit/cost	Total of 20
0.0–0.49	0
0.50–0.99	3
1.00–1.49	5
1.50–1.99	8
2.00–2.99	10
3.00–4.99	15
>4.99	20
Air quality/energy conservation, $ per pound of NO_2 reduction	Total of 20
>$99.99	0
50.00–99.99	5
10.00–49.99	10
5.00–9.99	15
<5.00	20
Local cost participation, percent commitment	Total of 20
0–20	0
21–25	3
26–30	7
31–35	10
36–40	13
41–45	17
>45	20
Intermodal/multimodal/social mobility	Total of 20
Automobile (occupancy = 1)	0
Goods movement, pedestrian, bicycle, TDM, transit, elderly/handicapped, intermodal	20
Congestion management plan/transportation control measure	Total of 20
Is project in congestion or air-quality plan?	
No	0
Yes	20
Surface Transportation Program Capacity Improvements: Total of 100 Points	
Current cost-effectiveness	24
Future cost-effectiveness	18
Air quality/energy conservation	18
Local cost participation	24
Intermodal/multimodal/social mobility	16

SOURCE: NCTCOG, 2000

The value of time for all project types was $8.92 per hour (1993 dollars). The estimated number of person hours of travel saved came from the regional travel forecasting model.

Massachusetts   In the late 1980s, the Massachusetts DOT was receiving numerous public complaints about the traffic noise from freeways. In response, the agency undertook a statewide study to identify the potential sites for noise barriers, the expected benefits, and the associated costs [Massachusetts DPW, 1988]. Not surprisingly, the cost for all of the potential projects far exceeded the budget. Accordingly, the DOT developed a prioritization method that assigned the following points for different types of impacts.

1. Five points for each year that a sensitive land use had been exposed to the noise effect.

2. For residences of all types:

Each residence now experiencing 68–72 decibels	1 point
Each residence now experiencing 73–77 decibels	5 points
Each residence now experiencing 77 or more decibels	25 points

3. For places of worship:

Each place of worship now experiencing 68–72 decibels	5 points
Each place of worship now experiencing over 72 decibels	25 points

4. For schools, hospitals, nursing homes, libraries, or recreational areas:

Each location now experiencing 68–72 decibels	10 points
Each location now experiencing over 72 decibels	50 points

The more points a project was assigned, the higher its priority. Table 9.4 shows the results of this evaluation. Note that a supplemental cost-effectiveness rating was used to provide more information on those projects having similar priority point levels. The effectiveness measure in this case was noise level reduction in decibels per sensitive receptor (e.g., households or schools).

Northern New Jersey   Because of ISTEA requirements for a financially constrained program, the North Jersey Transportation Planning Authority, Inc. (NJTPA) developed "prioritization procedures that evaluated and ranked proposed projects based on technical measures of how well the projects fulfill regional transportation goals" [NJTPA, 1998]. The MPO advisory committee developed a prioritization process that assigned points to each project as it related to each of six goals established for the regional transportation plan. Table 9.5 shows the possible points associated with the goal of enhancing system coordination, efficiency, and intermodal connectivity. Two themes run throughout this goal's priority categories—enhancing system linkages and maximizing/optimizing existing capacity (so much so that there appears to be double weighting with the level of service *and* the volume-to-capacity ratio criteria).

Table 9.4   Priority setting for noise barrier projects, Massachusetts

		Barrier Information				
Priority	Points	Location	Height	Length	Est. Cost (000s)	Supplementary Rating ($/dB/unit)
1	842	Milton/Quincy I-93	12	3400	600	1000
2	794	Milton I-93	12	3900	700	1300
3	756	Milton/Quincy I-93	18	6300	1600	1500
4	745	Boston I-93	14	2700	500	700
5	743	Boston I-93	18	1700	500	2400
6	576	Lynnfield I-95	14	5500	1100	1700
7	545	Woburn I-93	14	3500	700	1000
8	407	Wellesley I-95	24	3100	1100	7400
9	397	Lynnfield I-95	18	2800	700	2200
10	374	Wakefield I-95	14	6700	1300	2800
11	351	Fall River I-95	16	3400	800	1100
12	349	Wellesley/Newton I-95	14	3700	700	1100
13	319	Medford I-93	18	2900	900	1000
14	311	Stoneham I-93	24	4200	1400	1300
15	288	Boston I-93	18	4300	1300	2000
16	286	Lowell I-495	16	3400	700	1700
17	277	Boston I-93	24	1900	600	3600
18	269	Wakefield I-93	10	1200	200	1500
19	261	Lynnfield I-95	16	4000	900	5100
20	259	Boston I-93	20	3200	1100	4700
21	254	Wakefield I-95	14	1200	200	2300
22	254	Boston I-93	14	4100	900	2700
23	252	Lynnfield I-95	26	2000	700	6500
24	250	Lynnfield I-95	24	4000	1400	76,400
25	216	Newton I-95	14	1400	300	1700
26	210	Woburn/Reading I-93	16	5000	1100	2300
27	208	Wakefield I-95	20	1700	500	1700
28	206	Lynnfield/Wakefield I-95	22	1300	400	14,900
29	198	Reading I-95	20	2000	600	3900
30	196	Chelmsford I-495	22	3100	1000	9200
31	189	Wakefield I-95	26	1800	600	2300
32	187	Wakefield I-95	20	2200	600	4000
33	187	Lynnfield/Wakefield I-95	18	1800	500	5300
34	185	Chelmsford I-495	26	7900	2900	5300
35	183	Medford I-93	12	2200	400	700
36	172	Lowell I-495	16	3600	800	2600
37	169	Wilmington I-93	20	3600	1000	4700
38	165	Wilmington I-93	18	4600	1100	3000
39	162	Wilmington I-93	18	2300	600	3600
40	161	Chelmsford I-495	20	1900	600	5500

SOURCE: Massachusetts DPW, 1988

Table 9.5:   Project prioritization criteria for goal of enhancing system coordination, northern New Jersey

Highway and State Bridge Projects		Local Bridge Projects		Transit Projects	
Will it provide linkages to other existing transportation systems?		*Will project remove weight or height restrictions?*		*Will it provide linkages to other existing transportation systems?*	
High: completion of missing links	30 pts	High: yes	40 pts	High: bus/rail links	25 pts
				Medium: stations	15 pts
Medium: grade-separated interchanges	15 pts	*Will it reduce congestion?*		Low: fare coordination	15 pts
Low: at-grade intersections	15 pts	High: Level of service E or F	20 pts		
		Medium: Level of service C or D	10 pts	*Will it reduce congestion?*	
Will it reduce congestion?				High: increase ridership	25 pts
High: Level of service E or F	15 pts			Medium: ITS projects	10 pts
Medium: Level of service C or D	10 pts	*Will it maximize/optimize existing capacity?*			
		High: Top 1/3 of v/c ratios	40 pts	*Will it maximize/optimize existing capacity?*	
Will it maximize/optimize existing capacity?		Medium: Middle 1/3 of v/c ratios	25 pts	High: signals, tracks	25 pts
High: Top 1/3 of v/c ratios	30 pts	Low: Bottom 1/3 of v/c ratios	5 pts	Medium: rolling stock	15 pts
Medium: Middle 1/3 of v/c ratios	15 pts				
Low: Bottom 1/3 of v/c ratios	5 pts			*Will it promote intermodalism?*	
				High: feeder service	25 pts
				Medium: park-and-ride lots	15 pts
Will it promote intermodalism?					
High: Access to intermodal facilities	25 pts				
Medium: Park-and-ride lots	15 pts				

| SOURCE: NJTPA, 1998

Columbus, Ohio   During the update of the 1994 transportation plan, more than 800 transportation improvement needs were identified in the Columbus metropolitan area [MORPC, 1998]. This was later refined to 540 projects that were actually considered in the transportation plan evaluation. Each of these projects was evaluated with a scoring system that assigned points in such categories as economic benefits, safety, social benefits, environmental benefits, transportation efficiency, accessibility/connectivity/mobility, and system preservation (see Sec. 8.6.2). In order to establish priorities among these 540 projects, the Mid-Ohio Regional Planning Commission (MORPC) defined a Likelihood Code (LC) for each project that was the sum of the points assigned in four categories

1.   If already listed in the previous TIP: 65 points.

2.   If identified in a local plan and/or regional study: 25 points.

3.   If identified in the MPO congestion management system: 5 points.

4. A weight of 10 is assigned to the final project score that was calculated during the planning process.

Likely funding sources for the TIP were identified (10 in all), and projects were prioritized within each funding source by giving higher priority to those projects with the largest Likelihood Codes.

9.2.4   Programming Evaluation Matrix

The programming evaluation matrix approach is similar to the evaluation matrix introduced in Chap. 8, except that the criteria relate to project priorities instead of project impacts. Each criterion can be estimated quantitatively or subjectively. Four examples illustrate this approach—Albany, New York; Phoenix, Arizona; Portland, Oregon; and Denver, Colorado.

Albany, New York   The 1997 TIP for Albany was the first transportation program for Albany to be developed subsequent to the adoption of the *New Visions Plan* (see Sec. 8.7.3). Discussions with decision makers early in this process led to two policy directions that guided TIP development. First, fulfilling existing commitments was a higher priority than adding new projects. Second, if the opportunity existed to consider new projects, all project sponsors wanted a chance to compete for the funds. Thus, the prioritization method had to be unbiased, focusing on the function and purpose of a proposed investment rather than who "owned" the facility. Of the $450 million in federal aid expected to be available during the 5-year investment period, $360 million was allocated to satisfy existing commitments, leaving $90 million that could be programmed for new projects [Younger and O'Neill, 1998]. A three-step process was used to determine project merit. The first step, project screening, was targeted on project consistency with the *New Visions* and local land-use plans. The projects also had to be constructed by 2002 and eligible for federal aid. This screening process decreased the number of projects to 154 with a value of $420 million, still too costly for the budget.

The second step, evaluating project merit, used benefit/cost ratios and system performance measures to present evaluation information to decision makers. A one-page evaluation form was used for each project that identified the most important characteristics of each project (Figure 9.2). Where possible, a benefit/cost ratio was calculated using as benefits accident reduction, travel time savings, energy/user cost savings due to pavement improvements, and other monetary benefits not captured by any of the above, but which were significant enough to affect system-level performance [CDTC, 1998]. The degree to which implementation would help achieve *New Vision* goals was also considered, focusing on whether it was an improvement to a priority network identified during the planning process.

The third step, building a program, followed a stepwise decision process that allocated first-year programming dollars to three types of projects. The first allocation of $55 million identified high-priority projects based on the evaluation process previously described. The main emphases of this decision process were highway/ bridge preservation, transit support, and projects relating to safety, eco-

Location:	Benefit/cost ratio:
Description:	Total benefits ($1,000/year)
Project type:	Safety
Cost: $(total), $(federal)	Travel time
Sponsor:	Energy/user
Class/AADT/condition:	Life cycle
	Other
Priority network(s):	Annualized cost:

Congestion relief:

Air-quality benefit:

Regional system linkage:

Land use compatibility (planned or existing):

Contribution to community or economic development:

Environmental issues:

Business or housing dislocations:

Facilitates

    Bicycling?

    Walking?

    Goods movement?

    Transit use?

    Intermodal transfers?

Screening issues:

Matching/maintenance:

Other considerations:

Figure 9.2   Evaluation form for project prioritization in Albany
| SOURCE: CDTC, 1998

nomic development, and bicycle/pedestrian mobility. The second allocation of $30.5 million emphasized program distribution, providing balance among project sponsors, jurisdictions, and project types. The final allocation of $4.5 million was reserved for projects that received strong public and elected official support.

Phoenix, Arizona   Unlike other metropolitan areas of its size, Phoenix did not construct an extensive urban freeway system during the 1960s and 1970s. By the late 1970s, however, constructing such a freeway system had become a major priority of the regional transportation plan. The Maricopa Association of Governments (MAG) used several quantitative and qualitative criteria to rank the freeway projects including travel demand, congestion relief, accident reduction, air-quality improvement, cost-effectiveness, joint funding, social and community impact, system continuity and mobility, rapid completion of a freeway system, linkage to regional needs, and connectivity [MAG, 1997]. Table 9.6 shows the evaluation

Table 9.6   Freeway/expressway priority data for Phoenix

Corridor	Segment	Miles	Cost (Millions)	ADT (000s)	Congested Intersections Per Mile	Accidents Per Mile	Average CO Concentration	Cost-Effectiveness	Joint Funding	Overall Rating
Agua Fria	Northern to I-10	6.64	$130	117 High	0 Med.	0.01 Med.	4.0 High	$0.02 High	0 Med.	Med.
Grand	Thomas to Camelback	2.83	119	42 Low	2.12 High	0.33 High	8.0 High	0.12 Low	0 Med.	Med.
Pima	I-17 to Squaw Pk.	6.02	107	125 High	0.17 Med.	0.03 Med.	2.0 Med.	0.02 High	1 High	High
Pima	Squaw Pk. to Scottsdale	5.08	47	121 High	0.59 Med.	0 High	2.0 Med.	0.01 High	1 High	High
Pima	Scottsdale to FLW	3.28	67	115 Med.	0.61 Med.	0.02 Med.	2.0 Med.	0.02 Med.	1 High	High
Pima	FLW to Shea	3.27	64	124 High	0.31 Med.	0.06 Med.	2.0 Med.	0.02 High	0 Med.	High
Price	Guadalupe to Warner	2.01	73	136 High	0 Med.	0.2 High	6.0 High	0.03 Med.	0 Med.	High
Price	Warner to Santan	2.93	71	104 Med.	0 Med.	0.1 High	2.0 Med.	0.03 Med.	1 High	Med.
Red Mtn.	Country Club to Gilbert	3.24	56	95 Med.	0 Med.	0.06 Med.	3.3 High	0.02 Med.	0 Med.	Med.
Red Mtn.	Gilbert to Bush Hwy.	6.54	90	76 Med.	0 Med.	0.01 Med.	2.0 Med.	0.02 Med.	0 Med.	Low
Red Mtn.	Bush Hwy. to Superstition	7.87	101	50 Med.	0.13 Med.	0.01 Med.	2.0 Med.	0.03 Med.	0 Med.	Low
Santan	I-10 to Price	6.14	115	77 Med.	0 Med.	0.04 Med.	2.0 Med.	0.03 Med.	0 Med.	Med.
Santan	Price to Arizona Ave.	3.04	58	114 Med.	0.66 High	0.03 Med.	2.0 Med.	0.02 High	0 Med.	Med.
Santan	Arizona Ave. to Gilbert	3.09	52	79 Med.	0.65 High	0.01 Med.	2.0 Med.	0.03 Med.	0 Med.	Med.
Santan	Gilbert to Power Rd.	7.94	103	52 Med.	0 Med.	0 High	2.0 Med.	0.03 Med.	0 Med.	Low
Santan	Power Rd. to Superstition	5.58	78	42 Low	0 Med.	0 High	2.0 Med.	0.04 Low	0 Med.	Low
Sky Hbr.	University Dr. to I-10	0.94	20	24 Low	2.13 High	0.13 High	8.0 High	0.12 Low	0 Med.	Med.
S. Mtn.	Papago to Baseline Rd.	6.09	96	44 Low	0.82 High	0.06 Med.	2.0 Med.	0.04 Low	0 Med.	Med.
S. Mtn.	Baseline Rd. to 7th St.	10.8	85	41 Low	0 Med.	0 High	2.0 Med.	0.02 Med.	2 High	Low
S. Mtn.	7th St. to Maricopa	2.36	56	46 Med.	0 Med.	0.06 Med.	2.0 Med.	0.06 Low	2 High	Med.
Squaw Pk.	Bell Rd. to Pima	2.24	55	95 Med.	0.45 Med.	0.13 High	2.0 Med.	0.03 Med.	0 Med.	Med.

measures associated with each freeway project. It is interesting to note that the quantitative estimates for each measure were transformed into a subjective rating of high, medium, or low so that decision makers would be better able to understand the relative ranking of the projects.

Several of these measures are quite unique. Because it was not feasible to estimate the reduction in congestion or accidents on freeways that did not exist, the positive benefits of a freeway in these important impact areas were related to corresponding measures in the surrounding community. For example, congestion relief was measured as the reduction in the number of congested intersections (volume-to-capacity ratio greater than 0.90) per mile during the P.M. peak hour within a 2-mile radius of the planned freeway in the horizon or target year. Similarly, the safety measure was the reduction in the number of accidents per day per mile within a 1-mile radius of the planned freeway. The cost-effectiveness measure was defined as the annualized cost of a freeway section divided by its total annual passenger miles of travel at the horizon year. A joint funding measure gave higher priority to those projects funded mostly by private groups (e.g., toll roads) or local governments. Figure 9.3 shows the freeway/expressway priority segments that resulted from this process.

Portland, Oregon   Portland used a programming process similar to Albany's, with the evaluation of priorities linked directly to the long-range transportation plan. The evaluation approach, shown in Figure 9.4, included the determination of a priority index score related to system performance (V/C ratio and accidents), support of the 2040 Plan, degree to which the project supported a multimodal transportation system, and a cost/benefit analysis [Portland METRO, 1995]. For the project shown in Fig. 9.4, the cost/benefit ratio was defined as the dollar cost per vehicle hour of reduced delay. For transit-oriented development and TDM projects, this ratio was defined as dollar cost per reduced vehicle mile traveled (VMT). Dollar cost per VMT avoided was used for bicycle projects, and a low/medium/high estimation was designated for pedestrian projects. The information presented also included issues such as the degree of multijurisdictional support, the level to which local governments provided more than the required share of local matching (called overmatch) to increase the project's priority, and whether the project was "implementable."

Denver, Colorado   The Denver Regional Council of Governments (DRCOG) had since 1988 developed a multiyear program for improving traffic signals in the region. DRCOG provided system planning and signal timing plans for local communities, while operating agencies were responsible for maintenance and ongoing system management. A traffic signal subcommittee of DRCOG consisting of signal operations personnel from the state DOT and local governments provided input to project prioritization. A system inventory indicated that there were 2900 traffic signals in the region operated by 30 different agencies [DRCOG, 1999]. The subcommittee defined needs in four areas

1. *Network coverage.* Approximately 33 percent of the signals on major regional roadways were not under a coordinated system control. Signals outside the Denver central business district, spaced less than one-half mile apart and on arterials with daily volumes greater than 20,000 vehicles, were given priority consideration.

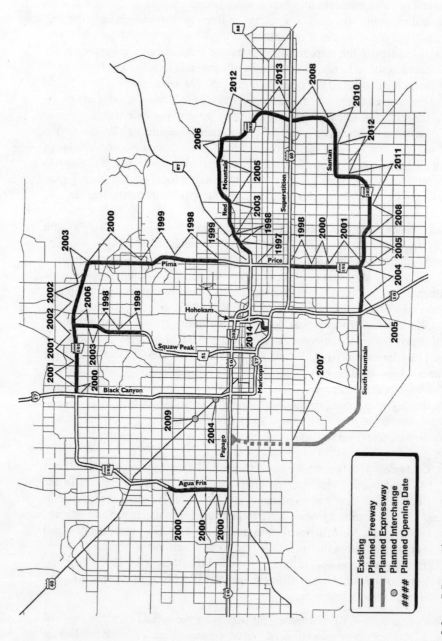

Figure 9.3   E Freeway/expressway priority segments in Phoenix
I SOURCE: MAG, 1997

PROJECT: Sunnyside Road (Sunnybrook to 122nd Avenue)
SPONSOR: Clackamas Co.
TECHNICAL RANK: 1st of 17
REQUESTED FUNDS: $5,000,000

Criteria	Data	Score	Max Score	Comments
1990 V/C Ratio	1.01	15	15	
2015 V/C Ratio	1.76	10	10	
Accident Rating	see comments	20	20	2.9 accidents/million vehicle miles. Points based on County staff analysis of relative hazards.
2040 Support	see comments	19	25	Project west terminus serves Regional Center, 2040 HCT Corridor
Cost/Benefit	$10,242/vhd reduced	15	15	Project eliminates 51 vehicle hours of delay that would occur in its absence.
Multi-Modal	bike/ped/transit factors	13	15	Extends regional bike syst; median design to enhance ped travel/safety, #71; #151 line & 2040 HCT route.
	TOTAL	92	100	

Project Description

Widen existing 3-lane road to accommodate 4 travel lanes including curbs, sidewalks, bike lanes. Additional ROW (design width of 115 ft.) also to be acquired for turn lanes, median pedestrian refuge and future HCT.

2040 Relationship

2040 Concept plan identifies corridor for future HCT. Project ROW acquisition would secure this objective. Project construction would help to facilitate Clackamas Town Center buildout although this is mostly expected to be driven by market conditions with or without additional public assistance. Congestion benefits are more strongly related to easing conditions associated with existing and planned residential/commercial development east of the Regional Center.

Administrative Criteria

• *Overmatch:* 47% @ total cost of $10.5 million, and regional provision of $5.6 million (includes $600,000 of Regional STP programmed for 30% PE/EIS).

• *Multi-jurisdictional financial support:* Significant private sector participation through system development charges and potential swap of LID funds for state funds related to the Sunnybrook Ext. project.

• *Implementable:* Qualified yes: Draft EIS starts summer '95 using programmed Reg. STP funds. Fin. Design in spring '97. PS&E possible by '98.

• *Future Projects:* Project would coordinate with construction of currently programmed Sunnybrook Extension and serve to minimize congestion expected at the Extension's juncture with Sunnyside Road at 108th. The Extension is, in turn, related to programmed construction of the Sunnybrook Split Diamond Interchange in FY 98. Coordinates with bike and ped improvements on the new Sunnybrook Extension.

Other Relevant Information

Bike and pedestrian multi-modal points should be made contingent on commitment to sensitive median design. Signal timing and intersection modifications have already been implemented. Shuttle service from 122nd to Sunnyside Transit Center funded. Capacity needed to accommodate easterly residential buildout. Priority project in the Sunnyside Area Transportation Master Plan, Nov., 1994.

Potential Phases

No feasible lesser construction phase. Reduced ROW would impede securing 2040 HCT alignment. ROW acquisition would achieve primary 2040 goal. Est. of $1 million for ROW; Final Design cost uncertain.

Figure 9.4 Priority assessment in Portland
SOURCE: Portland METRO, 1995

2. *Reliability/expandability.*This area of need related to providing consistent signal system software and communications infrastructure. The subcommittee placed emphasis on projects that permitted local communities to partner with telecommunications companies to place fiber optics cable for signal communications.

3. *Functional capabilities.* Signal systems have several basic functions, for example, upload/download timing parameters, monitor system and intersection operation, and report remote errors. The subcommittee wanted to fund functional expansions in the areas of detection, incident management, transit priority, and ITS integration.

4. *Control strategy.* The majority of signal systems are pretimed for time-of-day operation and are unable to respond to unexpected fluctuations in volumes. DRCOG wanted to begin the evolution toward more advanced control strategies.

The subcommittee recommended an improvement program based on three criteria: (1) criticality of the need, (2) resulting benefits, and (3) participation and commitment of operating agencies to implement and support the project.

Table 9.7 shows the type of information that was considered in the prioritization process. Such data were collected from previously implemented projects so that the subcommittee had an understanding of the likely benefits of similar projects. Daily user benefits were calculated from the reduction in vehicle hours of travel and the savings in fuel consumption.

Table 9.7   Traffic signal improvement summary in Denver

Project	Reduction in Vehicle Hours of Travel	Fuel Consumption Reduction	Air Pollutant Emission Reductions	Daily User Benefits ($)
Capital Improvement Projects				
Sheridan	888	337	1104	4850
Parker	1370	646	1504	7650
Havana	741	268	845	4050
Buckley	594	342	720	3400
Chambers	488	164	560	2650
East Alameda	340	141	400	1850
East Iliff	786	319	923	4300
Signal Timing/Coordination Projects				
88th/92nd	583	2552	669	3200
Quebec	1120	480	1294	6200
104th	1625	652	1905	8900
84th/88th	716	308	867	3950
Washington	741	307	906	4050
East Quincy	330	152	327	1850
Santa Fe	1270	383	1558	6800
Total	11,592	7,051	13,582	63,700

SOURCE: DRCOG, 1999

9.2.5   Systems Analysis Techniques

Much of the early technical work in supporting a programming process focused on methods that would result in an "optimal," time-staged sequencing of projects over the investment period [North Carolina Department of Transportation, 1972; U.S. DOT, 1973; Ontario Ministry of Transportation and Communications, 1974; U.S. DOT, 1976; Bellomo et al., 1979]. The analysis tool used most often was a linear programming model, designed to maximize discounted net benefits subject to a series of constraints. However, a large amount of data was necessary to conduct the analysis, a requirement that reduced the attractiveness of such an approach. Most of the applications of these methods have occurred at the state or provincial level, especially for targeted prioritization of well-defined project types such as pavement rehabilitation. Generally, similar types of analysis techniques have not been used at the metropolitan level for several reasons.

First, the goals and objectives of transportation policy in urban areas tend to be much more diverse than at higher levels of government. As disagreements over goals multiply, it becomes more difficult to develop a technical approach that reflects such divergence of opinion.

Second, most state and provincial transportation agencies are responsible for only one or two types of transportation projects, such as highways and airports. MPOs, however, must consider several different types of transportation investment, including highway projects, transit services, ride sharing, and operational strategies. Determining trade-offs among these types of investment in a quantitative manner is difficult.

Third, the limited number of funding sources usually provides state and provincial officials with a good estimate of the funds that will be available for the construction of projects. At the metropolitan level, many funding sources support the transportation program, and each often includes great uncertainty (e.g., voter approval is sometimes a prerequisite for regional taxes that support transportation investment). As an example, Table 9.8 shows the many different sources of funding for transportation projects/services in the Sacramento metropolitan area. Estimation of many of these revenues, beyond 1 or 2 years, can be very difficult.

Fourth, the network interdependence between project alternatives at the state or provincial level is minimal. Projects are located far apart and can thus be considered independent of one another. Technically, this allows the use of a linear formulation, more specifically, the assumption that different project scores can be added in the objective function. In a metropolitan context, however, the interdependence between several projects creates significant analysis problems.

Finally, the number of transportation projects considered in a large metropolitan area is typically much greater than at the state or provincial levels. In large areas, over 1000 projects could be considered in a typical year. Collecting the data needed to satisfy the requirements of a linear programming analysis package for each project would be beyond the resources of most government agencies.

Multiobjective optimization methods, however, could provide some useful information (see, for example, [El Dessouki et al., 1998]). A multiobjective application in Provo, Utah, for example, evaluated more than 1.9 million different

Table 9.8   Different funding sources for the Sacramento TIP

Benefit assessment district	Other local funds
Congestion mitigation/air quality (CMAQ)	Other state funds
CMAQ transfer	Petroleum violation escrow account
Developer funded	Private funds
FTA 5307 carryover	Proposition 116
FTA 5309 carryover	Recreational trails program
FTA 5310 carryover	Redevelopment funds
Fair share/in lieu	Regional surface transportation program
Farebox revenue	Road fund
Federal demonstration funds	Sacramento County measure a sales tax fund
Federal lands program	Sacramento Housing and Redevelopment Authority
FTA Section 3037	State matching funds
FTA Section 5303	State surface transportation program
FTA Section 5307	State transit assistance
FTA Section 5309	State/local partnership program
FTA Section 5310	Surface transportation program safety fund
FTA Section 5311	Surface transportation program/RR crossing fund
Highway bridge replacement/rehabilitation	Traffic impact fees
Interstate maintenance	Transit capital investment program
Local transportation fund	Transportation development act
National highway system fund	Transportation enhancement activities
Other federal funds	Transportation system management funds

SOURCE: SACOG, 1999

program combinations based on three objectives—to minimize travel time, per capita cost, and land-use change [Taber et al., 1999]. Of these 1.9 million combinations, the model found 195 that provided optimal achievement of the objective function. However, as noted by the authors, "choosing among 195 optimal plans and their different ramifications is still too difficult for decision makers to assimilate in a reasonable amount of time."

9.2.6   Summary

Transportation planners use several methods to prioritize projects, and the result of the prioritizing is an investment program. Although the prioritization process will be unique to the decision-making context of a region, several characteristics of good practice arise from the preceding examples. First, and perhaps most importantly, a direct linkage exists between the goals of the transportation planning process and the programming criteria. This was best illustrated by the examples from Albany and Portland.

Second, even in those cases where substantial effort was made to produce quantifiable scores, the information presented to decision makers also included qualitative and/or subjective interpretation of a project's value. This type of information is important to those with limited knowledge of the technical approach to planning and often is the only way of assessing the value of a project as it relates to impacts that are difficult to quantify.

Third, the fiscal constraint requirement has increased the importance of project prioritization. With only limited funds available, and given the mandate to produce only financially realistic programs, technical analysis can play a critical role in helping to define the most appropriate investment strategy for a region. Such analysis could be the deciding factor in debates on which projects to pursue in the regional transportation program.

Finally, in each case previously described, the *process* of prioritization was as critical for the credibility of the result as were the technical methods. The weights included in the scoring methods were always selected by decision makers or through a public process. The definitions of the criteria used to prioritize resulted from extensive public participation, and the decision steps that produced a final transportation program included the participation of stakeholders, decision makers, interest groups, and the general public.

9.3   FINANCIAL ANALYSIS AND FUNDING AVAILABILITY

Predicting the amount and time stream of future funding is critical to the development of a transportation program and a financially constrained transportation plan. In addition, given the often large costs associated with new transportation facilities, determining the financial impacts of these costs becomes an important criterion for the decision-making process. For example, project or plan evaluation criteria often have the following as part of the set of evaluation measures

$\dfrac{Total\ capital\ cost}{Financial\ Capacity}$: Percentage of the region's (or agency's) financial capacity required to cover the capital cost of each alternative.

Valuation of annual tax burden: Per capita or per household annual tax requirements to pay annualized capital, OMM, and debt costs.

Revenues required for debt service: Dollars needed to pay principle and interest balance over life of project and finance bond (if used to fund the project).

Percent revenues for OMM: Proportion of tax revenues required to subsidize the operation, maintenance, and management of each alternative [Seattle METRO, 1993].

Many factors outside the control of planners can influence future funding levels, including changes in funding commitments from higher levels of government, reduced revenues from gas or sales taxes due to economic conditions, and changes in public funding priorities due to newly elected leaders. Each of these factors can significantly influence the amount of funds that will be available for transportation investment, but each is very difficult to predict. As a result, the determination of future funding is often a loosely structured process, based on the expectations and intuition of the planner.

> It [fund forecasting] requires a knowledge of transportation: trends in priorities and funding on the part of Congress, state legislatures, county boards, and city councils; departmental and executive-branch priorities and trends; and project development in various modes for different types of projects and the way projects might be affected by everything from environmental laws and citizen opposition to design delays [Campbell and Humphrey, 1978].

Financial analysis consists of two major steps—assessment of the financial condition of the region to afford a transportation program and an assessment of the financial capability to afford new debt.

Financial Condition   The point of departure for a financial analysis is determining the current financial condition of the region or of an agency proposing to fund a project. Historical and forecasted trends of economic variables, service revenues, and expected program expenditures provide a snapshot of the financial health of a region. For example, an assessment of population growth and employment trends by sector provides a basis for estimating income levels, sales tax receipts, and general government revenues. Operating cost and revenue data are compared in an annual cash flow statement of revenues and expenses that can identify likely end-of-year cash positions [Portland METRO, 1996; 1997]. Sensitivity analyses are usually conducted on the key input variables and assumptions. Different business cycle growth rates and different federal appropriation levels, for example, could be modified to determine a range of future year revenues.

Financial Capacity   Once the financial condition is established, the ability to fund and/or finance future operating and capital liabilities must be determined. This is called financial capacity. Financial capacity analysis differs from that for financial condition in that it focuses on supporting new services, whereas financial condition analysis examines the financial wherewithal to support existing services. To the data collected in the financial condition analysis must be added expected construction costs and schedules, OMM costs, capital revenues and financing techniques, OMM sources of revenues, bond issuance costs, and other borrowing requirements. Important to this process is the determination of future funding availability, which consists of four steps.

1. *Forecast future funds from higher levels of government.* This step is probably the most difficult, as the politics and legislative processes that secure funding are far removed from the local funding context of the planner. In the United States, for example, this task would involve forecasting the level of transportation funds coming from federal and state governments. At the federal level, transportation funds are allocated by funding categories that disburse funds in different ways. The major types of federal programs are described in the following

- Some formula funding is apportioned to each state on the basis of population, highway mileage, minimum return on contributions to the Highway Trust Fund, and so on.

- Some formula funding is apportioned based on need to address a national problem. For example, federal funds may be distributed on the basis of the severity of air-quality problems.

- Flexible funding expands the eligibility for the use of formula funds so that state and local officials can shift portions of one program category to another to address state or local priorities.

- Set-aside funding is a special funding allocation taken before funds are distributed by formula in order to fund high-priority initiatives. Federal programs emphasizing job access, scenic byways, and recreational trails have been funded in this manner.

- Pilot programs provide funding for several years to encourage states and MPOs to test innovative concepts. ISTEA, for example, provided such funds to encourage demonstrations of congestion pricing.

- Incentive grants provide funds to reward states and MPOs that have achieved certain national objectives, such as increasing seat-belt usage or reducing drunk driving [ITE, 2000].

Program funding levels are easiest to predict when based on formulas. These formulas seldom change during the 5- to 6-year authorization time frame of the enabling legislation, allowing determination of future allocations with some certainty over this time period.

At the state level, which is the greatest source of transportation revenues in the United States, 60 percent of the transportation revenue comes from taxes on gasoline consumption. Predicting future levels of these funds requires an understanding of issues such as the impact of more fuel-efficient and alternative fueled automobiles on fuel sales and the trends affecting vehicle miles traveled. State officials are often able to estimate what funds they will have available, at least in the near term. At other times, these officials will be unable to provide this information because the appropriate legislative body has not yet decided on the level of funds to be allocated or the tax rates to be levied. Planners might thus find themselves estimating levels of funding from state government that, at best, would be a rough approximation of what action the legislative body will take.

2. *Estimate future funds from local government.* Local governments rely more on property taxes than any other source to fund transportation services. The planner should therefore be able to determine with some certainty the level of funding that will be provided for transportation purposes by local government. This can be done by closely monitoring the political and budgetary process of local governments or, in the case of dedicated funding sources (e.g., 1 percent sales tax for local transportation purposes), estimating the likely trends associated with the taxed activity. In the case of revenue projections from transit fares or highway tolls, the level of analysis can be relatively simple (e.g., trend projections) or sophisticated (e.g., transit

ridership demand models). The specific analysis technique used in estimating revenues would depend on the characteristics of the transportation system and on the operating environment. Clearly, adding a new line to a subway network or having the price of gasoline increase dramatically would affect ridership and revenues on a transit system, requiring a more sophisticated revenues analysis than a simple trend projection. Estimating the impacts of selected projects on budget availability thus becomes an important task for transportation planners concerned with programming.

3.   *Determine cash flow requirements.* Although estimating the total level of funding available for transportation purposes is an important element of programming, an equally important task is determining the flow of those funds among project categories and over time. Often, project costs are divided among different agencies. For example, the interstate highway program in the United States was constructed with the federal government covering 90 percent of the project cost and state governments providing the rest. Where matching costs are required, the planner must determine if sufficient funds are available at the required time to match funds from other agencies. Another cash-flow problem can be related to the reimbursement policies of government agencies. For example, the costs of project design and right-of-way acquisition typically need to be paid when the work is being done, whereas for construction projects the costs might not have to be paid until the contractor reaches certain stages of the construction work.

4.   *Bring funding categories together to determine overall budget and deficit.* Once the funding levels have been estimated for specific program areas, they should be combined with estimates of operating and maintenance expenditures to provide a clear picture of the financial situation over the programming time horizon. Such a synthesis not only gives decision makers an indication of the absolute levels of funding available, but also provides an "early warning" system for identifying future areas with uncertain funding.

Several examples illustrate the process of developing funding estimates for the financially constrained transportation plan and program.

Denver   The financial analysis of the 2020 transportation plan included "the participation of the Colorado DOT, the regional transit agency, local governments, special districts and authorities, paratransit operators, and various special funding agencies" [DRCOG, 1998]. Over the 22-year period of the plan, close to $14 billion in revenues were expected to be raised by local governments. The largest funding source was the regional transit agency, which was expected to generate $6.2 billion in nonfederal revenues—$4.6 billion from a sales and use tax, $1.4 billion from fares, and $0.2 billion from other sources. An approximate $4.21 billion was to be contributed by the Colorado DOT.

The major assumptions used in this financial analysis included the following:

- For transit revenues, the base sales tax rate did not change, but transit fares were expected to be increased in line with inflation.

- Colorado motor fuel tax receipts were to increase at a rate reflecting the increase in vehicle miles traveled.

- Vehicle registration and drivers' license fees were to increase in line with the forecasted rate of Colorado's population growth.

- The Denver metropolitan area would receive 41.4 percent of the state's transportation resources.

- There would be no growth in federal aid highway funds over the authorized levels, and only 90 percent of the authorized funds will be obligated and made available.

- Federal revenues for interstate maintenance and national highway system and bridge projects were expected to decrease in purchasing power over the 22-year period.

- Surface transportation and congestion mitigation/air-quality funds would be increased to keep up with inflation and be separately allocated to Denver by formula.

- Federal Transit Administration formula funds and $0.39 billion in discretionary funds would be available.

The summary of expected revenues and expenditures for the Denver regional transportation plan is shown in Table 9.9.

San Francisco   A 20-year transportation revenues projection for the San Francisco metropolitan area resulted in an estimated $90 billion budget for the regional transportation plan [MTC, 1998]. The bulk of these revenues came from local sources, primarily transit fares, property taxes, dedicated sales tax programs, and state gas tax disbursements to local jurisdictions. Of this $90 billion, $83 billion was allocated to operations, maintenance, and rehabilitation of the existing system. The major assumptions used in this financial analysis included the following:

- Federal and state revenues were to increase 2.2 percent per annum.

- Annual inflation would increase at a rate of 2.2 percent from 1999 to 2004 and 3.5 percent from 2005 to 2018.

- Forecasted project revenues and costs were inflated to year of expenditure dollars.

- Transit fares would keep pace with inflation.

- The state fuel tax was projected to increase by 5 cents per gallon in fiscal year 2009/2010.

- County transportation taxes were not assumed to continue after their sunset dates.

- No new revenue sources were assumed to become available.

Sacramento   The Sacramento Area Council of Governments (SACOG) conducted a financial analysis of its 1999 transportation plan that included the results of a survey of all its member jurisdictions, a review of existing TIPs showing funds committed previously, and projections of uncommitted state and federal funds over the planning time horizon. An approximate $12.8 billion in revenues was projected for

Table 9.9   Expected transportation plan revenues and expenditures in Denver

2020 RTP Surface Transportation Expenditures (Millions of 1996 Dollars)	
Improvement Type	**Expenditures**
Rapid transit improvements	$1338.9
Nonrapid transit improvements	1083.9
Bus replacement and expansion vehicles	578.3
Park-n-ride lots (nonrapid transit)	40.8
ADA Fleet	73.0
RTD maintenance facilities and equipment	391.8
Roadway capacity	2721.5
1997–2002 TIP committed improvements	433.7
W/E/SE MIS highway improvements	414.8
Freeway and arterial roadway improvements	1543.0
Other interchange improvements	330.0
Reserved for undetermined MIS corridor recommendations	636.8
Operational improvements on emphasis corridors/ITS	442.5
Other	181.5
Regional traffic signal improvement program	47.3
Ride arrangers program	30.1
US-36 transportation management organization	0.3
Pedestrian and bicycle program	88.2
Specialized services for elderly and disabled (federal portion)	15.6
2020 RTP federal state, and regional subtotal	$6405.1
100% locally funded improvements	640.3
Total	$7045.4

Summary of Revenues Available to Denver Region FY 1999–2020, Eight-County Total (Millions of Constant 1996 Dollars)	
Funding Source	**Total**
Interstate maintenance and reimbursement	$194.7
NHS and discretionary	199.7
Bridge	179.0
STP-non Metro	343.8
STP-Metro	341.0
CMAQ	77.7
Federal highway subtotal	$1335.9
HUTF, SB 97-01, HB 98-1202	$3499.0
Other new revenues	631.7
Miscellaneous	77.5
State subtotal	$4208.2

Table 9.9   Expected transportation plan revenues and expenditures in Denver (continued)

Summary of Revenues Available to Denver Region
FY 1999–2020, Eight-County Total (Millions of Constant 1996 Dollars)

Funding Source	Total
RTD sales tax	$4223.2
RTD use tax	401.2
Farebox revenues	1392.7
Other RTD revenues	165.2
RTD subtotal	$6182.3
FTA Capital Section 5307 and Fixed Guideway Modernization	$325.1
FTA discretionary	356.4
Specialized service programs	15.6
FTA subtotal	$ 697.1
Yet to be determined combination of I-70 mountain corridor FHWA dollars and FTA discretionary dollars	$511.3
Local match for STP-Metro	$227.3
Local match for enhancement	9.2
Denver match for east corridor rail	68.0
Local match for DUT	7.0
Local match subtotal	$311.5
Southwest corridor remaining funds	$57.9
USDOT, RTD, CDOT subtotal	$13,334.7
Local project revenues	$840.3
Total	$14,175.0

SOURCE: DRCOG, 1998

the region. The recommended use of these funds was presented in the plan categorized by mode and by county. The revenue projections were based on SACOG's growth projections for the region and on the local government's assumptions regarding future impact fees, assessment districts, and other funding sources. A comparison of recommended expenditures and expected revenues showed a shortfall of $2.274 billion (Table 9.10). This shortfall became the focus of a comprehensive strategy to raise more transportation funding for the region.

San Diego   The 2020 transportation plan for the San Diego metropolitan area estimated total transportation expenditures of $29.38 billion over the 21-year planning time horizon. However, a financial analysis showed that only $17.865 billion in revenues were likely to occur, thus leaving a shortfall of $11.515 billion. Table 9.11 shows the plan's cost and revenue estimates by time period and the assumptions underlying many of the estimates.

Table 9.10   Planned expenditures and revenues in Sacramento

Recommended Expenditures	Totals	Percentage of Total Expenditures
Capital Improvements		
Bicycle/pedestrian	103	1%
Bus/light rail/heavy rail	1174	8%
HOV lanes	306	2%
Freeway interchanges	690	5%
Local roads and bridges	2112	14%
Total capital improvements	4385	30%
Operations and maintenance		
Transit operations/maintenance	3074	20%
Road maintenance	6477	43%
Road maintenance backlog	909	6%
Total operations and maintenance	10,460	69%
Other		
Transportation management	104	< 1%
Studies	8	< 1%
Air quality	90	<1%
Total other	202	1%
Total expenditures	15,047	100%
Projected existing revenues	12,773	85%
Surplus (shortfall) based on revenues from existing sources	(2274)	15%

SOURCE: SACOG, 1999

The projected shortfall in the San Diego plan led to the identification of several potential revenue sources that could help make up the difference, including such strategies as raising the gas tax, extending local sales taxes, increasing truck weight fees, instituting regional development impact fees, building more toll roads, and implementing congestion pricing. The transportation plan also illustrated the impact of not having sufficient funds to build all of the improvements identified in the transportation plan. Figure 9.5 shows the predicted level of service on the freeway system resulting from the revenue-constrained plan. A similar map was provided for the transit system.

Southeastern Wisconsin   The recommended 2010 transportation plan for the Milwaukee metropolitan area was estimated to cost $521.9 million annually [SEWRPC, 1994]. The expected average annual public revenues were expected to total $299.9 million, thus resulting in an annual shortfall of $220.0 million. Several possible options were examined to make up the difference, including increased state funds and gas taxes. SEWRPC planners presented the information to decision makers in a way that was easy to understand and that highlighted possible implementation

Table 9.11 San Diego transportation plan's costs and revenues
(in millions constant 1999 dollars)

	FY 1999–2004	FY 2005–2010	FY 2011–2020	FY 1999–2020
Highway Costs				
Admin/operations[1]	$110	$115	$200	$425
Maintenance[2]	180	185	335	700
Construction/ROW/eng'g[3]	1700	1985	5580	9265
Total costs	1990	2285	6115	10,390
Revenues				
State/federal (STIP)[4]	430	325	410	1165
SHOPP/minor[5]	300	300	500	1100
Federal STP/CMAQ[6]	205	150	155	510
TransNet[7]	220	120	0	340
Local/privatization[8]	545	0	95	640
Admin/opns/other[9]	110	115	200	425
Maintenance[10]	180	185	335	700
Total Revenues	1990	1195	1695	4880
Total surplus (deficit)	$0	($1090)	($4420)	($5510)

SOURCE: SANDAG, 1999
NOTES: [1] FY 1999 base is $18 million; assumes 1 percent increase per annum starting in 2005
[2] FY 1999 base is $30 million; assumes 1 percent increase per annum starting in 2005
[3] Includes state DOT's major rehabilitation projects at 10-year plan levels
[4] Assumes continuation of state transportation programs and funding levels based on 1998 STIP
[5] Assumes funding will occur at least at historical levels
[6] Highway projects assumed to receive 50 percent of future STP and CMAQ funds
[7] Highway TransNet revenues reflect sales tax receipts plus interest bond proceeds minus bond debt service
[8] Local privatization includes state routes 125 and 241 as toll roads
[9] Assumes state will fund all state highway administration, operations, and maintenance costs

mechanisms. This was done by estimating the equivalent motor fuel tax in cents per mile or the equivalent general sales tax in percent that would be necessary to fund the shortfall. Table 9.12 shows this information categorized by modal investment and by county. During the planning process, a proposal was made to stop funding highways and transit service with local property taxes, a $60 million-per-year use of property tax revenues. A table similar to Table 9.12 was provided in the plan to show what level of gas tax or sales tax would be needed to offset the loss of these funds.

In order to obtain these additional funds, the plan recommended that an area-wide authority be created with the power to raise funds to support the regional transportation program. The importance of this step was noted in the plan:

> Southeastern Wisconsin, like most large metropolitan areas in the nation, must come to grips with this uncertainty by identifying and securing a dedicated nonproperty-tax revenue source for county and local transportation purposes. This uncertainty is perhaps greatest of all in terms of plan implementation and will need to be carefully monitored as the implementation period proceeds [SWRPC, 1994].

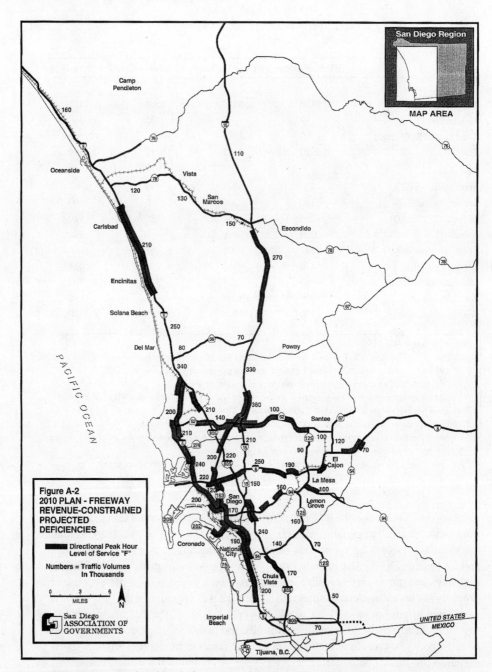

Figure 9.5   Predicted level of service on freeways in San Diego with constrained plan
| SOURCE: SANDAG, 1999

The successful implementation of the regional plan not only needed a financial component, but also an institutional capability to manage the transportation system.

Table 9.12   Southeastern Wisconsin cost and revenue estimates

County	Average Annual Local Funding Shortfall (millions)			Equivalent Motor Fuel Tax Required to Fund Shortfall (cents per gallon)			Equivalent General Sales Tax Required to Fund Shortfall (percent)		
	Highways	Transit	Total	Highways	Transit	Total	Highways	Transit	Total
Kenosha	3.9	0.2	4.1	6.8	0.4	7.2	0.5		0.5
Milwaukee		13.2	13.2		3.3	3.3		0.2	0.2
Ozaukee	3.4	0.2	3.6	7.8	0.4	8.2	0.7		0.7
Racine	3.0	1.2	4.2	4.1	1.6	5.7	0.3	0.1	0.4
Walworth	4.8		4.8	10.9		10.9	1.0		1.0
Washington	4.8	0.1	4.9	9.1	0.2	9.3	0.8		0.8
Waukesha	8.2	2.4	10.6	4.6	1.4	6.0	0.4	0.1	0.5
	28.1	17.3	45.4	43.3	7.3	50.6	3.7	0.4	4.1

SOURCE: SEWRPC, 1994

9.4   PROGRAM/PROJECT IMPLEMENTATION AND INNOVATIVE FINANCING

The completion of a transportation plan and TIP is just the beginning of a successful strategy for investing in a metropolitan transportation system. The projects and services proposed in the transportation plan must be implemented, which requires funding and institutional capability. For example, the 2010 regional transportation plan for southeastern Wisconsin, discussed in the preceding section, outlined extensive changes to the structure of transportation system management in the region [SEWRPC, 1994]. These ranged from recommended changes in the jurisdictional responsibility for portions of the arterial street and highway network to the creation of a regional transportation authority. The mechanisms for effecting these changes similarly ranged from changes to local statutes to enactment of state-enabling legislation. Without adequate capital and operations/maintenance funding, the ability of transportation officials to preserve and enhance the transportation system would be severely constrained. Without institutional capability and flexibility, transportation organizations might be unable to respond to a changing policy environment. Funding and institutional capability are clearly prerequisites for any improvements to the transportation system.

9.4.1   Nontraditional Funding

Transportation needs in a metropolitan area often far exceed the level of available funding. Traditional sources of funding such as motor fuel, sales, and property taxes do not provide enough revenues to meet these needs. Many states and metropolitan areas have begun to use a variety of nontraditional funding strategies that fall under the label of "innovative funding." As shown in Figure 9.6, the identification of inno-

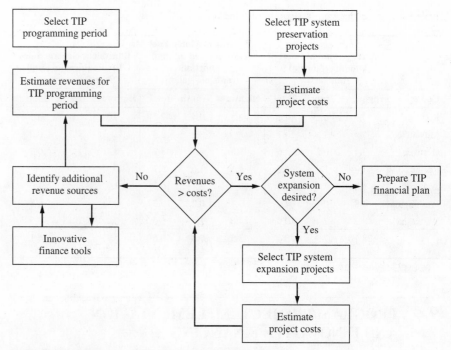

Figure 9.6   Role for innovative financing in the TIP process
| SOURCE: FHWA, 1996

vative financing opportunities becomes necessary when a metropolitan area needs additional resources to support a program. In very broad terms, innovative financing includes any strategy that does not rely solely on conventional highway user fees and taxes [FHWA, 1998]. Over the past decade, such strategies have included toll roads, privatization of services, infrastructure banks, revolving loan funds, federal loan programs, advanced construction financing, and cross-border leasing for transit investments [FTA, 1998]. It is beyond the scope of this book to describe all of the innovative finance strategies that are available. However, the approaches that occur primarily in metropolitan areas and that include a substantial role for private sector participants are described in the following paragraphs.

Public-private transportation financing strategies commonly found in metropolitan areas include toll roads; development fees, exactions, and value-added taxation; and privatization.

Toll Roads   The United States currently has 37 toll roads and 44 toll bridges totaling 5000 miles and accounting for about 5 percent of total highway revenues. An estimate 1600 miles are being planned. The number of toll facilities having electronic toll collection capability has increased from 49 in 1995 to 118 in 1999 [FHWA, 1999]. Toll roads can be financed in several ways—general obligation bonds, revenue bonds, revenue bonds supplemented by income other than that paid by users, private financing, and combinations of the above. Toll revenues are then used to pay off the principal and interest of these bonds. The toll project could also be part of a much larger leveraged loan program that supports other projects [Lockwood, 1995].

The current public–private partnership models for toll facilities include

Build-Own-Operate: A private consortium finances, builds, owns, and operates a facility.

Build-Operate-Transfer: A private consortium receives a concession to finance, build, own, and operate a facility for a specified time period, after which the facility is transferred to the responsible government agency.

Build-Transfer-Operate: Similar to the preceding, but the ownership of the facility is transferred to the government upon completion, and the consortium then leases the facility and collects revenues for some limited time period.

Buy-Build-Operate: A private consortium buys an existing facility from the government, upgrades it, and then operates it, collecting tolls.

Lease-Develop-Operate: A facility is leased, upgraded, and operated by a private consortium for the duration of the lease. Ownership is continuously held by the government.

Temporary Privatization: A firm takes over operation and maintenance of an existing facility, improves it, and collects tolls until the cost of repair plus a reasonable return on capital is attained. Ownership is continuously held by the government.

Many states already have legislation that authorizes toll roads, and others are putting laws in place (see [FHWA 1994]).

Development Fees, Exactions and Value-Added Taxation This model of public–private partnership requires private groups to contribute to the cost of the transportation improvements needed to handle the additional demand associated with a development site. Development fees, exactions, or taxation of the added value of land due to enhanced accessibility are the most common ways of collecting these revenues. The enabling institutional mechanisms take the following forms

Assessment Districts A special tax is levied on all property owners in a designated area to pay for an improvement that benefits primarily those owners. The tax must be approved by a majority of the property owners. Usually a district is supported by a municipality that issues revenue bonds and assesses property owners to repay the bonds. For example, Chatham County, Georgia, has established a special finance district to support transit services in Savannah, Georgia, by applying an additional millage to tax rates on property in the district [FTA 1993].

Special Districts Special districts are similar to assessment districts, except their governing body is separate from local government. Special districts have authority to tax, issue bonds, and provide services within a specified area. Portland, Oregon, uses Local Improvement Districts (LIDs) authorized by state statute to finance road and other improvements. A developer can initiate a LID, but 50 percent of the voters in the proposed district and in the governmental jurisdiction must approve it.

Once approved, the LID can issue tax-exempt general obligation bonds that are backed by the faith and credit of the city or county and are also supported by a lien on the properties benefited by the bond revenues.

Development Agreements In this agreement, developers pay for transportation improvements in return for a public commitment to remove as many administrative impediments and delays in approval as possible. For example, in Montgomery County, Maryland, the county pays for required road improvements and is then reimbursed either in cash per lot, payable upon conveyance of title to buyers of new houses, or by a deferred payment plan that places so-called "road club" charges on homeowners' tax bills, amortized over 5 to 10 years. Developers choose the repayment method. Montgomery County requires the "road clubs" be included in formal documents, called public works agreements, that legally identify the participants and that set forth the responsibilities of the county and the developers.

Development or Impact Fees Developers pay fees to compensate the community for the extra costs of public facilities needed for a development site. Paid at the time of the building permit, the fees are placed in a fund designated for the construction of facilities that "solve" the problems caused by the development. For a good description of impact fees, see [Nelson, 1988]. For example, Ft. Collins, Colorado, requires developers to provide all streets internal to their project and, in addition, pay a street oversizing fee for collector and arterial streets. These fees were set at $584 per residential and multifamily dwelling unit, $5252 per acre of light industrial, $7003 per acre of heavy industrial, $10,504 per acre of office/general commercial, and $14,005 per acre of retail/commercial (1996 dollars).

Broward County, Florida, imposes road impact fees that are representative of the fair share of the service or facility costs associated with the development site. A travel demand model is used to estimate the number of trips generated by the proposed development. If the development will significantly increase traffic volumes over the existing road capacity, the developer is required to pay a proportionate share of the costs required to increase the capacity of the road. The developer is not required to pay for existing deficiencies in the road.

Tax Increment Financing (TIF) Increased tax revenues that occur as a result of new development are earmarked for financing the public improvements that supported this growth. Improvements are paid with public funds or bonds, then repaid with property tax revenues from the new development. Prince George's County, Maryland, for example, established a TIF to finance a parking garage for an AMTRAK station.

Privatization Privatization of transportation facilities and services has the highest level of private involvement of public–private partnerships. Privatization simply means shifting the responsibility of providing transportation services from the public sector to private providers. This can include the complete takeover of public programs, sales of public assets to private investors, contracting of services, and deregulation of activities previously treated as a public monopoly [Starr, 1987].

Privatization is not a new phenomenon to transportation; many early roads and transit services in the United States were operated by private operators.

Privatization has received a great deal of attention in recent years, especially with regard to the distribution of public benefits. To the extent that private equity or debt replaces public investment, taxpayers will gain from not having to make such an uncompensated contribution. More efficient operation of the facility or service might also provide economic gains to taxpayers. Investors would gain if they were able to capture the user fees and other revenues in sufficient quantity to provide a return on investment. The largest societal benefits from privatization will occur where efficiency gains are the greatest and where the private operator faces effective competition.

Implementation Considerations Public–private partnerships created to finance transportation projects can encounter significant barriers. Many relate to the assumption of the investment risk or to the appropriate roles of government agencies and private investors in projects that can provide public benefits while resulting in private gain. Some of the major issues that can confront these partnerships include

1. *Financial barriers*: Financial viability is a basic enabling criterion for private-sector involvement. Risks associated with this financial viability include start-up financing problems, unsure demand levels and resultant uncertain income streams, fluctuating construction costs, general exposure to liability, and the uncertainty of project scope as the project wends its way through the environmental review process.

2. *Equity capital*: The private equity capital market has its own lending operating procedures and constraints that could mitigate against a private investor obtaining investment dollars. Partnering with a government agency, for example, is often considered a risky investment, or at least one that will not provide much return on investment.

3. *Concession or franchise agreements*: If a private entity is to operate a service or facility for public use, the government agency that oversees the operation will most likely require minimum design and operation characteristics, risk assignment, regulatory oversight, provisions for public funding participation, tort liability guarantees, right-of-way provisions, and an agreement on default conditions.

4. *Constitutional powers of state agencies*: State constitutions provide very specific limitations in the mission and authority of state agencies. Thus, constrained state contractual or police powers and the limited flexibility in using federal or state funds could be important issues in fulfilling the governmental role of a public–private partnership.

5. *Procurement*: Government agencies are subject to stringent conditions on how goods and services are procured. Private organizations often do not have such limitations. For example, a choice of contractor is often competitively bid for government proposals but can be sole sourced in a private construction setting. Other issues could include business set-asides, protection of intellectual property, and use of the design/build contracting process.

6. *Permitting process*: Environmental clearances can often take a long time and potentially lead to project delay. The environmental analysis process could add a great deal of risk to a private investment.

7. *Tax structure*: The combination of public tax-exempt financing with taxable private investment financing can add a high level of complexity to a project. Given that most costs for a large project are incurred up front and most benefits occur over a longer time frame, high tax rates on private investment early in the project, with the promise of revenues occurring in the future, is a significant disincentive to invest.

8. *Community and government support*: Providing opportunities for public involvement is a requirement for most transportation projects. Given the unusual nature of the financing arrangements, public participation could be either a positive or negative influence on the ultimate outcome. Such support could be critically important in gaining the necessary approvals from government officials and agencies [FHWA, 1994]. However, if public groups perceive that this partnership is not good for the community or is deliberately avoiding the safeguards that have been put in place for public projects, the public involvement process could become a source of frustration and delay.

Private investment is usually targeted on a particular project, except in the case of a jurisdictionwide assessment district. Private financial contributions are therefore not a reliable or stable source of funding for a regional program of transportation improvements. User fees will likely remain the most important source of transportation funding for the foreseeable future. With user fees, however, any change to the perceived cost of travel is likely to meet strong opposition. Implementing a regional toll system or congestion pricing, for example, would be viewed by many as just another form of taxation. A survey of southern California voters indicated strong opposition to any possibility of changing the road pricing system. A large majority of the respondents felt that the existing system of motor fuel taxes was fair and appropriate. The assessment report accompanying the survey results offered some suggestions on what would be necessary for success if the price system were to be changed.

1. *The problem must be understood and imminent*: Voters must clearly understand the nexus between the pricing program and the problem it addresses. Without some motivation to solve a "problem," voters were significantly inclined to keep the motor fuel tax.

2. *The program must work*: Assuming voters can be motivated to accept a change in finance, the new approach must be viewed as being effective in actually accomplishing congestion reduction and mobility goals. In addition to effectiveness, the survey showed that the financing program must be viewed as being simple, fair, seamless, and user friendly. Most importantly, it must show that it is better than the current system.

3. *Tread slowly and carefully toward implementation*: Successful implementation of a new financing scheme must increase voter intensity/awareness of the basic problem being addressed; explain to voters the nexus between the solution and the problem; and clearly describe to voters all of the program elements [Southern California Association of Governments, 1996].

The report concluded that the environment currently for changing financing strategies in general "must be mindful of the following basic voter predilection." Voters are skeptical that government can effectively implement a program that will affect millions of drivers without great inconvenience. This was especially important given voters' strong support for the existing auto-dominant transportation system.

9.4.2 Institutional Capability

Institutional capability is a key ingredient to successful implementation. Such capability is not limited to appropriate organizational structures for carrying out project implementation. It also requires the types of skills and analytical capabilities needed to implement, operate, and maintain transportation projects and systems. This section identifies several important institutional issues that can either facilitate project implementation or serve as barriers to implementing a project or program. Many of these issues are common to any type of transportation project or program, whereas some might be very specific to the particular project being considered. Three institutional contexts are used to illustrate the strategies that can successfully establish a capability for project and program implementation—integrating ITS actions into system operations, coordinating land-use and transit service planning, and creating private associations to advocate and direct transportation investment to solve activity center mobility problems.

Institutional Challenges In Implementing ITS   Successful implementation of ITS strategies must address two key issues—(1) technical integration of the electronic systems that make up the different components and (2) institutional integration, which involves the coordination among various agencies and jurisdictions to achieve systemwide operations. Many of the early efforts in implementing ITS programs have experienced significant issues with regard to this institutional integration [U.S. DOT, 1994]. These challenges related to transportation agencies refusing to accept ITS strategies as solutions to problems that had been traditionally "solved" with new construction. Given the significant role that private companies have had in many ITS deployments, constraints arise from the issues discussed earlier regarding the appropriate role for government agencies and the regulatory, financial, and legal constraints associated with contracting and purchasing of services.

A survey of the institutional issues and barriers to ITS planning found that these concerns could be grouped into six major categories—organizational, leadership and management, legal and regulatory, technological, ITS impacts and benefits, and financial [JHK and Assocs., 1996]. Table 9.13 shows the different types of issues that fall within each category. Many of these nontechnical issues can be dealt with by incorporating ITS into the regional transportation planning process, thus providing a forum for all interested parties to understand and resolve such issues. However, others must necessarily be the product of legal and negotiated compromises. Note in the table that many of these issues are common to all types of transportation projects, not just those having a technology component.

Table 9.13   ITS Institutional Issues

Barrier	Breadth	Impact	Resolution Potential	Unique to ITS
Organizational				
Perception of loss of control	Moderate	Moderate	High	U
Public agency not accepting ITS	Moderate	Moderate	High	U
Differing agency objectives	Moderate	High	Moderate	A
Thinking limited on solutions	Moderate	High	High	A
Difficulties cooperating cross border	Moderate	High	Moderate	C
Cultural differences between public and private organizations	Moderate	High	Moderate	A
Lack of trust	Moderate	Moderate	High	C
Unclear definition of goals and roles	Low	Moderate	Moderate	A
Resistance to change	Low	Low	High	A
Lack of staff continuity	High	Moderate	High	A
Leadership and Management				
Lack of advocates at the staff level	Moderate	High	High	C
No advocates at top management	High	Moderate	Moderate	C
Failure to provide leadership	Moderate	High	Moderate	C
Lack of interagency communication	High	Moderate	High	C
Lack of intra-agency communication	Moderate	Moderate	High	C
Too large steering committee	Moderate	Low	Low	A
Inadequate committee representation	Moderate	Moderate	High	A
Unempowered steering committee	Moderate	High	Moderate	C
Inability to maintain interest	Low	Moderate	High	C
Devoting planning resources to low-priority areas	Low	Low	High	C
Difficulties in agencies' contracting	Low	High	Moderate	C
Failure to show progress	High	Moderate	Moderate	A
Ignoring maintenance and operations staff	Moderate	Moderate	High	U
Trying to accomplish too much	Low	Moderate	High	A
Lack of written documentation	Moderate	Moderate	High	C
Uncertainty over long commitment	High	High	Low	A
Overly complex procurement	High	Moderate	Moderate	A
Resources: Personnel and Facilities				
Lack of specialized technical skills	High	High	Moderate	U
Lack of familiarity with ITS	Moderate	Moderate	High	U
No knowledgeable ITS spokesperson	Moderate	Moderate	Moderate	C
No knowledgeable public/private partnership expertise	High	Moderate	Moderate	A
Unacceptable business risk	Moderate	High	Low	A
Technologies				
Fear of technological obsolescence	Moderate	Moderate	Moderate	U
Inattention to details	Moderate	High	High	A

Table 9.13   ITS Institutional Issues (continued)

Barrier	Breadth	Impact	Resolution Potential	Unique to ITS
Technologies (continued)				
Lack of technical standards	Moderate	Moderate	Moderate	U
Limited communication frequencies	High	Moderate	Moderate	U
Its Impacts and Benefits				
Lack of information on ITS impacts	High	Moderate	Low	A
Redistribute traffic in communities	Low	Moderate	Low	U
Leaning to capacity increases	Moderate	High	Moderate	C
No explanation to officials	Moderate	Moderate H	High	U
Public reaction against technologies that compromise privacy	Low	High	Low	U
Environmental impacts	Moderate	Low	Low	C
Low visibility of ITS benefits	High	High	Moderate	U
Inequitable allocation of benefits	Moderate	Moderate	High	A
Perception that ITS doesn't solve problems	Low	Moderate	Moderate	U
Legal and Regulatory				
Liability concerns	Low	High	Moderate	A
Regulatory limitations	High	Moderate	High	A
Difficulty in intellectual property rights	Low	High	Moderate	U
Regulations on cross-border projects	Low	High	Moderate	C
Organizational conflict of interest	Low	Moderate	High	A
Financial				
No state/local funding	High	High	Moderate	C
Lack of funding for operations/ maintenance	High	High	Moderate	A
Justifying expense of ITS	Moderate	High	Low	A
Low-priority allocation to ITS	Moderate	High	Low	A

U = Unique to ITS
A = Amplified, not unique but more prominent in ITS
C = Common to all transportation projects
SOURCE: JHK and Assocs., 1996

The survey of ITS experience also looked at the lessons learned from early ITS projects. These included the following

- *Customer orientation*: The delivery of ITS projects and programs focused on the needs of the customer, including commuters, travelers, transit riders, and goods transporters. Long-term success depended on the perception that a useful service was being provided.

- *Problem-solving emphasis*: The focus of successful ITS applications was on addressing problems or on improving travel convenience and safety to the public. Such successes encouraged elected officials to look upon ITS strategies as ways of gaining the favor of their constituents.

- *Integration*: ITS was considered as just one of a broad array of techniques to address transportation problems. ITS was used to best advantage when integrated with other techniques.

- *Partnerships*: Substantial ITS deployment was never achieved by a single agency. Partnerships brought agencies together across geographic boundaries and lines of functional responsibility. Long-term support of the ITS program most often included public/private partnerships.

- *Communication with elected and appointed officials*: Elected official support was essential to long-term success. Developing this support required continual attention, information, and education [Lappin et al., 1998].

- *Maintaining credibility*: The public mistrusted projects that did not work predictably and consistently. Faiiure to maintain credibility can ultimately erode support for ITS. Maintaining credibility was clearly linked with customer orientation.

Many metropolitan areas have conducted strategic assessments of ITS deployment to identify those strategies that are most appropriate for implementation. As was noted in the review of the early experiences with ITS, the more successful assessments provided a realistic estimate of the funding needed and provided guidance on the partnerships and agreements that were necessary to sustain implementation. Importantly, these assessments often focused on the institutional issues related to ITS deployment that had a side benefit of improving interagency communication and coordination [JHK and Assocs., 1996].

Coordinating Land-Use and Transit Service Planning   One of the important linkages in multimodal transportation planning is the symbiotic relationship between the transportation system and urban development. Some agencies and metropolitan areas have adopted policies that encourage development patterns that are conducive to transit use by providing easy and convenient access to transit facilities or by providing other services that are supportive of transit. Transit-oriented or pedestrian pocket development includes dense, mixed-use development that includes residential, commercial, and employment use, all of which would be focused within a 5- or 10-minute walk from a transit station [Carlson, 1995].

The major benefits of transit-oriented development relate primarily to how the automobile is used and the resulting impact on other modes of transportation. For example, Cervero found that mixed-use developments in suburban environments yielded four main benefits [Cervero, 1988]: walk trips increase, trip making is more evenly distributed throughout the day and week, shared-use parking is possible, and car pooling is more likely. A study in the Puget Sound area established a positive relationship between land-use density and nonauto trips—the higher density of population and employment, the more nonauto trip making [Frank and Pivo, 1995].

Figure 9.7 shows the different types of planning and design guidelines that could be incorporated into a transit supportive development program. The successful implementation of such a program would require the coordinated efforts of

1. Mixed use development
   - Supports increased densities
   - Integrates surrounding development and neighborhoods
   - Incorporates public and civic space
   - Encourages walking and biking

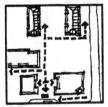

2. Strong connections
   - Improves transit accessibility and promotes use
   - Emphasizes bike use and walking over auto use
   - Increases efficient transportation
   - Links uses and activities

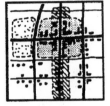

3. Concentrated development
   - Extend the hours of activity
   - Supports increased densities and commercial traffic
   - Consolidate trips
   - Makes walking and biking a convenient option

4. Buildings fronting the street
   - Encourages street activity and walking
   - Makes pedestrian areas more interesting
   - Removes parking to a side or rear location

5. Good vehicle circulation
   - Incorporates improvements in street design, pedestrian amenities, bicycle access, parking and multimodal connections
   - Efficiently connects the transit center and the surrounding area
   - Provides visual connections to the transit center
   - Utilizes various street types to carry different levels of vehicular traffic

6. Pedestrian environment
   - Prioritizes pedestrians over auto uses and space
   - Includes sidewalks, paths, trees, benches, and usable public open spaces

7. Bicycle environment
   - Makes biking easy, efficient, and safe
   - Requires adequate locking facilities such as racks, lockers, or other storage facilities located conveniently near the entrances to buildings or transit stops
   - Provides necessary connections with pedestrian and transit modes

8. On-street parking
   - Increases the safety of pedestrians by establishing a buffer between cars and people walking
   - Provides convenient parking
   - Reduces parking lot requirements

9. Structured parking
   - Decreases the amount of land required for parking
   - Makes parking more convenient, closer to the buildings

Figure 9.7   Strategies for transit-oriented development
SOURCE: Regional Transit District, 1996

numerous government agencies as well as the cooperation of the development community. A growing number of land-use guidelines and model ordinances are being adopted that favor transit use [TRI-MET, 1993; Municipal Research and Services Center, 1995; Center for Urban Transportation Research, 1994; Beimborn and Rabinowitz, 1991; Jarvis, 1993]. TRI-MET, the transit service provider in Portland, Oregon, provides a good example of what a regional implementation strategy for transit-oriented development entails. Figure 9.8 shows the approach, beginning with an interim strategy of identifying corridors and working with local governments to establish multimodal transportation ordinances that give priority to transit accessibility. A longer-term strategy, shown as track two, would take place concurrently and would develop a metropolitan transit strategy within the context of the region's 2040 vision. Figure 9.9 illustrates the institutional component of this strategy. The specific actions of key actors in the region are identified in this table, providing a regionally coordinated approach toward linking transit service and land-use policy. TRI-MET's planning manual also provided a prototype multimodal transportation zoning ordinance that could be considered by local communities.

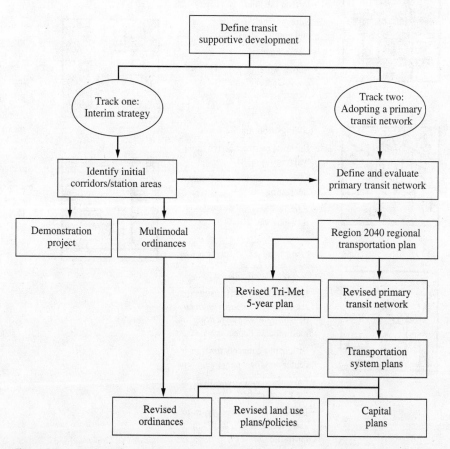

Figure 9.8   Strategy framework for transit-oriented development in Portland, Oregon
SOURCE: TRI-MET, 1993

The Business Community and Transportation Management Associations Employers and developers have a critical role to play in improved management of a community's transportation system. Not only are employment and development sites the location of significant employee travel, but the sites often attract large numbers of visitors and goods deliveries that create demand on the transportation system. Transportation management associations (TMAs) are partnerships between business and local government designed to help solve local transportation problems associated with rapid growth. TMAs give the business community a voice in local transportation decision making, build local constituency for better transportation, and serve as a forum for public/private consultations on issues of transportation planning, financing, and implementation (Ferguson and Davidson, 1995; Ferguson et al., 1992). They have become an important institutional forum for dealing with mobility, in particular in increasing commuting options to suburban employment centers that are poorly served by public transportation. TMAs offer a forum for public/private consultations on such varied issues as establishing highway funding priorities, restructuring of public transit routes, improving transit service, minimizing disruption caused by road reconstruction, and mitigating traffic congestion. Some TMAs have been instrumental in launching innovative programs to help entry-level workers gain access to suburban jobs. Table 9.14 shows the types of activities that could be found in a TMA.

Transportation management associations often fill an institutional void characteristic of fast-growing activity centers. Many of these centers often have no institutional structure to articulate public needs and concerns. TMAs can act as a surrogate for traditional institutions and serve as advocates for underrepresented interests. Entrepreneurial in nature, a TMA offers the promise of maturing into an instrument of advocacy well-suited to the realities of contemporary metropolitan areas.

9.5 CHAPTER SUMMARY

1. Programming is the process of matching available or expected resources with transportation needs to achieve planning goals. A transportation program is a document that lists all of the projects to be implemented in an investment period and the agency responsible for each. In the United States, the federal government requires every metropolitan area to prepare a transportation improvement program (TIP), which is a staged, multiyear program describing all regionally significant projects that will be implemented during the programming period.

2. The transportation plan and the TIP must be financially constrained, which means that only projects having realistic funding sources can be included in the program. This requirement has increased the importance of prioritization as part of the planning process.

3. Several methods have been used by transportation planners to establish project priorities. These include the use of goals achievement, numerical ratings,

	Local governments	**TRI-MET**
1. Define transit supportive development	• Generate local definition	• Outreach planning and design for transit • Research development supportive transit • Evaluate land use/transit performance
2. Target specific areas	• Identify mixed-use centers and PTN • Identify potential pedestrian districts	• Establish PTN • Service improvements to PTN
3. Adopt policies and regulations	• Identify PTN and pedestrian districts in comp plan and transportation plan maps and policies • Adopt code revisions—overlays, development regulations, specific plans • Incorporate bike, pedestrian, transit facility standards in code	• Adopt strategic plan • Adopt policies supporting PTN • Adopt 5-year plan and land use based service policies and standards • Refine HOV corridor analysis and selection process • Service demonstration projects in developing corridors of PTN
4. Development incentives	• CIP funding priority to PTN • Incorporate transit access into urban renewal goals	• Priorities capital investments into PTN • Create a land use based capital plan • Priority service expansion to PTN
5. Advocacy education	• Neighborhood outreach and workshops • Planning commission/council briefings • Distribute TSD brochures • Notify Tri-Met of major projects	• Planning and development technical assistance • Expand role in development review • Sponsor conferences/workshops • Promote ridership and development in PTN

Legend:
CIP: Capital improvement plans
PTN: Primary transit network
TPR: Transportation planning rule
TSD: Transit supportive development

Figure 9.9   Implementation matrix for transit-oriented development in Portland, Oregon
| SOURCE: TRI-MET, 1993

Development community	Metro	State agencies
• Provide development expertise to public agencies	• Define mixed-use centers and PTN in Region 2040	• TSD concepts into TPR evaluation
• Assist in identifying pedestrian districts	• Identify specific locations for mixed use centers and PTN	• Identify state-owned corridors and sites within PTN
• Incorporate TSD concepts into development proposals	• Adopt mixed-use centers and PTN in Region 2040 plan and regional transportation plan	• Incorporate TSD concepts into transportation planning rule review • Require transportation demand management programs for major developments outside of PTN
• Assist government agencies in identifying effective capital investments and zoning incentives • Participate in public/private demonstration projects	• Specify incentives in Westside Station area planning • Identify in fill incentives for PTN • Investigate reduced development impact fees in PTN	• PTN of pedestrian districts as a priority for state housing dollars • Target PTN as priority for 6-year plan capital improvements • Support legislative changes for development assistance in station areas
• Sponsor seminars for design/development communities • Include TSD in professional conferences	• Research and provide data for PTN • Sponsor conferences and workshops on TSD	• Sponsor workshops/seminars

Table 9.14   Typical TMA activities

Offer a Forum for Public/Private Consultation On

Highway funding priorities

Minimizing disruption from road repairs

Transit service improvements

Traffic engineering improvements

Represent and Advocate Needs and Interest of TMA Members before Public Agencies, Legislative Bodies, and in the Planning Process

Monitor traffic conditions, and recommend appropriate "quick fixes"

Conduct employee travel surveys, assess commuter travel needs, and recommend appropriate changes in transit routing and level of service

Monitor development and employment trends and assess their impact on future road and transit needs

Advise on alignment and location on new transportation facilities.

Build Local Constituency for Better Transportation and Raise Funds for Local Transportation Improvements

Promote and Coordinate Demand Management Actions Designed to Reduce Peak-Hour Demand on Transportation Facilities and Help TMA Members Comply with Local Traffic Mitigating Requirements (Trip Reduction Ordinances, Conditions of Development Permits, Proffers, etc.)

Ride sharing

Variable work hours to spread peak-hour traffic

Parking management

Transit marketing and promotion

Facilitate Commuting and Provide Internal Circulation within the Area Through

Daytime circulators

Subscription vans/buses

Short-term car rentals

Shuttles to commuter rail stations and fringe parking lots

Emergency transportation for employees without cars

"Reverse commute" services for service employees

Provide Specialized Membership Services to TMA Members

Conduct employee "travel audits"

Provide relocation to newcomers

Train in-house transportation coordinators

Manage shared tenant services (i.e., daycare centers, security, sanitation, etc.)

SOURCE: Ferguson, Ross, and Meyer, 1992

priority indexes, programming evaluation matrices, and systems analysis techniques. No matter which method is used, the prioritization process should provide a direct linkage between the goals of the transportation planning process and the programming criteria. In addition, nonquantifiable or subjective information should be part of the assessment process. Finally, the process of prioritization is as critical to the credibility of programming as the technical approach used.

4. An effective programming process should

   - Ensure that investment decisions are related to established transportation policy and planning goals.

   - Provide a general direction for transportation investment by integrating projects over time and location.

   - Set priorities among policy and agency objectives.

   - Include credible estimates of funding sources.

   - Provide decision makers with a timely review of the effectiveness of transportation policy and planning in the region.

   - Be open to interested groups to ensure consistency between the program and community goals.

5. Another basic component of the programming process is the estimation of future funding availability. Several aspects merit special note:

   - Funds to be made available by higher levels of government are the most difficult to forecast, as allocations can be determined both by changing economic trends and by legislative decisions.

   - Monitoring the budgetary processes of local governments and estimating the impacts of selected projects on future local budgets are crucial elements to developing accurate estimates of funding from local sources.

   - The cash flow requirements of different project categories over time and the shares to be paid by different levels of government have to be determined beyond the short-term programming horizon.

   - Estimates of funding sources and availability have to be related to projected operating and capital expenditures in order to provide a complete picture of the financial situation over the programming horizon.

6. Project funding and institutional capability are two critical elements of successful program implementation. Because of the limited level of funding that is available from traditional sources such as gas or property taxes, transportation decision makers have experimented with innovative financing strategies. Most often described under the general label of public–private partnerships, these strategies have included toll roads, development fees/tax increment financing, assessment districts, and privatization. In each case, significant implementation issues will most likely need to be addressed.

QUESTIONS

1. For your urban area, obtain a copy of a transportation programming document and discuss the following:

   (a) What are the major sources of funding for transportation projects in your urban area?

(b) Which agencies seem to be actively involved in the programming process, based on the documentation?

(c) What type of project seems to be receiving priority in the document?

(d) For the major sources of funding, what factors are likely to hinder or help forecasting future contributions?

2. A multiyear framework for programming allows the planner to integrate projects over time and location. What problems are likely to arise in developing such a multiyear program?

3. Establishing priorities for resource allocation is an important element of the programming process. Most efforts to establish such priorities have focused on projects found in one funding category or in one modal group. Thus, for example, highway projects are often compared and ranked with other highway projects but rarely with transit projects. Outline a procedure for establishing priorities between highway and transit projects. What are the advantages of such a procedure?

4. Divide the class into groups and develop a process to establish the importance of transportation problems facing university students. Establish a ranking scheme to prioritize possible solutions. How does the relative importance of the identified problems differ from one group to another?

5. Select a funding source for transportation programs sponsored by a local agency. Forecast the level of funds that will be available from this source next year, in 3 years, in 5 years, in 10 years. What are some of the problems associated with such forecasting?

6. Outline a public participation program that can be used to provide input into a regional transportation programming process.

REFERENCES

Beimborn, E. and H. Rabinowitz. 1991. *Guidelines for transit sensitive suburban land use design.* Washington, D.C.: Federal Transit Administration.

Bellomo, S. et al. 1979. Evaluating options in statewide transportation planning/programming. *National Cooperative Research Program Report 199.* Washington, D.C: National Academy Press.

Campbell, B. and T. Humphrey. 1978. *Methods of cost-effectiveness analysis for highway projects.* NCHRP Synthesis 142. Washington, D.C.: National Academy Press.

Capital District Transportation Committee. 1998. *FY 1999–2004 transportation improvement program.* Albany, NY.

Carlson, D. 1995. *At road's end: Transportation and land use choices for communities.* Washington, D.C.: Surface Transportation Policy Project. Island Press.

Center for Urban Transportation Research. 1994. *Development incentives that support transit.* Tampa: University of South Florida. June.

Central Florida Regional Transportation Authority. 1995. *Central Florida mobility design manual.* Orlando, FL.

Cervero, R. 1988. Land use mixing and suburban mobility. *Transportation Quarterly.* Vol. 42. No. 3. Washington, D.C.: Eno Foundation.

Chicago Area Transportation Study. 1976. *Transportation improvement program FY76–FY80: A documentation of the programming process.* Chicago, IL. June.

Davidson, D. (ed.). 1993. *The TMA summit.* Proceedings of a conference sponsored by the Association for Commuter Transportation and the Federal Highway Administration. October.

Denver Regional Council of Governments. 1998. *Metrovision 2020: Regional transportation plan, the fiscally constrained element.* Denver, CO.

_____. *Traffic signal system improvement program, 1999 update summary report.* Denver, CO. December.

East–West Gateway Coordinating Council. 1999. *Transportation redefined II: Building on a solid foundation for 2020.* St. Louis, MO. March 24.

El Dessouki, W. et al. 1998. Multiperiod highway improvement and construction scheduling: Model development and application. *Transportation Research Record 1441.* Washington, D.C.: National Academy Press.

Federal Highway Administration. 1980. *Programming projects.* Washington, D.C.: U.S. DOT.

_____. 1993. *Guidance for state implementation of ISTEA toll provisions in creating public–private partnerships.* Report No. FHWA-PL-93-015. Washington, D.C. November.

_____. 1994. *Summary of the Federal Highway Administration's symposium on overcoming barriers to public–private partnerships.* Report No. FHWA-PL-94-026. Washington, D.C. September.

_____. 1998. *Toll facilities in the U.S.* Report FHWA-PL-99-01. Washington, D.C. February.

Federal Transit Administration (FTA). 1998. *Innovative financing techniques for America's transit systems.* Washington, D.C. May.

Ferguson, E. and D. Davidson. 1995. Transportation management associations: An update. *Transportation Quarterly.* Vol. 49. No. 1. Washington, D.C.: Eno Foundation.

Ferguson, E., C. Ross, M. Meyer. 1992. *Transportation Management Associations in the United States: Final Report.* Office of Technical Assistance and Safety. Washington, D.C.: Federal Transit Administration. May.

Frank, L. and G. Pivo. 1995. Impacts of mixed use and density on utilization of three modes of travel: Single occupant vehicle, transit, and walking. *Transportation Research Record 1466.* Washington, D.C.: National Academy Press.

Geotechnical Engineers. 1985. *Hobbs Brook reservoir sodium chloride study.* Report prepared for the Massachusetts Department of Public Works and City of Cambridge Water Department. Boston, MA.

Institute of Transportation Engineers. 2000. *Funding strategies.* National Committee on Transportation Operations White Paper. Washington, D.C.

Jarvis, F. 1993. *Site planning and community design for great neighborhoods.* Washington, D.C.: Home Builder Press.

JHK and Associates, Inc. 1996. *Integrating ITS with the transportation planning process: An interim handbook.* Washington, D.C.: Federal Highway Administration. August.

Lappin, J. et al. 1998. *Marketing ITS infrastructure in the public interest.* Report prepared for the FHWA. Washington, D.C. May.

Lloyd, E. and M. Meyer. 1984. Strategic approach to transportation project implementation: The Boston auto-restricted zone. *Transport Policy and Decision-making. Vol. 2.* 335–349. Martinus Nijhoff Publishers.

Lockwood, S. 1995. Public–private partnerships in U.S. highway finance: ISTEA and beyond. *Transportation Quarterly.* Washington, D.C: Eno Foundation.

Maricopa Association of Governments. 1997. *Long-range transportation plan. Summary.* Phoenix, AZ. September.

Massachusetts Department of Public Works. 1988. *Massachusetts type II noise attenuation study.* Final Report. Boston, MA.

Metropolitan Transportation Commission. 1998. *Regional transportation plan.* Oakland, CA.

Mid-Ohio Regional Planning Commission. 1998. *Vision 2020 transportation plan.* Columbus, OH: May 21.

Municipal Research and Services Center. 1995. *Creating transit supportive regulations.* King County, WA. August.

Nelson, A. (ed.) 1988. *Development impact fees.* Washington, D.C.: Planners Press, American Planning Association.

Neumann, L. 1976. *Integrating transportation system planning and programming: An implementation approach.* Unpublished Ph. D. dissertation. Department of Civil Engineering, MIT. Cambridge, MA.

_____. 1997. Methods for capital programming and project selection. *Synthesis of Highway Practice 243.* Washington, D.C.: National Academy Press.

North Central Texas Council of Governments. 2000. *Mobility 2025: Metropolitan transportation plan.* Arlington, TX. February 16.

North New Jersey Transportation Planning Authority, Inc. 1998. *Regional transportation plan for northern New Jersey, 1998 update.* Newark, NJ. January.

Pennsylvania Department of Transportation. 1988. *Transportation partnerships guidelines manual.* Harrisburg, PA. May.

Politano, A. 1983 *Financing urban transportation improvements, Report 1: Cost-effectiveness considerations in corridor planning and project programming.* Federal Highway Administration. U.S. Department of Transportation Report FHWA/PL/83/001. Washington, D.C. April.

Portland METRO. 1995. *Short list of technical rankings and assessment of administrative criteria.* Portland, OR. June 28.

_____. 1996. *Financial analysis: Methods report.* Portland, OR. March 22.

_____. 1997. *1996 financial assessment profiles. South/north corridor light rail transit.* Portland, OR. September.

Puget Sound Regional Council. 1999. *Policy framework for the 1999 TEA-21 TIP process.* Seattle, WA. February 25.

Regional Transit District. 1996. *Transit-oriented development guidelines.* Denver.

Sacramento Area Council of Governments. 1999. *Metropolitan transportation plan.* Sacramento, CA. July.

San Diego Association of Governments. 1999. *2020 regional transportation plan.* San Diego, CA. November.

Seattle METRO. 1993. Evaluation criteria for high capacity transit. Seattle, WA.

Southeastern Wisconsin Regional Planning Commission. 1994. *A regional transportation systems plan for southeastern Wisconsin: 2010.* Planning Report No. 41. Milwaukee, WI. December.

Southern California Association of Governments. 1996. *Pricing transportation congestion: The voters' view.* Los Angeles, CA. March.

Starr, P. 1987. *The limits of privatization.* Washington, D.C.: Economic Policy Institute.

Taber, J. et al. 1999. Optimizing transportation infrastructure planning with a multiobjective genetic algorithm model. *Transportation Research Record 1685.* Washington, D.C.: National Academy Press.

Transmanagement, Inc., M. Coogan and M. Meyer. 1998. Innovative practices for multimodal transportation planning for freight and passengers. *NCHRP Report 404.* Washington, D.C.: National Academy Press.

Transportation Research Board. 1980. *National Cooperative Highway Research Program Synthesis of Highway Practice 72.* Washington, D.C.

_____. Priority Programming and Project Selection." 1978. *National Cooperative Highway Research Program Synthesis of Highway Practice 48.* Washington, D.C.

TRI-MET. 1993. *Planning and design for transit.* Portland, OR. March.

U. S. Department of Transportation. 1994. *Nontechnical constraints and barriers to implementation of intelligent vehicle-highway systems.* Report to Congress. Washington, D.C. June.

_____. 1997. *The status of the nation's highway bridges: Highway bridge replacement and rehabilitation program and national bridge inventory.* Report to Congress. Washington, D.C.: Federal Highway Administration. May.

_____. Transportation improvement program. 1975. *Federal Register.* September 17.

Wilson, D. I. and J. L. Schofer. 1978. Decision-maker-defined cost-effectiveness framework for highway programming. *Transportation Research Record 677.* Washington, D.C.: National Academy Press.

Younger, K. and C. O'Neill. 1998. Making the connection: The transportation improvement program and the long-range plan. *Transportation Research Record 1129.* Washington, D.C.: National Academy Press.

Zimmerman, K. and ERES Consultants, Inc. 1995. Pavement management methodologies to select projects and recommend preservation treatments. *Synthesis of Highway Practice 222.* Washington, D.C.: National Academy Press.

Chronology of Selected Federal Actions Related to Urban Transportation Planning

Year	Action	Impact
1962	Federal Aid Highway Act	• Encouraged development of comprehensive transportation systems • Directed states to develop long-range highway plans coordinated with other modes • Required that all federally funded highway projects be based on a continuing, comprehensive, and cooperative (3C) planning process involving states and local communities • Defined planning focus as the urban area
1963	Guidelines for implementing the 3C planning process	• Resulted in development of planning process in all urbanized areas
1964	Urban Mass Transportation (UMT) Act	• Encouraged planning of areawide urban mass transportation systems • Established federal support match for acquisition and construction of transit facilities at two-thirds of cost; federal share was limited to 50 percent when no comprehensive plan existed • Required that all funds be channeled through public agencies to projects initiated locally • Established programs of mass transportation research, development, and demonstrations
1964	Civil Rights Act (esp Title VI)	• No person shall on the grounds of race, color, or national origin be excluded from participation in, be denied the benefits of, or be otherwise subjected to discrimination under any program or activity receiving federal financial assistance

Year	Action	Impact
1965	Housing and Urban Development Act	• Authorized grants for comprehensive planning to regional organizations
1966	Amendments to UMT Act	• Established program for two-thirds support for planning, engineering, and design of local transit projects that would lead to a federal grant application • Established grants for transit management training • Established program to develop new system technology
1966	Department of Transportation Act	• Created U.S. DOT and provided focal point for coordinated federal transportation policy • Provided for protection of publicly owned wildlife refuges, recreation areas, parks, and public or private historic sites from highway impacts (Section 4[f])
1966	Demonstration Cities and Metropolitan Development Act	• Required review of federal aid applications (Section 204) by an areawide agency for coordination with long-range comprehensive development plans
1967	U.S. DOT Policy and Procedure Memorandum 50-9	• Specified the 3C process for urban transportation planning, including jurisdictions covered, plan elements, scope of studies, and citizen participation
1968	Federal Aid Highway Act	• Required that Secretary of Transportation not approve a program or project requiring use of publicly owned park, recreation area, or wildlife or waterfowl refuge unless (1) there is no prudent and feasible alternative and (2) all possible planning has been done to minimize them • Required public hearings
1968	Intergovernmental Cooperation Act	• Required that national, state, regional, and local viewpoints be taken into account (to the extent possible) in planning of federally assisted development programs and projects
1968	Reorganization Plan No. 2 from the President to Congress	• Established the Urban Mass Transportation Administration (UMTA) within DOT and transferred existing urban mass transportation programs from the Department of Housing and Urban Development to DOT
1968	U.S. DOT Instructional Memorandum 50-4-68	• Required operations plans defining organizational structures and provided further specification of procedures for urban transportation planning • Provided additional guidance for technical approaches and scheduling of plan reviews and reevaluation

Year	Action	Impact
1969	U.S. Office of Management and Budget Circular A-95	• Encouraged creation of project notification and review systems • Required areawide comprehensive planning agencies to comment on the relationship of proposed projects to the planned development of the area • Required that federal agencies notify governors of awards within their state
1969	FHWA Policy and Procedure Memorandum 20-8	• Required two public hearings for highway projects • Required consideration of a full range of social, economic, and environmental impacts in the assessment of projects
1969	National Environmental Policy Act	• Required the preparation of environmental impact statements (EIS) for major federal actions including a discussion of alternatives and unavoidable adverse effects • Required a systematic interdisciplinary approach for planning and decision making • Created Council of Environmental Quality to implement policy
1970	Amendments to Clean Air Act	• Created Environmental Protection Agency (EPA), authorized to set ambient air-quality standards • Required development of state implementation plans (SIPSs) to meet these standards • Set deadline nonattainment areas to meet standards • Encouraged consideration of low-capital and traffic management actions to meet air-quality goals
1970	Uniform Relocation Assistance Act	• Ensured owners of property acquired for and persons displaced by federally aided projects that they will be treated fairly, consistently, and equitably • Assured that those affected will not suffer disproportionate injuries
1970	Federal-Aid Highway Act	• Established federal-aid highway urban systems (FAUS) program • Authorized expenditure of highway funds on bus transit projects • Required DOT regulations to assure adverse economic, social, and environmental effects are fully considered in highway projects • Increased federal share for non-interstate projects to 70 percent

Year	Action	Impact
1970	Federal-Aid Highway Act (continued)	• Required consultation with local officials before any highway project is built in urban areas with populations of 50,000 or more
1970	National Historic Preservation Act	• Required Secretary of the Interior to publish a national register of historic places; once selected, a site is subject to Section 106 study that identifies and determines impacts of projects on historic properties
1970	Fish and Wildlife Coordination Act	• Required consultation and coordination between relevant agencies for any project that modifies any water body of the U.S. in which federal aid is used
1972	FHWA Policy and Procedure Memorandum 90-4 "Process Guidelines"	• Required states to develop their own action plans to describe organization and procedures for highway planning and allowed different procedures for different categories of projects • Topics to be covered included social, economic, and environmental impacts; alternative courses of action; involvement of other agencies and the public; responsibility for implementation; and fiscal and other resources
1973	Rehabilitation Act	• Section 504 of this Act required that no person should be discriminated against due to handicap in any program funded with federal support; subsequent UMTA regulations established guidelines and criteria for all bus and rail systems to be accessible to handicapped persons
1973	Federal-Aid Highway Act	• Allowed expenditures of FAUS funds on mass transportation projects • Allowed withdrawal of interstate segments and substitution of mass transit projects • Urban transportation planning to be funded as separate program with dedicated funds; planning funds to be made available to a metropolitan planning organization designated by state as responsible for 3C process in urban areas • Required that projects selected on the urban system be in accordance with required federally mandated planning procedures
1973	Endangered Species Act	• Prohibited federal activity that may further jeopardize the existence of a plant or animal on the federal endangered species list; by 2000, over 700 animal and plant species on this list

Year	Action	Impact
1974	Emergency Highway Energy Conservation Act	• FHWA funds can be used for ride-sharing demonstration programs • 55-mile-per-hour speed limit established on major highways
1974	DOT Order 5610.1B	• Formalized DOT procedures for preparation of environmental impact statements
1974	National Mass Transportation Assistance Act	• Authorized federal operating assistance for urban transit systems • 80 percent federal share for transit capital projects; 50 percent for operating assistance
1975	FHWA–UMTA Joint Regulations on Urban Transportation Planning	• Required as condition for continuing federal assistance the designation by the governor of a metropolitan planning organization (MPO) in each urban area • MPO must develop a unified planning program and a prospectus of the planning process • Transportation plan must consist of a long-range element and a transportation system management (TSM) element • MPO must develop a transportation improvement program (TIP) and an annual element detailing the following year's projects • Special efforts needed to plan for needs of elderly and handicapped
1976	Federal-Aid Highway Act	• Allowed funds from interstate highway projects to be used for other highways and busways
1976	UMTA Major Mass Transportation Investments Policy	• Areawide transportation improvement programs should be multimodal • Major mass transportation investments should be planned and implemented in stages, with full consideration given to initial improvements to existing services and systems • Required and analysis of transportation alternatives and final environmental impact statement for new start projects in order for a project to be eligible for federal assistance • Established principles and procedures for UMTA review of alternatives analysis
1977	FHWA Mass Transit and Special Use Projects Policy	• Authorized use of highway funds for projects relating to transit and high-occupancy vehicles

Year	Action	Impact
1977	Federated Water Pollution Control Act	• Prohibited dumping of dredged or fill material into wetlands unless a permit is obtained from the Army Corps of Engineers (Section 404 permits)
1977	Clean Air Act Amendments	• Required revisions to state implementation plans (SIPs) for areas not in attainment of national air-quality standards • In many urbanized areas, SIPs were required to develop transportation control plans that included strategies to reduce mobile (i.e., from transportation sources) emissions • 1981 regulations issued that required transportation plans, programs, and projects to conform with the approved SIPs; priority to be given to transportation control measures
1978	Surface Transportation Assistance Act	• First Act that combined highway, transit, and highway safety authorization in one piece of legislation • Set September 30, 1986 as date by which all remaining portions of interstate system must be under contract or be designated; raised federal share to 85% for interstate projects • Required DOT to create guidelines for interstate maintenance • Created highway bridge replacement and rehabilitation programs; made funds available for interstate system resurfacing • Created formula transit capital and operating assistance grant program for nonurbanized areas (80% federal share for capital and 50% federal share for operating) • Energy conservation added as a new goal for transportation planning and TSM alternatives required to be included in planning process
1978	National Energy Act	• Required states to undertake conservation actions, including the development of car-pool and vanpool programs • All phases of transportation planning and project development were to encourage fuel conservation
1981	Federal-Aid Highway Act	• Established early completion and preservation of interstate systems as highest priority for federal investment
1982	Surface Transportation Assistance Act	• Provided additional funds for highway and transit projects and increased federal gas tax by $0.05 per gallon • Established separate mass transit account in highway trust fund to support capital transit investment; $0.01 of the $0.05 gas tax to go to this account

Year	Action	Impact
1982	Surface Transportation Assistance Act (continued)	• Capped operating assistance and established formula for distributing planning, operating, and capital funds; formula based on urban population, density, vehicle miles, and route miles • Required DOT to publish regulation for minimum criteria for transportation service to elderly and handicapped, public participation, and monitoring of transit system performance
1983	FHWA/UMTA Final Planning Rules	• Simplified planning process for urbanized areas under 200,000 population • Increased state and local role in directing urban transportation planning; federal role diminished • Reduced number of required products of planning process (e.g., TSM element no longer needed), but still required transportation improvement program (TIP), transportation plan, and unified planning work program
1984	UMTA Major Capital Investment Policy	• Federal decision on supporting new project starts would be based on local planning studies and determination of cost-effectiveness • Alternatives analysis and draft EIS stage of project development would be the important decision point for proceeding with project • Capital alternatives must be compared to TSM alternatives
1987	FHWA/UMTA Environmental Impact Related Procedures Final Rule	• Established specific NEPA requirements to be followed by FHWA and UMTA • Specified three classes of actions that prescribe necessary level of documentation—environmental impact statements, environmental assessments, and categorical exclusions • Outlined scooping process for agency review
1987	Surface Transportation and Uniform Relocation Assistance Act	• Codified grant criteria of new fixed guideway transit projects • Discretionary grant program for transit to be funded from the mass transit account: 40% for rail modernization, 40% for new rail starts, 10% for major bus projects, and 10% discretionary • Required long-term financial plans for regional mass transit improvements • Allowed states to raise speed limit to 65 miles per hour outside urbanized areas

Year	Action	Impact
1990	Americans with Disabilities Act	• Accessible paths of travel must be available to individuals with disabilities (e.g. curb ramps at intersections) • New buses purchased after 1990 must be accessible • Transit authorities must provide comparable transit service to individuals with disabilities who cannot use fixed route bus services • New rail and bus stations must be accessible
1990	Clean Air Act Amendments	• Each state must submit a state implementation plan to EPA that outlines steps to be taken to meet air-quality standards • In addition to showing attainment of standards, areas in nonattainment of ozone standards must demonstrate reasonable further progress toward attainment for certain milestone years • Expanded scope and concept of conformity provisions to mean that attainment strategies must conform to an SIPs purpose of eliminating or reducing the severity and number of violations of the standards • The projected emissions associated with transportation projects and programs must be reconciled with the required emission reductions of the SIP
1991	Intermodal Surface Transportation Efficiency Act (ISTEA)	• Required the consideration of 15 planning factors in metropolitan transportation planning, relating to mobility and access for people and goods, system performance and preservation, and environment and quality of life • Transportation plans and programs must be financially constrained so that there is a reasonable chance of funding those projects being proposed • The transportation plan and improvement program must conform to the state implementation plan • Major investment studies are to be conducted to address significant transportation problems in corridors or subareas where federal transportation funds might be used • Strong emphasis on proactive public involvement in the transportation planning process • Promoted the implementation of intelligent transportation system (ITS) technologies • Changed name of Urban Mass Transportation Administration to Federal Transit Administration

Year	Action	Impact
1993	EPA Criteria and Procedures for Determining Conformity to Transportation Plans Rule	• Established interagency consultation procedures for determining plan and program conformity • Criteria for conformity determination include the following: – Plan, program, or project must be based on latest planning assumptions and latest emission estimation model available – Plan, program, or project must provided timely implementation of transportation control measures – There must be a conforming plan and TIP at time of project approval and project must come from them – Plan, program, or project must not cause or contribute to new pollutant violations or increase frequency or severity of existing problems – Plan, program, or project must be consistent with SIP emission targets – A project must eliminate or reduce in severity and number carbon monoxide (CO) violations (if area is in violation of CO standard)
1993	FHWA/FTA regulations on statewide and metropolitan planning; management systems	• Implemented requirements of ISTEA listed previously • Six management systems relating to intermodal transportation, congestion mitigation, public transit facilities, pavement management, bridge management, and traffic safety are to be part of the planning and decision-making process • Required states to develop a statewide transportation plan that is intermodal, covers at least 20 years, contains a plan for bicycle/pedestrian transportation, and is coordinated with transportation plans
1994	Executive Order 12898 on Environmental Justice (implemented by FHWA in 1998)	• Federal agencies should identify and address disproportionately high and adverse human health and environmental effects of their programs on low-income and minority populations • FHWA will identify and evaluate such effects of its programs; propose measures to avoid, minimize and/or mitigate disproportionately high adverse effects; consider alternatives to proposed programs, policies, and activities where such effects would result; and provided proactive public involvement opportunities for minority and low-income populations

Year	Action	Impact
1995	National Highway System Designation Act	• Designated 160,955 miles of nation's road system as the National Highway System (NHS) • Suspended requirement for management systems (except for congestion management system in nonattainment areas) • Allowed 10 states to establish infrastructure banks as demonstration of innovative financing
1996	FTA New Starts Criteria	• Revised new starts criteria to broaden project criteria including: alternatives analysis and preliminary engineering provide data for analysis; justified based on a comprehensive review of project mobility improvements, environmental benefits, cost-effectiveness, and operating efficiencies; and supported by an acceptable degree of local financial commitment
1997	FTA Policy on Transit Joint Development	• Jointly created transit improvements and adjacent land development will be considered mass transportation projects and eligible for funding under FTA capital programs • Revenues generated from the development can be used by the transit property for capital or operating expenses
1998	Transportation Equity Act for the 21st Century (TEA-21)	• New budget categories established for highway and transit spending • States guaranteed a return of 90.5% of state's contributions to highway trust fund • Innovative financing strategies proposed using federal funds • Up to 50% of NHS apportionments may be transferred to other categories of projects • Eliminated transit operating assistance, although areas under 200,000 population are still eligible for such assistance • Funds provided for development of intelligent transportation systems projects • Continued provisions for long-range comprehensive transportation planning under umbrella of the MPO • Consolidated previous planning factories into seven broad areas to be considered in planning process—support for economic vitality (including global competitiveness, productivity, and efficiency), safety and security, accessibility and mobility for people and freight, environmental protection, integration and connectivity of the transportation system, efficient system management and operation, and emphasis on preservation of the existing transportation system

Year	Action	Impact
1998	Transportation Equity Act for the 21st Century (TEA-21) (continued)	• Initiated a process for "streamlining" project planning/development and environmental review procedures; major investment studies no longer a separate requirement • Added several considerations to the federal project evaluation process of transit capital projects—cost of sprawl, infrastructure cost savings due to compact land use, population density and current transit ridership in corridor, and technical capacity of the grantee to undertake project

NOTE: There are numerous other laws and regulations that might be relevant to a particular transportation project. Interested readers should check the following websites: http://www.dot.gov; http://www.fhwa.dot.gov; http://www.epa.gov; http://www.bts.dot.gov; http://www.fta.dot.gov. See also: E. Weiner. 1999. *Urban transportation planning in the U.S.: A historical overview.* Westport, CT: Praeger.

Appendix

B

Determination of Sample Size

The determination of sample size involves two major steps:

1. Because the purpose of sampling is to gain information about the nature or distribution of elements in a particular population, one must make assumptions about the form of the underlying distribution of these elements when selecting a sample size. One common assumption is that the population is normally distributed (Fig. B.1a). An important characteristic of the normal distribution is that, no matter what the mean μ or standard deviation σ of a particular normal distribution, the same proportion of observations will always lie between the mean and a specified number of standard deviations. Thus, as shown in Fig. B.1a, 68.26 percent of the area under the distribution will always be within one standard deviation of the mean, and 95.46 percent within two standard deviations for any normally distributed variable.

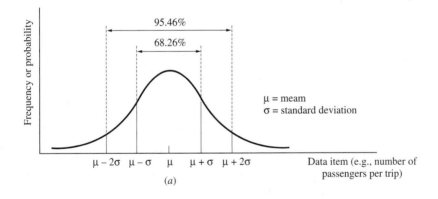

(a)

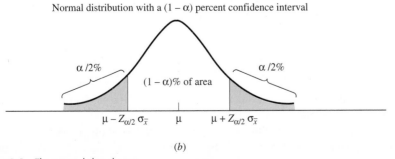

(b)

Figure B.1 The normal distrubution

2. Some decision must be made about the acceptable limits of error for the sample. This is usually done by specifying that the population mean for a data item should be within some confidence interval d around the sample mean for a certain percentage of samples. This latter percentage (i.e., the percentage of samples falling within the desired limit of error) is called the *level of confidence*. This level of confidence is denoted as $100(1 - \alpha)$, where α is the fraction of the area under the normal distribution falling outside the confidence limits (Fig. B.1b). For example, one could specify that the precision of the sample mean of normally distributed random variables should be within plus or minus two units of the true population mean at a 95 percent confidence level. Specifically, plus or minus two standard deviations around the sample mean will include the population mean 95 percent of the time (note in this case that $d = 2$ and $\alpha = 0.05$). It should be noted that in any particular trial, the 95 percent confidence interval either does or does not contain the true population mean. Thus, it is never known which outcome has materialized, merely that in 95 percent of repeated trials, the 95 percent confidence interval will contain the population mean.

Once an assumption about the distribution of the elements of a population has been made and the desired sample precision has been chosen, the required sample size can be determined. For example, if one assumes that the elements of a particular population are normally distributed, it can be shown that the sample size required to achieve a precision of d units with $100(1 - \alpha)$ percent confidence is

$$n = \left[\frac{Z_{1-(1/2)\alpha}\sigma}{d} \right]^2 \qquad \text{[B.1]}$$

where

n = sample size

d = tolerable margin of error of mean value

σ = standard deviation of population distribution

α = fraction of area under normal curve representing events *not* within confidence level (thus, $1 - \alpha$ is desired level of confidence)

$Z_{1-(1/2)\alpha}$ = standard normal statistic corresponding to the $(1 - \alpha)$ confidence level (found in tables in any statistics book)

For example, assume an agency wanted to estimate the required proportion of some vehicle type (e.g., trucks) within a tolerance of ± 0.025 with a 95 percent level of confidence. Also assume that a previous survey has shown the standard deviation of the underlying vehicle type distribution to be 0.04. The sample size can thus be calculated as

$$n = \left[\frac{Z_{1-0.05/2}(0.04)}{0.025} \right]^2 = 9.8 \text{ or } 10$$

Thus, vehicle classification data should be collected on 10 randomly selected days to meet the required precision.

One of the problems with using Eq. [B.1] is that σ is often unknown. Existing records or preliminary samples can be used to estimate σ and thus determine the appropriate sample size. Another distribution, the t distribution, allows the population standard deviation to be replaced with an estimated value based on the sample. As sample size increases, the t distribution approaches the shape of the normal curve. The t distribution is used, for example, in transit sampling efforts where the data item examined is number of bus trips per route per hour and where the sample size is small (e.g., less than 30 trips) [Attanucci et al., 1981].

Another approach, based on an estimate of the proportion of the data item in the population, is also used by planners. The equation for this sample size determination is

$$n = \frac{\left[Z_{1-(1/2)\alpha}\right]^2 (p)(1-p)}{d^2} \qquad \text{[B.2]}$$

where p is the observed value of the proportion of the data item in the population (e.g., percentage of trip types of all trips made in an area), and the remaining symbols are the same as in Eq. [B.1].

One modification of Eq. [B.2] can occur when planners are interested in the relative error r from the true mean value instead of the absolute error d. For example, planners might want to estimate a variable with an error not exceeding 5 percent of the true mean value. In order to make this estimation, Eq. [B.2] is modified by substituting $(r \times p)$ for d. The resulting formula is:

$$n = \frac{\left[Z_{1-(1/2)\alpha}\right]^2 (p)(1-p)}{(rp)^2} = \frac{\left[Z_{1-(1/2)\alpha}\right]^2 (1-p)}{r^2 p} \qquad \text{[B.3]}$$

where r is the margin of error or precision, expressed as a fraction of the mean value, and the remaining symbols are the same as above. Instead of using Eq. [B.3], planners can use tables or figures that show the relationship between proportions (frequencies), confidence intervals, margins of error (precision), and sample sizes. Table B.1 and Fig. B.2 are examples of such aids.

For instance, suppose one wished to estimate the total number of trips for one trip type within an urban area, and it was estimated that 50 percent of all trips in the area ($p = 0.50$) were indeed of this trip type. Assume that planners require an estimate of this number within ± 5 percent ($r = 0.05$) of the real value 95 percent of the time ($\alpha = 0.05$). Using Table B.1, one can see that the level of precision for a value of p of 0.50 is 8.85 percent at a sample size of 500, 6.26 percent for n equal to 1000, and 4.43 percent for n equal to 2000. Thus, for the required level of precision, a sample size of between 1000 and 2000 would be necessary. To determine the exact value, Eq. [B.3] can be used; that is

$$n = \frac{(1.96)^2 (0.50)}{(0.05)^2 (0.50)} = 1537$$

Thus, 1537 observations are needed to achieve the specified level of precision.

Because some survey methods rely on voluntary questionnaire return, and given that many individuals contacted will not respond, the sample size in such

Table B.1 Sample size related to frequency, precision, and confidence interval

Sample Size

Frequency	90% Confidence Interval					95% Coonfidence Interval					99% Confidence Interval				
	500	1000	2000	5000	10,000	500	1000	2000	5000	10,000	500	1000	2000	5000	10,00
0.10	22.14	15.65	11.07	7.00	4.95	26.65	18.78	13.28	8.40	5.94	34.35	24.29	17.17	10.86	7.68
0.11	20.99	14.84	10.49	6.64	4.69	25.19	17.81	12.59	7.96	5.63	32.57	23.03	16.28	10.30	7.28
0.12	19.98	14.13	9.99	6.32	4.47	23.98	16.96	11.99	7.58	5.35	31.00	21.92	15.50	9.80	6.93
0.13	19.89	13.50	9.54	6.14	4.27	22.91	16.20	11.05	7.24	5.12	29.62	20.94	14.81	9.37	6.62
0.14	18.29	12.93	9.14	5.78	4.09	21.95	15.52	10.97	6.94	4.91	28.38	20.06	14.19	8.97	6.34
0.15	17.57	12.42	8.78	5.55	3.93	21.18	14.90	10.54	6.67	4.71	27.25	19.27	13.63	8.62	6.09
0.16	16.91	11.96	8.45	5.35	3.78	20.29	14.35	10.14	6.42	4.54	26.23	18.55	13.12	8.30	5.87
0.17	16.30	11.53	8.15	5.15	3.65	19.57	13.84	9.78	6.19	4.38	25.30	17.89	12.65	8.00	5.66
0.18	15.75	11.14	7.87	4.98	3.52	18.90	13.36	9.45	5.98	4.23	24.44	17.28	12.22	7.73	5.46
0.19	15.24	10.77	7.62	4.82	3.41	18.28	12.93	9.14	5.78	4.09	23.64	16.71	11.82	7.48	5.29
0.20	14.76	10.44	7.38	4.67	3.30	17.71	12.52	8.85	5.60	3.96	22.90	16.19	11.45	7.24	5.12
0.21	14.31	10.12	7.16	4.53	3.20	17.17	12.14	8.59	5.43	3.84	22.21	15.78	11.10	7.02	4.97
0.22	13.89	9.82	6.95	4.39	3.11	16.67	11.79	8.34	5.27	3.73	21.56	15.24	10.78	6.82	4.82
0.23	13.50	9.55	6.75	4.27	3.02	16.20	11.46	8.11	5.12	3.62	20.95	14.81	10.47	6.62	4.68
0.24	13.13	9.29	6.57	4.15	2.94	15.75	11.14	7.88	4.98	3.52	20.37	14.41	10.19	6.44	4.56
0.25	12.78	9.04	6.39	4.04	2.86	15.34	10.84	7.67	4.85	3.43	19.83	14.02	9.91	6.27	4.43
0.26	12.45	8.80	6.22	3.94	2.78	14.94	10.56	7.47	4.72	3.34	19.31	13.66	9.66	6.11	4.32
0.27	12.13	8.58	6.07	3.84	2.71	14.56	10.30	7.28	4.60	3.26	18.82	13.31	9.41	5.95	4.21
0.28	11.83	8.37	5.92	3.74	2.65	14.20	10.04	7.10	4.49	3.18	18.36	12.98	9.18	5.81	4.11
0.29	11.55	8.16	5.77	3.65	2.58	13.85	9.80	6.93	4.38	3.10	17.91	12.67	8.96	5.66	4.01
0.30	11.27	7.97	5.54	3.56	2.52	13.53	9.56	6.76	4.28	3.02	17.49	12.37	8.74	5.53	3.91
0.31	11.01	7.78	5.50	3.48	2.46	13.21	9.34	6.51	4.18	2.95	17.08	12.08	8.54	5.40	3.82
0.32	10.76	7.61	5.38	3.40	2.41	12.91	9.13	6.45	4.08	2.89	16.69	11.80	8.34	5.28	3.73
0.33	10.51	7.43	5.26	3.32	2.35	12.62	8.92	6.31	3.99	2.82	16.31	11.54	8.16	5.16	3.65
0.34	10.28	7.27	5.14	3.25	2.30	12.34	8.72	6.17	3.90	2.76	15.95	11.28	7.98	5.04	3.57
0.35	10.06	7.11	5.03	3.18	2.25	12.07	8.53	6.03	3.82	2.70	15.60	11.03	7.80	4.93	3.49
0.36	9.84	6.95	4.92	3.11	2.20	11.81	8.35	5.93	3.73	2.64	15.26	10.79	7.63	4.83	3.41
0.37	9.63	6.81	4.81	3.04	2.15	11.55	8.17	5.78	3.65	2.58	14.94	10.56	7.47	4.72	3.34

(continues)

Table B.1 Sample size related to frequency, precision, and confidence interval

Sample Size

Frequency	90% Confidence Interval					95% Coonfidence Interval					99% Confidence Interval				
	500	1000	2000	5000	10,000	500	1000	2000	5000	10,000	500	1000	2000	5000	10,00
0.38	9.43	6.66	4.71	2.98	2.11	11.31	8.00	5.66	3.58	2.53	14.62	10.34	7.31	4.62	3.27
0.39	9.23	6.53	4.61	2.92	2.06	11.07	7.83	5.54	3.50	2.48	14.32	10.12	7.16	4.53	3.20
0.40	9.04	6.39	4.52	2.86	2.02	10.84	7.67	5.42	3.43	2.42	14.02	9.91	7.01	4.43	3.14
0.41	8.85	6.26	4.43	2.80	1.98	10.62	7.51	5.31	3.36	2.38	13.73	9.71	6.87	4.34	3.07
0.42	8.67	6.13	4.34	2.74	1.94	10.41	7.36	5.20	3.29	2.33	13.45	9.51	6.73	4.25	3.01
0.43	8.50	6.01	4.25	2.68	1.90	10.19	7.21	5.10	3.22	2.28	13.18	9.32	6.59	4.17	2.95
0.44	8.32	5.89	4.16	2.63	1.86	9.99	7.06	4.99	3.16	2.23	12.92	9.13	6.46	4.08	2.89
0.45	8.16	5.77	4.08	2.58	1.82	9.79	6.92	4.89	3.10	2.19	12.66	8.95	6.33	4.00	2.83
0.46	7.99	5.65	4.03	2.53	1.79	9.59	6.78	4.80	3.03	2.15	12.48	8.77	6.20	3.92	2.77
0.47	7.84	5.54	3.92	2.48	1.75	9.40	6.65	4.70	2.97	2.10	12.16	8.60	6.08	3.84	2.72
0.48	7.66	5.43	3.84	2.43	1.72	9.22	6.52	4.61	2.91	2.06	11.92	8.43	5.96	3.77	2.66
0.49	7.53	5.32	3.76	2.38	1.68	9.03	6.39	4.52	2.86	2.02	11.68	8.26	5.84	3.69	2.61
0.50	7.38	5.22	3.69	2.33	1.65	8.85	6.26	4.43	2.80	1.98	11.45	8.10	5.72	3.62	2.56
0.51	7.23	5.11	3.62	2.29	1.62	8.63	6.14	4.34	2.74	1.94	11.22	7.94	5.61	3.55	2.51
0.52	7.09	5.01	3.54	2.24	1.59	8.51	6.02	4.25	2.69	1.90	11.00	7.78	5.50	3.48	2.46
0.53	6.95	4.91	3.47	2.20	1.55	8.34	5.90	4.17	2.64	1.86	10.78	7.62	5.39	3.41	2.41
0.54	6.81	4.82	3.41	2.15	1.52	8.17	5.78	4.09	2.58	1.83	10.57	7.47	5.28	3.34	2.36
0.55	6.67	4.72	3.34	2.11	1.49	8.01	5.66	4.00	2.53	1.79	10.36	7.32	5.18	3.27	2.32
0.56	6.54	4.63	3.27	2.07	1.46	7.85	5.55	3.92	2.48	1.76	10.15	7.18	5.07	3.21	2.27
0.57	6.41	4.53	3.20	2.03	1.43	7.69	5.44	3.85	2.43	1.72	9.94	7.03	4.97	3.14	2.22
0.58	6.28	4.44	3.14	1.99	1.40	7.54	5.33	3.77	2.38	1.68	9.74	6.89	4.87	3.08	2.18
0.59	6.15	4.35	3.08	1.95	1.38	7.38	5.22	3.69	2.33	1.65	9.54	6.75	4.77	3.02	2.13
0.60	6.02	4.26	3.01	1.91	1.35	7.23	5.11	3.61	2.29	1.62	9.35	6.61	4.67	2.96	2.09

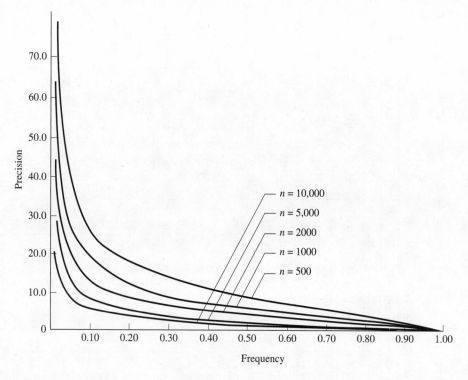

Figure B.2   Precision versus frequency for 95 percent confidence-level sampling

cases must be adjusted by the expected response rate. Incorporating this change, Eq. [B.3] becomes

$$n = \frac{\left[Z_{1-(1/2)\alpha}\right]^2 (1-p)}{r^2(p)(s)}$$ **[B.4]**

where s is the expected response rate from the survey. Suppose in the previous example the sampling method used was a mail-back survey in which planners expected to have a 60 percent response rate. The required sample size would be

$$n = \frac{(1.96)^2(0.50)}{(0.05)(0.50)(0.60)} = 2561$$

Equations [B.2] through [B.4] assumes that the sample n is small relative to the total population size. However, in those cases where the estimated sample size is a substantial percentage of the total population (i.e., n/N is greater than some standard, usually set at 0.10), the following formula should be used to modify the sample size estimate:

$$n_1 = \frac{n_0}{1 + n_0/N} \qquad \text{[B.5]}$$

where

n_0 = number of sample observations originally estimated
n_1 = adjusted number of observations
N = total population

For example, assume that an original estimate of the size of a household sample to estimate mode of travel was 2561 households. Further assume that these 2561 households were part of a district containing 8537 households. Because 2561/8537 = 0.30 exceeds our standard of 0.10, Eq. [B.5] should be used. Thus,

$$n_1 = \frac{2561}{1 + 2561/8537} = 1970$$

In this case, 1970 instead of 2561 households need to be surveyed for the requisite level of precision.

For those interested in further reading on sample size determination as it relates to transportation data collection, see the following references:

REFERENCES

Attanucci, J., I. Burns, and N. H. M. Wilson. 1981. *Bus transit monitoring manual*, Vol. 1, Report UMTA-IT-09-9008-81-1. Washington D.C: Aug.

Freund, R. and W. Wilson. 1997. *Statistical methods*. Chestnut Hill, MA: Academic Press.

Levy, P. and S. Lemeshow. 1999. *Sampling of populations, methods and applications*. Third edition. New York: Wiley Interscience.

Ott, W. 1995. *Environmental statistics and data analysis*. Boca Raton, FL: CRC Press.

Patel, J. and C. Read. 1996. *Handbook of the normal distribution*. Second edition. New York: Marcel Dekker.

Vardeman, S. and J. M. Jobe. 2001. *Data collection and analysis*. Pacific Grove, CA: Duxbury.

INDEX

NAME & SUBJECT